Lecture Notes in Physics

The Editorial Policy for Proceedings

The series Lecture Notes in Physics reports new developments in physical research and teaching – quickly, informally, and at a high level. The proceedings to be considered for publication in this series should be limited to only a few areas of research, and these should be closely related to each other. The contributions should be of a high standard and should avoid lengthy redraftings of papers already published or about to be published elsewhere. As a whole, the proceedings should aim for a balanced presentation of the theme of the conference including a description of the techniques used and enough motivation for a broad readership. It should not be assumed that the published proceedings must reflect the conference in its entirety. (A listing or abstracts of papers presented at the meeting but not included in the proceedings could be added as an appendix.)

When applying for publication in the series Lecture Notes in Physics the volume's editor(s) should submit sufficient material to enable the series editors and their referees to make a fairly accurate evaluation (e.g. a complete list of speakers and titles of papers to be presented and abstracts). If, based on this information, the proceedings are (tentatively) accepted, the volume's editor(s), whose name(s) will appear on the title pages, should select the papers suitable for publication and have them refereed (as for a journal) when appropriate. As a rule discussions will not be accepted. The series editors and Springer-Verlag will normally not interfere with the detailed editing except in fairly obvious cases or on technical matters.

Final acceptance is expressed by the series editor in charge, in consultation with Springer-Verlag only after receiving the complete manuscript. It might help to send a copy of the authors' manuscripts in advance to the editor in charge to discuss possible revisions with him. As a general rule, the series editor will confirm his tentative acceptance if the final manuscript corresponds to the original concept discussed, if the quality of the contribution meets the requirements of the series, and if the final size of the manuscript does not greatly exceed the number of pages originally agreed upon.

The manuscript should be forwarded to Springer-Verlag shortly after the meeting. In cases of extreme delay (more than six months after the conference) the series editors will check once more the timeliness of the papers. Therefore, the volume's editor(s) should establish strict deadlines, or collect the articles during the conference and have them revised on the spot. If a delay is unavoidable, one should encourage the authors to update their contributions if appropriate. The editors of proceedings are strongly advised to inform contributors about these points at an early stage.

The final manuscript should contain a table of contents and an informative introduction accessible also to readers not particularly familiar with the topic of the conference. The contributions should be in English. The volume's editor(s) should check the contributions for the correct use of language. At Springer-Verlag only the prefaces will be checked by a copy-editor for language and style. Grave linguistic or technical shortcomings may lead to the rejection of contributions by the series editors.

A conference report should not exceed a total of 500 pages. Keeping the size within this bound should be achieved by a stricter selection of articles and not by imposing an upper limit to the length of the individual papers.

Editors receive jointly 30 complimentary copies of their book. They are entitled to purchase further copies of their book at a reduced rate. As a rule no reprints of individual contributions can be supplied. No royalty is paid on Lecture Notes in Physics volumes. Commitment to publish is made by letter of interest rather than by signing a formal contract. Springer-Verlag secures the copyright for each volume.

The Production Process

The books are hardbound, and quality paper appropriate to the needs of the authors is used. Publication time is about ten weeks. More than twenty years of experience guarantee authors the best possible service. To reach the goal of rapid publication at a low price the technique of photographic reproduction from a camera-ready manuscript was chosen. This process shifts the main responsibility for the technical quality considerably from the publisher to the authors. We therefore urge all authors and editors of proceedings to observe very carefully the essentials for the preparation of camera-ready manuscripts, which we will supply on request. This applies especially to the quality of figures and halftones submitted for publication. In addition, it might be useful to look at some of the volumes already published.

As a special service, we offer free of charge LATEX and TEX macro packages to format the text according to Springer-Verlag's quality requirements. We strongly recommend that you make use of this offer, since the result will be a book of considerably improved technical quality.

To avoid mistakes and time-consuming correspondence during the production period the conference editors should request special instructions from the publisher well before the beginning of the conference. Manuscripts not meeting the technical standard of the series will have to be returned for improvement.

For further information please contact Springer-Verlag, Physics Editorial Department V, Tiergartenstrasse 17, W-6900 Heidelberg, FRG

U. Heber C. S. Jeffery (Eds.)

The Atmospheres of Early-Type Stars

Proceedings of a Workshop Organized Jointly by the UK SERC's Collaborative Computational Project No. 7 and the Institut für Theoretische Physik und Sternwarte, University of Kiel
Held at the University of Kiel, Germany
18-20 September 1991

Springer-Verlag Berlin Heidelberg GmbH

Editors

U. Heber
Institut für Theoretische Physik und Sternwarte
Olshausenstraße 40, W-2300 Kiel, FRG

C. Simon Jeffery
Department of Physics and Astronomy
University of St. Andrews
North Haugh, St. Andrews KY16 9SS
United Kingdom

DOI 10.1007/978-3-540-46997-1

Originally published by Springer-Verlag Berlin Heidelberg New York in 1992
MyCopy version of the original edition 1992

Typesetting: Camera ready by author

58/3140-543210 - Printed on acid-free paper
www.springer.com/mycopy

Preface

The modern history of research into hot-star atmospheres began with the work of Unsöld in Kiel, in the 1930s. Having laid down the basic theory of stellar atmospheres, Unsöld's rigorous application of theory to the spectroscopic observations has been continued by his successors to the present day.

Established more recently, in 1980, the UK SERC's Collaborative Computational Project No. 7 (CCP7) exists to support the UK astronomical community by providing the computational radiative transfer tools necessary for the analysis of astronomical spectra. That it has been successful has been due in part to the close links established between CCP7 and astronomers in Germany.

It was therefore appropriate that the Institut für Theoretische Physik und Sternwarte, University of Kiel, and CCP7 should jointly organise a workshop in Kiel, both to continue Unsöld's tradition by bringing together observers and theoreticians, and to encourage further international collaborations. As a result, these proceedings provide a wide-ranging picture of current research into the atmospheres of hot stars, which reflects the stimulating contributions and discussions which took place in Kiel.

Fittingly, Prof. Dr. Dr. (h.c.) Kurt Hunger celebrated his seventieth birthday during the workshop. As one of Unsöld's students, and his successor in Kiel, Kurt Hunger's ambition and encouragement ensured that the institute kept its place as one of the leading institutes in stellar atmospheres research. Kurt Hunger played a prominent rôle in German and European astronomy, latterly as president of the ESO council. The scientific organising committee is pleased to dedicate this workshop to him on the occasion of his birthday.

The broad divisions into which the subject of hot-star atmospheres naturally fall have been followed in these proceedings. Thus the first two sections deal with the analyses of O, B and A stars in our own and other galaxies, in which methods similar to those first developed by Unsöld for τ Sco have been employed. The more complex situations created by the presence of mass outflows, magnetic fields, diffusion and pulsation are described in the following two sections. These first four sections deal mainly with upper main-sequence stars. The large variety of evolved low-mass stars with early-type spectra, from blue horizontal branch stars to PG 1159 stars and white dwarfs are discussed in section five. Two final sections deal with the derivation of accurate atomic data, and the modern methods used in radiative transfer calculations, both vital ingredients for the modelling of stellar spectra.

We are grateful for the help we received from our colleagues of the Scientific Organizing Committee, Hartmut Holweger and Volker Weidemann. Our particular thanks go to Inge Schmidt and many other colleagues and students from the Kiel Institute, without whose efforts the workshop could not have been so successful. Financial support from CCP7, the Deutsche Forschungsgemeinschaft and Die Ministerin für Bildung, Wissenschaft, Jugend und Kultur des Landes Schleswig-Holstein is gratefully acknowledged. But foremost, we thank the delegates for their contributions, formal and informal, which provided the foundation for a memorable occasion.

U.Heber and C.S.Jeffery
Kiel, December 1991

To Kurt Hunger:

on the occasion of his seventieth birthday and in recognition of his contributions to research on hot star atmospheres.

Contents

I. "Normal" Early-Type Stars

V. Evolved Stars

VI. Atomic Data

VII. Model Atmospheres and Radiative Transfer

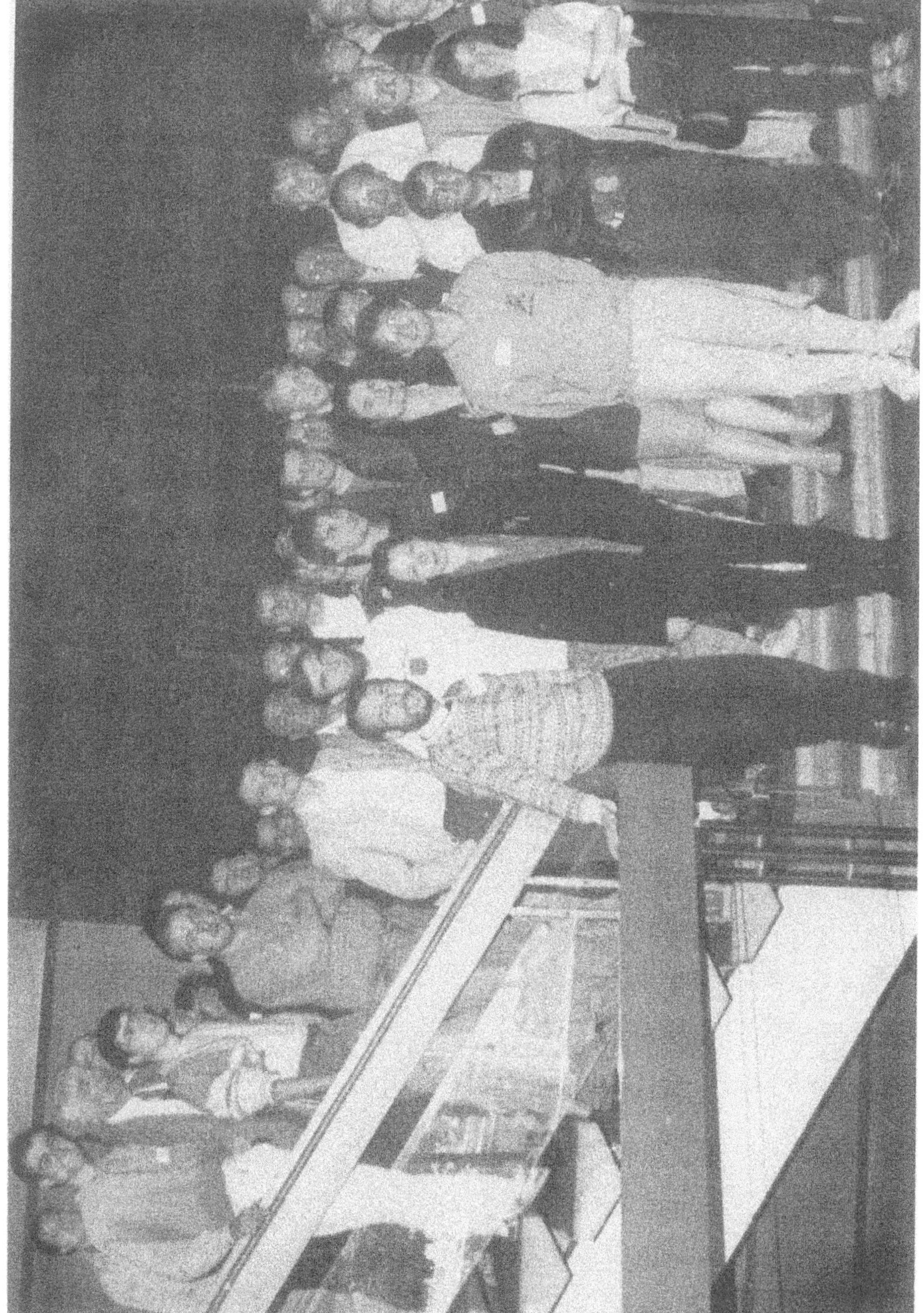

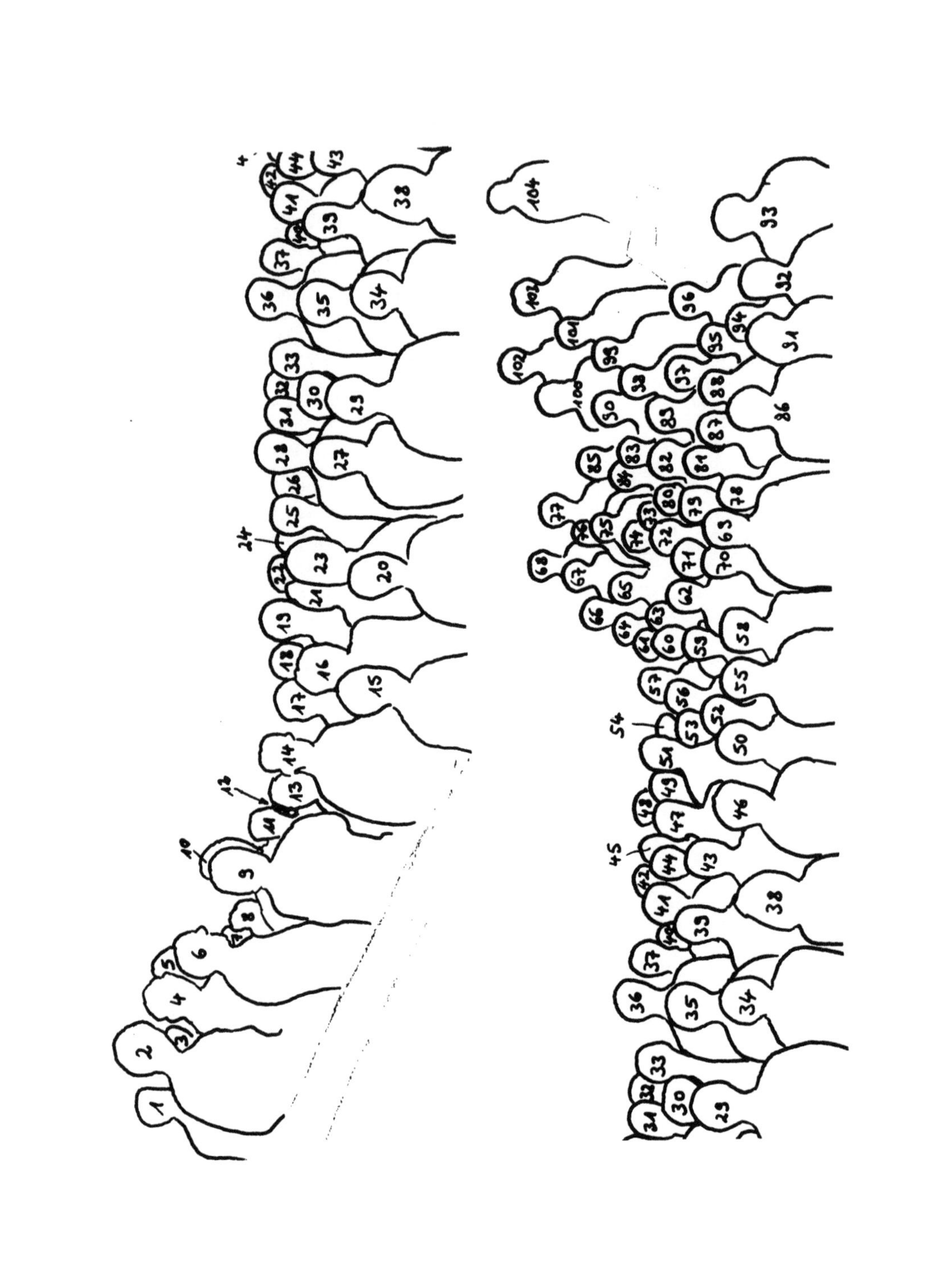

List of Participants

Numbers refer to location in photograph

Aerts, Conny	Astronomisch Instituut Leuven, Belgium [38]
Allard, Nicole	Observatoire de Paris Meudon, France
Baade, Dietrich	ESO Garching, Germany [101]
Barstow, Martin	University of Leicester, UK [14]
Baschek, Bodo	Institut für Theoretische Astrophysik Heidelberg, Germany [81]
Baum, Ekkehard	Inst. f. Theor. Physik u. Sternwarte Kiel, Germany [41]
Becker, Sylvia	Universitätssternwarte München, Germany [91]
Bianchi, Luciana	Observatorio Astronomico di Torino, Italy [59]
Bikmaev, Ilfan	Special Astrophysical Observatory Nizhnij Arkhyz, USSR [43]
Blomme, Ronny	Koninklijke Sterrewacht van Belgie Bruxelles, Belgium [47]
Bohlender, David	University of British Columbia Vancouver, Canada
Blöcker, Thomas	Inst. f. Theor. Physik u. Sternwarte Kiel, Germany [82]
Brown, Paul J. F.	Queens University Belfast, UK [10]
Bues, Irmela	Dr.-Remeis-Sternwarte Bamberg, Germany [74]
Burkhart, Claude	Observatoire de Lyon, France [70]
Butler, Keith	Universitätssternwarte München, Germany [92]
Chentsov, Eugene L.	Special Astrophysical Observatory Nizhnij Arkhyz, USSR [44]
Clausen, Jens Viggo	Copenhagen University Observatory, Denmark [84]
Conlon, Elisabeth	Queens University Belfast, UK [20]
Cugier, Henryk	Astronomical Observatory Wroclaw, Poland
de Groot, Mart	Armagh Observatory, UK [37]
de Koter, Alex	Laboratory for Space Research SRON Utrecht, Netherlands [80]
de Vos, Roland	Vrije Universiteit Brussel, Belgium [65]
Dimitrijevic, Milan	Astronomical Observatory Belgrade, Yugoslavia
Dreizler, Stefan	Inst. f. Theor. Physik u. Sternwarte Kiel, Germany [2]
Drilling, John S.	Louisiana State University Baton Rouge, USA
Dudley, Richard E.	St. Andrews, UK [66]
Dufton, Phillip	Queens University Belfast, UK [58]
Engelhardt, Dieter	Dr.-Remeis-Sternwarte Bamberg, Germany [19]
Farthmann, Michael	Inst. f. Theor. Physik u. Sternwarte Kiel, Germany
Field, David	School of Chemistry Bristol, UK
Finley, David	University of California Berkeley, USA [31]
Freytag, Bernd	Inst. f. Theor. Physik u. Sternwarte Kiel, Germany
Friedrich, Susanne	Theoretische Astrophysik Tübingen, Germany [21]
Gesicki, Krzysztof	Copernicus Astronomical Center Torun, Poland [85]
Groote, Detlef	Sternwarte Hamburg, Germany [75]
Groth, Hans-Günter	Universitätssternwarte München, Germany [4]
Gussmann, Ernst-August	Zentralinstitut für Astrophysik Potsdam, Germany [64]
Hamann, Wolf-Rainer	Inst. f. Theor. Physik u. Sternwarte Kiel, Germany [49]
Heap, Sarah	NASA Goddard Space Flight Center Greenbelt, USA [71]
Heber, Ulrich	Inst. f. Theor. Physik u. Sternwarte Kiel, Germany [55]

Hensler, Gerhard	Inst. f. Theor. Physik u. Sternwarte Kiel, Germany [73]
Herrero Davo, Artemio	Instituto de Astrofisica de Canarias La Laguna, Spain [34]
Hillier, John	Universitätssternwarte München, Germany [41]
Holweger, Hartmut	Inst. f. Theor. Physik u. Sternwarte Kiel, Germany [83]
Houziaux, Leo	Liege, Belgium [62]
Howarth, Ian D.	UCL London, UK [29]
Hubeny, Ivan	NASA Goddard Space Flight Center Greenbelt, USA [95]
Hubrig, Swetlana	Zentralinstitut für Astrophysik Potsdam, Germany [8]
Hummel, Wolfgang	Astronomisches Institut Ruhr-Universität Bochum, Germany [22]
Hunger, Kurt	Inst. f. Theor. Physik u. Sternwarte Kiel, Germany [86]
Husfeld, Dirk	Universitätssternwarte München, Germany [24]
Jeffery, C. Simon	St. Andrews, UK [15]
Jordan, Stefan	Inst. f. Theor. Physik u. Sternwarte Kiel, Germany [77]
Jüttner, A.	Landessternwarte Heidelberg, Germany [76]
Kaper, Lex	Astronomical Institute Amsterdam, Netherlands [57]
Kilian, Judith	Universitätssternwarte München, Germany [79]
Kirchgesser, Ulrike	Universitätssternwarte München, Germany [78]
Klochkova, Valentina	Special Astrophysical Observatory Nizhnij Arkhyz, USSR [52]
Koester, Detlev	Louisiana State University Baton Rouge, USA [60]
Koesterke, Lars	Inst. f. Theor. Physik u. Sternwarte Kiel, Germany [23]
Kolka, Indrek	Tartu Astrophysical Observatory, Estonia [16]
Krüß, Andrea	Inst. f. Theor. Physik u. Sternwarte Kiel, Germany
Kügler, Lutz	Inst. f. Theor. Physik u. Sternwarte Kiel, Germany
Kunze, Dietmar	Universitätssternwarte München, Germany [69]
Landstreet, John D.	University of Western Ontario London, Canada [99]
Lehmann, Holger	Karl-Schwarzschild-Observatorium Tautenburg, Germany
Lemke, Michael	University of Texas Austin, USA [68]
Lennon, Danny J.	Universitätssternwarte München, Germany [87]
Leuenhagen, Uwe	Inst. f. Theor. Physik u. Sternwarte Kiel, Germany
Ludwig, Hans-Günter	Inst. f. Theor. Physik u. Sternwarte Kiel, Germany
Lynas-Gray, Anthony E.	Department of Astrophysics Oxford, UK [89]
Marten, Holger	Inst. f. Theor. Physik u. Sternwarte Kiel, Germany [39]
Mathys, Gautier	Geneva Observatory, Switzerland [18]
Mc Causland, Robert	Queens University Belfast, UK [9]
MacFarlane, Josef	University of Wisconsin Madison, USA [98]
Michaud, Georges	Universite de Montreal, Canada [90]
Moehler, Sabine	Sternwarte der Universität Bonn, Germany [46]
Möller, Rolf	Institut für Polarökologie Kiel, Germany
Monier, Richard	IUE-Observatory VILSPA Madrid, Spain
Napiwotzki, Ralf	Inst. f. Theor. Physik u. Sternwarte Kiel, Germany [30]
Niedzielski, Andrzej	Copernicus Astronomical Center Torun, Poland [63]
Owocki, Stan	Bartol Research Institute Newark, USA [93]
Perinotto, Mario	Dipart. di Astronomia e Scienza d. Spazio Firenze, Italy [33]
Rauch, Thomas	Inst. f. Theor. Physik u. Sternwarte Kiel, Germany [103]
Reimers, Dieter	Sternwarte Hamburg, Germany [48]
Rieschick, Andreas	Inst. f. Theor. Physik u. Sternwarte Kiel, Germany [1]
Rolleston, Robert	Queens University Belfast, UK [6]
Rons, Nadine	Vrije Universiteit Brussel, Belgium [61]

Runacres, Marc	Vrije Universiteit Brussel, Belgium [53]
Schaerer, Daniel	Geneva Observatory, Switzerland [26]
Schlüter, Dieter	Inst. f. Theor. Physik u. Sternwarte Kiel, Germany
Schmidt, Jelena	Sternwarte der Universität Bonn, Germany [50]
Schönberner, Detlef	Inst. f. Theor. Physik u. Sternwarte Kiel, Germany [56]
Schöneich, Werner	Zentralinstitut für Astrophysik Potsdam, Germany [51]
Schmutz, Werner	ETH Zürich, Switzerland [13]
Simon, Klaus Peter	Universitätssternwarte München, Germany [100]
Smith, Keith C.	University College London, UK [11]
Solheim, Jan-Eric	Mathematical and Physical Sciences Tromsoe, Norway [97]
Stahlberg, Jürgen	Zentralinstitut für Astrophysik Potsdam, Germany
Steenbock, Wolfgang	Inst. f. Theor. Physik u. Sternwarte Kiel, Germany [94]
Steffen, Mathias	Inst. f. Theor. Physik u. Sternwarte Kiel, Germany [45]
Stift, Martin	Institut für Astronomie Wien, Austria [96]
Stolzmann, Werner	Inst. f. Theor. Physik u. Sternwarte Kiel, Germany
Stürenburg, Sven	Inst. f. Theor. Physik u. Sternwarte Kiel, Germany [72]
Theissen, Armin	Sternwarte der Universität Bonn, Germany [28]
Thejll, Peter	NORDITA Copenhagen, Denmark
Tobin, William	University of Canterbury Christchurch, New Zealand [32]
Unglaub, Klaus	Dr.-Remeis-Sternwarte Bamberg, Germany [3]
Van Rensbergen, W.	Vrije Universiteit Brussel, Belgium [5]
Van Winckel, Hans	Astronomisch Instituut Leuven, Belgium [25]
Vranken, Myriam	Vrije Universiteit Brussel, Belgium
Waelkens, Christoffel	Astronomisch Instituut Leuven, Belgium [35]
Weidemann, Volker	Inst. f. Theor. Physik u. Sternwarte Kiel, Germany
Wenske, Volker	Inst. f. Theor. Physik u. Sternwarte Kiel, Germany [36]
Werner, Klaus	Inst. f. Theor. Physik u. Sternwarte Kiel, Germany [102]
Wessolowski, Ulf	Inst. f. Theor. Physik u. Sternwarte Kiel, Germany
Wolf, Bernhard	Landessternwarte Heidelberg, Germany [17]
Zboril, Milan	Slovak Academy of Sciences Tatranska Lomnica, CSFR [67]
Zelwanowa, Elia	Zentralinstitut für Astrophysik Potsdam, Germany [27]

I. “Normal” Early-Type Stars

Galactic B-type Stars

Philip Dufton

Department of Pure and Applied Physics, Queen's University of Belfast
Belfast BT7 1NN, United Kingdom

Abstract: Some recent analyses of the spectra of apparently normal B-type stars are discussed. These include studies of relatively nearby stars, distant stars in the plane of our galaxy and stars at high galactic latitudes.

1 Introduction

Normal B-type stars provide very useful probes of our own galaxy and also of other nearby galaxies. Their advantages include

– their brightness, which allows studies over considerable distance scales;

– their evolutionary short lifetimes with the implication that they provide information on the current state of the galaxy;

– the belief that their atmospheres are relatively well understood and hence model atmosphere analyses will provide reliable results;

– they provide useful tests of the reliability of calculations of stellar evolution and in particular the effects of mass loss.

Although I will discuss some work with which I have been involved, I have deliberately tried to include other studies. I, therefore, apologize if I have mis-understood or mis-represent these studies. Areas that will not be considered are atomic data requirements for model atmosphere calculations, different theoretical or numerical methods, peculiar (for example, subluminous stars) or stars in external galaxies, as these will be discussed by others.

Before considering recent results, it is instructive to briefly consider earlier studies. As an example, one of the first analyses of a B-type star was by Unsöld (1942) for τ Sco. The chemical composition, which he deduced from a curve-of-growth analysis is summarized in table 1, together with results from the comprehensive LTE model atmosphere analysis of Hardorp and Scholz (1970) and from recent sophisticated non-LTE model atmosphere calculations (Becker and Butler 1988,1989; Becker 1991). Despite all the improvements in observational and theoretical techniques, it is astonishing how closely the abundances from Unsöld's pioneering analysis agree with the best results currently available. This should warn us of the danger of completely ignoring earlier studies.

Table 1. Comparison of analyses of τ Sco by Unsöld (1942), Hardorp and Scholz (1970) and Becker and Butler (1988,1989), Becker (1991). Abundances are on a logarithmic scale with that of hydrogen being 12.

Element	Unsöld	HS LTE	BB nLTE
He	11.0	11.0	-
C	8.1	8.0	-
N	8.3	8.3	8.2
O	8.7	8.7	8.8
Ne	8.6	8.8	-
Mg	7.5	7.5	-
Al	6.2	6.3	-
Si	7.6	7.5	7.7
Fe	7.3	7.4	-

Returning to the present day, the analyses discussed here have been arbitrarily divided into three groups, namely nearby stars, distant stars in the galactic plane and high galactic latitude objects.

2 Nearby stars

In this context 'nearby' is a relative term and implies a distance of typically less than 2 kpc. Such stars are bright enough ($V \leq 10 - 11$) for their spectra to be observed with some precision using large telescopes and modern detectors. Such an observation is shown for τ Sco in figure 1 and is based on data discussed by Kilian and Nissen (1989), who used the CASPEC spectrograph on the 3.6m telescope at La Silla. These authors obtained spectra from 4060 to 5060Å for 21 (near) main sequence early-type stars ranging in spectral-type from O9 to B3 and luminosity classes III to V. Besides line identifications and equivalent width measurements, they collated the relevant atomic data (Kilian et al 1990a) and then deduced atmospheric parameters using non-LTE model atmosphere calculations for hydrogen, helium and silicon (Kilian et al 1990b). Results for the analysis of the metal line spectra will be presented at this meeting.

Spectra of comparable signal-to-noise, resolution and wavelength coverage have recently been presented by Gies and Lambert (1991) for 39 B-type stars, including 7 supergiants. The principal impetus for this study was to search for abundance variations with evolutionary status and in particular to investigate the surprising discovery of Lyubimkov (1989 and references therein) that nitrogen abundances systematically increased with stellar age. Gies and Lambert had some targets in common with Kilian and Nissen and it is disconcerting that despite the very high quality of both observational datasets, there appears to be a systematic difference of approximately 12% between the two set of equivalent widths (the differences between Gies and Lambert and Kane et al (1980) is less worrying given the relatively low signal-to-noise of the latter's photographic spectra). Gies and Lambert undertake a comprehensive LTE analysis of their

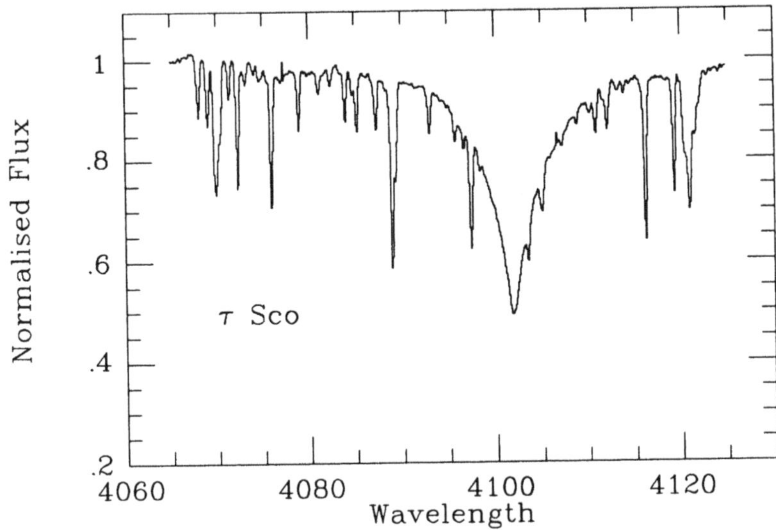

Fig. 1. Spectra of τ Sco discussed by Kilian and Nissen (1989)

stars and then use the results of non-LTE calculations to compensate for possible errors due to the assumption of LTE. They find no evidence for the variation of nitrogen abundance with age found by Lyubimkov, which they ascribe to errors in his analysis. However they discuss a number of other topics, including the good agreement between their CNO abundances and those found from H II region studies. Another interesting result is the variation in nitrogen abundances with evolutionary status. They find enhanced nitrogen abundances in some but not all of the evolved stars, together with marginal variations in the helium (correlated with nitrogen) and carbon (anti-correlated with nitrogen) abundances. They conclude that this is evidence for partial processing by the CN cycle and that the stars have not yet evolved through a red supergiant stage (as then the materially should be fully processed by the CNO cycle). Why only some of these objects should have such material at their surface is unclear.

We at Queen's have also been investigating the chemical composition of B-type stars. Our approach differs from those discussed above as we have deliberately restricted out observations to a small number of weak metal lines (equivalent widths of typically 5mÅ). This choice was made in order to minimise theoretical uncertainties and also to ensure that the lines were on the linear part of the curve of growth where the equivalent width is sensitive to the adopted element abundance. Also by only considering a restricted set of lines we have been able to undertake new calculations of their oscillator strengths. We believe that by combining high resolution and signal-to-noise spectra, with reliable atomic data and a non-LTE radiative transfer equation, abundances accurate to 0.1 dex should be obtainable (Keenan et al 1989, Holmgren et al 1990).

The evolutionary status of galactic B-type supergiants has also been considered by Lennon et al (1991a) who have observed 43 targets ranging from O9 to B9 and with luminosity classes I and II. The quality of their observational data is illustrated in figure 2. They speculate that the morphologically nitrogen weak objects (the OBC stars) may have normal nitrogen abundances with the other supergiants showing nitrogen enriched atmospheres possibly after passing through a red-supergiant phase. Non-LTE abundance analyses for these stars are currently underway.

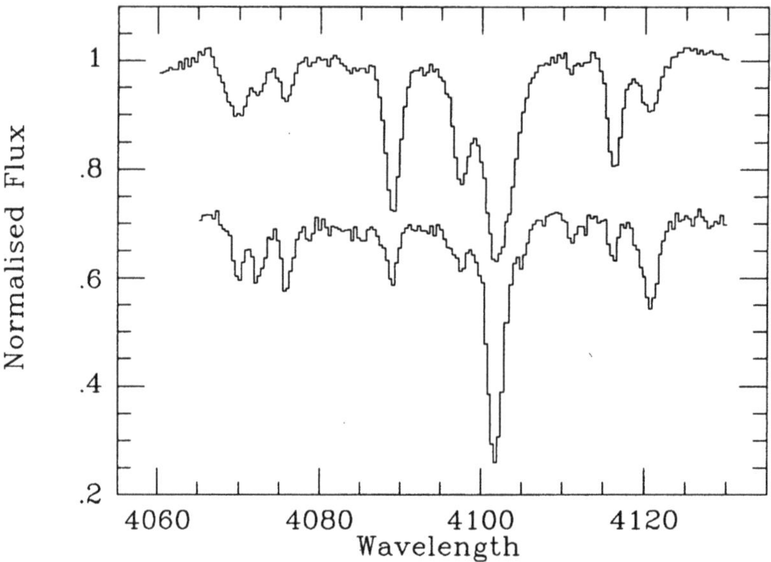

Fig. 2. Spectra of two supergiants in the region of Hδ discussed by Lennon et al (1991a). The upper plot is for HD167264 (B0Ia) and the lower for HD14956 (B2Ia)

3 Distant stars

Groups in both Munich and Belfast have been using distant young clusters to probe large scale abundance variations in our galaxy. Unfortunately main sequence B-type stars in such clusters are faint ($V \simeq 12 - 14$) and to obtain spectra of sufficient quality for a significant number of stars is observationally very demanding. Previous studies of H II regions (see for example Shaver et al. 1982, Fich and Silkey 1991) have implied that the logarithmic abundance of elements such as nitrogen and oxygen decrease approximately linearly with galactocentric distance (R_g). In contrast the results of Gehren et al (1985), Brown et al (1986) and Fitzsimmon et al (1990) indicate that the abundance variations are very small within a few kiloparsecs of the sun. Table 2 list the abundance gradients found by Fitzsimmons et al for nitrogen, oxygen, magnesium, aluminium and silicon. In all cases the results are compatible with no significant abundance variations within 2 kpc of the sun. However for the distant cluster Dolitze 25, Lennon et al (1990) have found a systematic depletion of approximately 0.6 dex in a number of elements, including carbon, nitrogen, oxygen, magnesium and silicon. Additionally current research at Munich (discussed by Sylvia Becker at this meeting) and Belfast implies that there may be variations in abundances even for clusters with similar galactocentric distances.

Table 2. Abundance gradients in the solar neighbourhood deduced by Fitzsimmons et al (1990). Also listed are the absolute abundances on a logarithmic scale with hydrogen being 12.

Element	Gradient	Abundance
N	$+0.045 \pm 0.028$	7.55 ± 0.19
O	-0.015 ± 0.014	8.65 ± 0.14
Mg	-0.017 ± 0.018	7.26 ± 0.13
Al	$+0.060 \pm 0.017$	6.17 ± 0.07
Si	$+0.008 \pm 0.010$	7.46 ± 0.08

Hence there is an apparent contradiction in the results from H II region studies and B-type stars. However a comparison shows that the latter results lie within the scatter of abundances found from the H II regions results (Fitzsimmons et al 1991). However what the cluster studies do clearly indicate is that the assumption of a unique and linear relationship between abundance and R_g is probably an over simplification.

4 Stars at high galactic latitude

Most B-type stars at high galactic latitude are highly evolved objects, for example subdwarfs, horizontal branch or post-asymtopic giant branch objects. These will be discussed elsewhere at this meeting and I will therefore concentrate on the small number of stars which appear to be either on or evolving away from the hydrogen burning main sequence. In the last ten years, approximately 20 such objects have been identified from model atmosphere analyses apparently at distances from the galactic plane of more than 1 kpc (see for example Heber and Langhams, 1986, Viton et al, 1991, Conlon et al 1990).

Most of these stars have evolutionary lifetimes (again assuming they are normal stars) that are consistent with them having been ejected from the galactic disk. Conlon et al (1990) considered the competing mechanisms and concluded that most of the stars were probably ejected by close gravitational interactions within a cluster (Leonard and Duncan 1990). However the Queen's group have identified a small number of stars that they believe could not have formed in the halo. The most extreme example is probably PG 0832+676 which has a visual magnitude of approximately 14.5 and a B1 V spectral type (see figure 3). If it is a normal hydrogen burning main sequence star it is at a distance of 31 kpc and lies approximately 18 kpc from the galactic plane (Brown et al, 1989). Another interesting example is SB 357, first studied at Kiel by Heber and Langham (1986) and re-observed by Conlon et al (1991). Both studies found emission in the Balmer line cores and deduced that this was a population I Be star at approximately

6 kpcs. Its very high latitude means that a reliable kinematical analysis is possible and again this object appears to have formed some distance from the plane.

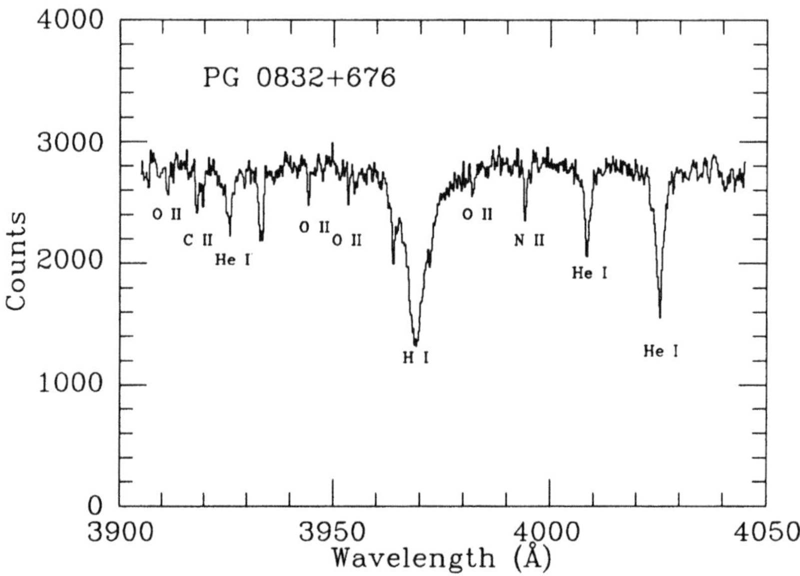

Fig. 3. Spectrum of PG 0832+676. This star appears to be a hydrogen burning main sequence object despite having a magnitude of approximately 14.5 and a galactic latitude of 35 degrees.

One complication in the study of such objects is the possibility that they are evolved subluminous stars whose spectra mimic those of normal stars even at high signal-to-noise and resolution. Last week two colleagues at Belfast, Liz Conlon and Robert McCausland, observed with the William Herschel Telescope faint blue targets ($B \simeq 20$) above and below the plane of the external galaxy, M31. Most turned out to be either quasars or subluminous stars. However one appears to be a normal B-type star with the same radial velocity as the galaxy, implying an absolute magnitude of approximately -4. Hence we believe that we have found a young star in the halo of M31; a result which provides significant support for such identifications in our own galaxy.

An example of the importance of quantitative model atmosphere calculations is given by the (in)famous high galactic latitude star HD 93521. The evolutionary status of this object has been a matter of controversy for over a decade. For example, while Hobbs et al (1982) classified it as a normal O9.5V star, Ramalla et al (1980) and Ebbets and Savage (1982) deduced that it was a low mass population II star on the basis of the low wind terminal velocity deduced from IUE spectra. However Irvine (1989) has questioned whether the relationship between terminal and escape velocities found by Abbott (1978) holds for late O-type stars. One serious constraint on model atmosphere analysis of its spectrum is the very large projected rotational velocity ($\simeq$ 400km s^{-1}). Hence although Hack and Yilmaz (1977) deduced a helium overabundance, they were unable to derive any metal abundances. Recently Lennon et al (1991b) have obtained very high signal-to-noise ($\simeq$ 500) spectra for HD 93521 and for the first time have identified and measured the strength of some of the metal lines (see figure 4). A non-LTE model atmosphere analysis confirms that helium is overabundant while its metal abundances are near normal. They believe that HD 93521 is probably a population I star whose main sequence lifetime has been extended by mixing in the core due to its large rotational velocity. This extended lifetime would then allow HD 93521 to reach its current position after ejection from the galactic disk.

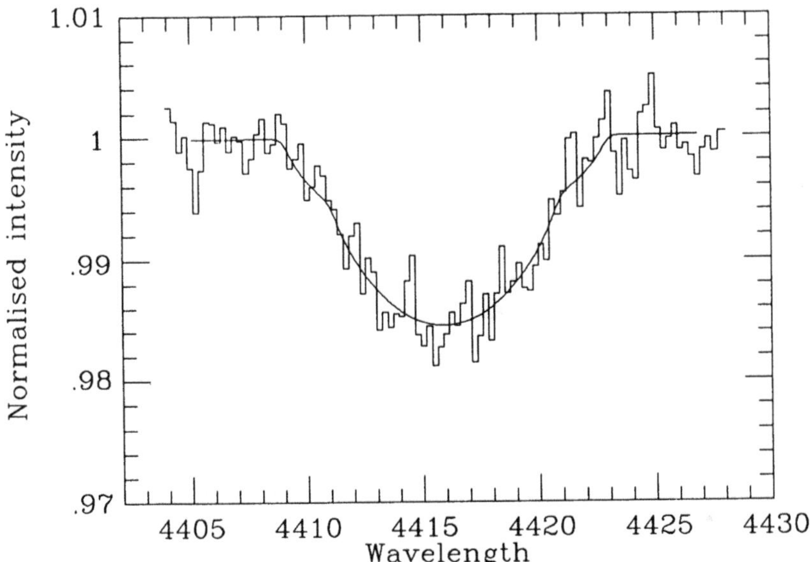

Fig. 4. Observed spectra of an O II doublet in HD 93521. Despite a central depth of less than 2%, a convincing fit can be obtained using rotationally broadened theoretical profiles

I am grateful to my colleagues at Queen's University and Munich for providing unpublished results and for advice.

References

Abbott, D.C. 1978. Astrophys. J., **225**, 893.

Becker, S.R 1991. Thesis.

Becker, S.R., Butler, K. 1988. Astr. Astrophys., **201**, 232.

Becker, S.R., Butler, K. 1989. Astr. Astrophys., **209**, 244.

Brown, P.J.F., Dufton, P.L., Lennon, D.J., Keenan, F.P. 1986. Mon. Not. R. Astr. Soc., **220**, 1003.

Brown, P.J.F., Dufton, P.L., Keenan, F.P., Boksenberg, A., King, D.L., Pettini, M. 1989. Astrophys. J., **339**, 397, 1989.

Conlon, E.S., Dufton, P.L., keenan, F.P., Leonard, P.J.T. 1990. Astr. Astrophys., **236**, 357.

Conlon, E.S., McCausland, R.J.H., Dufton, P.L., Keenan, F.P. 1991. Mon. Not. R. Astr. Soc., submitted.

Ebbets, D.C., Savage, B.D. 1982. Astrophys. J., **262**, 234.

Fich, M., Silkey, M. 1991. Astrophys. J., **366**, 107.
Fitzsimmons, A., Brown, P.J.F., Dufton, P.L., Lennon, D.J. 1990. Astr. Astrophys., **232**, 437.
Fitzsimmons, A., Dufton, P.L., Rolleston, W.R.J. 1991. in preparation.
Gehren, T., Nissen, P.E., Kudritzki, R.P., Butler, K. 1985. Proc. ESO workshop on production and distribution of CNO elements, European Southern Observatory, Garching, p. 171.
Gies, D.R., Lambert, D.L. 1991. Astrophys. J., in press.
Hack, M., Yilmaz, N. 1977. Astrophys. Sp. Sc., **48**, 483.
Hardorp, J., Scholz, M. 1970. Astrophys. J. Suppl., **13**, 353.
Heber, U., Langhams, G. 1986. ESA SP-263, 279.
Hobbs, L.M. Morgan, W., Albert, C.A., Lockmann, F.J. 1982. Astrophys. J., **263**, 690.
Holmgren, D.E., Brown, P.J.F., keenan, F.P., Dufton, P.L. 1990. Astrophys. J., **364**, 657.
Irvine, N.J. 1989. Astrophys. J., **337**, L33
Kane L.G., Dufton,P.L., McKeith, C.D. 1981. Mon. Not. R. Astr. Soc., **184**, 537.
Keenan F.P., Bates B., Dufton, P.L., Holmgren, D.E., Gilheany, S. 1989. Astrophys. J., **348**, 322.
Kilian, J., Nissen, P.E. 1989. Astr. Astrophys. Suppl., **80**, 255.
Kilian, J., Montenbruch, O., Nissen, P.E. 1990a. Astr. Astrophys. Suppl., **88**, 101.
Kilian, J., Becker, S.R., Gehren, T., Nissen, P.E. 1990b. Astr. Astrophys., **244**, 419.
Lennon, D.J., Dufton, P.L., Fitzsimmons, A., Gehren, T., Nissen, P.E. 1990. Astr. Astrophys., **240**, 349.
Lennon, D.J., Fitzsimmons, A., Dufton, P.L. 1991a, Ast. Astrophys. Suppl., submitted.
Lennon, D.J., Dufton, P.L., Keenan, F.P., Holmgren, D.E. 1991b. Astr. Astrophys., **246**, 175.
Leonard, P.T.J., Duncan, M.J. 1990., Astr. J., **99**, 608.
Lyubimkov, L.S. 1989. Astrofizika, **30**, 99.
Ramalla, M., Morossi, C., Santin, P. 1980. Astr. Astrophys., **197**, 209.
Shaver, P.A., McGee, R.X., Newton, L.M., Danks, A.C., Pottasch, S.R. 1982. Mon. Not. R. Astr. Soc., **204**, 53.
Unsöld, A. 1942. Zeit. f. Astrophys., **21**, 22.

NON–LTE ANALYSES OF B STARS

Sylvia R. Becker

Institut für Astronomie und Astrophysik der Universität München, Scheinerstraße 1, D–W–8000 München 80, Germany.

Abstract: In this paper we present a spectroscopic method for parameter determination in B stars and applications thereof. The derived abundances are considered in the context of the existence or non-existence of a radial gradient of galactic abundances.

1 Introduction

It has been known for a number of years that fine analyses of B stars require the corresponding model calculations to be carried out in non–LTE. Meanwhile a set of non–LTE line formation calculations for various ions has been published (Becker and Butler, 1990 a and references therein) These can be used for abundance analyses in the parameter region of early B stars. Such analyses are generally carried out for given values of the gravity and effective temperature of a star. Thus the determination of the stellar parameters is the key step to any further analysis.

Given the excellent observational material a method for parameter determination based on these spectra is of interest. Such a method is the use of the ionization balance of ions of one atomic species, *e.g.* silicon in B stars to determine the effective temperature coupled with the hydrogen Balmer lines as gravity indicators. (Becker and Butler, 1990 a and b) provide a set of silicon lines which can be used reliably.

This paper sketches the silicon calculations and the method applied for parameter fitting, presents some examples of the reliability of this scheme and finally shows some results of parameter and abundance analyses and discusses their implications for a radial galactic abundance gradient.

2 The Calculations

The complete description of the non–LTE model atom for silicon (Si II, III and IV) and a grid of non–LTE equivalent widths is given in Becker and Butler (1990 a, b).

All lines between 4000 Å and 5100 Å are reliable for temperature determination except for $\lambda\lambda$ 4716/4813/4819/4828 Å. The discrepancy for these lines can be removed by creating a significantly larger model atom and taking dielectronic recombination into account. The resulting model atom is too big for production runs and there is still a small discrepancy at the high temperature end of the grid. Thus the published grid is used for parameter determinations.

3 Parameter Determination

The method of parameter determination to be described only uses features in the part of the spectrum most frequently observed with the ESO CASPEC, no further data are necessary.

The theoretical data available are a grid of non–LTE equivalent widths of Si II, III and IV and a set of non–LTE line profiles of hydrogen Balmer lines based on the same model atmosphere grid. In the course of the calculations it became obvious that the ratios of silicon lines of different ionisation stages depend on the silicon abundance. Thus, a determination of T_{eff} and $\log g$ using the silicon ionisation balance must include the determination of the silicon abundance and possibly a microturbulent velocity. The following procedure is adopted here:

1. H profile fits: comparing the theoretical profiles to the observed spectra results in a set of values $T_{eff}/\log g$ for which the two profiles match. In a $T_{eff}/\log g$ diagram these points will be on a curve which mainly determines $\log g$ of the object.

2. Si ionisation balance: comparing theoretical and observed equivalent widths of silicon lines in several ionisation stages results in sets of values $T_{eff}/\log g$ for any given silicon abundance (and microturbulence). Plotting $T_{eff}/\log g$ diagrams for all values of $\log \varepsilon_{Si}$ and v_{turb} including the profile fits and fit curves for a number of silicon lines from more than one ionisation stage leads to a set of diagrams out of which there will be one where all fit lines meet in a small region. This region defines T_{eff} and $\log g$ the choice of plot determines $\log \varepsilon_{Si}$ and v_{turb}. Fig. 1 shows the fit diagram for τ Sco.

Since this method is based solely on spectral lines it does not depend on the reddening of the object.

The model atom has been designed to match the observations in a small wavelength region in the optical part of the spectrum. Fig. 2 shows the resonance lines of Si III and Si IV in τ Sco. The theoretical profiles have been calculated using the stellar parameters derived from the analysis in the optical. The UV lines obviously match the observations. The slight asymmetry of the observed Si IV lines is due to wind effects which our hydrostatic models cannot account for.

4 Comparison with other methods

The method outlined in the previous section has been applied to several objects and shown to produce reliable results. In cases where results from other authors were in the literature these were in reasonable agreement. For τ Sco we find: this work: T_{eff}= 30900 K±250 K, $\log g$= 4.18, Gulati *et al.* (1989): T_{eff}= 30000 K and Remie and Lamers (1982): T_{eff}= 31800 K±1500 K, $\log g$= 4.09.

Since τ Sco is a relatively nearby star its spectrum is barely reddened. Thus the agreement between the various methods is to be expected. However, applying the

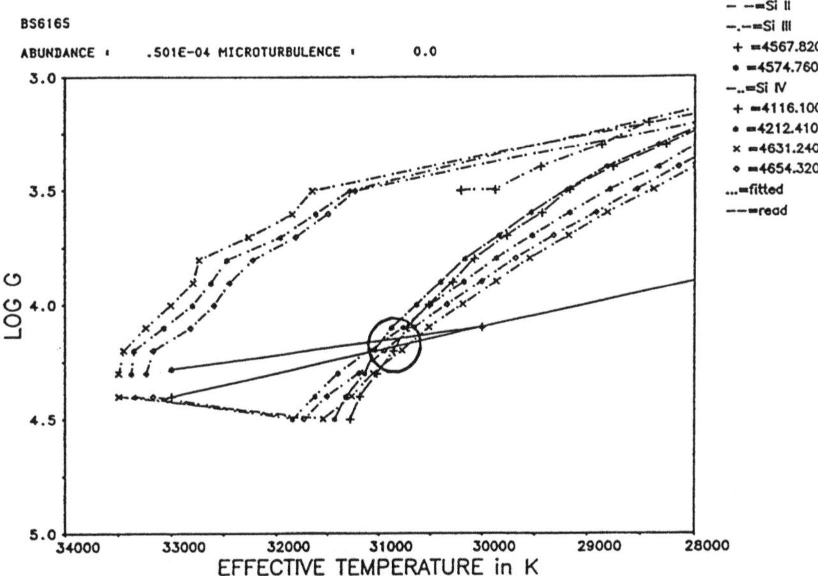

Figure 1: Fit diagram for τ Sco. The circle marks the fit region for T_{eff} and $\log g$.

energy distribution method to early B stars leads to a serious problem (Hummer *et al.*, 1988). The slope of the continuum in the UV does not depend on temperature to any large degree. This can be seen in figure 3, where the observed dereddened spectrum (IUE) of S289-13 is plotted together with Kurucz (1979) fluxes for 25000 K, 30000 K and 35000 K (all $\log g = 4$). All these theoretical fluxes match the observations. In such cases spectroscopic methods provide more reliable results.

5 Applications

5.1 Abundance Analyses

Given the stellar parameters a fine analysis of abundances can be carried out easily. Results of such analyses are or will be published elsewhere (Becker, 1991a, Becker, 1991b, Kilian *et al.*, 1992 a, b, Becker and Butler, 1992). Figure 4 shows a collection of results from abundance analyses by various authors. The logarithm of the number density relative to hydrogen divided by the solar value is plotted against radial galactic distance R_G. The 'x' symbols mark the current non–LTE results for stars at large galactocentric distances. Note that the results from stellar analyses show no evidence of an abundance gradient which is still in contrast to work on H II regions (Shaver *et al.*, 1983).

5.2 Spectral Classification

The line ratios of silicon lines depend on the abundance. The classification criteria for late O and early B stars include line ratios of silicon and of silicon with helium. A comparison of these line ratios for different values of the silicon abundance shows that

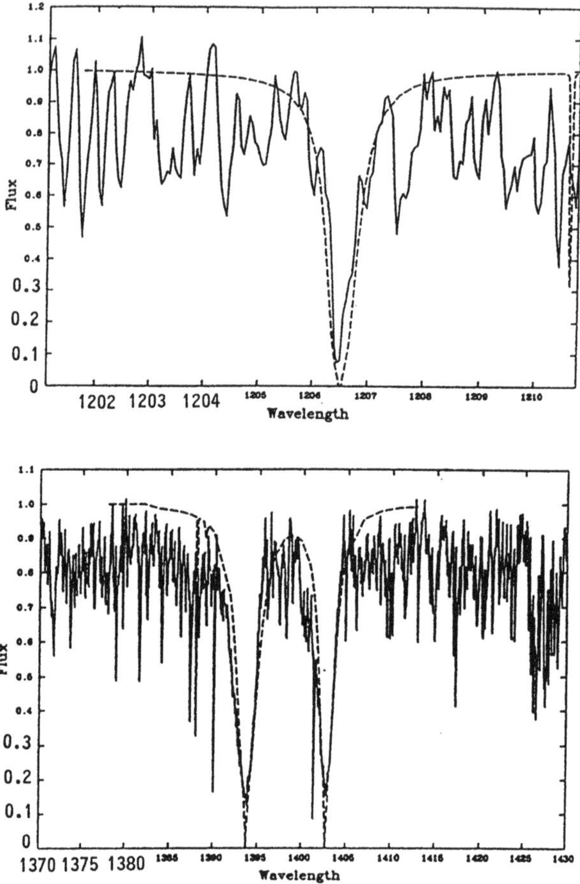

Figure 2: Resonance lines of Si III (top) and Si IV (bottom) in τ Sco.

any attempts at a finer classification system than the one currently in use (Walborn, 1971, Walborn and Fitzpatrick, 1990) cannot be based on silicon line ratios due to abundance effects. A change in the abundance by a factor of two means that a unique spectral class cannot be assigned.

6 Summary and Outlook

A spectroscopic method for parameter determination in B stars has been introduced and applied to a sample of stars. This method can be used whenever high resolution spectra of an object are available. It does not depend on reddening but only on the quality of the observations.

Abundance analyses show no evidence of a radial galactic gradient of abundances in contrast to results from H II regions. It is however necessary to carry out a larger set of analyses to be able to make a definite statement as to the existence or nonexistence of such a gradient. This work is under way.

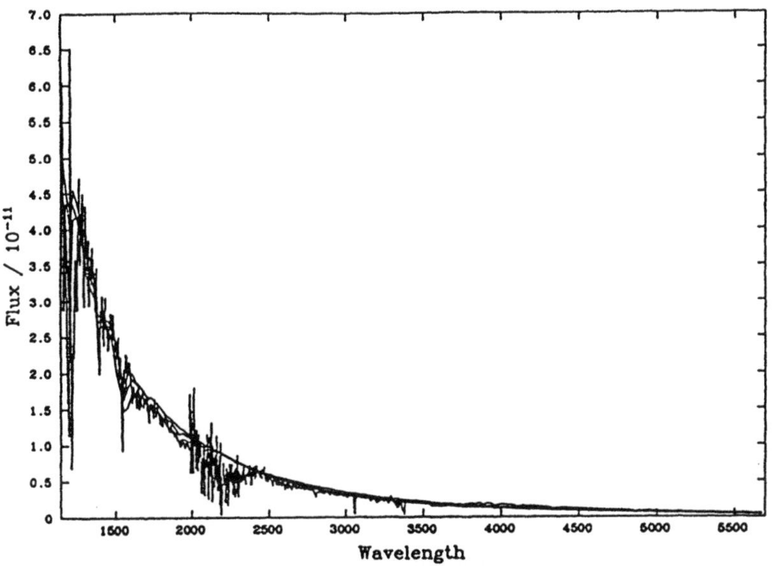

Figure 3: Energy distribution of S289-13 and theoretical fluxes, see text.

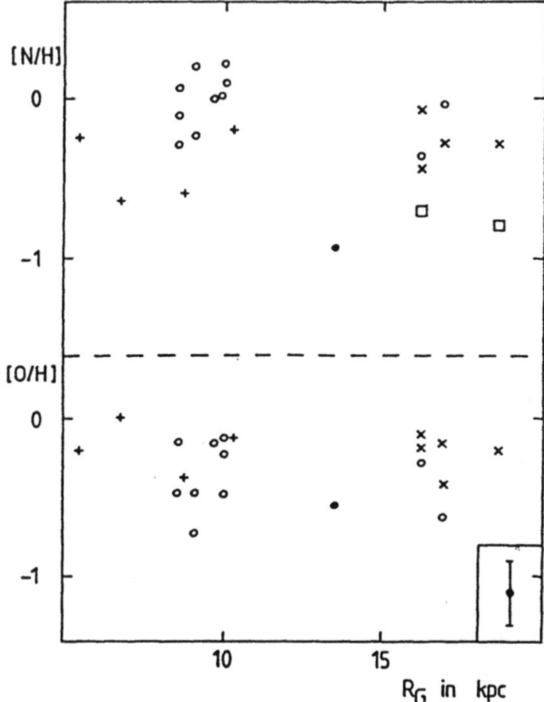

Figure 4: Abundance versus galactocentric distance. Symbols: ×=this work, o=Gehren *et al.*, 1985, •=Lennon *et al.*, 1990, +=Fitzsimmons *et al.*, 1990 und □=Fich and Silkey, 1991. Bottom right corner: typical error bar.

Acknowledgements

This work has been supported by the Deutsche Forschungsgemeinschaft under grants Ku 474/13-2 and Ku 474/16-1. The calculations were carried out on the Cyber machines of the Leibniz Rechenzentrum in München and the Cray of the Bayerische Akademie der Wissenschaften. Thomas Gehren kindly provided the spectra.

References

Becker, S.R., 1991 a, in *"Stellar Atmospheres: Beyond Classical Models"*, Ed. L. Crivellari, I. Hubeny and D.G. Hummer, NATO ASI series, Series C, No. **341**,353.
Becker, S.R., 1991 b, *thesis*, Ludwig-Maximilians-Universität München.
Becker, S.R. and Butler, K., 1990a, *Astron. Astrophys.*, **235**, 326.
Becker, S.R. and Butler, K., 1990b, *Astron. Astrophys. Supp. Ser.*, **84**, 95.
Becker, S.R. and Butler, K., 1992, *Astron. Astrophys., in preparation.*
Fich, M. and Silkey, M., 1991, *Astrophys. J.*, **366**, 107.
Fitzsimmons, A., Brown, P.J.F., Dufton, P.L. and Lennon, D.J., 1990, *Astron. Astrophys.*, **232**, 437.
Gehren, T., Nissen, P.E., Kudritzki, R.P. and Butler, K., 1985, in *"Production and Distribution of C, N, O Elements", ESO Conf. Workshop Proc.* No. **21**, Ed. I.J. Danziger, F. Matteucci, K. Kjär.
Gulati, R.K., Malagnini, M.L. and Morossi, C., 1989, *Astron. Astrophys. Supp. Ser.*, **80**, 73.
Hummer, D.G., Abbott, D.C., Voels, S.A. and Bohannan, B., 1988, *Astrophys. J.*, **328**, 704.
Kilian, J., Becker, S.R., Gehren, T. and Nissen, P.E., 1992 a, this book.
Kilian, J., Becker, S.R., Gehren, T. and Nissen, P.E., 1992 a, *Astron. Astrophys.*, , in preparation.
Kurucz, R.L., 1979 *Astrophys. J. Supp. Ser.*, **40**, 1.
Lennon, D.J., Lynas-Gray, A.E., Brown, P.J.F. and Dufton, P.L., 1986, *Monthly Notices Roy. Astron. Soc.*, **222**, 719.
Lennon, D.J., Dufton, P.L., Fitzsimmons, A., Gehren, T. and Nissen, P.E., 1990, *Astron. Astrophys.*, **240**, 349.
Remie, H. and Lamers, H.J.G.L.M, 1982, *Astron. Astrophys.*, **105**, 85.
Shaver, P.A., McGee, R.X., Newton, L.M., Danks, A.C. and Pottasch, S.R., 1983, *Monthly Notices Roy. Astron. Soc.*, **204**, 53.
Walborn, N.R., 1971, *Astrophys. J. Supp. Ser.*, **198**, 257.
Walborn, N.R. and Fitzpatrick, E.L., 1990, *P.A.S.P*, **162**, 379.

Galactic B-type Supergiants

D.J. Lennon [1]

[1]Universtäts-Sternwarte München, Scheinerstr. 1, DW-8000 München 80

Abstract: The chemical composition of the atmospheres of the B-type supergiants is an important diagnostic of their evolutionary history, in particular with regard to the question of whether or not their atmospheres become contaminated by the products of the CNO-cycle due to mixing during a previous red-supergiant evolutionary phase. There is some circumstantial evidence for this scenario; i)a number of analyses of OB-supergiants have implied a helium enrichment of their atmospheres and ii)there is evidence that the precursor of SN1987A (a B3 I supergiant) in the LMC was surrounded by a nitrogen rich cloud thought to have been ejected while the star was a red-supergiant. With regard to i), there is little convincing quantitative evidence for the expected CNO abundance anomalies which should accompany a helium enrichment. There are a number of difficulties in this area, for example, curve-of-growth and LTE abundance analyses of B-supergiants have been obliged to adopt rather large microturbulent velocities (values range between 10km/s and 25km/s). There are also inconsistencies in the reproduction of the He I line profiles and these problems are not yet adequately resolved by the use of plane-parallel non-LTE model atmospheres and line formation calculations. A current survey of 46 bright galactic B-supergiants does imply a correlation between the strength of the N II lines and the weakness of the C II lines but there is also evidence that this trend is also correlated with increasing luminosity. It is also clear that the He I lines are also much stronger than non-LTE model predictions for all B-supergiants in the sample. It is possible that at least part of this effect is due to inadequacies in the current models for these stars, a possibility that is currently under investigation. A more complete description of this work may be found in;

Lennon, D.J., Kudritzki, R.-P., Becker, S.R., Butler, K., Eber, F., Groth, H.G., Kunze, D., 1991, A& A, in press.

Lennon, D.J., Dufton, P.L., Fitzsimmons, A., 1991, A& A, in press.

Lennon, D.J., Dufton, P.L., Fitzsimmons, A., Kudritzki, R.-P., in preparation

On the determination of effective temperature and surface gravity using Strömgren $uvby\beta$ photometry

R. Napiwotzki*, D. Schönberner*, and V. Wenske
Institut für Theoretische Physik und Sternwarte der Universität Kiel
Olshausenstr. 40, W-2300 Kiel, Germany

Abstract: We compare different calibrations of the Strömgren $uvby\beta$ system with independently derived values of T_{eff} and g. Additionally we present new temperature calibrations of the $[u-b]$ index for normal and Ap/Bp stars.

Introduction

The photometric $uvby\beta$ system as designed by Strömgren and Crawford in the sixties is a powerful instrument to determine stellar parameters as effective temperature and surface gravity. Numerous calibrations have been published since then. By applying different recommendations we noticed

1. inconsistencies between these calibrations, and
2. discrepancies between photometrically and spectroscopically determined values of T_{eff} and g.

Thus we decided to investigate the nature of these differences and to remove them as far as possible (and necessary). In particular we investigated the calibrations of Moon & Dworetsky (1985), Lester et al. (1986), and Balona (1975).

For this purpose we selected sets of stars with well known T_{eff} from the literature and derived surface gravities from fitting theoretical Balmer profiles to new spectra obtained at the Calar Alto observatory, Spain. We give further color-temperature calibrations, with special attention to metal-rich Ap stars.

Temperature calibrations

We decided to use effective temperatures derived from the integrated flux given by Code et al. (1976), Beeckmans (1977), and Malagnini et al. (1986), because these temperatures are only little dependent on the employed model atmospheres. We didn't use temperatures derived by the infrared flux method for normal type stars, because it was shown by Mégessier (1988) that these temperatures differ systematically from those of Code et al. For Ap/Bp stars however, we had to rely on the infrared flux method (Mégessier, 1988, and Glushneva, 1987) owing to the lack of other temperatures. We will show that the differences between Bp/Ap stars and normal type A and B stars is larger than possible systematic errors.

Normal type stars: We divided the stars into cool ($T_{eff} < 8500\,K$), intermediate ($8500\,K < T_{eff} < 11000\,K$), and hot groups ($T_{eff} > 11000\,K$) according to Moon & Dworetsky, and applied the different temperature calibrations.

Moon & Dworetsky: This calibration fits our temperature scale best: there are no systematic differences exceeding 1.5%. Originally the calibration was limited to $T_{eff} < 20000\,K$. We extended it up to $T_{eff} = 30000\,K$ and found still good agreement.

*Visiting astronomer, German-Spanish Astronomical Center, Calar Alto, operated by the Max-Planck-Institut für Astronomie Heidelberg jointly with the Spanish National Commission for Astronomy

Lester et al.: This calibration fits the temperature scale well for the A stars but for hotter and cooler stars the agreement is less satisfactory (systematic differences up to 5 % at 20000 K).

Balona: The comparison reveals a strong trend with temperature: at $T_{\rm eff} = 10000$ K the temperatures are too high by 3 %, and too low by 5 % at 20000 K.

From our temperature standards we derived a new $[u-b]$-calibration. With $\Theta = \frac{5040\rm K}{T_{\rm eff}}$ we obtained:

$$\Theta = 0.1723 + 0.2792[u-b] - 0.018[u-b]^2$$

Ap/Bp stars: For these stars we got a different calibration. Let us note that we, in contrast to Mégessier, use the modern relation as given by Crawford & Mandwewala: $[u-b] = (u-b) - 1.53(b-y)$. The result is

$$\Theta = 0.2127 + 0.2399[u-b]$$

The temperature scale of Kilian et al.: Kilian et al. (1991) derived temperatures from the ionization equilibria of helium and silicon. Their NLTE line formation calculations are based upon the LTE models of Gold (1984). They compared their temperature scale to that of Lester et al. (1986) and found significant differences, which were assigned to NLTE effects.

We compared the temperature scale of Kilian et al. with that of Moon & Dworetsky, which is identical with that of our temperature standards. Again significant differences between LTE and NLTE temperatures occur, but astonishingly these are larger ($\approx$ 2000 K) in the region of 20000 K than in the 30000 K region. We infer that these differences are not caused by NLTE effects but by the insufficient line blanketing in the LTE models (only about 100 important lines were included) combined with NLTE effects for the hottest stars. We strongly recommend to use the fully blanketed model atmopsheres of Kurucz (1979).

Gravity calibrations

Our determination of surface gravities is based on the fitting of Balmer lines (mostly H_γ, in some cases also H_β) with theoretical profiles calculated by Kurucz (1979) from his fully blanketed LTE model atmospheres. Spectroscopy was carried out at the German-Spanish-Astronomical Center, Calar Alto, in two observing runs during Jan. 1987 and May 1990. We used the 2.2 m telescope in connection with the Coudé spectrograph and a CCD detector. The (reciprocal) dispersion was 8 Å/mm and 17 Å/mm.

We used bright field stars from the list of photometric β standards of Crawford & Mander (1966) supplemented by the well observed stars α Lyr and α CMa. To get surface gravities from the Balmer line fits, one firstly has to fix the effective temperature. We used the temperatures derived from Moon & Dworetsky (1985). The results were tested using the binary systems α CMa and AR Aur which showed excellent agreement. We also found no difference between the results from the fits of H_γ and H_β: the mean difference amounts to only 0.01 dex.

Moon and Dworetsky: There are small but significant deviations: $\log g_{\rm MD} - \log g_{\rm fit} = +0.036 \pm 0.015$ (s.e.m) for the intermediate group and $\log g_{\rm MD} - \log g_{\rm fit} = -0.069 \pm 0.016$ for the hot group.

Lester et al.: This calibration strongly deviates for the hotter stars (up to 0.4 dex). It is certainly caused by a wrongly adopted gravity of their only B-type standard star (η UMa): for $T_{\rm eff} = 17000$ K the profile fit gives $\log g = 4.2$ instead of 3.9 as assumed by Lester et al. (see Fig. 1).

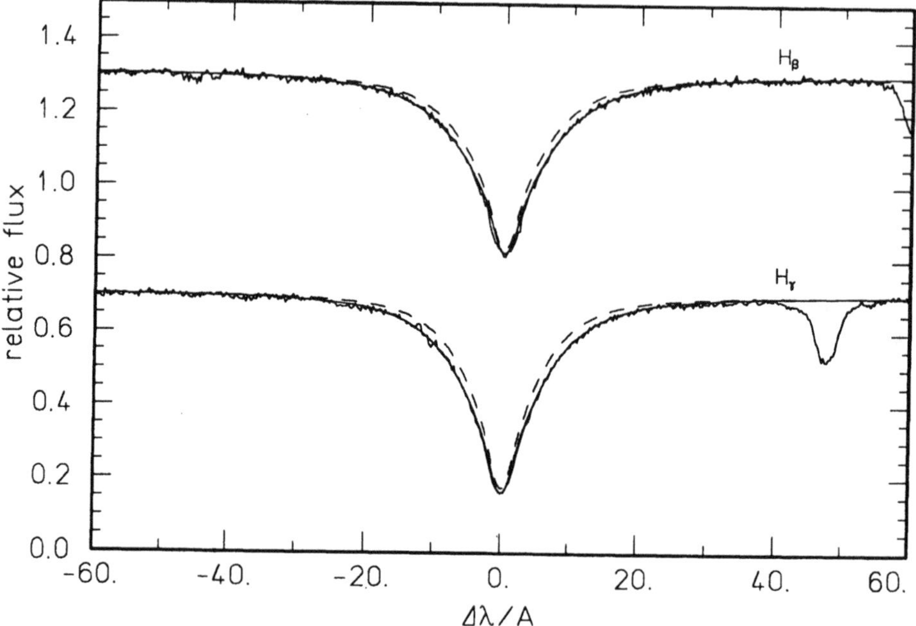

Fig. 1: H_β and H_γ lines of η UMa compared to theoretical profiles with $T_{\rm eff} = 17000$ K and $\log g = 3.9$ (dashed line) and $\log g = 4.2$ (solid line).

Balona: This calibration is at complete variance with our spectroscopic gravity determinations: the deviation varies from about 0.2 dex at 10000 K to 0.4 dex at 20000 K.

Conclusions

We recommend to use the Moon & Dworetsky calibration together with our small gravity corrections for the determination of effective temperatures and surface gravities (of main sequence stars) by means of the *uvbyβ* photometric system.

Acknowledgements: This work was supported by DFG travel grants.

References:

Balona, L.A.: 1975, MNRAS 172, 191
Beeckmans, F.: 1977, A&A 60, 1
Code, A.D., Davis, J., Bless, R.C., Hanbury Brown, R.: 1976, ApJ 203, 417
Crawford, D.L., Mander, J.: 1966, AJ 71, 114
Crawford, D.L., Mandwewala, N.: 1976, PASP 88, 917
Gold, M.: 1984, Diplomarbeit, Ludwig Maximilians Universität, München
Kilian, J., Becker, S.R., Gehren, T., Nissen, P.E.: 1991, A&A 244, 419
Lester, J.B., Gray, R.O., Kurucz, R.L.: 1986, A&AS 61, 509
Malagnini, M.L., Morossi, C., Rossi, L., Kurucz, R.L.: 1986, A&A 162, 140
Mégessier, C.: 1988, A&AS 24, 159
Moon, T.T., Dworetsky, M.M.: 1985, MNRAS 217, 305

The Mass and Helium discrepancy in massive young stars

A.Herrero [1], R.P. Kudritzki [2,3], J.M. Vilchez [1], D. Kunze [2],

K. Butler [2] and S. Haser [2,3]

[1]Instituto de Astrofísica de Canarias, E-38200 La Laguna, Tenerife, Spain
[2]Universität-Sternwarte München, Scheinerst. 1, D-8000 München 80, FRG
[3]Max-Planck Institut für Astrophysik, D-8064 Garching bei München

Abstract: We present the results of the spectroscopic analysis of 25 galactic luminous OB stars. Using these data, spectroscopic and wind masses are derived. Both masses agree within the errors and disagree with evolutionary masses. Helium abundances are higher than normal for many objects, which cannot be explained by evolutionary calculations. We believe that the differences are significant, and call them the *helium discrepancy* and the *mass discrepancy* respectively. We indicate some possibilities of improving the physics in both theories in order to reduce the discrepancies.

1 Introduction

Some indications may be found in previous work about discrepancies between the stellar masses derived from spectroscopical analyses using the stellar atmospheres and radiatively driven wind theories and those predicted by stellar evolution calculations (Groenewegen et al., 1989; Herrero et al, 1991). Also the helium abundances of some OB stars are larger than normal (Kudritzki et al., 1983, 1989a; Voels et al., 1989; Lennon et al., 1991) and cannot be explained by present evolutionary calculations (f.e., Maeder, 1990).

For that reason have observed single massive OB stars in open clusters (except ζ Oph). The observations were carried out with the 2.5m INT of the Observatorio del Roque de los Muchachos on La Palma. The spectral region covered ranges from 4000 to 5000 Å, with an effective resolution of 0.6 Å and S/N > 200. A model grid (H/He NLTE models with 43 NLTE levels) with $\Delta T_{\rm eff}$= 2500 K and $\Delta \log g$= 0.2 was calculated for two different values of the helium abundance ϵ= N(He)/ (N(H)+N(He)), ϵ= 0.09 and $\epsilon = 0.20$.

For the analysis we plot the fit curves of the H and He lines in the log g–$T_{\rm eff}$–ϵ diagram and choose the stellar parameters giving the smallest dispersion. The results are given in Table 1.

Table 1. Results of the spectroscopic analysis and mass determination. Masses are given in solar units

HD number	Sp. Clas.	T_{eff} (kK)	$\log g$	ϵ	$V_r \sin i$	log M_{spec}	Δ log M_{spec}	log M_{wind}	log M_{evol}
193 682	O5 III(f)	43.5	3.50	0.30	200	1.32	0.26		1.69
46 150	O5 V((f))	43.0	3.70	0.09	120	1.49	0.35	1.62	1.69
227 018	O6.5 III	41.0	3.70	0.09	80	1.27	0.22		1.56
190 864	O6.5 III (f)	41.0	3.55	0.17	105	1.41	0.22	1.46	1.67
191 612	O6.5 III f	40.0	3.60	0.09	80	1.55	0.22		1.68
217 086	O7 Vn	40.0	3.60	0.17	375	1.35	0.22	1.32	1.54
47 839	O7 V((f))	39.5	3.70	0.07	80	1.25	0.22	1.26	1.51
34 656	O7 II(f)	39.0	3.50	0.17	85	1.06	0.38	1.22	1.50
192 639	O7 Ib (f)	38.5	3.35	0.20	125	1.49	0.22	1.51	1.77
46 966	O8 V	38.0	3.90	0.09	80	1.46	0.22	1.24	1.47
193 514	O7 Ib(f)	38.0	3.35	0.12	105	1.50	0.22	1.56	1.77
214 680	O9 V	37.5	4.00	0.10	50	1.40	0.22		1.42
203 064	O7.5 III:n((f))	37.5	3.50	0.12	315	1.48	0.22	1.43	1.60
34 078	O9.5 V	36.5	4.05	0.09	40	1.31	0.38		1.36
227 757	O9.5 V	36.0	4.00	0.09	50	1.27	0.22		1.35
24 912	O7 I	36.0	3.25	0.18	250	1.71	0.38	1.67	1.78
13 268	O8 III	35.0	3.30	0.20	320	1.21	0.22	1.20	1.44
191 423	O9 III:n*	34.0	3.40	0.20	450	1.46	0.22		1.45
207 198	O9 Ib-II	34.0	3.30	0.12	80	1.22	0.22	1.25	1.50
210 809	O9 Iab	33.0	3.10	0.12	100	1.33	0.22	1.48	1.61
149 757	O9 V	32.5	3.50	0.16	400	1.19	0.38		1.29
209 975	O9.5 Ib	32.5	3.20	0.09	100	1.23	0.22	1.29	1.50
18 409	O9.7 Ib	31.5	3.10	0.12	160	1.14	0.22		1.45
228 199	B0.5 V	30.0	3.90	0.09	120	1.24	0.22		1.21
227 634	B0 Ib	28.5	3.20	0.18	115	1.20	0.22		1.38

2 The Helium and Mass Discrepancies

From Tab. 1 we see that: 1) All main sequence stars, except fast rotators, have normal helium abundances; 2) Most objects classified as fast rotators, Ia, Iab, f or (f) (9 out of 12) have abundances of 0.15 by number or more. All of them but one (HD 191 612) show some indication of helium enhancement ($\epsilon \geq 0.12$); 3) All objects not classified as Ia, Iab, f, (f) or fast rotators, except one (HD 227 634, which is the coolest and latest object) show helium abundances of 0.12 or below.

Our opinion is that most O giants and supergiants are showing CNO processed material and that this enrichement is reinforced when the star rotates rapidly or has a low gravity and high luminosity. This coincides with previous suggestions (Walborn, 1976; Voels et al., 1989) but our conclusion is based in the detailed analysis of 25 objects.

This helium enrichment is not explained by the evolutionary calculations (Maeder, 1990), which predict: a) if the stars are in their redwards tracks, they should show normal helium abundances; b) if they are in their bluewards tracks, they should show helium abundances much larger than the observed; and c) if they are in their bluewards tracks they should have mass loss rates one or two orders of magnitude larger than observed.

We have determined the stellar mass by three different methods (see Herrero et al., 1991): spectroscopically, using the theory of radiatively driven winds (Kudritzki et al., 1989b) and from the evolutionary calculations (Maeder, 1990).

The results are given in Tab. 1. Some stars have large errors due to distance uncertainties (for these objects nothing can be stated) and there is no discrepancy for main sequence objects. For all other objects not included in that categories there is a *large discrepancy* between M_{spec} (or M_{wind}) and M_{evol} and a *small discrepancy* between M_{wind} and M_{spec}. If the stars are in their bluewards tracks the sign of the mass discrepancy is inverted (as is the case also for the helium discrepancy).

We have tried to minimize the mass discrepancies, but these are in most cases larger than the error bars (note that it is not possible to take simultaneously the largest T_{eff} and the smallest $\log g$).

We conclude that the discrepancy between evolutionary masses and wind and spectroscopic masses is real and due to defficiencies in the physics of the calculations.

3 Discussion

If we represent the mass discrepancy (evolutionary minus spectroscopic mass) versus the distance to the Eddington limit in $\log g$ we find a correlation, which indicates that the mass discrepancy is probably related to mass loss, atmospheric extensions and winds.

Both kind of calculations (evolutionary/ wind-atmospheric) may be improved: on the spectroscopic side, wind contamination in H_{γ} could explain the small discrepancy between M_{wind} and M_{spec}. We do not include wind blanketing in our models, but we estimate that this effect does not solve the discrepancies. On the evolutionary side, the models of Maeder (1990) include mass loss and overshooting, but our results indicate that the mass loss rates near the ZAMS could be higher than adopted. From the models in the literature, only the homogeneous models (Maeder, 1987) could account for the enhanced helium abundances observed in some stars. For a comparison, however, detailed calculations are needed.

References

Groenewegen, M.A.T., Lamers, H.J.G.L.M., Pauldrach, A.W.A.: A&A **221** 78 (1989)

Herrero, A., Vilchez, J.M., Kudritzki, R.P., Butler, K., Kunze, D. in "Stellar Atmospheres: Beyond Classical Models", ed. L. Crivellari, I. Hubeny and D.G. Hummer, NATO ASI Series C, Vol. 341, pg. 409 (Kluwer Ac. Pub.) (1991)

Kudritzki, R.P., Cabanne, M.L., Husfeld, D., Niemela, V., Groth, H.G., Puls, J., Herrero, A.: A&A **226** 235 (1989a)

Kudritzki, R.P., Pauldrach, A., Puls, J., Abbott, D.C.: A&A **219** 205 (1989b)

Kudritzki, R.P., Simon, K.P., Hamman, W.R.: A&A **118** 245 (1983)

Lennon, D.J., Kudritzki, R.P., Becker, S.R., Butler, K., Eber, F., Groth, H.G., Kunze, D.: A&A, in press (1991)

Maeder, A.: A&A **178** 159 (1987)

Maeder, A.: A&AS **84** 139 (1990)

Voels, S.A., Bohannan, B., Abbott, D.C., Hummer, D.G.: ApJ **340** 1073 (1989)

Walborn, N.R.: ApJ **205** 419 (1976)

A COMPARISON BETWEEN THE ORBITAL MASSES OF EARLY TYPE BINARY COMPONENTS AND MASSES PREDICTED BY STELLAR EVOLUTION.

M. Vrancken [1], W. van Rensbergen [1], D. Vanbeveren [2].

(1): Astrophysical Institute,Vrije Universiteit Brussels, Pleinlaan 2, 1050 Brussels,

(2): I.H. Group T Leuven, Vuurkruisenlaan 2, 3000 Leuven.

ABSTRACT.

In this paper we show that for eight components of early type binary systems (including three supergiants) the orbital mass corresponds reasonably well with masses resulting from evolutionary computations.

INTRODUCTION.

The masses of early type single stars can be determined in two indirect ways:

1. Using T_{eff} resulting from a comparison of an observed and theoretically predicted photospheric spectrum of stars with a known distance and luminosity and comparing the HRD position with evolutionary tracks.

2. Using T_{eff} and $\log g$ resulting from a comparison of an observed and theoretically predicted photospheric spectrum of stars with a known distance and luminosity.

The masses which are obtained from the first method using a non-LTE code (Hubeny, 1988) are about 40% larger than the results predicted by the second method (see e.g. Kudritzki et al., 1986).

In this paper we consider a number of binary components for which an independent orbital mass can be computed and we will compare these masses to the theoretically predicted evolutionary masses.

BINARY SYSTEMS.

The number of massive binary components of which the evolutionary mass can be compared to the orbital mass is obviously small. The systems to which these components belong have to be detached (non-evolved), they should eclipse so that the orbital plane is known to be about in the line of sight and the distance (luminosity) must be known. A search through literature revealed four non-evolved candidates, i.e. CWCep, V453Cyg, V478Cyg and YCyg.

The orbital masses, luminosities and effective temperatures were taken from Popper (1980). Reminding that prior to the Roche lobe overflow phase, both components of a binary system evolve like single stars, we used evolutionary tracks of single stars of Maeder (1991).

In table 1 we compare the orbital and evolutionary mass of the four systems.

Table 1:
A comparison between orbital and evolutionary masses of primaries of four eclipsing non-evolved binaries.

star	spectral type	orbital mass range	mass range from evolution
YCyg HD 198846	O9.8	16.2 - 17.2	13.80 - 17.25
V478 Cyg HD 193611	O9.8	14.9 - 16.3	15.58 - 18.03
V453 Cyg HD227696	B0.4	13.3 - 15.7	15.77 - 17.94
CWCep HD 218066	B0.4	1.6 - 12	12.4 - 15.54

There are four X-ray binaries with an O or B type component (Joss et al., 1984). We compared the orbital mass of these components with the mass determined from evolution using computations of accretion stars of Vanbeveren (1990). These accretion stars resemble closely normal single stars. They are slightly overlumineous (max. 0.5 mag). Table 2 compares the orbital and evolutionary mass of these X-ray binaries.

CONCLUSIONS.

The orbital masses of the components of four massive non-evolved binaries and the optical companions in four massive X-ray binaries are in good agreement with the masses obtained from a comparison with evolutionary tracks even for the three supergiants in the sample.

Table 2:
A comparison between orbital mass and evolutionary mass of four massive X-ray binaries.

star	spectral type	orbital mass range	mass from evolution
SMC X-1	B0.5 I	13 - 21.5	≈ 14
Cen X-3	06.5 IIIe	16.5 - 25	≈ 24.5
Vela X-1	B0.5 I	21.5 - 26.5	≈ 21
4U1538 - 52	B0I	12 - 30.5	≈ 20

FINAL REMARK.

New LTE models with a large number of lines predict log g values which are 0.3 dex larger for early type stars compared to non-LTE models with a restricted amount of lines (Kurucz, 1991). This largely removes the discrepancy between the masses determined by the two methods given in the introduction.

REFERENCES.

Hubeny, I., 1988, Comp. Phys. Comm. 52, 103.
Joss, P.C., Rappaport, S.A., 1984, Ann. Rev. Astron. Astrophys. 22, 537.
Kudritzki, R.P. and Hummer, D.G., 1986, in Luminous Stars and Associations in Galaxies, editors C.W.H. De Loore et al.
Kurucz, 1991, private communication.
Maeder, A.,1991, Astron. Astrophys. (in press).
Popper, D.M., 1980, Ann. Rev. Astron. Astrophys. 18, 115.
Vanbeveren, D.: 1990, Astron. Astrophys. 234, 243.

B–TYPE STARS IN YOUNG CLUSTERS

D. SCHÖNBERNER AND V.WENSKE
Institut für Theoretische Physik und Sternwarte der Universität Kiel
Olshausenstr. 40, W-2300 Kiel, Germany

We studied B–type stars of two young stellar clusters, viz. NGC 2264 and α Per (Mel 20), in order to address the following questions:

1. Is star formation in clusters coeval, meaning that all stars lie on one single isochrone?
2. Does some mixing of processed matter (helium) to the surface occur during the main–sequence phase?
3. What is the influence of rotation on the stellar loci in the $T_{\rm eff}$,g plane?

Therefore, high–quality spectrogrammes centered at H_γ were obtained with the Coudé–spectrograph of the Calar Alto 2.2 m telescope, in conjunction with a CCD detector. The resolution is 0.25 Å, with $\frac{S}{N} \geq 50$. These observations were supplemented by *uvbyβ* photometry from the literature, and by IUE spectrogrammes from the VILSPA database if available.

The effective temperatures were derived from the photometric c_0 index (dereddened) according to the calibrations of Moon & Dworetsky (1985). For those stars with available IUE spectrogrammes, we determined $T_{\rm eff}$ also from the total flux. The mean difference between both temperature determinations is only $(10 \pm \frac{150}{\sqrt{14}})$ K. Except for $c_0 \geq 0.8$, $T_{\rm eff}(c_0)$ is insensitive to gravity.

For given temperature, surface gravity, $v \sin i$, and $\frac{He}{H}$ were determined by matching theoretical line profiles based on Kurucz's fully line–blanketed model atmospheres to the observations, viz. H_γ for gravity and *He* I 4387 Å for $\frac{He}{H}$.

NGC 2264:

Fig. 1 shows the analysed objects of the young cluster NGC 2264, together with the ZAMS ($X = 0.70, Z = 0.03$) from Hejlesen (1980). Since that particular cluster is very young, (age $\approx 5 \cdot 10^6$ yr) one expects that its stars would define an observational ZAMS, hopefully close to the theoretical one. This is not the case. Closer inspection, however, shows that only stars with large $v \sin i$ deviate from the theoretical ZAMS, whereas the slow rotators ($v \sin i < 200 \frac{\rm km}{\rm s}$), except one, behave as expected.

In order to quantify the effects of stellar rotation on the loci of these stars, we employed the extensive calculations for rigidly rotating stellar models by Collins and Sonneborn (1977). Assuming that the distorted surface of a rotating star, viewed under the aspect angle i with regard to the axis of rotation, can be characterized by a mean surface temperature and gravity, we can transform Collins & Sonneborn's $\Delta c(v \sin i)$, $\Delta\beta(v \sin i)$ into the corresponding values $\Delta T_{\rm eff}(v \sin i)$, $\Delta \log g(v \sin i)$ with the help of Moon & Dworetsky's (1985) photometric grid. Stellar rotation reduces, in general, gravity *and* temperature, and this effect is clearly seen in Fig. 1 where the displacement of the ZAMS for $i = 60°$ but different $w = \frac{\omega}{\omega_{\rm crit}}$ is indicated by the dashed lines. It appears evident that rotation is responsible for the "ZAMS widening" in NGC 2264. The effect is, however, only significant if $w > 0.5$!

We tried to correct for the effects of rotation using $i = 60°$, which corresponds closely to mean value for random orientation. Only for a few objects (3 out of 10) we got better results by choosing $i < 45°$ or $i \approx 90°$ resp. The final result is presented in Fig. 2.

α Per:

Fig. 3 shows the $T_{\rm eff}$, g plane for the α Per cluster, as it results directly from our analysis, together with isochrones from Maeder & Meynet (1991). This figure seems to support two distinct age groups, namely $\approx 6 \cdot 10^7$ yr and $2 \cdot 10^8$ yr, as proposed by Eggen & Iben (1988).

Fig. 4, same as Fig. 3, but after corrections for the rotational effects have been applied in the same manner as discribed above. The ambiguity about the cluster age disappeared, and one can assigne one single age for α Per, viz. $(6 \pm 1) \cdot 10^7$ yr, based on the isochrones of Maeder & Meynet (1991).

From our analysis of the helium lines, it appears that α Per is marginally enriched in helium compared to NGC 2264: $\frac{He}{H} = 0.108 \pm \frac{0.008}{\sqrt{8}}$ vs. $0.094 \pm \frac{0.015}{\sqrt{7}}$. In this context it is interesting to note that Lyubimkov (1988) contended that helium is mixed to the photosphere during central hydrogen burning, and that this mixing is more effective for larger stellar masses. For the brightest α Per members one can estimate from his figures that $\Delta n_{\rm He} \approx 0.02...0.04$. Our derived helium abundances do not indicate any differential helium enrichement in α Per.

In conclusion, we found by means of accurate spectroscopic determination of stellar parameters and application of simple rotating stellar models:

- Fast rotation explains the relatively wide main sequence of the young cluster NGC 2264.
- If the effects due to rotation are corrected for, the B–type stars in α Per indicate *one single* cluster age.
- Differential enrichement of surface helium is not detectable in α Per.

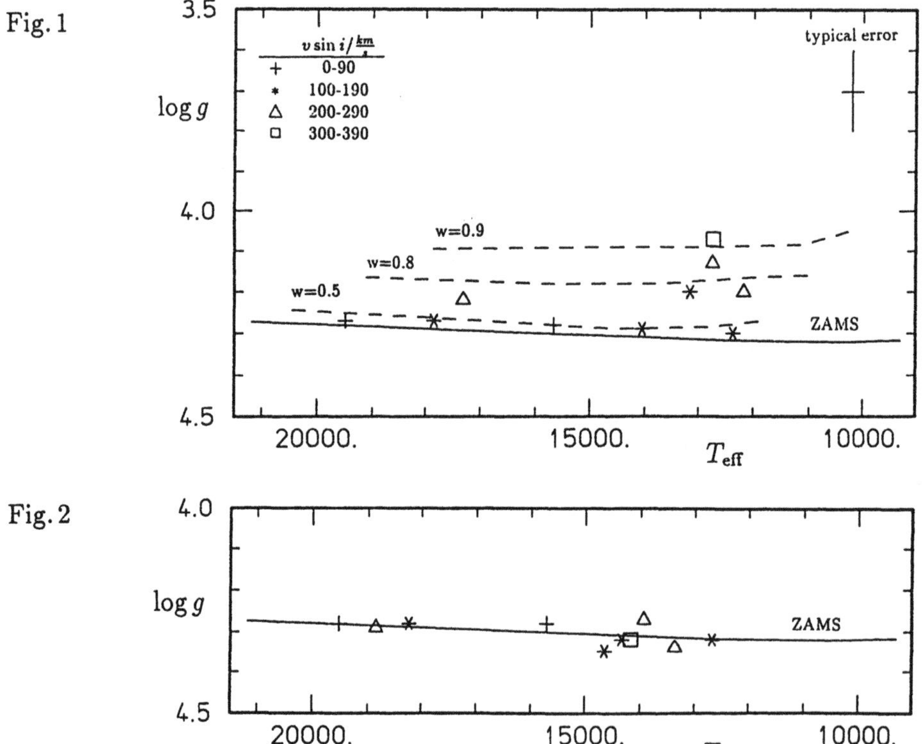

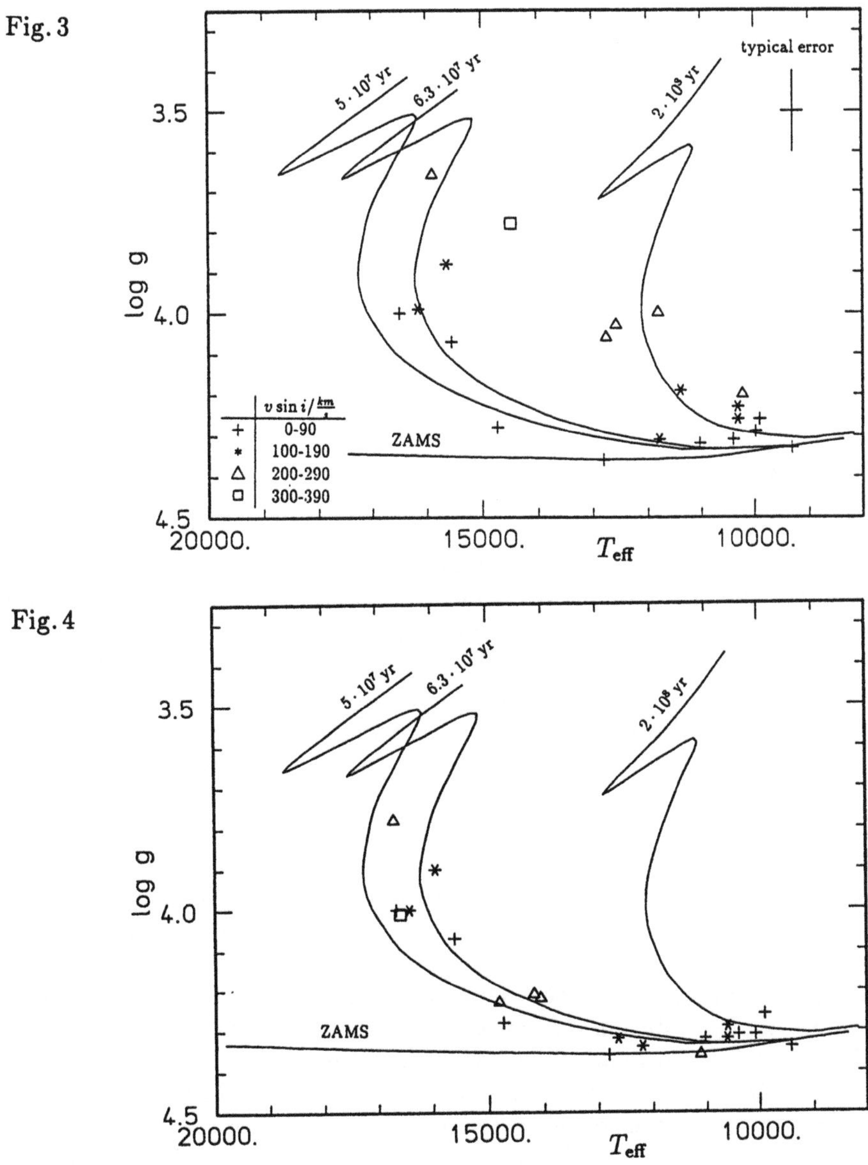

References:

Collins, G.W., Sonneborn, G.: 1977, *Astrophys. J. Suppl.* **34**, 41

Eggen, O.J., Iben, I.Jr.: 1988, *Astron. J.* **96**, 635

Hejlesen, P.M.: 1980, *Astron. Astrophys. Suppl.* **39**, 347

Lyubimkov, L.S.: 1988, *Astrofizika* **29**, 479

Maeder, A., Meynet, G.: 1991, *Astron. Astrophys. Suppl.* **89**, 451

Moon, T.T., Dworetsky, M.M.: 1985, *Monthly Notices Roy. Astron. Soc.* **217**, 305

Chemical abundances in early B-type stars

J. Kilian, S.R. Becker, T. Gehren, and P.E. Nissen

By means of high resolution Echelle spectra of ESO/Chile in the wavelength range from 4060 Å to 5060 Å with in total about 470 identified transitions of H, He, C, N, O, Ne, Mg, Al, Si, S, Ar, and Fe (Kilian et al., 1991a) observed by one fo us (P.E. Nissen) we determined chemical abundances (He, C, N, O, and Si) for 21 main sequence B-stars in the local field, the Ori OB1 association, and the Sco Cen association ($r \approx 1000$ pc). The aim of our work is to get some information about the evolutionary state of these stars and the chemical composition of the interstellar matter out of which the stars are.

Instead of using photometric quantities the stellar parameters were determined by fitting theoretical NLTE hydrogen line profiles ($H\beta$ and $H\gamma$) and equivalent width ratios of NLTE silicon lines from different ionization stages to the observed spectra (Kilian et al. 1991b). The microturbulent velocity was estimated using oxygen lines and the abundance analysis was made by comparing theoretical and observed equivalent widths (Kilian & Nissen 1989). NLTE model spectra of hydrogen and helium (Herrero 1987, Becker, Butler & Husfeld private communication) and carbon, nitrogen, oxygen, and silicon (Eber & Butler 1988, Becker & Butler 1988, 1989, 1990) made it possible to determine reliably effective temperatures, gravities, microturbulent velocities, and absolute abundances.

We preferred line blanketed LTE atmospheres (Gold 1984) to unblanketed NLTE atmospheres (Mihalas 1972) since inspection shows that the effect of line blanketing on the atmospheric structure in the formation region of silicon lines and hydrogen line wings is more pronounced than that of deviations from local thermodynamic equilibrium (Montenbruck 1987).

Using $T_{\rm eff}$ and $\log g$ from the detailed analysis of hydrogen line profiles and the silicon ionization equilibrium we made a new calibration for the relation between the reddening corrected photometric index $[c_1]$ and the effective temperature. Comparing these results with the $[c_1] - T_{\rm eff}$ calibration of Lester et al. (1986) shows up to 2000 K higher temperatures. Due to the restricted set of reference stars used to establish the $[c_1] - T_{\rm eff}$ relation of Lester et al. the calibration is subject to high systematic errors for B-stars in the temperature range from 25000 K to 30000 K.

An essential result is the fact that the microturbulent velocity is not negliable although the microturbulence was mainly explained by NLTE effects. The value of the microturbulence ($v_{\rm turb} \leq 14$km/s) is less than in LTE analyses but not at all zero. Additional there is an increase of $v_{\rm turb}$ during the main sequence as the evolution increases.

The so estimated chemical abundances for the members in the different groups of stars show deviations up to 0.8 dex whereas the abundance error is about 0.1 dex. The mean abundances of He, C, N, and O of the field stars and both association are rather equal: He is about solar and C, N, and O are about half the solar abundance. This leads to the conclusion that the interstellar matter is very well inhomogen within short distances while the mean values of the abundances of different associations show a large scaled homogen distribution of the interstellar matter.

Using evolutionary models from Maeder & Meynet the HRD for the observed stars with $T_{\rm eff}$ and $\log g$ determined here shows that all stars are nearby main sequence stars. Their typical age is 10 million years and their masses going from 8 to 20 solar masses.

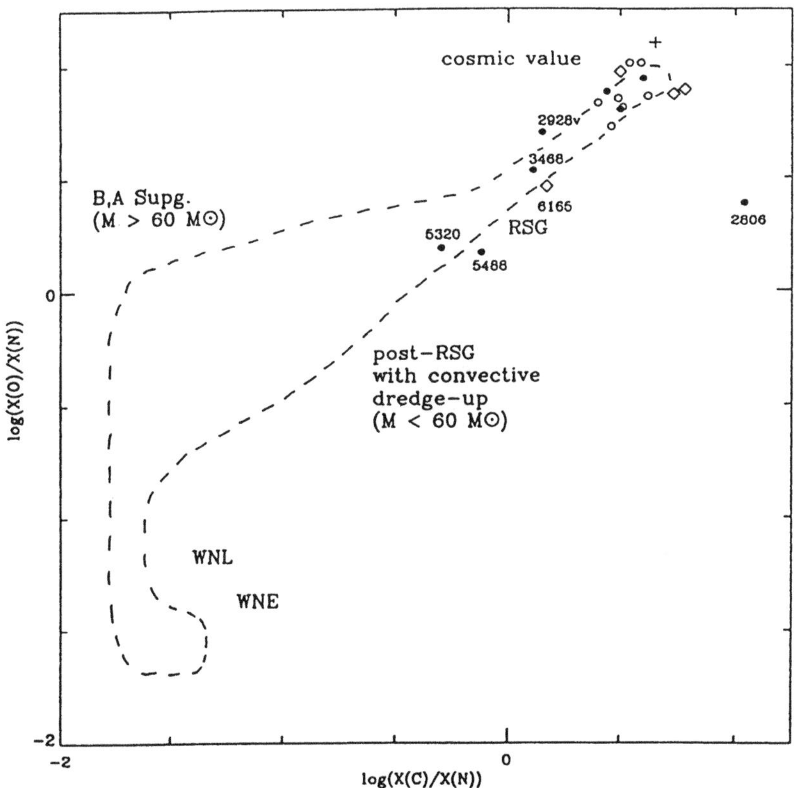

Figure 1: Correlation of O/N to C/N. The marked area gives the abundance relation for evolved stars by Maeder (1985). The symbols mark field stars (•) and stars of the Ori OB1 (∘), Sco Cen (⋄), and Sgr OB1 (+) associations.

During the main sequence the energy of B-stars in the observed mass range is produced due to the CNO cycle. Although the reactions in this cycle are calalytic the relative abundances of C, N, and O are changing. Whereas C/N and O/N are going from 1/10 to twice the solar the C/O mass fraction is nearly the same in all stars (about 1:3). The correlation of C/N to O/N (see Fig. 1) shows that almost all stars lie in a small band indicating the solar C/O abundance. This means that carbon and oxygen already existed in the interstellar matter out of which the stars evolved. That speaks well for an uniform formation of these elements by the helium burning phase of earlier star generations. Only in the atmospheres of HR 5320 and HR 5488 a nitrogen abundance above the average was detected and the matter seems to be mixed. One explanation of this phenomenon could be the assumption of quick rotation. Models of fast rotating rich mass stars by Maeder (1987) show that during the main sequence turbulent diffusion of processed matter from the core to the surface is possible.

References:

Becker S.R., Butler K.: 1988, *Astron. Astrophys.* **201**, 232
Becker S.R., Butler K.: 1989, *Astron. Astrophys.* **209**, 244
Becker S.R., Butler K.: 1990, *Astron. Astrophys.* **235**, 326
Eber F., Butler K.: 1988, *Astron. Astrophys.* **202**, 153
Herrero A.: 1987, *Astron. Astrophys.* **171**, 189
Kilian J., Nissen P.E.: 1989, *Astron. Astrophys. Suppl. Ser.* **80**, 255
Kilian J., Montenbruck O., Nissen P.E.: 1991a, *Astron. Astrophys. Suppl. Ser.* **88**, 101
Kilian J., Becker S.R., Gehren T., Nissen P.E.: 1991b, *Astron. Astrophys.* **244**, 419
Lester J.B., Gray R.O., Kurucz R.L.: 1986, *Astrophys. J. Suppl. Ser.* **61**, 509
Maeder A.: 1985, in *Production and Distribution of C, N, O Elements*, ESO Conference and Workshop Proceedings No. 21, eds. I.J.Danziger, F.Matteucci, K.Kjaer, p. 187
Maeder A.: 1987, *Astron. Astrophys.* **178**, 159
Maeder A., Meynet G.: 1988, *Astron. Astrophys. Suppl. Ser.* **76**, 411
Mihalas D.: 1972, *Astrophys. J.* **177**, 115
Montenbruck O.: 1987, *Diplomarbeit Universität München*

The Determination of Accurate Cosmic Abundances from B-type Stellar Spectra

P.J.F. Brown, P.L. Dufton, F.P. Keenan,

D.E. Holmgren and G.A. Warren

Department of Pure and Applied Physics, Queens University of Belfast, Belfast BT7 1NN, United Kingdom

Abstract: We have obtained high-resolution, high-signal-to noise observations of weak A II and O II absorption lines formed in the atmospheres of main-sequence early-type stars. The observed line strengths have been combined with equally well determined oscillator strengths, in an analysis which used both LTE and non-LTE model atmosphere techniques. Cosmic argon and oxygen abundances (on a logarithmic scale with hydrogen = 12) of [A] = 6.49 and [O] = 8.93 have been determined; these should have an accuracy of ±0.05 dex.

1 Introduction

We are undertaking an extensive programme to derive accurate cosmic abundances using weak stellar absorption lines (with typical equivalent widths $\leq$5 mÅ) observed in the high resolution optical spectra of slowly rotating main-sequence B-type stars. Due to their weakness, these lines lie on the linear part of the curve of growth and hence have equivalent widths which are sensitive to the element abundance, but less so to the assumptions made in the model atmosphere analysis. As main-sequence stars normally have atmospheres uncontaminated by the products of interior reactions, these abundances should reflect those of the interstellar material from which the stars formed some $10^6 - 10^7$ yrs ago, and therefore reliable contemporary cosmic abundance estimates are available using this method. Such estimates are essential in, for example, the study of element depletions in the interstellar medium, where the cosmic abundance values employed have usually been determined from the sun and meteorites, and are hence strictly applicable only to the interstellar medium as it was some 5×10^9 yrs ago.

Here we discuss results for argon and oxygen both elements that have been extensively observed in the interstellar medium.

2 Observations

We have obtained high resolution ($\simeq$0.09 Å FWHM), high signal-to-noise ($\simeq$200) observations of weak A II and O II absorption lines using the coudé spectrograph with a CCD detector on the Coudé Feed Telescope at the Kitt Peak National Observatory. Figures 1 illustrate the high quality of the observational data for an A II line in the B2V star, γ Peg. These data were reduced on the Northern Ireland STARLINK node (Bromage 1984) using the packages FIGARO and DIPSO

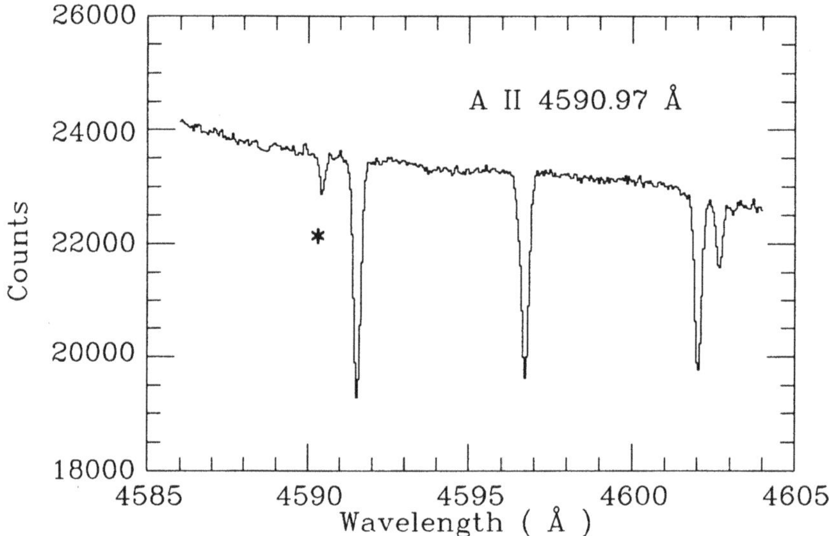

Fig. 1. The spectrum of γ Peg from 4585 to 4605 Å, which clearly shows the weak A II line at 4590 Å (marked with an asterisk). This feature has an equivalent width of $W_\lambda = 6.3 \pm 0.5$ mÅ.

3 Analysis

Oscillator strengths for the A II 4590 and 4658 Å transitions were taken from Garcia and Campos (1985), who measured A II lifetimes (and hence f-values) to an accuracy of better than 10% using the delayed-coincidence method. The O II oscillator strengths are those of Bell et al (1991) calculated using the CIV3 code of Hibbert (1975) and again should be accurate to better than 10%. Initially an LTE abundance analysis was undertaken but for A II this was supplemented by further calculations (Holmgren et al 1990) which showed non-LTE effects to be insignificant. Similarly previous calculations (Brown et al 1988) indicate that non-LTE effects in O II should also be relatively small.

The equivalent widths and abundance estimates for argon and oxygen are summarized in tables 1 and 2 respectively.

Table 1. Equivalent widths (W_λ) and abundances ([A]) for argon on a logarithmic scale with hydrogen =12.

Star	λ (Å)	W_λ (mÅ)	[A]	ΔW_λ	ΔT_{eff}	$\Delta \log g$	Δv_T
γ Peg	4590	6.3 ± 0.5	6.50	$^{+0.04}_{-0.01}$	± 0.00	$^{+0.03}_{+0.02}$	$^{-0.01}_{+0.02}$
γ Peg	4658	5.0 ± 1.0	6.43	$^{+0.06}_{-0.11}$	$^{+0.02}_{-0.00}$	$^{+0.04}_{-0.03}$	$^{-0.01}_{+0.02}$
HR 1765	4590	6.8 ± 0.5	6.53	$^{+0.04}_{-0.03}$	$^{-0.02}_{+0.05}$	$^{+0.08}_{-0.04}$	$^{-0.03}_{+0.06}$
δ Cet	4590	6.3 ± 0.5	6.49	$^{+0.01}_{-0.03}$	$^{+0.01}_{-0.02}$	± 0.01	∓ 0.01

Table 2. Equivalent widths (W_λ) and abundances ([O])for oxygen on a logarithmic scale with hydrogen =12..

Star	λ (Å)	W_λ (mÅ)	[O]	ΔW_λ	ΔT_{eff}	$\Delta \log g$	Δv_T
τ Sco	4752	6.1 ± 0.5	8.91	± 0.04	$^{+0.02}_{-0.09}$	$^{-0.05}_{+0.01}$	$^{-0.02}_{+0.01}$
τ Sco	4845	6.1 ± 1.0	8.95	$^{+0.07}_{-0.08}$	$^{+0.06}_{-0.05}$	$^{-0.00}_{+0.05}$	∓ 0.01
ξ^1 CMa	4752	10.6 ± 0.5	8.93	$^{+0.03}_{-0.02}$	± 0.00	$^{+0.04}_{-0.01}$	$^{-0.02}_{+0.05}$

ΔW_λ, ΔT_{eff}, $\Delta \log g$ and Δv_T show the effects of varying the equivalent width by the error estimate, and the effective temperature, gravity and microturbulent velocity by ± 1000 K, ± 0.3 dex and $\pm 5\, km\, s^{-1}$, respectively.

4 Results

The main results can be summarized as follows

– The cosmic argon abundance is [A] $= 6.49 \pm 0.05$. This is similar to that estimated by Veck and Parkinson (1981; [A] $= 6.38^{+0.18}_{-0.30}$) from an analysis of solar flare data, but has a much smaller observational uncertainty.

– Adoption of the present argon abundance determination rather than the often used Withbroe (1971) value of [A] $= 6.65$, leads to effectively zero depletions for unreddened interstellar sightlines, such as those toward α Vir and λ Sco. This is consistent with such sightlines containing few interstellar grains.

– The cosmic oxygen abundance is [O] $= 8.93 \pm 0.05$ This is in excellent agreement with the solar value of [O] $= 8.91 \pm 0.02$ (Lambert 1978).

– Our derived oxygen abundance is substantially larger than that obtained for H II regions in the solar vicinity (8.70 ± 0.04; Grevesse 1984). This supports the suggestion of Rubin (1989) that abundance ratios relative to ionized hydrogen may be underestimated in the presence of a varying density environment.

References

Bell, K.L., Hibbert, A., Stafford, R.P. 1991. J. Phys. B, to be submitted.
Bromage, G.E. 1984. Proc. 4th. European IUE meeting, ESA SP-218, 473.
Brown, P.J.F., Dufton, P.L., Lennon, D.J., Keenan, F.P. 1988.????
Garcia, G., Campos, J. 1985. J. Quant. Spectr. Rad. Trans., **34**, 85.
Grevesse, N. 1984. Phys. Scripta, **T8**, 49.
Hibbert, A. 1975. Computer Phys. Comm., **9**, 141.
Holmgren, D.E., Brown, P.J.F., Dufton, P.L., Keenan, F.P. 1990. Astrophys. J., **364**, 657.
Lambert, D.L. 1978. Mon. Not. R. astr. Soc., **182**, 149.
Rubin, R.H. 1989. Astrophys. J. Suppl., **69**, 897.
Veck, N.J., Parkinson, J.H. 1981. Mon. Not. R. astr. Soc., **197**, 41.

The Origin of Distant B-type Stars in the Galactic Halo

E.S. Conlon, R.J.H. McCausland, P.L. Dufton,

F.P. Keenan and D.E. Holmgren

Department of Pure and Applied Physics, The Queen's University of Belfast, Belfast BT7 1NN, United Kingdom

Abstract: Using model atmosphere analyses of high resolution optical spectra, we have identified a group of young B-type stars at large distances from the galactic plane. A study of the kinematics and evolutionary ages of these objects reveals two groups of stars: those that could have formed in the disc and travelled to their present locations in their lifetimes and those that could not. The kinematics of the first group are in agreement with results from N-body simulations of dynamical ejection from young galactic star clusters. For the stars that could not have formed in the disc, star formation in the halo via collisions within intermediate and high velocity clouds appears the most likely explanation.

1 Introduction

In a series of studies, we have identified a substantial number of blue stars at high galactic latitudes which appear to be normal Population I objects at large distances from the galactic plane (see Conlon et al., 1990 and references therein). In order to distinguish these stars from subluminous, nearby objects with similar effective temperatures and surface gravities - for example the evolved post-AGB stars (McCausland et al., 1991) - a quantitative abundance analysis of high resolution spectra must be performed. Once the Population I nature of the stars is confirmed, their kinematics are studied to establish if they could have formed in the disc and been ejected out into the halo or if they must have formed in situ. Apart from the intrinsic interest in the origin of this group of stars, they are also relevant to studies of the interstellar medium towards galactic halo targets (York, 1982).

2 Observations

In this paper we present results for a group of eight stars observed with the 3.9m Anglo-Australian Telescope and the Image Photon Counting System (IPCS) over a two year period from July 1986 to August 1988. The Intermediate Dispersion and University College London échelle spectrographs were employed resulting in resolutions of 0.1 – 0.3 Å (FWHM of reference arc lines) over the wavelength region 3800 – 4700 Å. Reduction and analyses of the objects was performed using the STARLINK supported packages FIGARO (Shortridge, 1986) and DIPSO (Howarth and Murray, 1988). As the stars appeared to be normal hydrogen-burning main sequence B-type stars, Kurucz (1979) LTE model atmospheres were used to derive atmospheric parameters and elemental abundances; however in the case of Si II lines, non-LTE methods were employed (Lennon et al., 1986). Comparison of the derived abundances with a normal Population I composition indicates that the present sample are indistinguishable from young disc objects, apart from their location. Further details of the observations and model atmosphere analyses may be found in Conlon et al., (1991).

3 Kinematics

With the Population I nature of these objects confirmed, their kinematics were investigated in order to test the hypotheses that these stars were formed

– as normal young B-type stars in the spiral arms and were subsequently ejected violently out into the halo, i.e. they are distant counterparts to the so-called "OB runaway stars";

– away from the plane of our galaxy.

From the derived atmospheric parameters and the evolutionary tracks of Hjelesen (1980), masses, ages (T_{evol}) and stellar distances are found. Flight-times are then calculated assuming the current z-component of the space velocity can be approximated by the radial velocity, corrected for differential galactic rotation (Kerr and Lynden-Bell, 1986). Using the z-distance and velocity and adopting the acceleration laws of House and Kilkenny (1980), the flight-times (T_{flight}) and initial z-velocity required to reach the present z-distance are calculated.

Comparison of flight-times with evolutionary ages has revealed two sets of stars: namely, stars that could have formed in the disc, i.e. $T_{flight} \leq T_{evol}$, and stars that could not have formed in the disc, i.e. $T_{flight} \gg T_{evol}$.

4 Runaway Stars

For the stars that could have formed in the disc, we have plotted the ejection velocities as a function of their masses (in units of solar mass) in Figure 1. Also plotted are results found for 32 previously identified halo B-type stars (Conlon et al., 1990). The sample is divided into four groups as follows

— $|b| \geq 60°$ and $|z| \geq 1$ kpc (A) or 400 pc $\leq |z| \leq 1$ kpc (C).
— $45° \leq |b| \leq 60°$ and $|z| \geq 1$ kpc (B) or 400 pc $\leq |z| \leq 1$ kpc (D).

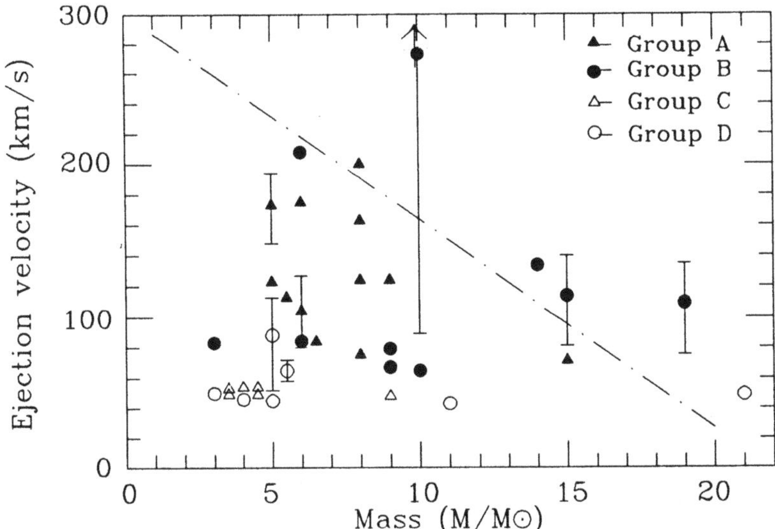

Fig. 1. Estimated ejection velocities plotted against stellar masses, together with the limits of simulations of gravitational encounters in a typical young star cluster (Leonard, 1989; dot-dashed line). Typical observational errors are $\pm$ 10 km s^{-1} and $\pm$ 1 $M_{\odot}$ respectively, unless otherwise indicated. The sample is divided into four groups with the results for A and C being the more reliable.

For Massive Close Binary ejection mechanisms, the runaway B-type star should be more massive than $\simeq$ 10 $M_{\odot}$ (Stone 1982). The eight stars with masses compatible with this limit show no evidence of the expected binary nature. Conversely, the limits of the results of N-body simulations of the dynamical ejection of runaway stars from young galactic star clusters (Leonard and Duncan, 1989), are generally in excellent agreement with our observations. One exception is TS 195, which requires an ejection velocity of 330 km s^{-1}; Leonard (1991) has shown that such velocities are possible through cluster ejection but they should be rare. Thus it appears that these halo stars are the most distant subgroup of B-type runaways produced by Cluster Ejection yet observed.

5 Star Formation in the Halo

A small group of stars require flight-times that are much longer than their evolutionary ages. In the present sample three stars fall into this category, namely BD -15° 115, SB 357 and BD -2° 3766. These are listed in Table 1, together with five stars found in previous studies. Also given are stellar distances from the Sun and z-distances from the galactic plane. It can be seen that these objects have been found over a large range of z-distances. In addition they are located both above and below the disc.

Table 1. Stars that appear to have formed in the halo

Star	Spectral Type	Distance (kpc)	z-distance (kpc)	T_{evol} (Myr)	T_{flight} (Myr)
PG0832+676	B1	31.0	17.8	13	145
PHL346	B1	9.5	8.0	16	60
BD-15° 115	B2	7.3	7.1	20	47
SB357	B2e	9.1	9.0	25	64
BD-2° 3766	B1	4.6	3.8	16	40
HD18100	B1	3.1	2.8	11	27
HD214080	B1	2.1	1.7	10	26
HD217505	B2	1.5	1.2	6	40

It therefore appears that these stars have formed in situ in the halo. Halo star formation is believed to be possible via collisions between cloudlets within intermediate- and high-velocity clouds (Dyson and Hartquist, 1983). The near-solar composition found for these objects implies that the interstellar clouds may have originated in the disc, possibly as condensed galactic fountain gas (Bregman, 1980).

Given the importance of these objects, a more detailed kinematical analysis is being performed and the results will be published in the near future. In addition, we are now involved in a systematic survey of a region of the sky ($\simeq$ 10,000 degree^2) to obtain the density distribution and hence production rate of young B-type stars in the halo.

References

Bregman, J.N. 1980, Astrophys. J., **236**, 577.
Conlon, E.S., Dufton, P.L., Keenan, F.P., Leonard, P.J.T. 1990, Ast. Astrophys., **236**, 357.
Conlon, E.S., Dufton, P.L., Keenan, F.P., McCausland, R.J.H., Holmgren, D.H., Pettini, M. 1991, Mon. Not. R. Astr. Soc., submitted.
Dyson, J.E., Hartquist, T.W. 1983, Mon. Not. R. Astr. Soc., **203**, 1233.
Hjelesen, P.M. 1980, Ast. Astrophys. Suppl., **39**, 347.
House, F., Kilkenny, D. 1980, Ast. Astrophys., **81**, 251.
Howarth, I.D., Murray, J.M. 1988, SERC Starlink User Note, No. 50.
Kerr, F.J., Lynden-Bell, D. 1986, Mon. Not. R. Astr. Soc., **221**, 1023.
Kurucz, R.L. 1979, Astrophys. J. Suppl., **40**, 1.
Lennon, D.J., Lynas-Gray, A.E., Brown, P.J.F., Dufton, P.L. 1986, Mon. Not. R. astr. Soc., **222**, 719.
Leonard, P.J.T. 1991, Astr. J., **101**, 562.
Leonard, P.J.T., Duncan, M.J. 1989, Astr. J., **96**, 222.
McCausland, R.J.H., Dufton, P.L. Conlon, E.S., Keenan, F.P. 1991, Astrophys. J., submitted
Shortridge, K., 1986, SERC Starlink User Note, No. 86.4.
Stone, R.C. 1982, Astr. J., **87**, 1.
York, D.G. 1982, Ann. Rev. Astr. Astrophys., **20**, 221.

Galactic Abundance Gradients from OB-type stars in Young Clusters and Associations

A. Fitzsimmons, P.J.F. Brown, P.L. Dufton and W.R.J. Rolleston

Department of Pure and Applied Physics, Queens University of Belfast,
Belfast BT7 1NN, United Kingdom

Abstract: High-resolution spectra of main sequence B-type stars in young galactic clusters have been analysed using LTE and (where appropriate) non-LTE model atmosphere calculations. A re-analysis of the O II data from 11 clusters indicates that within 2 kpc of the sun there appears to be no obvious abundance gradient with respect to galactocentric distance. However, the cluster Dolidze 25 lying at $R_g = 13.5$ kpc has been found to be oxygen deficient, implying that the abundance gradient may be non-linear. A comparison with studies of H II regions indicates that within the uncertainties the two methods yield consistent abundances. We are currently analysing clusters with 10 kpc $\leq R_g \leq$ 12kpc and will shortly be obtaining data on clusters with $R_g \leq 16kpc$.

1 Introduction

For a number of years we have been undertaking abundance analyses for main-sequence B-type stars in young galactic clusters and associations. The principal aim is to deduce the present-day chemical composition of the interstellar medium as a function of position in our galaxy. Concentrating on clusters has two advantages over observing field stars. Firstly previous photometry of clusters will generally give an accurate distance determination, enabling their position in the galaxy to be precisely placed. Secondly, if we assume that the progenitor interstellar cloud was homogeneous in composition, and that the formation time of the cluster was less than the lifetime of the most massive stars (*i.e.* no enrichment of the cloud took place), then abundances from cluster members may be combined to reduce random errors. Such studies will complement those of H II regions (see for example Shaver et al 1982, Fich and Silkey 1991) by increasing the numbers of elements that can be studied. Additionally the accuracy of the cluster studies (potentially better than ±0.1dex in relative abundances) compares well with that available from H II region analyses.

2 Observations and Analysis

High and medium resolution spectra have been obtained for stars in eleven clusters using the 2.4m Isaac Newton, 3.9m Anglo-Australian and 1.9 South African Astronomical Observatory telescopes. The distribution of clusters is illustrated in figure 1. Spectra were reduced on the Northern Ireland STARLINK node (Bromage 1984) using the DIPSO and FIGARO packages. The line strength and profiles were then analysed using LTE and non-LTE model atmosphere techniques (see Brown et al, 1986; Fitzsimmons et al, 1990; and Lennon et al, 1990 for details).

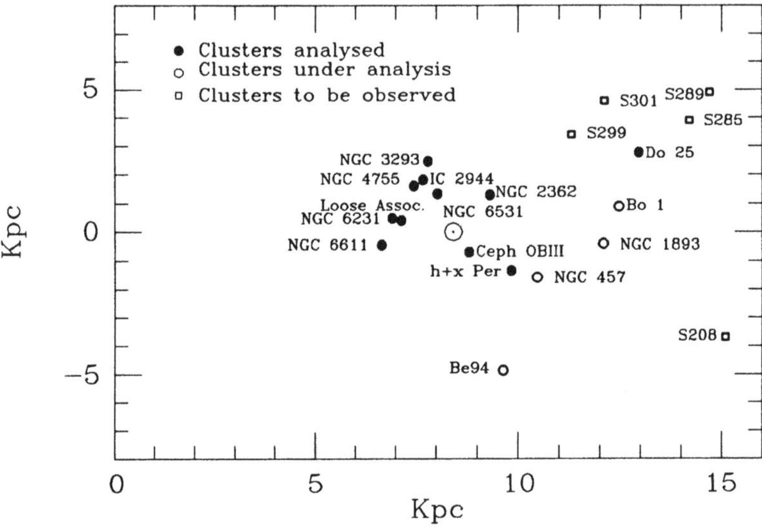

Fig. 1. Positions of young clusters in our observational program projected onto the galactic plane. The galactic center lies at $(0,0)$ and the sun ($\odot$ symbol) at $(8.5,0)$.

For the current study, we have re-analysed all the available O II observational data to deduce oxygen abundances. This should ensure a homogeneous set of results, which should be particularly reliable for determining abundance variations. Following the methods discussed in Fitzsimmons et al (1990), a subset of the targets were chosen in order to minimise uncertainties due to possible errors in the atmospheric parameters.

3 Results

The oxygen abundances (on a logarithmic scale with hydrogen having an abundance of 12) are given in table 1 and shown diagrammatically in figure 2. There appears to be no well defined decrease in [O/H] over the range of 6 kpc $\leq R_g \leq$ 10 kpc, assuming $R_\odot = 8.5$ kpc. Indeed least squares fits over this range of distances leads to an effectively zero abundance gradient. The corresponding oxygen abundance in the solar neighbourhood is [O/H] = 8.87 ± 0.12 dex, in good agreement with the solar value of 8.91 dex (Lambert 1978).

Table 1. Oxygen abundances in 11 young galactic clusters and associations. n refers to the number of stars used for the determination of [O/H].

Cluster	n	Heliocentric Distance (kpc)	Galactocentric Distance	$[\frac{O}{H}]$
NGC 6611	4	2.6 ± 0.3	6.1 ± 0.3	8.80 ± 0.06
NGC 6231	2	2.1 ± 0.3	6.9 ± 0.3	9.14 ± 0.16
NGC 6531	1	1.2 ± 0.3	7.3 ± 0.3	8.87 ± 0.15
NGC 4755	5	1.9 ± 0.1	7.6 ± 0.1	8.80 ± 0.14
IC 2944	3	2.0 ± 0.2	7.9 ± 0.1	8.86 ± 0.07
Loose Assoc.	2	1.4 ± 0.3	8.1 ± 0.1	8.96 ± 0.09
NGC 3293	7	2.8 ± 0.3	8.2 ± 0.1	8.87 ± 0.07
Cepheus OBIII	5	0.7 ± 0.2	8.8 ± 0.1	8.68 ± 0.06
NGC 2362	2	1.5 ± 0.3	9.4 ± 0.2	8.81 ± 0.11
h + χ Per	2	1.9 ± 0.1	9.9 ± 0.1	8.92 ± 0.09
Dolidze 25	3	5.2 ± 1.3	13.2 ± 1.2	8.32 ± 0.19

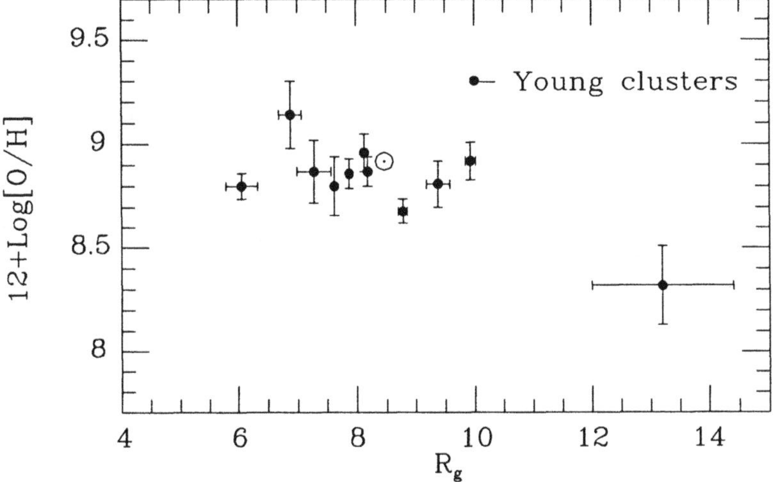

Fig. 2. Oxygen abundances as derived from young clusters and associations. ⊙ shows the solar photospheric abundance.

For the cluster Dolidze 25, we do find qualitative agreement with H II region results, in that the metallicity is lower at larger R_g. Indeed, including this cluster in a least-squares fit lead to an abundance gradient of -0.08 dex kpc^{-1}, in agreement with H II region studies. However, from this limited dataset it appears that the abundances may not be a simple linear function of R_g.

The above results appear to differ from those of Shaver et al (1982) who deduced a linear relationship between logarithmic abundance and R_g from studies of H II regions. In figure 3, we plot both our own results and those of Shaver et al and Fich and Silkey (1991). It is clear that within the uncertainties the two sets of abundances are consistent. Hence we conclude that the oxygen abundance gradient is effectively zero in the solar neighbourhood but for larger galactocentric distances there is a decrease in abundance.

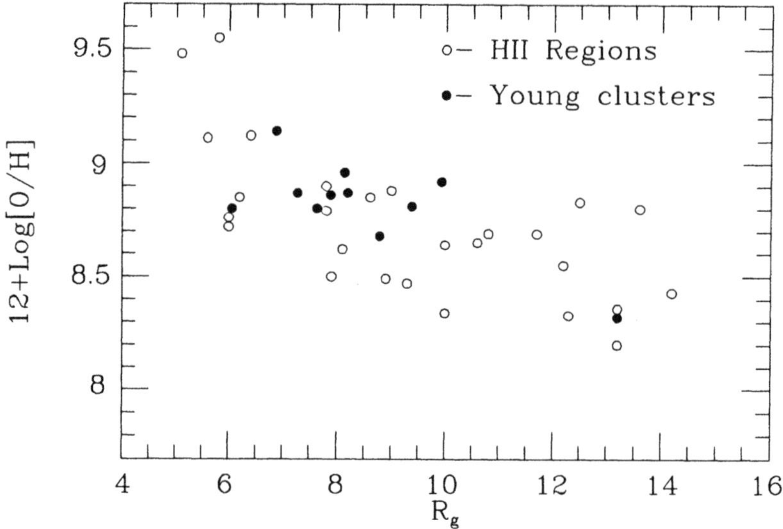

Fig. 3. Comparison of abundance determinations from B-type stars and H II regions. Within the uncertainties the two datasets are consistent.

We are currently analysing further clusters with $R_g > 10$ kpc and in forthcoming observations we will obtain spectra for galactocentric distances upto $R_g = 16$ kpc. The position of these clusters are also shown in figure 1.

References

Bromage, G.E. 1984. Proc. 4th. European IUE meeting, ESA SP-218, 473.

Brown, P.J.F., Dufton, P.L., Lennon, D.J., Keenan, F.P. 1986. Mon. Not. R. Astr. Soc., **220**, 1003.

Fich, M., Silkey, M. 1991. Astrophys. J., **366**, 107.

Fitzsimmons, A., Brown, P.J.F., Dufton, P.L., Lennon, D.J. 1990. Astr. Astrophys., **232**, 437.

Lambert, D.L. 1978. Mon. Not. R. astr. Soc., **182**, 149.

Lennon, D.J., Dufton, P.L., Fitzsimmons, A., Gehren, T., Nissen, P.E. 1990. Astr. Astrophys., **240**, 349.

Shaver, P.A., McGee, R.X., Newton, L.M., Danks, A.C., Pottasch, S.R. 1982. Mon. Not. R. Astr. Soc., **204**, 53.

Ionizing Radiation from Early-type Stars

Dietmar Kunze, Rolf-Peter Kudritzki, Joachim Puls

Institut für Astronomie und Astrophysik der Universität München
Scheinerstraße 1, D–8000 München 80, Germany

Abstract: A grid of NLTE model atmospheres for early-type stars is presented. Model atoms of eleven different elements with various abundances and, for a subset of the models, effects of wind blanketing, are taken into account. These model atmospheres supply realistic emergent fluxes which can be used in ionization calculations for circum- and interstellar gas. The dependence of the ionizing radiation of these stars on spectral type, luminosity class, evolutionary state and stellar environment is discussed.

1 Introduction

The radiation from hot and luminous early-type stars causes the gas in various objects, like planetary nebula, extended HII-regions and the wide-spread gas of starburst galaxies, to radiate. The understanding of the physical conditions and the interpretation of the observational properties of these circum- and interstellar gas clouds depend strongly on the knowledge of the spectrum of the ionizing radiation source. Excitation of circum- and interstellar gas takes place due to successive photoionization of atoms, starting from the ground state of the neutral ion. The photoionization edges of the ground states of the neutral and moderate charged ions lie short of 3000 Å, most of them between 2000 Å and 228 Å, where the radiation of hot early-type stars has its maximum. These stars are at the same time very luminous and, depending on their evolutionary stage, drive a moderate to strong stellar wind. The expanding envelope strongly influences certain parts of the emergent spectrum, like resonance lines or the ground state continuum of ionized Helium, but fortunately hardly the optical and UV continuum ($\lambda > 228$ Å), which is formed deep in the photosphere. Therefore a photospheric (i.e. plane-parallel and hydrostatic) approach is appropriate to compute ionizing fluxes form early-type stars. Because NLTE effects play an important role in the atmospheres of these stars, they must be included in the model atmospheres. As the emergent flux short of the Lyman edge of Hydrogen is strongly influenced by metal opacities, NLTE model atmospheres are needed which take opacities of heavier elements into account.

With modern approximate Λ-operator techniques it is possible nowadays to overcome the limitations of the classical complete linearization method by Auer and Mihalas (1969) and to include detailed model atoms of many elements besides Hydrogen and Helium with a large number of NLTE levels for every element and all radiative and collisional transitions between them. Additionally the backreaction of the expanding envelope onto the photospheric stratification can be summarized in a frequency-dependent albedo, which treats the backscattering and backward emission in the stellar wind as a partial reflection of the emergent flux in the upper boundary condition of the model atmosphere. Computation of such detailed NLTE model atmoshperes can supply reliable emergent UV fluxes for ionization calculations for circum- and interstellar gas.

2 NLTE Model Atmosphere Calculations

This work bases on plane-parallel and hydrostatic stratified NLTE model atmospheres in radiative and statistical equilibrium computed with the approximate Λ-operator technique according to Werner (1986) using the linear approximate Λ-operator proposed by Olson, Auer and Butchler (1987). A grid of model atmospheres was calculated for different values of effective temperature (T_{eff} = 30000 − 51000 K), surface gravity ($\log g$ = 4.00−close to Eddington limit), Helium-abundance ($\epsilon_{He} = 0.10, 0.20$) and metal-abundances ($\epsilon_{met} = 0.00*, 0.01*, 0.1*, 0.3*, 1*, 3*, 10*solar$), representing B- and O-stars of various spectral subtype and luminosity class in different evolutionary states and different environments like SMC, LMC or our Galaxy. Besides Hydrogen and Helium, model atoms of the most abundant, heavier elements C, N, O, Ne, Mg, Al, Si, S and Ar were included and treated completely in NLTE. For every element two to five ionization stages were considered. Overall 150 to 200 NLTE levels were taken into account and about 500 radiative and 1000 collisional transitions were linearized. Additionally about 1000 LTE levels were included to satisfy particle conservation.

For a subset of the models the backreaction of the stellar wind on the photosphere was taken into account in form of a frequency-dependent albedo as introduced by Abbott and Hummer (1985). This albedo was computed with the Munich wind code basing on the theory of radiative driven stellar wind by Castor, Abbott and Klein (1975) in the formulation of Pauldrach, Puls and Kudritzki (1986). It solves the problem of a spherical, hydrodynamical stratified, radiative driven stellar wind under NLTE conditions. Emergent fluxes of the detailed NLTE model atmospheres without wind blanketing were used as photospheric input fluxes for the stellar wind models. Frequency-dependent albedos at the sonic point were evaluated during the stellar wind calculations and used in the upper boundary condition of the model atmospheres. This process was iterated to convergence.

3 Results

Comparing the emergent spectra of different models of the grid, several conclusions can be drawn about the ionizing radiation of early-type stars. Generally there is a strong increase of the flux and a decrease in the Lyman jump of hydrogen with the effective temperature for all luminosity classes. But for giants and especially supergiants the flux in the Balmer continuum is smaller and the flux in the Lyman continuum is higher compared with main sequence stars with the same effective temperature. The Lyman jump for these stars is therefore much smaller and the UV spectrum somewhat flatter.

For supergiants above 50000 K, the Lyman jump even goes into emission. The ground state edge of neutral Helium shows the same NLTE effect at 35000 K. Because of the ionization of He I to He II this edge disappears totally for higher effective temperatures.

Metallicity effects can be studied by comparing the emergent fluxes of models with different metal contents. They depend strongly on the spectral type and appear with increasing effective temperature at higher frequencies, because the ionization equilibrium tends towards more highly charged ions, which have their bf-edges at shorter wavelengths. For the stars looked at here, the Balmer and Lyman continua up to the ground state edge of neutral Helium are therefore hardly affected, if metals are included. Only the cooler models show significant changes in the Lyman continuum. But the flux between the Lyman edge of neutral and singly ionized Helium falls dramatically (up to several orders of magnitude), if metal opacities are taken into account. This effect is non-linear with the metal contents and less dramatic for metal abundances greater than three times solar. For (super-)giants metallicity effects are in general smaller and concentrated at shorter wavelengths compared with main sequence stars with the same effective temperature.

Ionization calculations often use the number of ionizing quanta for different continua as input. These numbers can be calculated from the model atmospheres for every bf-edge $\lambda > 228$ Å. They increase strongly with the spectral type and also with the luminosity class but differently for every edge. The number of ionizing quanta for the He I ground state and bf-edges at shorter wavelengths show a decrease of one to some orders of magnitude with increasing metal abundances. For the cooler models, they decrease with increasing Helium abundance as well.

For stars with moderate bolometric albedo $A_{bol} < 0.1$, like metal poor stars or main sequence stars with not too high effective temperature, albedo effects are of minor importance. But for stars with a strong wind the frequency-dependent albedo changes the emergent spectrum in a significant way and the albedo should therefore be taken into account. A comparison of the ionizing fluxes from the wind blanketed NLTE model atmospheres and the stellar wind calculations show in general a good agreement.

In conclusion it should be noted that the ionizing radiation of early-type stars depends not only on spectral type, but also on luminosity class, chemical composition and wind blanketing. A grid of NLTE model atmospheres that provides reliable emergent fluxes for early-type stars as a function of this paramters is available.

References:

Abbott, D.C., Hummer, D.G.: 1985, *Astrophys. J.*, **294**, 286.
Auer, L.H., Mihalas, D.: 1969, *Astrophys. J.*, **158**, 641.
Castor, J., Abbott, D.C., Klein, R.: 1975, *Astrophys. J.*, **195**, 157.
Olson, G.L., Auer, L.H., Butchler, J.R.: 1987, *J.Q.S.R.T.*, **35**, 431.
Pauldrach, A., Puls, J., Kudritzki R.P.: 1986, *Astron. Astrophys.*, **164**, 86.
Werner, K.: 1986, *Astron. Astrophys.*, **161**, 177.

Abundance Patterns in A Stars: Carbon and Silicon

Hartmut Holweger

Institut für Theoretische Physik und Sternwarte
Universität Kiel
Olshausenstrasse 40, 2300 Kiel
Federal Republic of Germany

Abstract

It is shown that the seemingly erratic abundance variation of carbon among normal A stars is in fact a tight anticorrelation with silicon and other elements that may be explained by gas-dust separation in the protostellar or circumstellar environment.

1. Observation and analysis of sharp-lined normal A stars

The present study is based on results of the Kiel A star program, which is primarily intended to establish a homogeneous set of abundance data for a well-defined sample of A-type main sequence stars with no obvious chemical or evolutionary peculiarities.

The observational prerequisite was a uniform set of high-resolution spectra. The spectra analysed have been recorded with the ESO Coudé Echelle Spectrometer and a Reticon detector at a resolving power of 50 000. In order to ensure the desired accuracy of equivalent widths the sample was restricted to stars with v sin i < 50 km/s; it comprises all *sharp-lined B9.5 - A2 stars* classified as normal in the Bright Star Catalogue (Hoffleit and Jaschek 1982) which are known to be single stars or widely separated binaries and are observable from La Silla, Chile. The entire sample consists of 17 objects, and includes also *Sirius*. In addition to the ESO data high-dispersion photographic spectra of *Vega* recorded with the Coudé spectrograph of the 100-inch reflector at Mount Wilson Observatory were kindly made available to us in reduced form by Drs. R. and R.E.M. Griffin, Cambridge.

An overview of the observations and a first assessment of the *chemical anomalies* obviously existing among this reputedly homogeneous group of stars has been given by Holweger et al. (1986a,b). Detailed model-atmosphere analyses, taking into account departures from LTE in certain elements, are now available (Gigas 1986, 1988; Lemke 1989, 1990). The non-solar composition of Vega and the other 'normal' A stars has become obvious.

The model-atmosphere analysis by Lemke (1989, 1990) has confirmed the earlier suspicion that a considerable range in metallicity exists, and that the variations from star to star of the iron group are accompanied by parallel variations of lighter (Mg, Si) and heavier elements (Sr, Ba). *Carbon*, on the other hand, apparently behaves erratically, and no correlation with other elements or stellar parameters could be found.

The origin of these surface anomalies remains to be identified. The interplay of radiative and gravitational forces in the stellar envelope and atmosphere, eventually in conjunction with meridional circulation and mass loss or accretion, obvioulsy can lead to a wide variety of abundance patterns. Some of the predictions of diffusion theory worked out by Michaud and co-workers have been discussed by Lemke (1990, see references therein). For the normal A stars a decisive case has yet to be made out.

The purpose of the present paper is to demonstrate that the apparently erratic behaviour of *carbon* is in fact a tight *anti*correlation with other elements, which may indicate that another type of chemical separation processes not normally considered important in this context contributes to the observed abundance anomalies.

2. Carbon and silicon in the atmosphere of A stars

The peculiar behaviour of carbon among the elements investigated in the A star program suggests to look out for other properties that make this element exceptional. The only such property I was able to identify links the stellar carbon problem to the equilibrium of grains and gas in interstellar or circumstellar matter. Among the elements considered, carbon has by far the lowest *condensation temperature*, T_c, because it can remain in the gas phase in molecular form (mainly as CO) rather than being locked into grains. For carbon, $T_c \approx 80$ K, while all other elements studied in our sample of A stars have T_c values in excess of 1200 K (see e.g. the review on interstellar matter by Jenkins (1987), an references therein). This volatility of C is reflected in the comparatively small depletion of carbon (and of N, O, S, and the rare gases alike) in the interstellar gas as compared to the more refractory elements, which are almost completely removed from the gas phase.

If - during star formation or in a more recent accretion episode - grains and gas are combined in their original proportions, the photosphere of the star will have normal population I composition (unless other separation processes occur). If, on the other hand, the star happens to acquire an excess of *gas* - as the result of gas-solid fractionation having occurred somewhere on the way onto the star - then one would expect gas phase elements like carbon to be enriched with respect to refractory elements like silicon or iron, i.e. $[C/Si] > 0$, while silicon will be deficient with respect to hydrogen, i.e. $[Si/H] < 0$. In this simple (and certainly oversimplified) picture C and H are left unseparated, and one should expect $[C/H] = O$. Alternatively, if there is an excess of *dust*, the same kind of anomaly will occur, but with reversed sign. In other words, a variation of [C/Si] from star to star may indicate a *variable gas/dust ratio*, and if gas-solid fractionation has really occurred, one would expect a concomitant variation of [Si/H] in the sense that [C/Si] and [Si/H] are *anticorrelated*.

The normal A stars appear to comply with this theoretical prejudice surprisingly well (Fig. 1). Even if the gas-dust separation model should turn out inacceptable for some reason, it has led to find a connection between carbon and the other elements.

Fig. 1 is based on the original data given in Table 1 of Lemke (1990), who also quotes results for

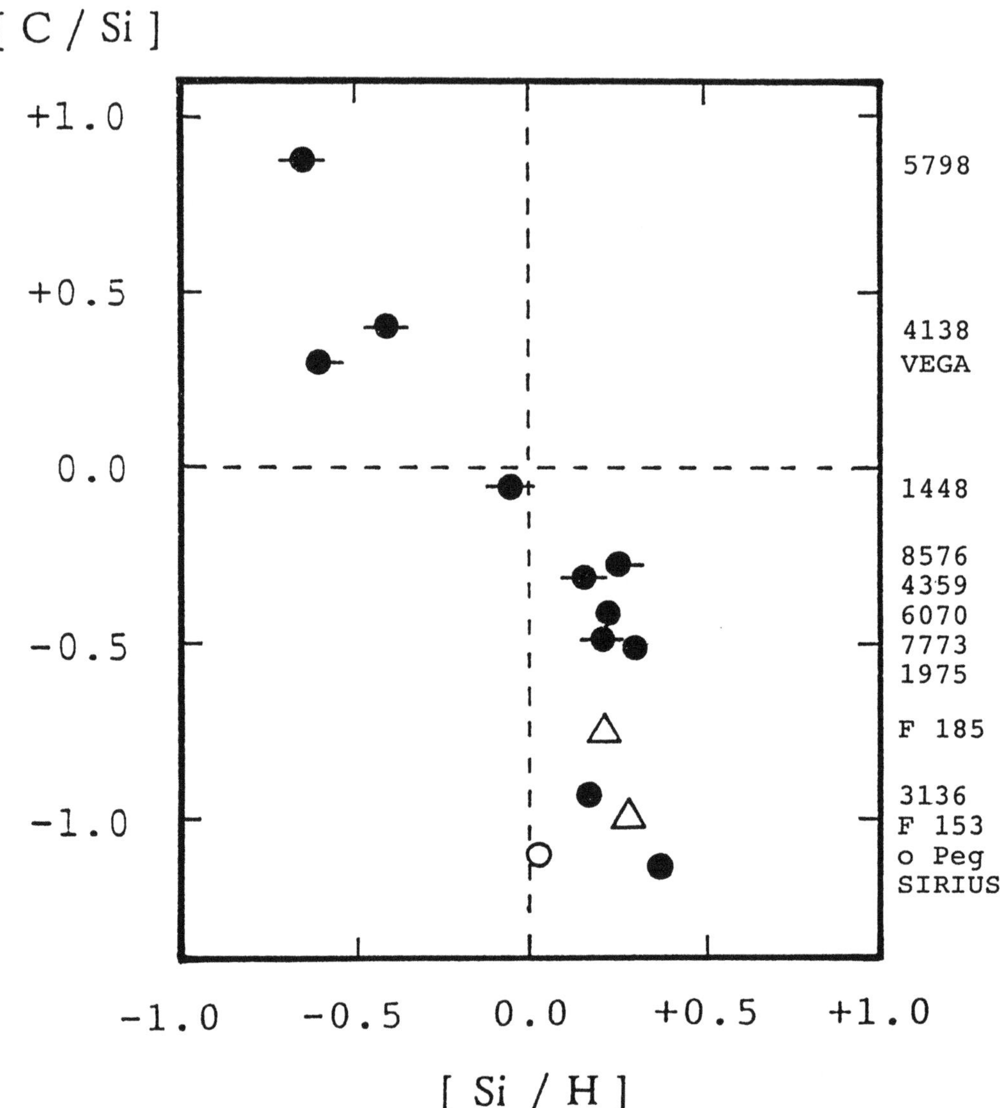

Fig. 1. Logarithmic abundance ratios with respect to the Sun. The filled circles designate 'normal' A stars studied in the Kiel A star program. The stars are identified by their names or HR numbers, respectively. Open circle: Omicron Pegasi, a hot Am star (Adelman 1988a,b). Open triangles: blue stragglers in the old open cluster M67 (Mathys 1991). Symbols with bars: see Sect. 3.

Vega obtained by Gigas (1986, 1988, and unpublished data), with the following exceptions. (1) For C I lines in Vega Non-LTE calculations are now available (Stürenburg and Holweger 1990). The Non-LTE corrections of Vega have been adopted for the other A stars of the program as well, taking into account their dependence on line strength. These corrections are small, typically -0.03 dex, and do not exceed -0.07 dex. Non-LTE corrections for Si are unknown but are likely to be even smaller because the abundance is derived from lines of Si II which is the dominant stage of ionization in A stars. (2) HR5959 has been omitted because the effective temperature adopted by Lemke is probably too low (Lemke 1991, private communication). (3) Owing to a typographical error a wrong value for the Si abundance of HR4138 (7.31 ± 0.09) appears in the Table. The correct value following from Lemke's Table 3 is 7.13. (4) The *solar* abundance of carbon has been changed from 8.69 to 8.58, in accordance with the recent solar analysis by Stürenburg and Holweger (1990). None of these modifications of Lemke's original data is of critical importance here.

Modern determinations of both C *and* Si in A-type stars are scarce. While I have not attempted to make an exhaustive survey of literature data, I have encountered two suitable recent sources and included them in Fig. 1. One is *Omicron Pegasi*, a typical *hot* Am star, which has been analysed by Adelman (1988a,b). Its effective temperature, 9600 K, fits well into the range spanned by the normal A stars shown. As can be seen, o Peg approximately conforms to the relation defined by the normal A stars. Perhaps more surprising in this context are the more recent results of Mathys (1991) for two *blue stragglers* in the old open cluster M67. Although blue stragglers are not believed to resemble normal A stars with regard to internal structure and evolution, their surface chemistry appears to be similar. This resemblance is not restricted to carbon and silicon; Mathys finds heavy elements like Sr and Ba to be overabundant - the same tendency is observed among normal A stars.

3. The C-Si relation: diffusion theory vs. gas-dust separation

The just mentioned overabundance of Sr and Ba in normal A stars that came to light in Lemke's (1990) work has prompted him to propose a connection with the Am phenomenon that is commonly explained within the framework of diffusion theory.

The obvious question now is: what is left of the observed C-Si variability (and of the gas-solid fractionation model) if the contribution of gravitational and radiative separation is subtracted? While at the moment probably nobody is able to perform this arithmetics theoretically because of the complexity of the physics involved, I suggest an empirical approach as a substitute.

As a measure of the extent to which diffusion processes have been active in a given star I propose to use the observed overabundance of Ba, more specifically, the overabundance relative to Ca, an element with similar condensation (and other) properties, but different response to diffusion (see the Am stars). Accordingly, two sets of stars have been marked in Fig. 1 by horizontal bars, one with [Ba/Ca] $\leq$ 0.8 (one bar), which spans the lower half of the observed range in [Ba/Ca], and a subset of five objects with closely coinciding overabundances, [Ba/Ca] = 0.6 ± 0.16 (two bars). All the other stars in Fig. 1, including o Peg and the blue stragglers of M67, have [Ba/Ca] values in excess of +1.0.

As can be seen, there is a clear tendency for 'Am-like' stars (those with more extreme Ba overabundances) to have particularly low [C/Si], about 0.5 dex below the others. This implies that radiative and/or gravitational separation that affects the heavy elements also affects the C/Si ratio.

At the same time it becomes evident that objects with a relatively mild excess of heavy elements nonetheless show extreme variations of [C/Si]; even those five stars with quite similar [Ba/Ca] values vary by 1.4 orders of magnitude. In addition, the anticorrelation of [C/Si] and [Si/H] predicted by the gas-dust separation model shows up more clearly among these stars.

This corroborates the working hypothesis that the observed variation of [C/Si] is indeed mainly due to a *variable gas/dust ratio in the interstellar or circumstellar matter from which the photosphere of normal A stars has been derived*, either during star formation of more recently through accretion. Gravitational or radiative separation of C and Si plays a minor but still important role.

4. The metallicity of A stars.

If the gas-dust scenario is correct then one would expect also other refractory elements, in particular the metals of the iron group, to show variations from star to star like silicon. Indeed Lemke (1989, 1990) finds a good correlation between [Fe/H] and [Si/H] on the one hand, and between [Ti/H] and [Fe/H] on the other. This then implies that the *range in metallicity* observed among population I A stars is, at least in part, caused by differences in the gas/dust ratio.

References

Adelman, S. 1988a, Mon.Not.R.astr.Soc. **230,** 671
Adelman, S. 1988b, Mon.Not.R.astr.Soc. **235,** 749
Gigas, D. 1986, Astron.Astrophys. **165,** 170
Gigas, D. 1988, Astron.Astrophys. **192,** 264
Hoffleit, D., Jaschek, C. 1982, *The bright star catalogue,* Yale Universityn Observatory, New Haven
Holweger, H., Gigas, D., Steffen, M. 1986a, Astron.Astrophys. **155,** 58
Holweger, H., Steffen, M., Gigas, D. 1986b, Astron.Astrophys. **163,** 333
Jenkins, E.B. 1987, in *Interstellar Processes,* eds. D.J. Hollenbach, H.A. Thronson, Jr., Reidel, Dordrecht, p. 533
Lemke, M. 1989, Astron.Astrophys. **225,** 125
Lemke, M. 1990, Astron.Astrophys. **240,** 331
Mathys, G. 1991, Astron.Astrophys. **245,** 467
Stürenburg, S., Holweger, H. 1990, Astron.Astrophys. **237,** 125

Quantitative Analysis of A-type Supergiants

H.G. Groth [1], R.P. Kudritzki [1 2], K. Butler [1], R.M. Humphreys [3]

[1]Universtäts-Sternwarte München, Scheinerstr. 1, DW-8000 München 80
[2]Max Planck Institut für Astrophysik, Karl-Schwarzschildstr. 1, DW-8046 Garching bei München
[3]University of Minnesota

Abstract: Spectra of α Cyg (A2 Ia), HD 92207 (A0 Iae), HD 33579 (A3 Ia-O, LMC) and HD 7583 (A0 Ia-O, SMC) have been analyzed using non-LTE atmospheric models calculated with a complete-linearization code in which opacities of C, N and Si were included. The stellar parameters were derived using the Balmer lines, Hγ, Hδ and the ionization equilibrium of Mg I/Mg II. The use of these diagnostics allow Teff and log g to be derived with high accuracy (ΔTeff $\sim \pm 100$ K, Δlog g $\sim \pm 0.1$). Abundances of He, C, O and Mg have been obtained by comparison with non-LTE line formation calculations. We derived the following parameters;

	α Cyg	HD 92207	HD 33579	HD 7583
Teff(K)	9100	9400	8500	8460
log g	1.27	1.10	0.85	0.75
$\varepsilon_i/\Sigma\varepsilon_i$				
He	0.09	0.09	0.09	0.11
C	$1 \cdot 10^{-4}$	$6 \cdot 10^{-4}$	$0.7 \cdot 10^{-4}$	$2 \cdot 10^{-4}$
O	$1 \cdot 10^{-4}$	$4.5 \cdot 10^{-4}$	$5.5 \cdot 10^{-4}$	$3 \cdot 10^{-4}$
Mg	$5.5 \cdot 10^{-5}$	$5 \cdot 10^{-5}$	$2 \cdot 10^{-5}$	$1 \cdot 10^{-5}$

Complete results will be published in Astronomy and Astrophysics.

Abundances in Rapidly Rotating A Stars

Michael Lemke

Astronomy Department, RLM 15.308, The University of Texas, Austin, TX 78712, U.S.A.

A sample of 21 rapidly rotating A stars ($50\,\mathrm{km/s} < v \sin i < 150\,\mathrm{km/s}$) was observed at McDonald Observatory, University of Texas, to investigate if the non-solar abundance patterns found in sharp lined, *i.e.*, slow rotating, A-type stars ([1], [2]) persist towards higher rotational velocities. Rapid rotation of the star can lead to meridional mixing of the stellar plasma counteracting the diffusion processes which are believed to be the most likely cause for abundance anomalies to occur.

Temperatures and Gravities: Evidence for Convective Overshoot?

Strömgren photometric data [3] were converted into T_{eff} and $\log g$ with the calibration of Moon&Dworetsky [4]. In addition, gravities were derived from H_β profiles by fitting synthetic spectra, computed with the photometrically determined T_{eff}. This procedure has an internal accuracy of $\pm\,0.02$ dex which makes it potentially much more precise than the photometric method. Unfortunately, the fitting also depends slightly on the adopted temperature making T_{eff} the main source of uncertainty.

Compared to the photometric data, the H_β fits generally give lower gravities, which also show a smaller spread yet extend to equal extreme values (see Fig. 1).

Comparison of the stars' positions in a $\log g$ – $\log T_{\mathrm{eff}}$ diagram with evolutionary tracks computed with and without convective overshoot ([5], [6]) give strong indication for overshoot parameters greater than 0 (Fig. 1). Similar results were found for other A-type stars ([7], [1]) and for 'blue straggler' stars in NGC 2301 [8].

The situation becomes less clear, however, when recent models by Castellani et al. [9] are considered which were calculated without overshoot but taking new opacities into account. Quite a few of the program stars are still above their TAMS (Fig. 1, left panel) implying either only a small amount of overshoot is required or the adopted temperature scale is in error. The latter cannot be ruled out from the current investigation: The temperatures from the Moon&Dworetsky calibration are approximately 150 K lower than those derived from least square fits to $T_{\rm eff}$ vs. c_0 or vs. a_0 [10]. This would translate into higher gravities by ~ 0.06 dex, which is marginally sufficient to move all stars (except HR 4194) onto Castellani's main sequence.

Abundances

Due to severe line blending caused by the rapid rotation of the stars the abundance determination is based on synthetic spectra. As a by-product rotational velocities were re-determined. We found fairly good agreement with the Bright Star Catalog except for two major discrepancies: HR 3711 and HR 4861 both rotate about 100 km/s faster.

Although locating the continuum in the observations with a semi-automatic procedure worked surprisingly well, both noise and non-unique spectroscopic solutions allowed only abundance estimates within the 0.5 dex level. We show those for the critical elements Ba and C (Tab. 1).

Most stars were found to have almost solar Ba and C abundances while a few might be enriched with Ba in a similar fashion like the slow rotators. HR 5023 might be the most likely candidate with a fairly high $v \sin i$ for such an enrichment. See Fig. 2 for details and a comparison with a typical example for the more solar-like stars (HR 6610). Variations of $T_{\rm eff}$ by 400 K and of $\log g$ by 0.4 dex have less influence on the spectra than varying Ba by 0.5 dex.

The crude estimates for C indicate a variation independent of Ba, which again is similar to the sharp lined stars.

Results for Ba and C with respect to $v \sin i$ are summarized in Fig. 3.

Preliminary results for Fe reveal no striking variation except for HR 4861, which is depleted between 1 and 1.5 dex relative to the sun. HR 3711 might also tend towards the same direction, putting both stars into the neighborhood of Vega.

References

1 Lemke, M. 1989: A&A 125, 225.
2 Lemke, M. 1990: A&A 240, 331.
3 Hauck, B., Mermilliod, M. 1985: A&AS 40, 1.
4 Moon, T.T., Dworetsky, M.M. 1985: MNRAS 217, 305.
5 Hejlesen, P.M. 1980: A&AS 39, 347.
6 Bertelli, G., Bressan, A., Chiosi, C., Angerer, K. 1986: A&AS 66, 191.
7 Andersen, J., Nordström, B., Clausen, J.V. 1990: ApJ 363, L33.
8 Napiwotzki, R., Schönberner, D., Weidemann, V. 1991: A&A 243, L5.
9 Castellani, V., Chieffi, A., Straniero, O. 1990: ApJS 74, 463.
10 Code, A.D., Davis, J., Bless, R.C., Hanbury Brown, R. 1976: ApJ 203, 417.

Table 1. Program stars, adopted parameters, and resultant Ba and C abundance estimates (relative to the sun). $T_{\rm eff}$ from Strömgren colors, $\log g$ from H_β profiles.

HR	Name	Sp. Type	V	$T_{\rm eff}$	$\log g$	$v \sin i$	[Ba]	[C]
830		A0 V	5.86	10480	3.69	65	0.5	0.0
945		A0 V	6.42	9850	3.73	80	0.0	0.0
1019		A0 V	5.61	11020	4.19	160	0.0	0.0
1251	3813 Tau	A1 V	3.91	9230	3.87	80	0.5	-0.5
1578		A0 V	6.50	9600	3.86	100	1.0	-0.5
1661		A0 V	6.05	10520	3.82	30	1.2	0.0
1718	18 Ori	A0 V	5.50	9840	3.70	85	0.0	-0.5
3711		A1 V	6.53	9770	3.74	230	0.0	0.0
4131		A1 V	6.45	9760	3.70	180	0.0	0.0
4194		A0 V	5.64	9730	3.55	90	0.0	0.0
4635	3 Crv	A2 V	5.46	9100	3.96	130	0.0	-0.5
4861	28 Com	A1 V	6.56	9340	3.92	225	0.0	0.0
5023	21 Cvn	A0 V	5.15	10750	3.93	100	1.0	0.0
5262		A2 V	5.99	8670	3.78	120	0.0	0.0
6195	37 Her	A1 V	5.77	10110	4.00	160	0.0	0.0
6610		A0 V	6.56	9890	3.96	135	0.0	0.0
6976		A1 V	6.40	9900	3.71	180	0.0	0.0
7351		A1 V	6.26	9520	3.77	120	0.6	0.0
8083		A0 V	6.17	10610	4.25*	170	0.0	0.0
8186		A1 V	6.63	9510	4.02*	160	0.0	-0.5
8844		A0 V	5.84	9900	3.77*	32	1.0	-0.1

*: from Strömgren colors

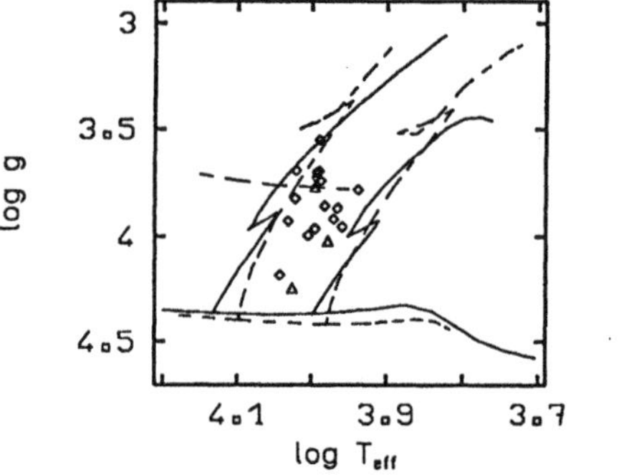

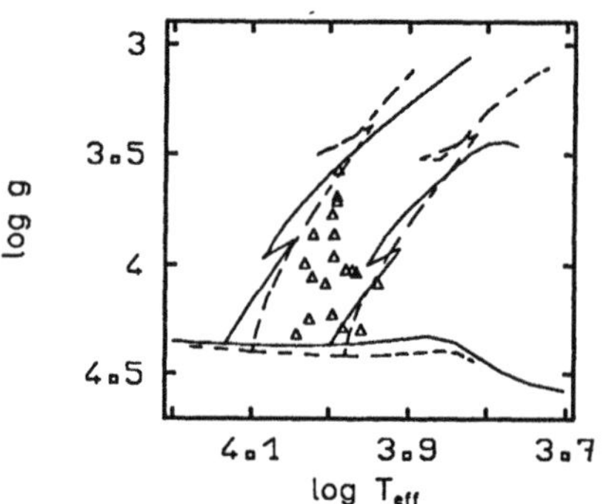

Fig. 1. $\log g - \log T_{\rm eff}$ diagrams. ◇: $\log g$ from H_β profiles; △: from photometry. Solid lines: Evolutionary tracks without overshoot [5], broken lines: with overshoot [6]; upper broken line (left panel): TAMS for new opacity calculations [9].

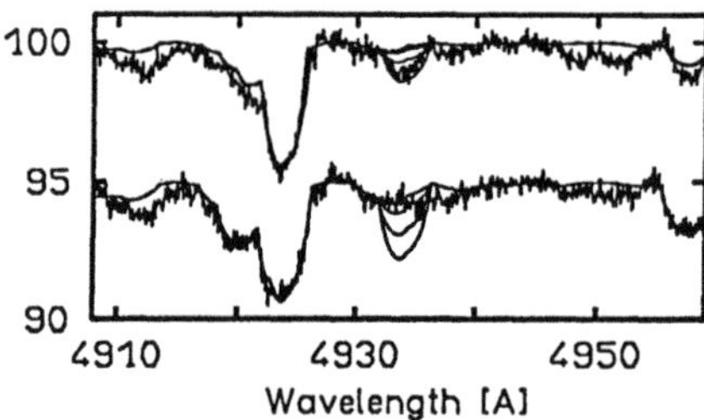

Fig. 2. Spectrum for HR 5023 (up) and HR 6610 (shifted by 5%) around Ba II λ4934 Å plus synthetic spectra. Ba abundance increased in steps of 0.5 dex, starting from 2.18 (solar).

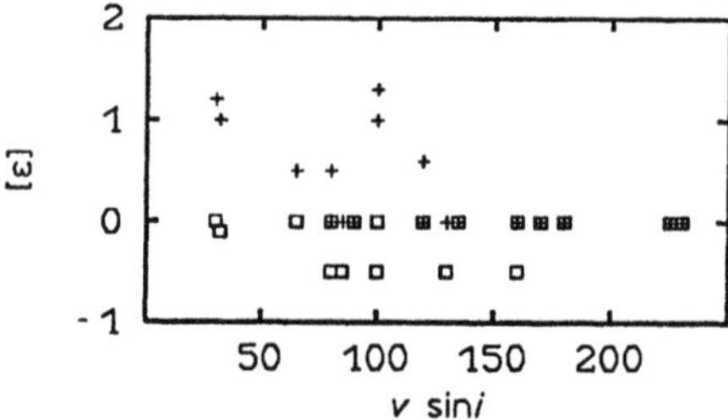

Fig. 3. Abundance estimates relative to the sun as a function of $v \sin i$. +: Barium; □: Carbon

Statistical equilibrium of Al I/II in A stars and the abundance of aluminium in Vega

W. Steenbock and H. Holweger
Institut für Theoretische Physik und Sternwarte der Universität Kiel, Olshausenstr. 40,
W-2300 Kiel, Germany

Introduction

In the last years there was growing evidence that normal A-type stars and especially the standard star Vega (A0V) have non-solar compositions. Recent studies of Vega revealed a non-uniform pattern of underabundances relative to the sun. The results of Sadakane et al. (1981), Gigas (1986, 1988) and Adelman et al. (1990) show nearly uniform underabundances of the iron group elements and magnesium of -0.6 dex, whereas eg. aluminium has a deficiency of about -1.0 dex. In two elaborate NLTE studies Gigas (1986, 1988) arrived at the conclusion that NLTE plays an important role for certain elements, eg. Fe I and Ba II, while other elements like Mg I are hardly effected by departures from LTE. In this poster paper we will present the results of NLTE computations for Al I in Vega.

Model-Atom and Atomic Data

We employ a newly developed model atom (Fig. 1). Details and its application to the sun can be found in Steenbock (1991). It follows a brief description:

```
Al I   26 levels        (each representing one term)
Al II   1 level         (continuum)
Al I   27 transitions   (each representing one multiplet)
```

The two resonance lines at 3944 Å and 3961 Å (multiplet 1) were treated separately. Because these two lines are situated on the wing of H_ϵ, the line opacity of H_ϵ was added to the underlying LTE continuum opacity in the NLTE computation.

Beside electron collisions we included collisions with neutral hydrogen by adopting the formulae of Drawin(1968, 1969). Since these are only order of magnitude estimates, we introduced a scaling factor of 1/3 as suggested by previous studies (see e.g. Steenbock et al. 1984). Fortunately, neutral hydrogen collisions are much less frequent than electron collisions in Vega in contrast to solar type stars.

The ionization balance of Al I is governed by the radiation field shortward of 2071 Å, where photoionization from the ground state occurs. Photoionizations rates were calculated with a theoretical UV-radiation field, where line opacities were taken into account following the distribution function scheme of Kurucz(1979). For line opacities an abundance of [M/H] = -0.5 dex was adopted.

Model Atmosphere

We adopted the model atmosphere used by Gigas(1986,88): $T_{\rm eff} = 9500$ K and $\log g = 3.9$ with an average metal underabundance of [M/H] = -0.6. The aluminium abundance was set to [Al/H] = -1.0 dex. A depth dependent microturbulence was chosen. Using a depth-independent microturbulence of 2 km/s does not change our calculated abundances. The temperature stratification was computed with the ATLAS6 program (Kurucz 1979).

Results

Fig. 2 shows the departure coefficients for selected levels. A significant underpopulation of low lying levels was found, a result of strong ground level photoionization. The exceptional large photoionization cross-section of the ground state (Kohl et al. 1973) is responsible for that overionization. The upper resonance line level (no. 2) exhibits a non-thermal excitation caused by pumping through the resonance lines and by the infrared line between level 2 and 5, making the source function of the resonance lines greater than in LTE. Both NLTE effects combined lead to weaker resonance lines as compared to LTE (Fig. 3, 4). Because the line at 3961 Å (Fig. 4) is situated on the deep wing of H_ϵ, it is formed higher in the atmosphere where the departures from LTE are larger, which leads to a more pronounced weakening of this line.

Abundances

Theoretical lines were convoled with a pure rotation profile whose $v \cdot \sin i$ corresponded to 22 km/s (Adelman et al. 1990) and fitted to the high-dispersion photographic spectrum of Vega kindly provided by R. and R.Griffin. The resulting LTE abundances are

$$[\mathrm{Al/H}] = -1.0\ (3944\ \text{Å})\ \text{and}\ [\mathrm{Al/H}] = -1.1\ (3961\ \text{Å}).$$

In NLTE the abundances become significantly larger,

$$[\mathrm{Al/H}] = -0.6\ (3944\ \text{Å})\ \text{and}\ [\mathrm{Al/H}] = -0.4\ (3961\ \text{Å}).$$

Because the line 3961 Å on the wing of H_ϵ is very insensitive against abundance changes, the abundance depends much on the details of the fitting procedure(rotation, local continuum, dispersion), and we give higher weight to the result of 3944 Å. Our result shows no significant dependence on neutral hydrogen collisions and microturbulence. If we increase the electron collisionrate by an arbitrary factor of 10 we obtain only a moderate change in NLTE abundances,

$$[\mathrm{Al/H}] = -0.72\ (3944\ \text{Å})\ \text{and}\ [\mathrm{Al/H}] = -0.6\ (3961\ \text{Å})$$

demonstrating the dominance of radiative processes in the statistical equilibrium.

Comparing our result with the average iron group and magnesium abundances of [M/H] = -0.6, aluminium does no longer stand out, and the underabundances pattern becomes more uniform. We come to the conclusion that in normal A-type stars, like Vega, the aluminium abundance is strongly effected by NLTE and that apparent abundance irregularities may disappear if proper account is made of departures from LTE.

References

Adelman, S.J., Gulliver, A.F., 1990, ApJ 348, 712
Drawin, H.W., 1968, Z.Phys. 211, 404
Drawin, H.W., 1969, Z.Phys. 225, 483
Gigas, D., 1986, A&A 165, 170
Gigas, D., 1988, A&A 192, 264
Kohl, J.L., Parkinson, W.H., 1973, ApJ 184, 641
Kurucz, R.L., 1979, ApJS 40, 1
Sadakane, K., Nishimura, M., 1981, PASJ 33, 189
Steenbock, W., Holweger, H., 1984, A&A 130, 319
Steenbock, W., 1991, submitted to A&A

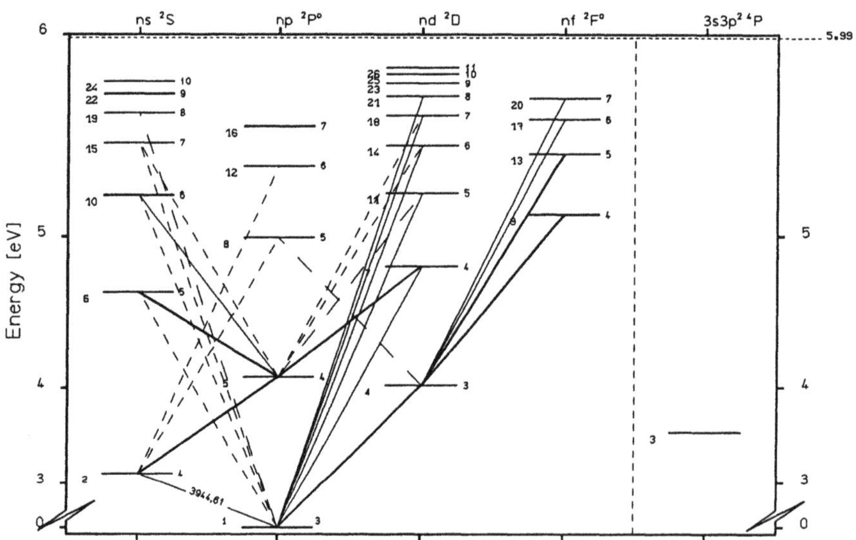

Fig. 1 Model atom for neutral aluminium. The levels are labelled with their respective principal quantum number (right) and level number in the model atom (left). Line transitions treated explicitely are indicated; heavy lines correspond to strong transitions, and vise versa.

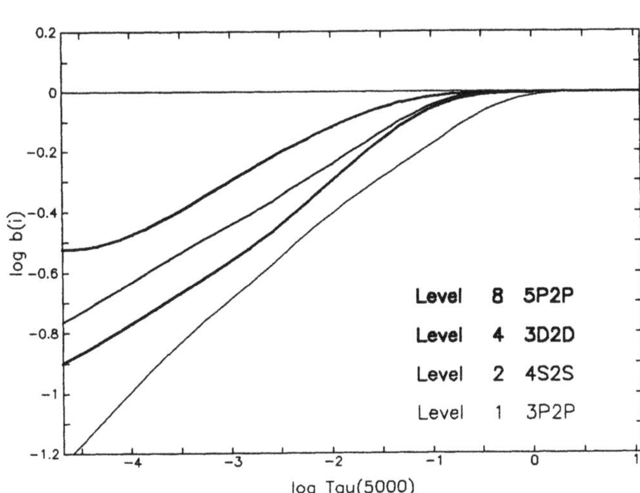

Fig .2
Selected departure coefficients of Al I in Vega. $\log b_i = \log(n_i^{NLTE}/n_i^{LTE})$, where n_i is the population number of level i.

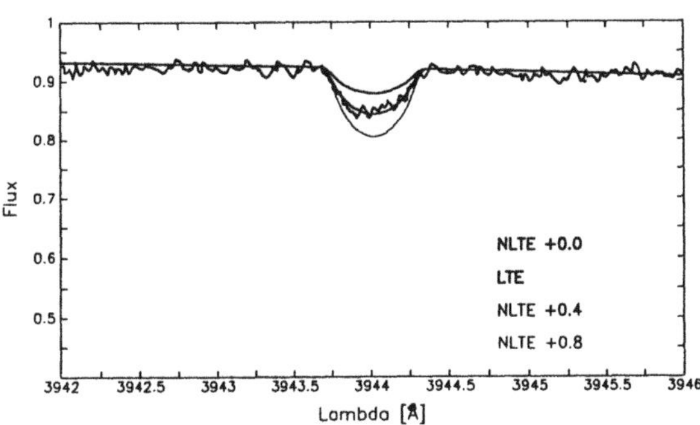

Fig. 3
Al I resonance line at 3944.01 Å. Theoretical line profile is convoluted with a rotation profile whose $v \cdot \sin i$ correspondes to 22 km/s. Shown are NLTE profiles with different logarithmic abundance corrections relative to LTE superimposed on the observation. The LTE profile is fitted to the observation and matches the NLTE profile with +0.4 dex.

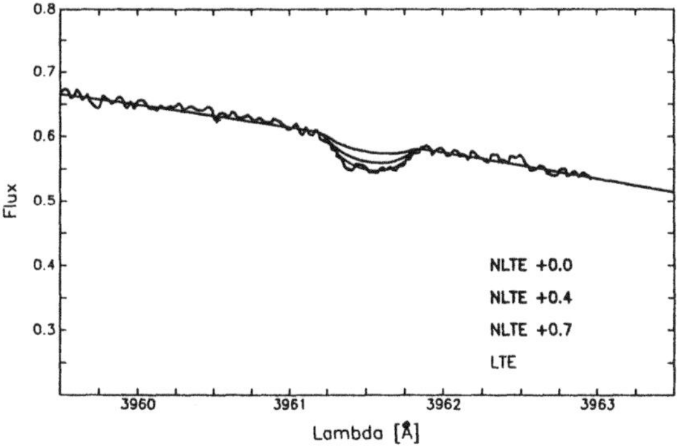

Fig. 4
Al I resonance line at 3961.52 Å. Theoretical line profile is convoluted with a rotation profile whose $v \cdot \sin i$ correspondes to 22 km/s. Shown are NLTE profiles with different logarithmic abundance corrections relative to LTE superimposed on the observation. The LTE profile is fitted to the observation and matches the NLTE profile with +0.7 dex.

II. Early-Type Stars in Other Galaxies

Analyses of B-Type Stars in the Magellanic Clouds

A. Jüttner, B. Wolf

Landessternwarte Königstuhl, D-6900 Heidelberg 1

Abstract: Model atmosphere analysis abundance determinations from B-type stars in the Magellanic Clouds are reviewed.

1 Introduction

The Magellanic Clouds differ in many respects from the Galaxy. They are of different Hubble type and therefore are expected to have a completely different chemical evolution.

B-type giants and main sequence stars are supposed to represent in their atmospheres the present day abundance pattern. Hence the major aim of abundance determinations from B-type stars is to learn more about the star formation history of the MCs and about the detailed production mechanisms for individual elements. Within the framework of the chemical evolution in the MCs it also important to analyse both stars in the general field and stars in clusters. The latter are important observational targets because the simultaneous knowledge of both ages and abundances should provide insight in chemical enrichment mechanisms of the MCs.

The very luminous B supergiants of the MCs are ideal for studying evolutionary scenarios of very massive stars formed in environments completely different from the Galaxy. The most luminous hot stars and the most luminous stars in general in the visual and photographic range in the MCs are A-type supergiants. Twenty years ago, therefore first model atmosphere analyses of members of the MCs have concentrated on the very brightest stars of both clouds (R76 of the LMC, $M_V = -9.5$ and R45 of the SMC, $M_V = -9.3$) due to technical problems in observing anything fainter at high enough dispersion (Wolf, 1972, Wolf, 1973). However, the abundance analyses of these extreme stars is very difficult due to the complicated physical structure of their atmospheres, i.e:

a. The atmospheres are very extended; the plane parallel approximation is presumably problematic.
b. Inspite of the comparatively low temperature ($T_{\rm eff} \approx 9000\,$K) NLTE effects are important since the gravity ($\log g \approx 0.7 - 0.9$) is so low.

c. Large scale hydrodynamic motions are important in these luminous stars and hydrostatic atmosphere models are not at all adequate. Recent high dispersion and high S/N-ratio spectra of R75 and R45 obtained with the Cassegrain-echelle spectrograph (CASPEC) attached to the ESO 3.6 m telescope evidence that even those layers where the bona fide photospheric absorption lines are formed expand with considerable velocities ($v > 100$ km/s; cf. Figure 1 where several FeII lines with blue extended wings are shown). Likewise the H_α-lines show strong stellar wind characteristics (Fig.2).

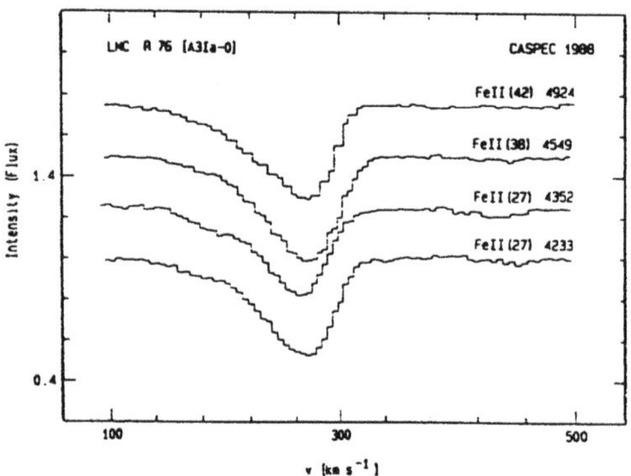

Fig. 1. FeII-lines of R76. As shown by the Figure the blue extended wings of the "photospheric" lines show a considerable mass flow of a terminal velocity of more than 100 km/s of this extreme LMC A supergiant.

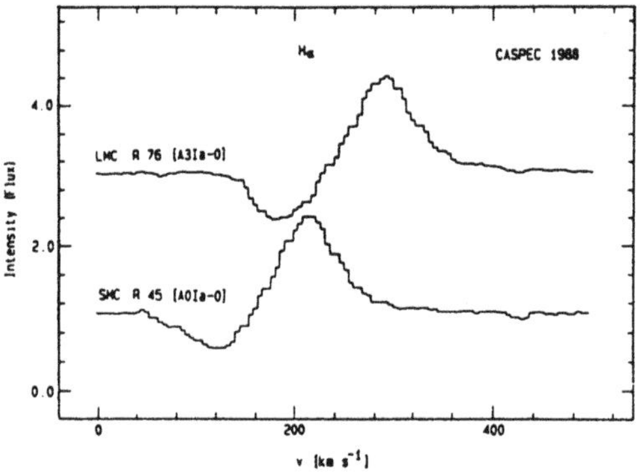

Fig. 2. H_α-P Cygni profiles of R76 and R45 indicating a strong mass loss of these luminous A supergiants.

As all these effects have not been taken into account in former analyses those abundances have to be considered with caution. Obviously very sophisticated NLTE models and a hydrodynamic code have to be applied for the interpretation of the spectra of those luminous A-type supergiants (see also Groth, 1991).

Since those early days dramatic progress was made both in model atmosphere codes and in instrumental developments.

I. Sophisticated codes to compute more realistic model atmospheres:
   a. line blanketed models (e.g. ATLAS 8, Kurucz, 1979)
   b. NLTE codes especially for hot stars (cf. Kudritzki, 1988 and literature quoted therein, Lennon, 1991)
   c. NLTE unified expanding model atmospheres (Puls, 1991, Lennon, 1991)

II. Instrumental developments:
   a. access to telescopes of the 3.6 m and 4.0 m class
   b. efficient echelle spectrographs
   c. new generation of digital recorders such as CCD or IPCS detector systems
   d. spectra in the UV range with IUE or with the HST

Using all these powerfull techniques MC stars over a wide range of luminosity and temperature are being studied by several groups (cf. Fig. 3).

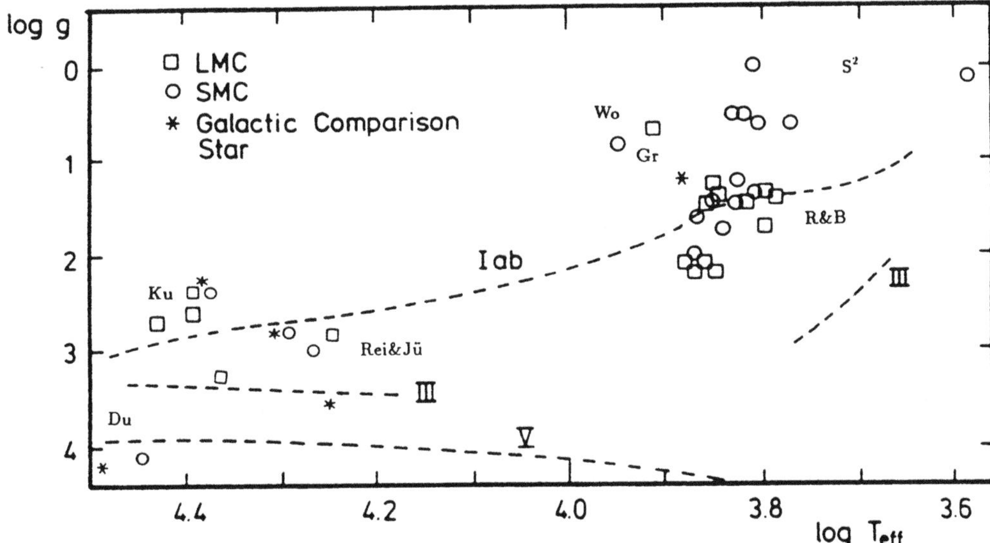

Fig. 3. $\log g - \log T_{\rm eff}$ **diagram of individual stars of the MCs for which abundance analyses have been carried out (Baschek, 1989). References: Du = Dufton et al. (1991), Gr = Groth (1991), Ku = Kudritzki et al. (1988,1991), R&B = Russell and Bessell (1989), Rei&Jü = Reitermann (1990) and Jüttner (1991), S² = Spite at al. (1989), Wo = Wolf (1972,1973).**

In the following the recent analyses of B-type stars of different luminousity classes of the MCs carried out by groups in Munich, Belfast and Heidelberg will be reviewed.

2 B supergiants

The analyses of B-type supergiants in the MCs are of paramount importance to test evolutionary scenarious of massive stars; of stars evolving in completely different environments (different Hubble types, different and/or late chemical evolutionary state).

Stimulated by the discovery that the progenitor of the Type II supernova SN 1987A in the LMC was the B3 I star Sk-69°202 (Walborn et al. 1987) Kudritzki and coworkers (cf. Kudritzki, 1988, Lennon, 1991) have started an extensive pogram of NLTE analyses of OB-supergiants in the MCs. The major aim of this program is to secure more information about the presence and the fraction of CNO-cycled material in the atmospheres of B supergiants in general and in MC B supergiants in particular. First results for O-type stars have been published by Kudritzki et al. (1988) while preliminary results for B- and A-type supergiants have been given previously by Kudritzki et al. (1987), Kudritzki et al.(1988), Lennon et al. (1990), Humphreys et al. (1991) and Groth (1991).

Recent NLTE analyses of MC supergiants are described in Lennon et al. (1991) and Lennon (1991). They obtained high resolution ($\Delta\lambda \approx 0.25$Å) spectra of the B supergiants Sk-68°41 (LMC) and Sk 159 (SMC) using the ESO 3.6 m telescope and the Cassegrain-echelle spectrograph (CASPEC). The abundance analyses are based on detailed NLTE model atmospheres and subsequent NLTE line formation calculations for all elements of interest. The principle of the analysis is to use the profile fit of the Balmer lines (H_γ, H_δ) for the gravity determination and the profiles of HeI and HeII lines for the simultaneous determination of $T_{\rm eff}$ and the helium abundances ($y = N_{\rm He}/(N_{\rm H} + N_{\rm He})$). On the other hand if the objects are to cool to show HeII lines, the ionization equilibria of SiII/III/IV or the energy distribution is necessary to deduce $T_{\rm eff}$. Since the gravity significantly influences the shape of the energy distribution here also $\log g$ has to be determined simultaneously. The adopted stellar parameters ($T_{\rm eff}$, $\log g$), the helium abundances and gravitational masses (derived from $\log g$ and luminousity) and their dereddened magnitudes are given in Table 1.

Table 1. Adopted stellar parameters.

Object	V	$T_{\rm eff}$	$\log g$	y	$M_{\rm grav}/M_\odot$	$M_{\rm ZAMS}/M_\odot$
κ Ori	2.06	25000	2.70	0.20	14	25
Sk-68°41	12.01	25000	2.60	0.23	15	35
Sk 159	11.90	25000	2.55	0.35	29	55

Note that all objects indicate a helium enrichment in their atmospheres. One explanation for the helium abundance, suggested by Lennon et al. (1991) is that these stars had evolved through a red supergiant phase, and had moved bluewards again in the HR-diagram to appear once again as blue supergiants. The helium overabundances could then be due to the influence of accumulated mass loss over the star's livetime and convective processes during the red supergiant phase of evolution. The results of the abundance analyses relative to κ Orionis are summarized in Table 2 and Table 3 respectively. Note, that the NIII lines imply a nitrogen rich atmosphere while the NII lines show a nitrogen underabundance.

Table 2. Metal abundances in Sk-68°41 realative to κ Ori.

Element	Sk-68°41	Late s'giants	HII regions
C	+0.06	-0.58	-0.77
N	$-0.12^{\rm NII}$,$0.14^{\rm NIII}$	-	-1.02
O	-0.33	-0.5	-0.49
Mg	-0.78	-0.13	-
Si	-0.14	0.38[1],-0.24[2]	-

([1]Russell & Bessell (1989), [2]Richtler et al. (1989))

For comparison the results for HII regions (Dufour, 1984) and for late type giants (Spite et al. 1989, Russell and Bessell 1989 and Richtler et al. 1989) are included. Considering Sk 159 one can see that the general picture of the SMC being the more metal deficient of the MCs is confirmed by this NLTE analysis. Note that Sk 159 exhibts a much lower oxygen abundance than found in any other SMC star. The fact that Sk 159 is nitogen rich and carbon and oxygen poor is at least qualitatively consistent with the idea of CNO-cycled material having been mixed into its atmosphere.

Table 3. Metal abundances in Sk 159 realative to κ Ori.

Element	Sk 159	Late s'giants	HII regions
C	-1.09	-0.9[1],-0.77[2]	-1.51
N	-1.00^{NII},-0.17^{NIII}	-	-1.53
O	-1.10	-0.8	-0.9
Mg	-1.00	-	-
Si	-0.75	-	-

([1]Spite et al. (1989), [2]Russell & Bessell (1989))

Important improvements in analysing B supergiants' atmospheres are under way; more sophisticated NLTE "unified" expanding model atmospheres are being developed by the Munich group (cf. Puls, 1991). However, apart from computational difficulties of deriving absolute abundances the finding that the atmospheres of B supergiants are contaminated by CNO-cycled material is even more serious. This makes B supergiants less suitable for deriving the abundances (He, C, N and O) characterizing the present evolutionary state of the MCs. To avoid these more principal difficulties analyses of less evolved stars are important.

3 B main sequence stars

First differential abundance analysis of a main sequence B-type star of the SMC was carried out by Dufton et al. (1990). They obtained high resolution spectra of the exceptionally sharp-lined star AV 304 (B0.5 V,V = 14.98, Garmany, Conti and Massey, 1987, Azzopardi et al. 1975) using the University College London Echelle Spectrograph and the Image Photon Counting System (IPCS) attached to the 3.9 m Anglo-Australian Telescope (AAT). The exposure time was 6.9 hr (split up into smaller exposures of approximately 1 hr) and the achieved signal-to-noise ratio is approximately 35.

The model atmosphere parameters of AV 304 were deduced using line- blanketed LTE model atmospheres of Kurucz (1979). The effective temperature ($T_{\rm eff}$ = 28000 K) was deduced from the SiIII/SiIV ionisation equilibrium and the surface gravity ($\log g = 4.1$) from the Balmer line profiles for H_γ and H_δ using the broadening theory of Vidal, Cooper and Smith (1973). From the OII lines they derived a microturbulence velocity of 5 km/s. The atmospheric parameters for the galactic comparison star τ Sco ($T_{\rm eff}$ = 31000 K, $\log g = 4.15$, $\xi_{\rm micro}$ = 5 km/s) were taken from Lennon et al. (1990).

The results of the differential abundance determinations are presented in Figure 4. For comparison the abundance pattern of τ Sco relative to the solar values is also shown in Figure 5.

The field star AV 304 exhibits a mean metal depletion of approximately 0.7 dex compared to τ Sco. Note that nitrogen is significantly more underabundant, while oxygen shows some evidence for being less depleted. Since AV 304 is located on the main sequence these results should not be affected by the contamination of nucleosynthetic material so that the derived chemical composition of AV 304 should reliably represent the interstellar material from which this star is formed. A similar analysis has recently

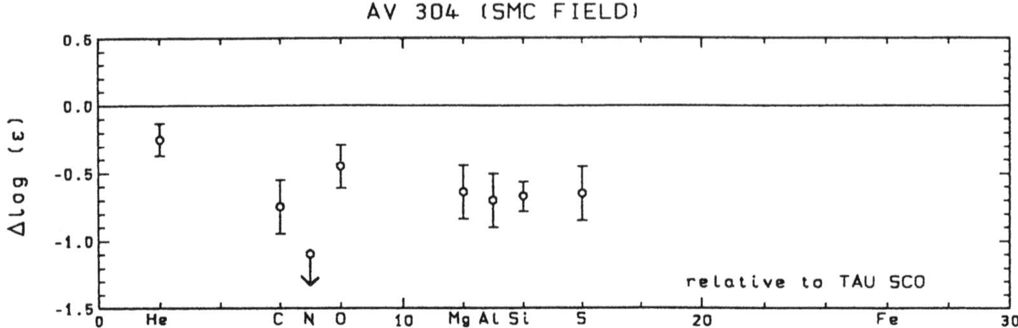

Fig. 4. Differential abundance pattern for the SMC field star AV 304, showing a mean abundance for all elements heavier than He of $\Delta \log \varepsilon = -0.70$ dex (Dufton et al. 1990).

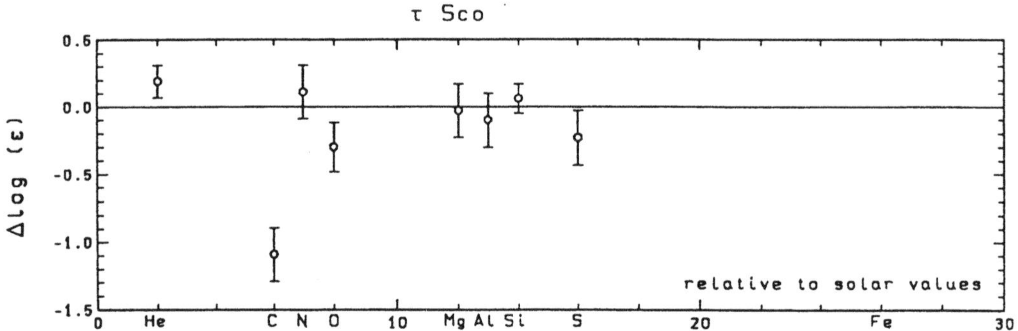

Fig. 5. Abundance pattern for the galactic comparison star τ Sco.

been carried out of an LMC main sequence field star; the results are given elsewhere in this volume (Dufton et al. 1991)

4 B giants

To find an answer to the question of how homogeneous are the MCs in metal content and in order to get a more detailed picture of the evolutionary status of the MCs a number of objects in different regions of the MCs has to be analysed.

Within the ESO-Key Programme "Coordinated Investigation of Selected Regions in the MCs" (de Boer et al. 1989) we obtained so far high resolution and high signal-to-noise spectra of 2 SMC stars and 6 LMC stars. For 4 of the above mentioned objects chemical abundance determinations are already available.

B-type giants exhibit some important advantages:

a. They are young stars and supposed to represent in their atmopheres the present day abundance pattern of the ISM.
b. B-type stars in this temperature range ($T_{\rm eff} \approx 18000$ - 25000 K) show apart from lines of He, C, N and O also lines of higher metals such as Mg, Al, Si and S up to Fe

so that results can be compared both to the abundances derived from HII regions and from late type stars.

c. Their atmospheres are presumably not contaminated by nuclear synthesis products.
d. Since we are working strictly differential the results on abundance determination are expected not to be seriously affected by NLTE effects.
e. Compared to B main sequence stars our objects are up to 2 mag brighter in the visual/photographic range so that we are able to obtain high signal-to-noise spectra within moderate exposure times and more objects can be analysed .

4.1 Strategy and Methods

One drawback is that B stars are frequently fast rotators. Stars with projected rotational velocities $v \sin i > 75\,\mathrm{km/s}$ are not suitable for fine analyses.

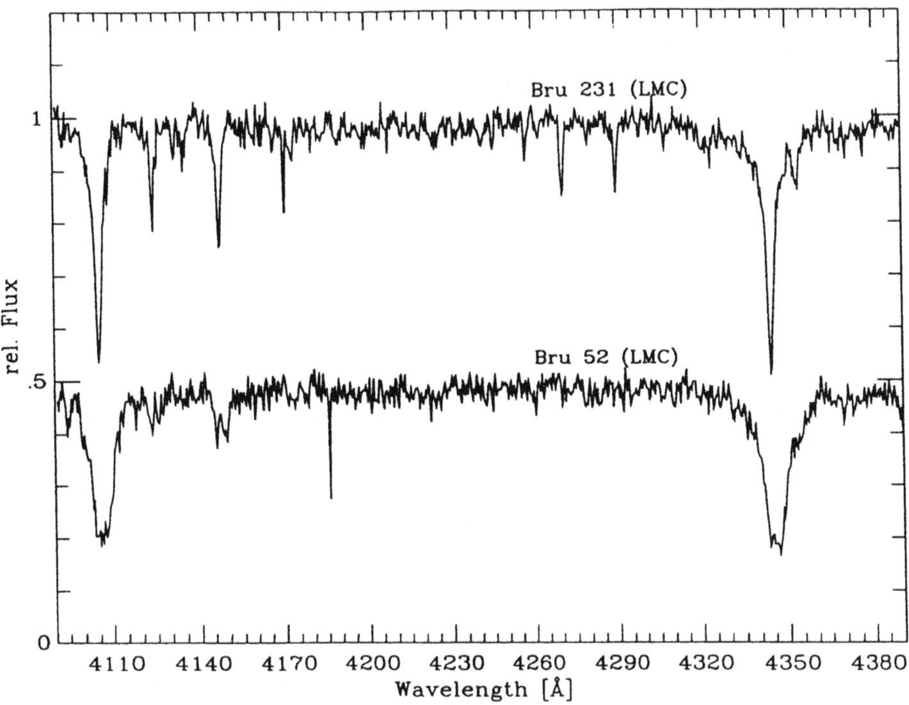

Fig. 6. Sections of B&C spectra obtained within the preselection program. Bru 231 has narrow lines and is a good candidate for further high resolution spectroscopy. Bru 52 is not useful for a fine analysis, it is a fast rotator.

For this reason we had first to carry out a preselection program, using the ESO/MPI 2.2 m telescope and the B&C spectrograph at a resolution of 30Å/mm. The typical exposure times are between one and two hours. The basis for this preselection program are the catalogues of Azzopardi-Vigenau (Azzopardi et al. 1975) and Brunet (Brunet et al. 1975) for the SMC and LMC, respectively.

Subsequent high resolution ($\lambda/\Delta\lambda \approx 20000$) and high signal-to-noise (S/N $\geq$ 75) spectra covering the wavelength range from 3900 Å to 5300 Å were obtained with CASPEC at the ESO 3.6 m telescope. In addition to the CASPEC data we obtained for each object low resolution IUE spectra in the SWP (1150-2000 Å) and LWP (1825-3300 Å) range.

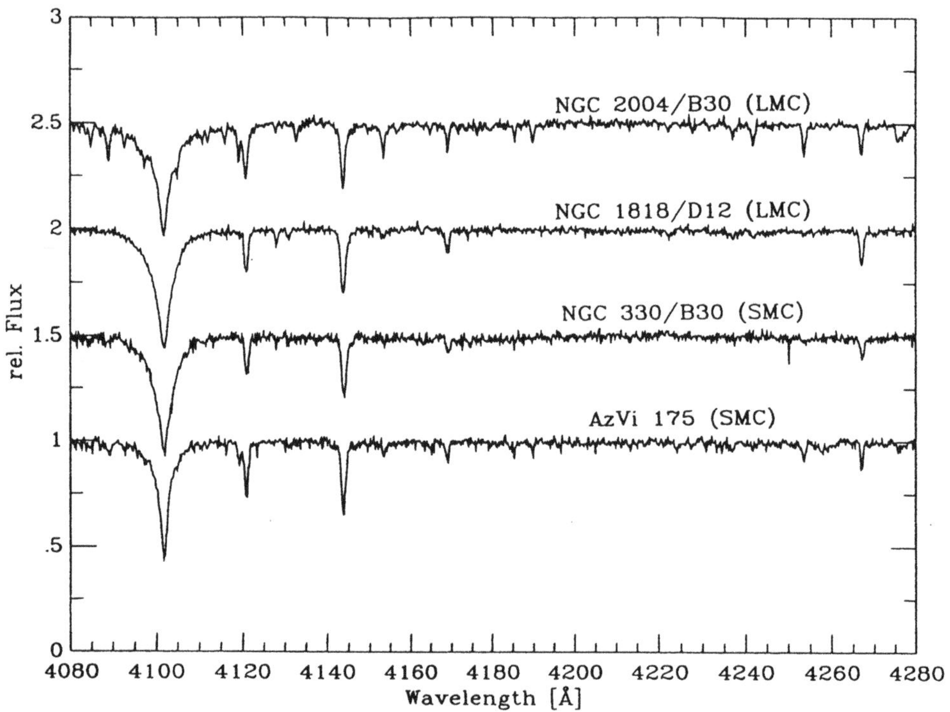

Fig. 7. Sections of CASPEC spectra obtained within the ESO-Key Programme

The analyses of our stars are based on Kurucz line blanketed LTE model atmospheres (Kurucz, 1979 and literature quoted therein) calculated with the ATLAS 8 code. The equivalent widths and line profiles are computed with a modified version of the BHT code (Baschek et al., 1966, Peytremann et al. 1967).

In principle it should be possible to derive uniquely the stellar parameters only from the Balmer line fits, however the observable variation of the profiles and equivalent widths with surface gravity and temperature usually leads to reasonably good fits over a wide range of values yielding in a curve in the $\log g$-$T_{\rm eff}$-diagram. This procedure is applied for each of the Balmer lines H_β, H_γ and H_δ using the broadening theory of Vidal, Cooper and Smith (Vidal et al., 1974) and results in a group of almost parallel fit curves with a scatter showing the accuracy expected for this kind of gravity determination.

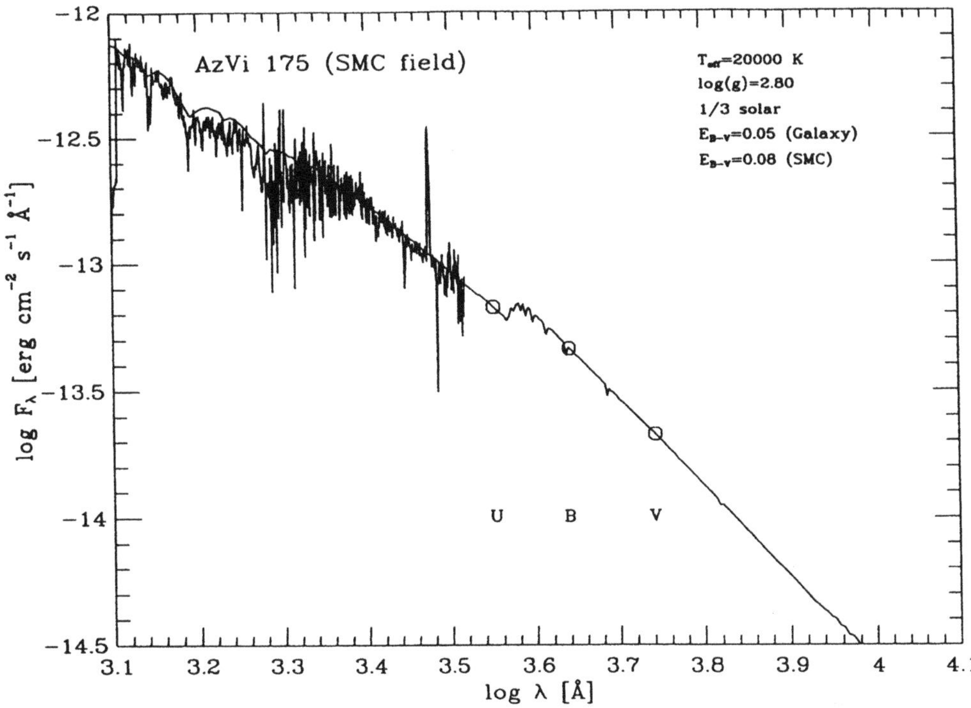

Fig. 8. Comparison of observed continuum (IUE and UBV) and computed energy distribution of AV 175. From the fit procedure we obatained reddening values of $E_{B-V} = 0.05$ for the galactic foreground and $E_{B-V} = 0.08$ for the SMC using the reddening law of Savage and Mathis (1979) for the galaxy and the law of Prevot et al. (1984) for the SMC.

The most temperature sensitive lines observed in B-type spectra are the lines of SiII, SiIII and SiIV. A second function in the $\log g$-T_{eff}-diagram with a steeper slope than produced by the Balmer line fit curves can be constructed by calculating the abundance ratios at given equivalent widths of all possible combinations of silicon lines from two neighbouring ionisation stages on a grid of model atmospheres. We use the ionisation equilibrium of SiII/SiIII and/or SiIII/SiIV. It should be noted that these equilibria depend strongly on the microturbulence and even on the abundances used in the model atmosphere code; both are initially unkown parameters. Therefore the above mentioned grid calculations are done for a set of microturbulence velocities ranging from 0 to 20 km/s in steps of 5 km/s. Since we are performing abundance analyses for stars in the MCs we usally use 1/3 solar abundance pattern in the model atmosphere code as an initial guess. The results of these grid calculations are two groups of fit curves, and various intersection points depending on the microturbulence parameter. The most prominent metal lines in B-type spectra showing both strong and weak equivalent widths are the OII lines. The microturbulence velocity is finally derived in such a way that there is no obvious dependence of the oxygen abundance on the line strength.

From the work of e.g. Kamp (1978) and Lennon et al. (1986), however, it is known that the temperatures derived from the SiII/SiIII ionisation equilibrium can be too high by approximately 2000 K for stars of spectral type B2 or later. In that case and if the reddening is known the temperature can be derived by fitting the observed IUE contiuum with the calculated continuum (cf. Fig. 8).

Otherwise if the derived temperature is based on the SiIII/SiIV ionisation equilibrium we are able to give reasonable reddening values at given reddening laws by comparing the observed continuum with the calculated continuum. The last point may provide the possibility to solve the reddening problem especially for the prominent blue globular cluster NGC 330 (SMC) (Bessell, 1991).

4.2 Results

So far four B giants have been analysed: AV 175 a field star near NGC 330 in the SMC and the blue globular cluster members NGC 330/B30 of the SMC and NGC 1818/D12 and NGC 2004/B30 of the LMC. The model atmosphere parameters of AV 175 are given in Table 4.

Table 4. Model atmosphere parameters derived from SiIII/SiIV equilibrium and the Balmer line fits.

Object	V	T_{eff}	$\log g$	ξ_{micro}
AV 175	13.65	20000 K	2.80	20 km/s
HR 2618	3.60	20100 K	2.88	20 km/s

Since our target stars are members of the MCs we have a reliable knowledge of the distance and hence luminosity. We adopt a distance modulus of 18.5 for the LMC and 18.85 for the SMC (Tammann, 1987). From the luminousity and the surface gravity $\log g$ we obtained $R = 26.1\,\text{R}_{\odot}$ and $M = 19.8\,\text{M}_{\odot}$.

Differential abundances of AV 175 are presented in Figure 9. For the abundance pattern of the galactic comparison star HR 2618 see Figure 10.

Note that the prominent CII (4267 Å) line yields a strong underabundance for this galactic star. The other elements show almost solar values. Since this particular CII line which is usually the only observable carbon line in MC stars is known for its strong NLTE dependence (i.e. Lyubminkov, 1989, Eber, 1988) it is important to work very differentially to derive reliable carbon abundances.

As shown by Figure 9 the metal abundance of AV 175 is down by $\Delta \log \varepsilon = -0.65$ dex. A more detailed discussion of the analysis of AV 175 is in preparation for Astron. Astrophys.. Comparing these results with the abundances of the B main sequence field star AV 304 (SMC) (Dufton et al. 1991, Fig. 4) we found a good agreement in both the pattern (apart from N !) and in the underabundance. The analyses of cool young field stars in the SMC carried out by Russell and Bessell (1989) and by Spite et al. (1989) also show an underabundance of approximately 0.60 dex compared to solar values. Similar values are found by photometric investigations of field stars in different parts of the SMC (Grebel et al. 1991).

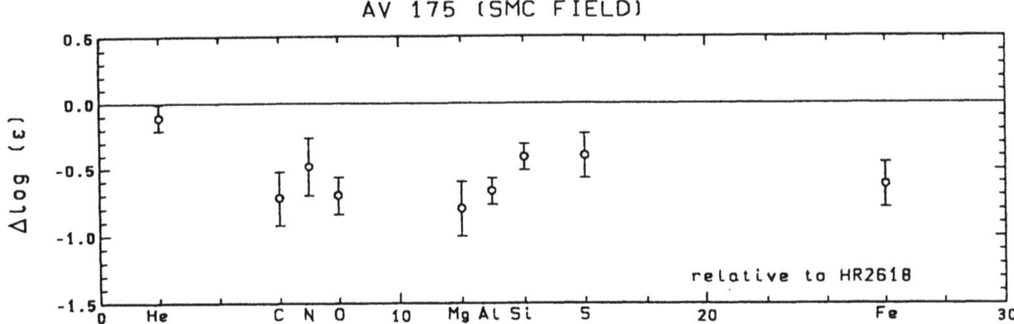

Fig. 9. Differential abundance pattern for the SMC field star AV 175, showing a mean abundance for all elements heavier then He of $\Delta \log \varepsilon = -0.65$ dex.

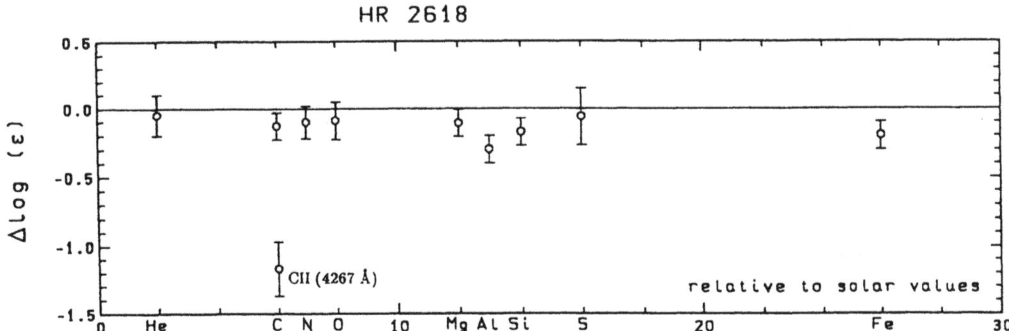

Fig. 10. Abundance pattern for the galactic comparison star HR 2618.

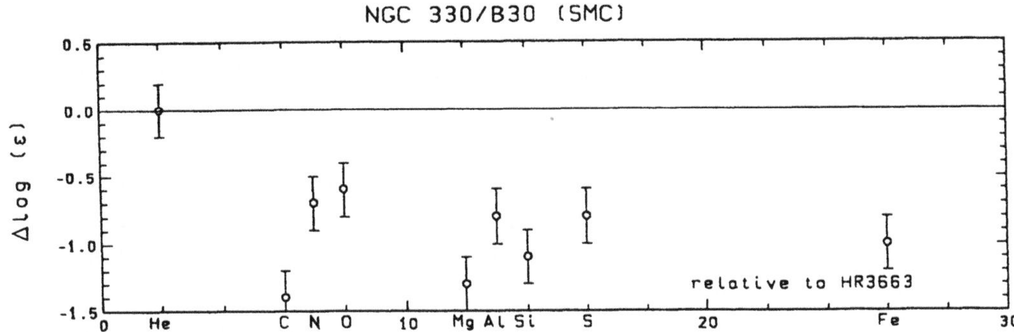

Fig. 11. Differential abundance pattern for the star B30 (Robertson, 1974) a member of the blue globular cluster NGC 330 in the SMC, showing a mean abundance for all elements heavier than He of $\Delta \log \varepsilon = -1.0$ dex.

The derived abundances of the blue globular cluster stars NGC 330/B30 (Reitermann et al. 1990), NGC 1818/D12 (Reitermann et al. 1990) and NGC 2004/B30 (Jüttner et al. 1991; preliminary results; a more detailed discussion is in preparation) are shown in Figures 11 to 13.

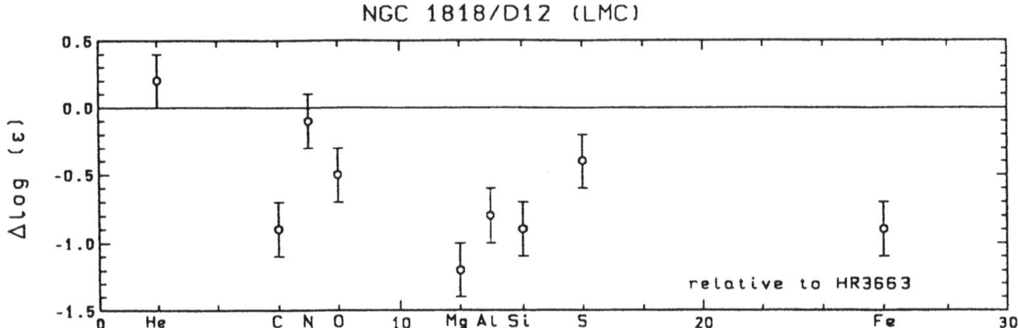

Fig. 12. Differential abundance pattern for the star D12 (Robertson, 1974) a member of the blue globular cluster NGC 1818 in the LMC, showing a mean abundance for all elements heavier than He of $\Delta \log \varepsilon = -0.8$ dex.

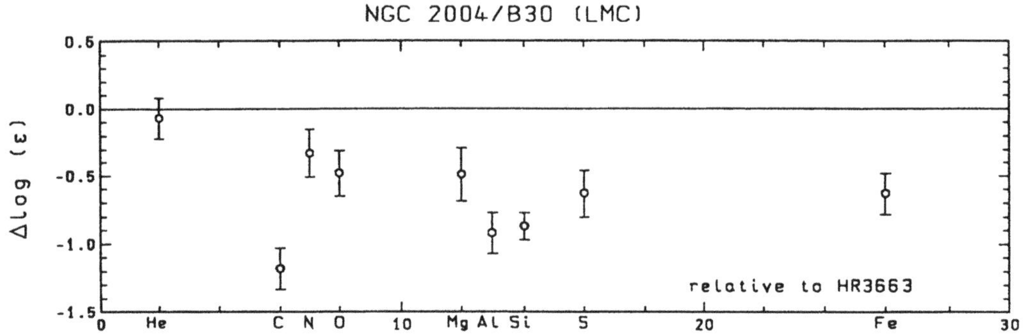

Fig. 13. Differential abundance pattern for the star B30 (Robertson, 1974) a member of the blue globular cluster NGC 2004 in the LMC, showing a mean abundance for all elements heavier than He of $\Delta \log \varepsilon = -0.7$ dex.

The scatter particularly of the CNO abundances is large. This might be partly due to the fact that the analyses were not strictly differential. More suitable galactic comparison stars are badly needed. Although bearing these difficulties in mind the following conclusions seem to be fairly safe:

a. The mean metallicity of B field stars in the SMC is down by about -0.65 dex compared to the solar values.
b. The members of the young blue globular clusters in the MCs show a underabunce in metallicity of about 0.3 dex compared to the field (cf. also de Boer, 1991)

This latter finding is in the case of NGC 330 in good agreement with Spite et al. (1989, 1991) who found a similar underabundances compared to the field from late type stars and with the photometry of Grebel et al. (1991).

5 Conclusions

The abundance analyses of luminous supergiants in the MCs provide a good test of evolutionary scenarious of very massive stars located and formed in different physical and chemical environments compared to the Galaxy. Yet the so far derived absolute abundances are more problematic and have to be treated with care due to the complex structure of the atmosphere and the mixing with CNO-cycled material.

The absolut abundances obtained from B giants and B main sequence stars representing the present evolutionary state and the abundances of the ISM can be derived with fair or even with good accuracy. The results are not seriously affected by NLTE effects when working stricity differential.

The overlap in C, N and O allows the comparison with the abundance determinations from HII regions. The reddening values of the MCs recently questioned by Bessell should be reexamined by analysing further suitable B stars observed with both CASPEC and IUE.

The metallicity in young globular clusters appears to be below that of the field indicating a chemical inhomogeneity in the MCs.

Acknowledgements: We would like to thank Dr. M. Bessell for the spectra of HR 2618. We thank the staff of the IUE Satellite Tracking Station operated by the European Space Agency at Villafranca (Spain) for kind and efficient support during the observations. This research was supported by the Deutsche Forschungsgemeinschaft (SFB 328).

References

Azzopardi, M., Vigeneau, J., Macquet, M.: 1975, *Astron. Astrophys. Suppl.* **22**, 285
Baschek, B.: 1989, Review presented at the "Ringberg Kolloquium"
Baschek, B., Holweger, H., Traving, G.: 1966, *Abhandl. Hamburger Sternwarte, VIII, Nr.1,* 26
Bessell, M.S.: 1991, *Astron. Astrophys.* **242**, L17-L20
Brunet, J.P., Imbert, M., Martin, N., Mianes, P., Prevot, L., Rebeirot, E., Rousseau, J.: 1975, *Astron. Astrophys. Suppl.* **21**, 109
de Boer, K.S., Azzopardi, M., Baschek, B., Dennefeld, M., Israel, F.P., Molaro, P., Segewiss, W., Spite, F., Westerlund, B.E.: 1989,*ESO Messenger* **57**, 27
de Boer, K.S.: 1991, *ESO Messenger* , in press
Dufour, R.J.: 1984, in *The Structure and Evolution of the Magellanic Clouds (IAU Symp. No. 108)*, eds. S. van den Bergh, K.S. de Boer, p.353
Dufton, P.L., Fitzsimmons, A., Howarth, I.D.: 1990, *Astrophys. J.* **362**, L59-L62
Dufton, P.L.: 1991, this volume
Eber, F., Butler, K.: 1988, *Astron. Astrophys.* **202**, 153
Garmany, C.D., Conti, P.S., Massey, P.: 1987, *Astron. J.* **93**, 1070
Grebel, E.K., Richtler, T.: 1991, *ESO Messenger* **64**, 56
Groth. H.G.: 1991, this volume
Humphreys, R.M., Kudritzki, R.-P., Groth, H.G.:1991 *Astron. Astrophys. Suppl.* in press
Jüttner, A., Reitermann, A., Stahl, O., Wolf, B.: 1989, *Astron. Astrophys. Suppl.* **81**, 93
Jüttner, A., Stahl, O., Wolf, B., Baschek, B.: 1991, in *The Magellanic Clouds (IAU Symp. No. 148)*, eds. R. Haynes and D. Milne, Kluwer, Dodrecht, p. 388

Kamp, L.W.: 1978, *Astron. Astrophys. Suppl.* **36**, 143
Kudritzki, R.-P., Groth, H.G., Butler, K., Husfeld, D., Becker, S.R., Eber, F., Fitzpatrick, E.: 1987, in *ESO workshop on SN1978A*, ed. I.J. Danziger, p.39
Kudritzki, R.-P., Gabler, A., Gabler, R., Groth, H.G., Pauldrach, A.W.A., Puls, J.: 1988, in *The Physics of Luminous Blues Variables (Proc. IAU Coll. 113)*, eds. K. Davidson, A.F.J. Moffat, H.J.G.L.M. Lamers, p. 67
Kudritzki, R.-P.: 1988, *The Atmospheres of Hot Stars: Modern Theory and Observations,* 18th Advanced Course, Swiss Soc. of Astron. and Astrophys., Leysin, MPA 380
Kurucz, R.L.: 1979, *Astrophys. J. Suppl.* **40**, 1
Lennon, D.J. Lynas-Gray, A.E., Brown, P.J.F, Dufton, P.L.: 1986, *Mon. Not. Roy. Astr. Soc.* **22 2**, 719
Lennon, D.J., Dufton, P.L., Fitzsimmons, A., Gehren, T., Nissen, P.E.: 1990, *Astron. Astrophys.* in press
Lennon, D.J., Kudritzki, R.-P., Becker, S.R., Eber, F., Butler, K., Groth, H.G: 1990, in *Properties of Hot Luminous Stars*, A.S.P. Conf. Ser., 7, 315
Lennon, D.J., Kudritzki, R.-P., Becker, S.R., Butler, K., Eber, F., Groth, H.G., Kunze, D.: 1991 *Astron. Astrophys.* submitted
Lennon, D.J.: 1991, this volume
Lyubimkov, L.S.: 1989, *Astrofizika* Vol.**30**, No.1, p.99
Peytremann, E., Baschek, B., Holweger, H., Traving, G.: 1967 Un Programme FORTRAN d' Analyse Quantitative de Spectre Stellaires, Obs. Geneve
Prevot, M.L., Lequeux, J., Maurice, E., Prevot, L., Rocca-Volmerage, B.: 1984, *Astron. Astrophys.* **132**, 389
Puls, J.: 1991 *Astron. Astrophys.* in press
Reitermann, A., Baschek, B., Stahl, O., Wolf, B.: 1990, *Astron. Astrophys.* **234**, 109
Richtler, T., Spite, M., Spite, F.: 1989, *Astron. Astrophys.* **225**, 109
Robertson, J.W.: 1974, *Astrophys. J. Suppl.* **15**, 261
Russell, S.C., Bessell, M.S.: 1989, *Astrophys. J. Suppl.* **70**, 865
Savage, B.D., Mathis,J.S.: 1979, *Ann. Rev. Astron. Astrophys.* **17**, 73
Spite, F., Spite, M.,Francois, P.: 1989, *Astron. Astrophys.* **210**, 25
Spite, M., Barbuy, B., Spite, F.: 1989, *Astron. Astrophys.* **222**, 35
Spite, F., Richtler, T., Spite, M.: 1991, *Astron. Astrophys.* submitted
Tammann, G.A.: 1987, in *Observational Cosmology (IAU Symp. No. 124)*, eds. A. Hewitt, G. Burbidge, Li Zhi Fang, Reidel, p. 151
Vidal, C.R., Cooper, J., Smith, E.W.: 1973, *Astrophys. J. Suppl.* **25**, 37
Walborn, N.R., Lasher, B.M., Laidler, V.G., You-Hua Chu : 1987, *Astrophys. J.* **21**, L41
Wolf, B.: 1972, *Astron. Astrophys.* **20**, 275
Wolf, B.: 1973, *Astron. Astrophys.* **28**, 335

The Chemical Composition of Main Sequence Stars in the Magellanic Clouds

I.D. Howarth[1], P.L. Dufton[2], A. Fitzsimmons[2],

W.R.J. Rolleston[2]

[1]Department of Physics and Astronomy, University College London,
Gower Street, London WC1E 6BT
[2]Department of Pure and Applied Physics, Queens University of Belfast,
Belfast BT7 1NN, United Kingdom

Abstract: High-resolution spectroscopic observations of main-sequence B-type stars in both the Small and Large Magellanic Clouds have been obtained using the UCL échelle spectrograph at the Anglo-Australian Telescope. Here we present a preliminary abundance analysis using LTE model-atmosphere techniques for two stars, AV 304 (in the SMC) and PS 34−16 (in the LMC). These stars lie on the hydrogen burning main-sequence; their surface abundances should reflect the current composition of the interstellar material in the Magellanic Clouds. Relative to the Galactic B0 V star τ Sco, AV 304 and PS 34−16 have have mean metal depletions of approximately 0.7 and 0.4 dex respectively.

1 Introduction

We are undertaking a programme to determine the chemical composition of main-sequence B-type stars in the Small (SMC) and Large (LMC) Magellanic Clouds. Analysis of spectra of these stars will yield reliable contemporary abundances of the interstellar medium, as their atmospheres should be uncontaminated by nuclear processing. Since both Clouds are stirred by the rotation of their central bars, the abundances of the local interstellar medium should in turn reflect global abundances within the Clouds. Those abundances are of interest for understanding the mass loss and evolution of early-type stars in environments different from those in the Galaxy.

In this paper we present preliminary results from an LTE model atmosphere abundance analyses of a subset of our observational data.

2 Observational Data

We have obtained high-resolution spectra of 6 main-sequence B-type stars in each Cloud by using the UCL échelle spectrograph at the AAT, during November 1989 and December 1990. Reduction of the data is still in progress; here we concentrate on two stars AV 304 (in the SMC) and PS 34–16 (in the LMC). The échelle data were reduced using the *Starlink* package, FIGARO (Shortridge, 1991); details of the procedures used can be found in Conlon *et al.* (1991). Line profiles and equivalent widths were measured by using the *Starlink* package, DIPSO (Howarth and Murray, 1990). The quality of the observational data is illustrated in figure 1 for both the stars discussed here.

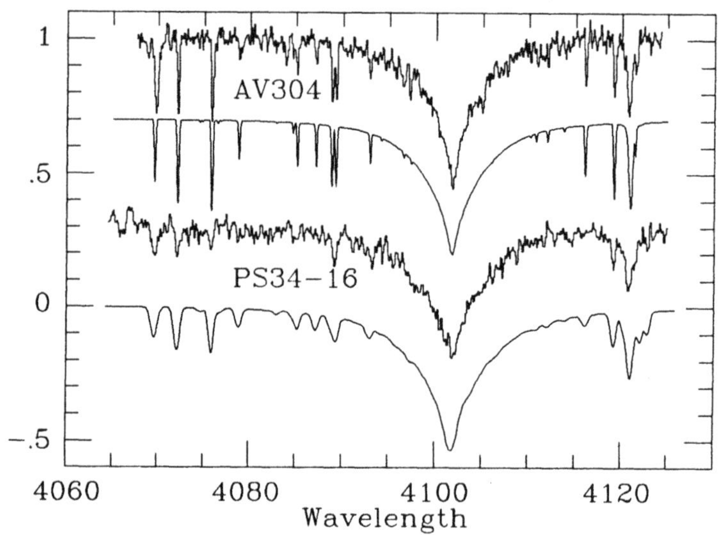

Fig. 1. Observed and theoretical spectra for AV 304 (in the SMC) and PS 34–16 (in the LMC). The theoretical spectra have been deduced using the abundances in Table 1. The agreement for the well-observed sharp-lined star AV 304 is excellent, while for the broader-lined PS 34–16 it is satisfactory, given the relatively poor signal-to-noise. Further observations of PS 34–16 are currently being reduced and should improve the quality of the dataset for this star.

3 Analysis and Results

The abundance analysis followed the procedures outlined in Dufton *et al.* (1990), where some results are presented for AV 304. Briefly, line-blanketted model atmospheres from Kurucz (1979) were used together with LTE radiative transfer codes; these methods should be adequate to provide reliable results for these relatively high-gravity stars. Theoretical spectra are shown in Figure 1 and, in general, agree well with the observations. However, to improve the accuracy of the abundances, a differential analysis was also performed with respect to τ Sco, the atmospheric parameters ($T_{\rm eff}$ = 31000 K,

$\log g = 4.15$, and $v_t = 5$ km s^{-1}) being taken from Lennon *et al.* 1990. The results of this differential analysis are summarized in Table 1.

Table 1. Logarithmic metal abundances (relative to those found for τ Sco) for the main-sequence B-type stars AV 304 (in the SMC) and PS 34–16 (in the LMC).

Element	Δ AV 304	Δ PS 34–16	
$T_{\rm eff}$		22000 K	25000 K
Carbon	−0.8	−0.4	−0.3
Nitrogen	< −1.1	−0.2	−0.4
Oxygen	−0.4	+0.3	−0.2
Neon	−0.4	...	...
Magnesium	−0.6	−0.9	−0.6
Aluminium	< −0.7	...	...
Silicon	−0.7	−0.4	−0.7
Sulphur	−0.6	...	...

The atmospheric parameters adopted for AV 304 were $T_{\rm eff} = 28000$ K, $\log g = 4.1$, and a microturbulent velocity of 5 km s^{-1}. AV 304 appears to have a normal helium abundance and a mean metal depletion of about 0.7 dex. However, nitrogen is significantly more underabundant while there is some evidence for oxygen being less depleted.

The effective temperature (22000 K) for PS 34–16 was deduced from the *uvby* photometry (Shobbrook and Visvanathan, 1987) and the surface gravity ($\log g = 4.0$) from the Hγ and Hδ profiles. However, adopting these atmospheric parameters and assuming a microturbulent velocity of 5 km s^{-1} implies a helium abundance of 10.7 dex and gives anomalous results for the metal abundances. Further observations of PS 34–16 are currently being reduced which should allow the effective temperature to be obtained from ionization equilibria. To test the sensitivity of our results to the adopted effective temperature, abundances have been deduced for a model atmosphere with $T_{\rm eff} = 25000$ K, $\log g = 4.2$, and $v_t = 5$ km s^{-1}; results are summarized in Table 1. For this hotter model the helium abundance is consistent with that found in other galactic main-sequence B-type stars, while a mean metal deficiency of about 0.4 dex is implied. Unlike AV 304, there is no evidence to suggest that nitrogen is more underabundant, but magnesium and silicon may be significantly more depleted. Due to the poorer quality of the PS 34–16 spectra, we were unable to obtain a reliable abundance for aluminium or sulphur, although both appear significantly depleted.

These results agree qualitatively with those deduced previously from analyses of late-type supergiants (Russell and Bessell, 1989) and emission-line objects (Russell and Dopita, 1990); both the LMC and the SMC are metal deficient compared with the Galaxy, with the SMC being significantly more heavily depleted.

References

Conlon, E.S., Dufton, P.L., Keenan, F.P., & McCausland, R.J.H., 1991. *Mon. Not. R. astr. Soc.,* **248**, 820.

Dufton, P.L., Fitzsimmons, A., & Howarth, I.D., 1990. *Astrophys. J.,* **362**, L59.

Howarth, I.D., & Murray, J.M., 1990. *SERC Starlink User Note,* No. 50.

Kurucz, R.L., 1979. *Astrophys. J. Suppl.,* **40**, 1.

Lennon, D.J., Dufton, P.L., Fitzsimmons, A., Gehren, T., & Nissen, P.E., 1990. *Astr. Astrophys.,* **240**, 349.

Russell, S.C., & Bessell, M.S., 1989. *Astrophys. J. Suppl.,* **70**, 865.

Russell, S.C., & Dopita, M.A., 1990. *Astrophys. J. Suppl.,* **74**, 93.

Shobbrook, R.R., & Visvanathan, N., 1987. *Mon. Not. R. astr. Soc.,* **225**, 947.

Shortridge, K., 1991. SERC Starlink Miscellaneous User Document, No. 1.

GHRS Observations of O Stars In Other Galaxies

Sara R. Heap[1], Dennis Ebbets[2], Eliot Malumuth[3]

[1]Code 681, NASA Goddard Space Flight Center, Greenbelt MD
[2]Ball Aerospace, Boulder CO
[3]Computer Sciences Corp., Lanham MD

Abstract: We apply modern wind theory to interpret ultraviolet spectra of O stars in other galaxies obtained by the Goddard High Resolution Spectrograph (GHRS).

With its unmatched ability to carry out ultraviolet spectroscopy of faint stars in crowded fields, the GHRS has brought in a new era of investigation of O stars in other galaxies. As examples, we will describe GHRS observations of three extragalactic objects made in the first year of HST:

- Melnick 42 (O3f), a single star in the LMC. It is one of the most luminous ($\mathcal{L} = 2.3x10^6$), massive ($\mathcal{M}$=100) stars known [1].
- R136a (O3f/WN), a compact cluster of luminous OB stars, also in the LMC. The inner 2" of R136a, the size of the GHRS aperture, contains 47 stars, with V magnitudes ranging from 12.8 to 19.3 [2].
- Knot #2, a giant H II region in the Seyfert galaxy, NGC 1068. Its overall spectral type is O8 II. Although only 1."1 in diameter, Knot #2 contains several thousand OB stars [3].

This sequence of observations forms stepping stones toward an understanding of OB stars in other galaxies. Melnick 42 is the most straight-forward case, being an apparently single star at a known distance. This O3f star makes an excellent case for testing the proposition [4] that the fundamental parameters of a star can be derived solely from its spectroscopic (photospheric and wind) parameters. R136a is a more difficult case. Here, we are forced to deal with a spectrum composed by the nearly 50 stars in the GHRS entrance aperture. On the other hand, the starfield is well resolved in WFPC pictures of the region, and we have photometry for each star [2]. Knot #2 in NGC 1068 is of course also composite. However, at 10 Mpc, NGC 1068 is too far away for the HST to resolve individual stars. We must use the techniques acquired from easier cases to learn about the stellar constituents of the knot.

Figure 1 shows the spectra of Melnick 42, R136a, and NGC 1068 Knot #2. All three spectra were obtained by the GHRS at low dispersion (G140L grating; 0.6 A resolution) with the object centered in the large (2"x2") entrance aperture. All show strong P Cygni features indicative of luminous O stars with strong winds. Besides the overall spectral appearance, there are three quantitative

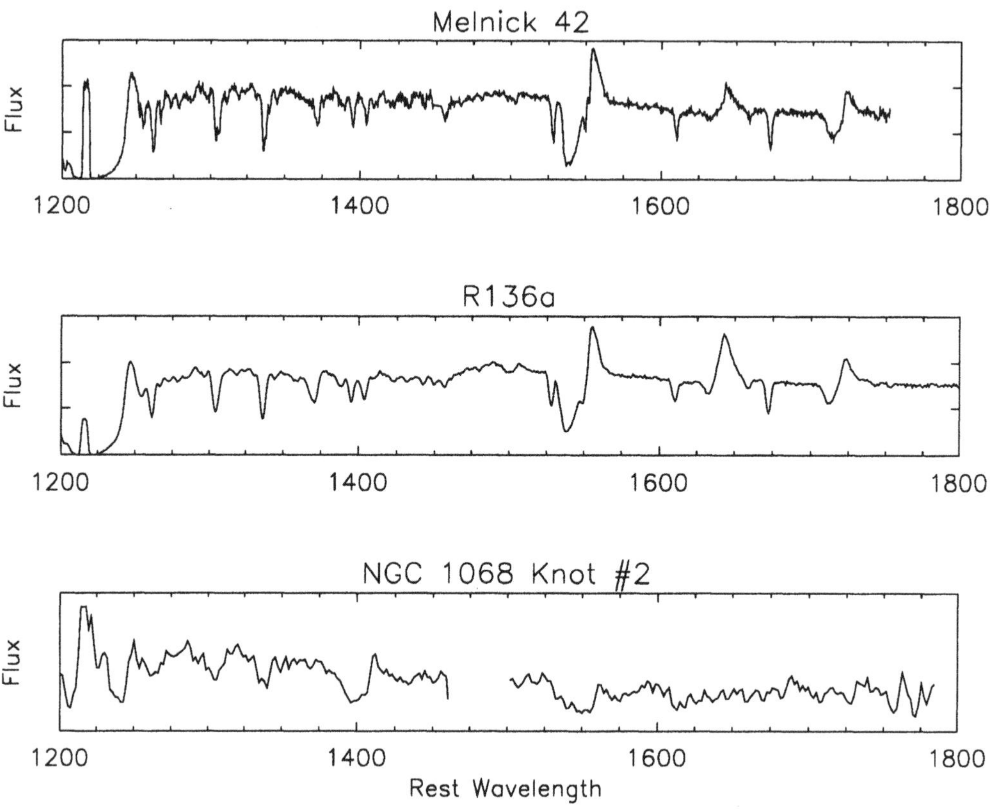

Figure 1: GHRS spectra of Melnick 42, R136a, knot in NGC 1068.

characteristics of a wind line that are particularly useful in revealing the properties of the underlying stars. The three features are: *(i)* the edge velocity of the absorption component of a wind line; *(ii)* the pedestal of a saturated resonance wind line; and *(iii)* the strength of the line after the pedestal has been subtracted out. The terminal velocity and mass-loss rate are well known determinants of the properties of a mass-losing star. The pedestal of a wind line gives the contribution of the lower-mass stars without winds.

Melnick 42. We were fortunate to obtain spectra of Melnick 42 with the star in the GHRS 0".25 aperture as well as in the 2" aperture. Saturated wind lines in the 0".25-aperture spectrum have measurably lower pedestals than in the 2"-aperture spectrum, even after convolution with the line-spread function. Upon checking pictures of Melnick 42 taken by the Planetary Camera we found up to three faint companions, so we believe that the higher pedestals of the resonance wind lines in the large-aperture spectrum are due to the presence of a nearby companion -- presumably a main sequence B star.

Contrary to intuition, the terminal velocity of wind models for Melnick 42 *increases* from 2400 to 2900 km s^{-1} when the metal abundances are reduced from solar to a quarter of the solar value [1]. That the terminal velocity of the wind is sensitive to metallicity but not necessarily in a monotonic way [5], serves as warning that analyses of distant OB stars should not be divorced from studies of metallicity.

R136a. Despite the fact that the large-aperture GHRS spectrum includes contributions from 47 stars, the spectrum of R136a is disturbingly similar to that of Melnick 42 (Fig. 1 top vs. middle). There is no overt evidence to indicate that the spectrum is composite. The edge velocity of the C IV λ1550 wind feature in R136a is only slightly higher than that of Melnick 42, convolved with the line spread function for R136a. More surprising, its pedestal is only slightly higher than in Melnick 42. Our photometry of R136a [2] indicates that about 65% of the flux is contributed by stars brighter than O5V stars, so there is no basic problem in understanding why the overall spectrum mimics that of a supergiant. Fainter stars, which are presumably main-sequence B stars that would contribute to the pedestal of the C IV wind line, make up less than 10% of the flux, consistent with the observed low pedestal.

NGC 1068. The spectrum of Knot #2 in NGC 1068 also has the appearance of an O supergiant. Unlike R136a, there are clear signs, *e.g.* the presence of Si III λ1300 absorption, of its composite nature. Hutchings *et al.* [3] suggest that the knot is a 3 million-year-old starburst. For a given age, IMF slope and mass range, we can calculate the properties of the stars [6] and their relative contribution to the observed UV continuous flux [7]. Since the minimum mass at which the Si IV λ1400 wind feature becomes noticeable is substantially higher than that for C IV or N V [8], the Si IV wind doublet should have a higher pedestal than should the C IV or N IV features. The actual values of the pedestals depend strongly on the IMF slope and lower mass-limit of the cluster. Similarly, the Si IV λ1400 wind feature should have a lower terminal velocity than would the C IV or N IV wind features. Unfortunately, the N V and C IV resonance wind lines are blended with interstellar lines, so that it is not possible to test this prediction. In any case, the computed terminal velocities are in qualitative agreement with the observed value, 1900 km s^{-1}, obtained from measurement of the Si IV wind line.

References

1. S.R. Heap et al.: Ap. J. **377** L29 (1991)
2. E. Malumuth et al.: B.A.A.S. **22**,1276 (1990)
3. J. Hutchings et al.: Ap. J. **377** L25 (1991)
4. R.-P. Kudritzki, D. Hummer: Ann. Rev. Ast. Ap. **28** 303 (1990)
5. K. Leitherer, H. Lamers: Ap. J. **373** 89 (1991)
6. A. Maeder: Ast. & Ap. Suppl. **84** 139 (1990)
7. R. Kurucz: priv. comm. (1991)
8. I. Howarth, R. Prinja: Ap. J. Suppl. **69** 527 (1989)

THE WINDS OF HOT STARS IN M31 AND M33

Luciana Bianchi [1], John B. Hutchings[2], Philip Massey[3]

[1] Osservatorio Astronomico di Torino, 10025 Pino Torinese, Italy
[2] Dominion Ap. Observatory, 5071 W.Saanich Rd., Victoria, B.C. V8X 4M6, Canada
[3] Kitt Peak National Observatory, NOAO, Box 26732, AZ 85726, USA

Abstract. Optical ground based (MMT, KFHT, KPNO) and UV *IUE* observations of bright blue stars in M31 and M33 have been collected over the past years. The spectral types, Teff and bolometric luminosities are derived from UV and visible spectra, and CCD optical photometry. The stars are all of O, B and WR type of high luminosity. Compared with galactic stars of similar types, the UV line spectrum and in particular the stellar wind indicators are weak, and the terminal velocities are low, in both galaxies. Early results were published in Massey et al. (1985) and in Hutchings et al (1987), and nine more stars are discussed, together with the previous data, in Bianchi et al (1991). The similarity of the wind features to MC's rather than to galactic counterparts, is hard to explain in M31, supposed to have a higher metallicity than Milky Way. The only possible explanation could be that we have so far studied the brightest stars, all in OB associations, that might be peculiar, or-else that a large metallicity gradient is present with radial distance in the galaxy. Both these hypoteses will be hopefully clarified by forthcoming HST observations.

References

Bianchi, L., Hutchings, J.B., Massey, P., 1991: Astron. Astrophys. **249** , 14
Hutchings, J.B., Massey, P., Bianchi, L., 1987: Astrophys. J., **322** , L79
Massey, P., Hutchings, J.B., Bianchi, L., 1985: Astron.J., **90** , 2239

III. Stellar Winds and Massive Stars

WOLF-RAYET STARS

Wolf-Rainer Hamann

Institut für Theoretische Physik und Sternwarte der Universität, Olshausenstraße, D-2300 Kiel, Germany

Abstract: Based on the so-called standard model for expanding atmospheres, non-LTE model calculations have been developed during the last years. Their application for the quantitative spectral analysis of WR stars is in progress. Recent results about the parameters and chemical composition of WR atmospheres are reviewed. From the achieved consistency between model predictions and observations, including the test case V444 Cygni, we conclude that the standard model is an adequate description of WR atmospheres to a satisfactory degree of accuracy. The empirical results of the spectral analyses obtained so far fit into the scenario of post-red-supergiant evolution, although there is not yet a detailed agreement with calculated tracks. The driving mechanism of WR winds remains an open question.

1. Introduction

More than a century ago, the two french astronomers Wolf and Rayet (1867) discovered three stars in the constellation Cygnus which exhibit very unusual spectra characterized by bright and broad emission lines. Stars of that spectral type are nowadays termed Wolf-Rayet (WR) stars. According to the dominant spectral lines, the nitrogen (WN) and carbon (WC) subclasses are distinguished. Subtypes (e.g. WN6, WN7 ...) were defined (cf. Smith 1968) in the order of decreasing degree of ionization. Subtype numbers of 6 and less form the group of "early" WN or WC stars (abbreviated WNE or WCE, respectively), while the "late" subtypes (WNL, WCL) start with WN7 and WC7.

The growing interest in WR stars, reflected by more than 100 publications per year (van der Hucht 1991), has several reasons. One is the key role of WR stars in the evolution of massive stars. It became evident (cf. the unexpected SN1987A progenitor) that the uppermost part of the HR diagram is poorly understood so far. A second reason concerns the physics of stellar winds, since UV spectroscopy has revealed that mass-loss is a very common phenomenon in astrophysics.

Beals (1929) was the first who recognized that the emission-line spectra of WR stars can be explained as the consequence of expanding stellar atmospheres. The so-called standard model for such stellar winds is described in the following section (Sect. 2). In Sect. 3 we review the various applications of that model for analyses of WR spectra. In Sect. 4 it is discussed to what degree real WR atmospheres are adequately described by the standard model. The results which have been obtained so far from spectral analyses of WR stars yield strong constraints for their evolutionary status (Sect. 5). However, the driving mechanism of their mass-loss is not yet understood (Sect. 6).

2. Basic assumptions of the standard model

The most-idealized model for expanding stellar atmospheres, the so-called standard model, is based on the following simplifying assumptions: spherical symmetry, stationarity, homogeneity, monotony of the velocity field, and radiative equilibrium. Models of that type are hence specified by only a few parameters: the stellar temperature T_* (i.e. the effective temperature, related to a well-defined radius, preferably to the stellar radius R_* of the hydrostatic stellar core), the luminosity L, the mass-loss rate $\dot{M}$, the terminal wind velocity v_∞ (and, less important, the run of $v(r)$), and the chemical composition of the atmosphere. Note that hydrodynamics are not self-consistently included.

3. Applications of the standard model

Based on these standard assumptions, Castor & Van Blerkom (1970) presented the first non-LTE calculations which were, however, restricted to one "representative point" in the atmosphere. The presence of the velocity field was accounted for by the escape probability method (Sobolev approximation). Calculations of that type were widely applied by various authors, mainly for abundance determinations (e.g. Willis & Wilson 1978; Smith & Willis 1982, 1983; Nussbaumer et al. 1982; Nugis 1991).

The weakness of those one-point models is that real WR atmospheres are extremely stratified, i.e. different lines may be formed in different layers. This can only be reflected adequately by "stratified models", which became available only since the last few years, due to the progress in the numerical techniques for multilevel non-LTE radiation transfer calculations. Presently there exist two independent codes, which fairly agree in their results.

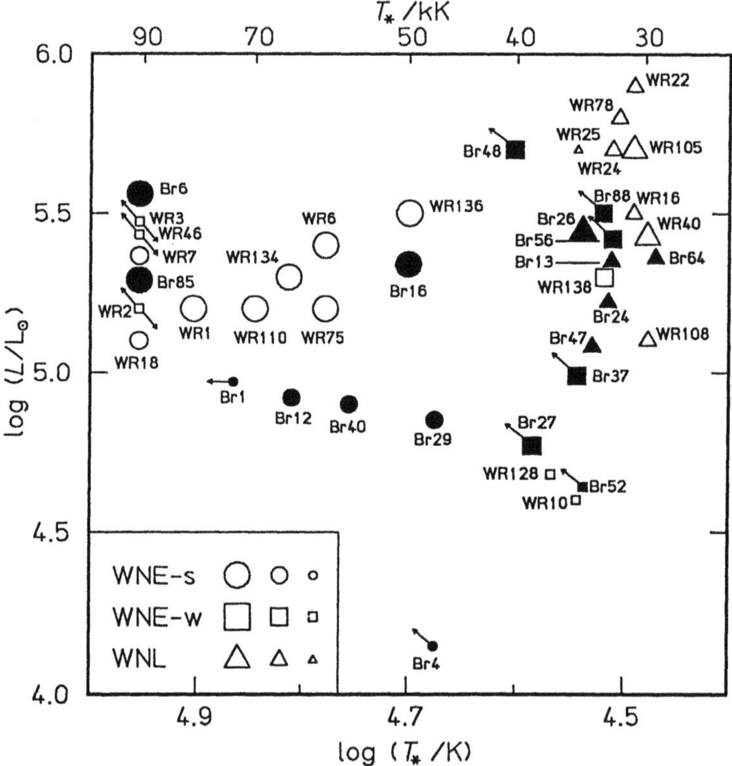

Fig. 1. Positions of WN stars in the HR diagram, discerning between Galactic and LMC members by open or filled symbols, respectively. The stars are identified by their catalogue number (Galaxy: van der Hucht et al. 1981; LMC: Breysacher 1981). The different subtypes are distinguished by different symbols, as explained in the inlet. The different sizes of the symbols indicate whether the rate of mass-loss is large ($\dot{M} > 10^{-4.2}$ $M_\odot$/yr), medium, or small ($\dot{M} < 10^{-4.8}$ $M_\odot$/yr). Uncertain positions are indicated by arrows. (From Koesterke et al. 1991)

The one has been written by John Hillier. It has been applied first for the modelling of the He and H spectrum of WR 6 (WN5) (Hillier 1987a,b). Subsequently the calculations were extended to nitrogen and carbon, again considering WR 6 (Hillier 1988). Recently the carbon spectrum of a WC star (resembling WR 111, WC5) has been simulated as well (Hillier 1989).

The other code was developed in Kiel. It takes advantage of the new method called "iteration with approximate lambda operators", which has been more and more improved and adapted to our purpose (Hamann 1985, 1986, 1987; Hamann & Wessolowski 1990; Koesterke et al. 1992). The Kiel code was used to establish a whole grid of pure-helium models (Wessolowski et al. 1988), which then was applied for the analyses of 30 Galactic WR stars (Schmutz et al. 1989) and of 19 WN stars in the LMC (Koesterke et al. 1991).

The resulting positions of the WN stars in the HR diagram (Fig. 1) reveal that WNL subtypes (i.e. WN7 and later) have stellar temperatures T_* around 35 kK, while WNE-s stars (early WN subtypes with strong lines) are hotter than 50 kK. The WNE-w stars (early subtypes with weak lines) do not form a uniform class. There is a tendency that LMC stars are less luminous than Galactic WN stars, but this must be further investigated with regard to selection effects.

In the next step hydrogen has been included into the model calculations. Four WN stars have been analyzed in detail with special regard to their hydrogen abundance by Hamann et al. (1991). The results confirm that WN atmospheres are hydrogen-deficient; for the four studied stars the mass fraction β_H ranges from 40% in WR 22 (WN7) to 10% in WR 136 (WN6-s). Although the numbers differ somewhat from the results obtained by Conti et al. (1983), their "semi-quantitative method"

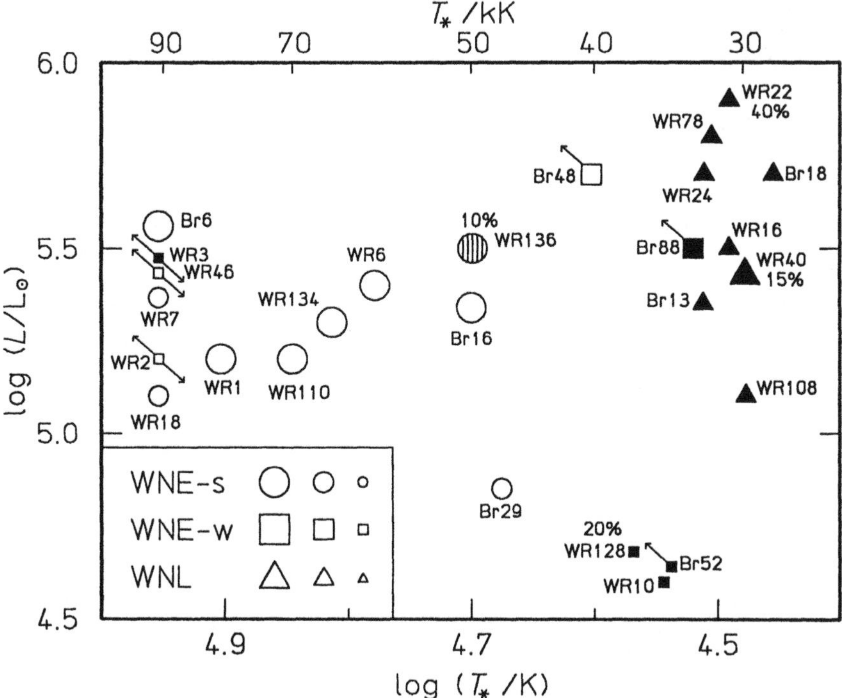

Fig. 2. Hydrogen in WN stars. Open symbols denote stars without hydrogen, filled symbols stars with hydrogen being detected (according to Conti et al. 1983). Percentages of hydrogen (mass fractions) are indicated for those four stars which have been analyzed in detail (Hamann et al. 1991). The symbol for WR 136 is hatched because of its small hydrogen content which Conti et al. (1983) could not detect. Form and size of the symbols and the identification of the stars have the same meaning as in Fig. 1. (From Hamann et al. 1991)

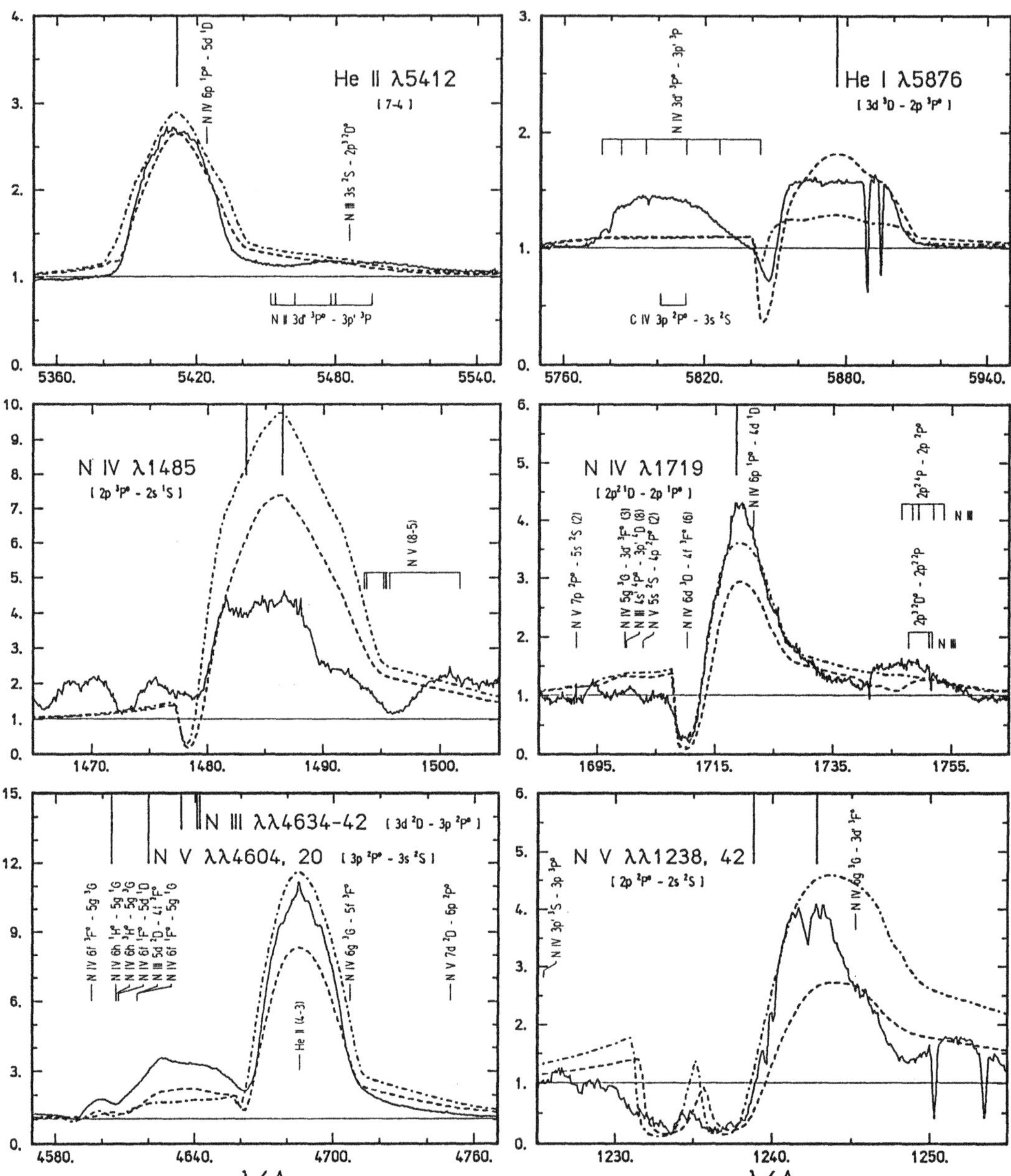

Fig. 3. Models with nitrogen. Theoretical profiles are given for two models with stellar temperatures of $T_* = 60$ kK (dashed-dotted) or 50 kK (dashed), while the other parameters are $L = 10^{5.5}\ L_\odot$, $\dot{M} = 10^{-3.85}\ M_\odot$/yr, $v_\infty = 1700$ km/s, and $\beta_N = 1.5\%$ (nitrogen mass fraction). Observed profiles are shown for WR 136 (WN6). (From Wessolowski, private communication)

is roughly confirmed and one can rely upon their discrimination between stars with or without hydrogen, respectively. In their large sample they found no clear correlation between the presence of hydrogen and the spectral subtype. However, there is a nearly perfect correlation between hydrogen abundance and the stellar temperature T_* as obtained from the spectral analyses (Fig. 2). Only the "cool" stars with $T_* < 40$ kK show hydrogen, while the hotter stars do not. WR 136 with

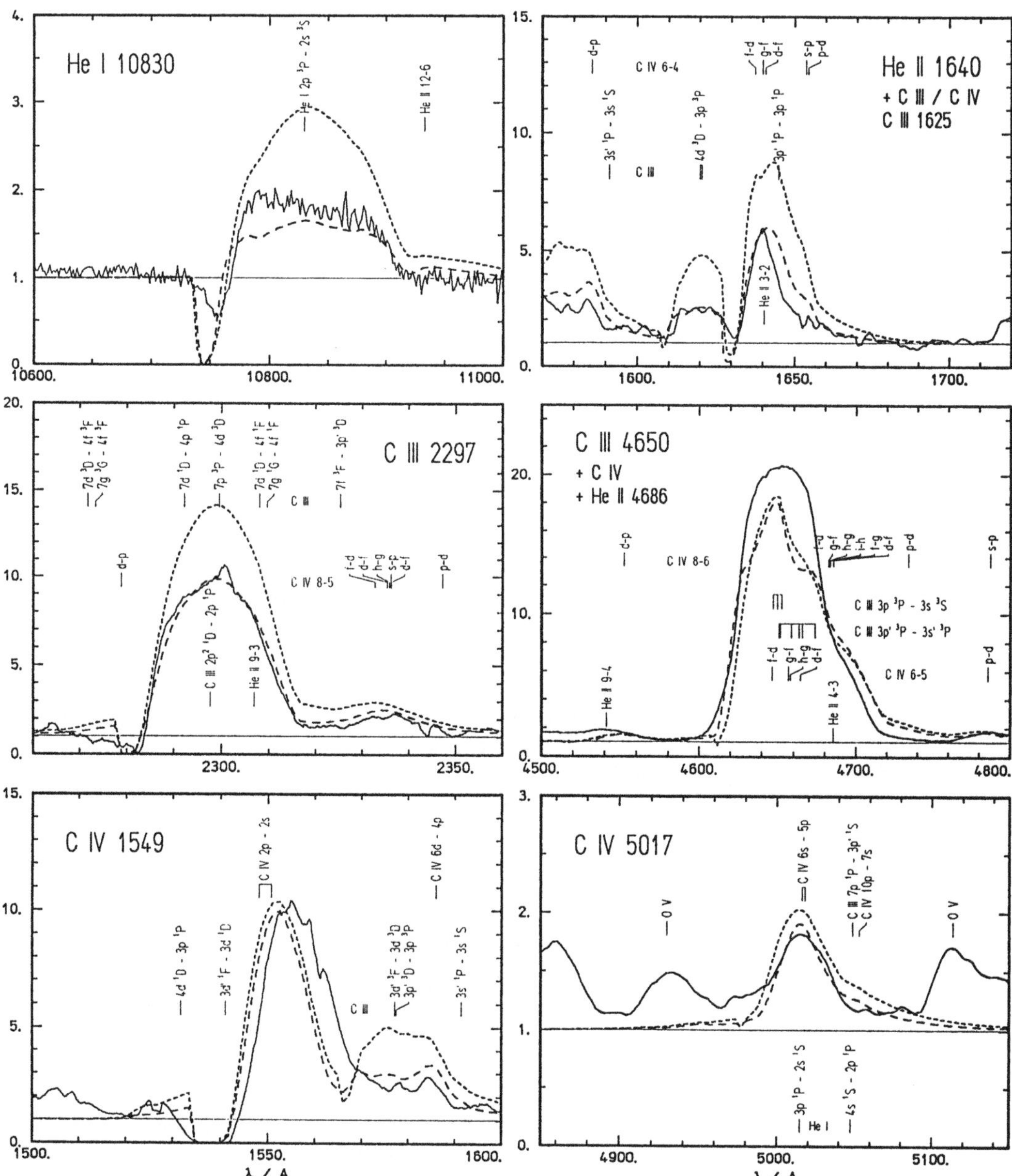

Fig. 4. Models with carbon. Theoretical profiles are given for two models with $\dot{M}$ = 4 10^{-5} $M_\odot$/yr (long-dashed) or a twice as large mass-loss rate (short-dashed), while the other parameters are L = $10^{5.0}$ $L_\odot$, T_* = 59 kK, v_∞ = 2400 km/s, and β_C = 60% (carbon mass fraction). Observed profiles are shown from WR 111 (WC5). (Combined from figures by Hamann et al. 1992)

β_H = 10% and T_* = 50 kK is intermediate in both parameters. This correlation gives strong evidence that evolution during the WN stage proceeds in the direction towards higher temperatures.

The inclusion of nitrogen into the models was a major step, due to the complexity of that atom. Hillier (1988) presented a model with T_* = 65 kK and β_N = 1% which could reproduce most equivalent widths of the N IV and N V lines (N III was not treated) observed in WR 6 (WN5) within a factor two or three. The Kiel group (Wessolowski et al., in preparation) developed model calculations which account for a complex model atom of nitrogen with 90 levels including N III. Low-temperature dielectronic recombination (LTDR) is treated in the way suggested by Mihalas & Hummer (1973). The formal integral allows for frequency redistribution of line photons by electron scattering.

In Fig. 3 two theoretical spectra are compared with observations of WR 136 (WN6). The agreement is reasonable, but N V 1240 requires $T_* = 60$ kK, while He I 5876 fits best with the cooler model (50 kK) where N III 4634/4642 is even still too weak. This might indicate that the temperature gradient in the models is a bit too shallow, due to the neglect of further trace elements as cooling agents. Nevertheless, Fig. 3 allows the important conclusion that the stellar temperatures derived from pure-helium analyses (Schmutz et al. [1989] obtained $T_* = 50$ kK for WR 136) are approximately confirmed. The nitrogen abundance of about $\beta_N = 1.5\%$ is obviously appropriate for that WN star.

The first model for a carbon-rich WC atmosphere was published again by Hillier (1989). He compared the result of one specific model ($T_* = 59$ kK, $L_* = 10^5 L_\odot$, $\dot{M} = 4\ 10^{-5} M_\odot$/yr, v_∞ = 1700 km/s, $\beta_C = 60\%$) with the observed equivalent widths (line profiles are not discussed) of WR 111 (WC5) and achieved a general agreement within a factor of two, with a few exceptions.

The Kiel group (Hamann et al. 1992) calculated models with similar parameters (cf. Fig. 4) and compared the resulting line profiles with the observed spectrum of WR 111. Although the stellar parameters were not yet optimized to yield the best fit, the agreement is rather good for almost all lines considered. Thus the stellar temperature of WR 111 is about $T_* = 60$ kK. Pure-helium analyses, which lead to 35 kK for that star (Schmutz et al. 1989), are obviously not adequate for WC spectra.

The carbon-to-helium abundance ratio in WCE atmospheres can be derived best from the neighbouring lines He II 5412 and C IV 5470. Fig. 5 displays two models, one with $\beta_C = 60\%$ (same model as in Fig. 4) and another model with $\beta_C = 20\%$ (the complementary part is always attributed to helium). The comparison with the observation of WR 111 suggests that an intermediate value ($\beta_C \approx 40\%$) would yield the best fit.

An extensive grid of WC models for the purpose of systematic spectral analyses will be published soon (Koesterke et al., in preparation).

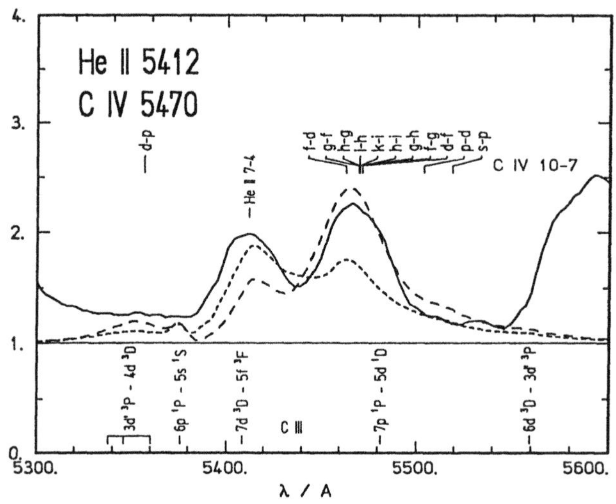

Fig. 5. Profiles of the neighbouring lines He II 5412 and C IV 5470 for the different carbon abundances $\beta_C = 60\%$ (long-dashed) or 20% (short-dashed). Other parameters have the same values as in Fig. 4. Obviously, a value of about 40% would fit best with the observation of WR 111. (From Hamann et al. 1992)

4. How adequate is the standard model?

The standard model for WR atmospheres relies on simplifying assumptions (cf. Sect. 2) which are certainly not perfectly fulfilled in reality. Only the achieved consistency between model predictions and observation can reveal whether that model can be considered as an adequate description of WR atmospheres. It should be mentioned that an entirely different model for WR atmospheres is promoted by Underhill and co-workers (cf. Underhill 1991), which however is not yet elaborated so detailed that line profiles can be predicted.

4.1. Spectrum synthesis

Strong support for the standard model comes from the achieved agreement between observed and synthetic spectra, concerning the line profiles and the continuum. Especially the fit of the WC spectrum (Fig. 4) is very encouraging. There are a few discrepancies, but we consider them as being of minor nature. Most of them can be probably resolved within the framework of the standard model. E.g., the problems with the WN spectra mentioned above might be due to the neglect of further trace elements causing cooling and blanketing effects. The form and strength of the blue-shifted absorption dips in He I, C III or N III lines, which are notoriously stronger in the model predictions than in the observations, might be due to the adopted velocity law. Problems with individual lines might be caused by deficiencies in the model atom and the atomic data, especially concerning the LTDR and the neglected HTDR (high-temperature dielectronic recombination).

Only a few remaining discrepancies possibly indicate the limitations of the standard model. Hillier (1991) pointed out that the electron-scattering wings of strong lines are weaker observed than predicted, which can be explained by inhomogeneities ("clumping") in the wind and a resulting over-estimate of the mass-loss rate by, say, a factor of two. The high "microturbulence" (typically 100 km/s) which is required for an optimum line fit, and the low-level time variabilities of the observed spectra suggest that further assumptions of the standard model (monotony of the velocity field, stationarity) are violated to some degree.

4.2. V444 Cygni

A test case for WR models is provided by the eclipsing binary V444 Cygni (WR+O). The light curve of that system has been analyzed by Cherepashchuk et al. (1984), and their solution (claimed as being unique) gave an effective temperature of the WR component (related to the radius of unity optical depth, which is not very different from our R_* in that case) of $T_{\rm eff}$ = 90 kK. Is that result compatible with a spectral synthesis based on the standard model? In a recent study (Hamann & Schwarz, in preparation) the Kiel models have been applied to both the composed spectrum and the light curve of V444 Cygni.

The analysis of the (helium) line spectrum alone cannot yield a unique solution, because the contribution of the O star component to the continuum level of the composite spectrum is not known a priori. Depending on the adopted brightness ratio between both continua (described by the flux ratio $Q_{\rm O:WR}$ at about 5000 Å), the helium lines can be reproduced with a WR temperature of $T_* \approx 35$ kK (if $Q_{\rm O:WR} \approx 1$ or less) as well as with $T_* \approx 60$ kK (if $Q_{\rm O:WR} \approx 5$), or even with 90 kK (if $Q_{\rm O:WR} \approx 7$), provided that the other free parameters are suitably adjusted. Once $Q_{\rm O:WR}$ is given, the line analysis also yields the "transformed radius" $R_{\rm t}$ as a combination of mass-loss rate and stellar radius (cf. Schmutz et al. 1989).

In order to synthesize the eclipse light curve, the following model is adopted. The O star revolves round the WR star within the outer regions of its extended atmosphere. Depending on the orbital phase, the O star shields a distinct volume of the WR atmosphere against the observer. On the other hand, the radiation which emerges from the surface of the O star is partly absorbed when passing through those parts of the WR atmosphere which lie in front.

The WR atmosphere is given by the standard models (pure-helium composition). Thus any disturbances caused by the O star are neglected. The intensity emitted by the O star is taken from non-LTE continuum models, accounting for limb-darkening.

The mass-loss rate has a very characteristic influence on the light curve, as it determines the width of the first minimum when the O star is occulted by the extended WR atmospheres. This width hardly depends on other parameters, and thus allows to derive $\dot{M}$ nearly independently. We find $\dot{M} = 10^{-4.9}\ M_{\odot}/yr$, remarkably consistent with the radio emission (Abbott et al. 1986) and the period change (Kornilov & Cherepashchuk 1979).

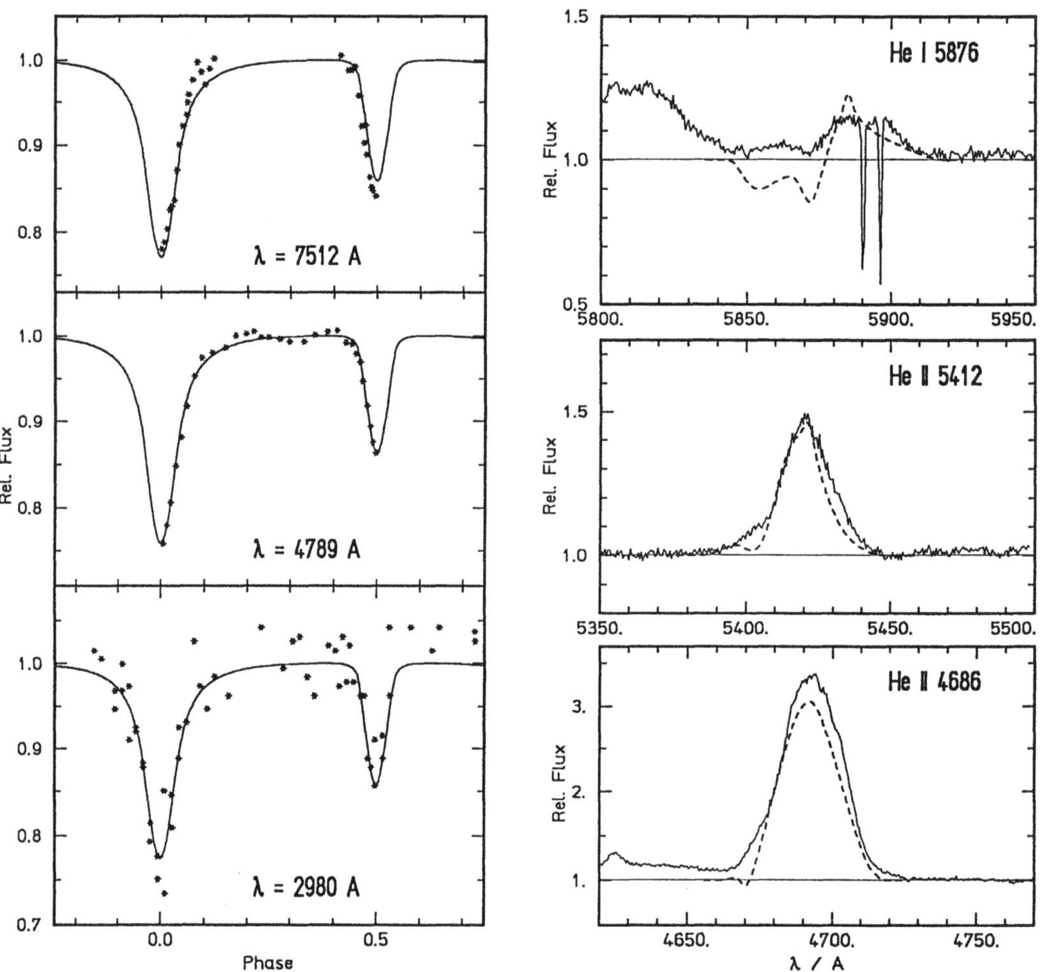

Fig. 6. Fit to the light curve (left) and to the helium lines (right) of V444 Cygni. Observations of the light curve (asterisks) are from Cherepashchuk et al. (1984) and references therein. Many photometric measurements have been averaged at the two visual wavelengths, while the UV data represent individual measurements. The line profiles have been observed at a phase of largest separation (0.24). The theoretical light curve (left, continuous lines) and the theoretical profiles (right, dashed lines) are calculated with the following, consistent set of parameters: WR star: $T_* = 34.8$ kK, $R_* = 6.8\ R_{\odot}$, $\dot{M} = 10^{-4.9}\ M_{\odot}/yr$, $v_{\infty} = 1900$ km/s; O star: $T_{eff} = 30$ kK, $R = 5.8\ R_{\odot}$; orbit: $a \sin i = 37.3\ R_{\odot}$, $i = 79°$. (After Hamann & Schwarz, in preparation)

The light curve now provides essentially two further properties, namely the depths of the two minima. But five further free parameters must be determined (temperatures and radii of both stars, and the inclination angle i; $a \sin i$ is known within error margins from the radial velocity curve). Hence it is not likely that the light curve fit alone can specify a unique solution. As a first result of our study we realized that the observed light curve can be reproduced with a whole three-dimensional subset of the parameter space (Hamann 1990). The solution presented by Cherepashchuk et al. (1984) is not unique, but only a particular one.

In a complicated procedure we have combined the light curve analysis with the constraints from the helium line analysis, which restricts the solution space by two dimensions. The effective temperature of the O star is taken as the remaining free parameter, which is varied between 30 kK and 40 kK as the range suggested from its (quite uncertain) spectral classification. Fig. 6 demonstrates the quality of the fit to both, the light curve and the helium line profiles, which is achieved with the consistent set of model parameters given in the figure caption. Nearly identical fits (not shown here) are obtained when T_{eff} = 35 kK or 40 kK is assumed; the corresponding sets of consistent parameters only differ in the O star radius (R = 5.8 $R_\odot$, 4.1 $R_\odot$ or 3.5 $R_\odot$ for T_{eff} = 30 kK, 35 kK or 40 kK, respectively) and the orbital inclination (i = 79°, 81° or 83°, respectively). The latter lie in the range derived from polarization measurements. Note that we obtain a WR temperature of $T_* \approx$ 35 kK, contradictory to the 90 kK given by Cherepashchuk et al. (1984).

4.3. X-ray emission

Many (even single) WR stars have been identified as X-ray sources (Pollock 1987). This type of emission is not predicted by the standard model for WR atmospheres. It is commonly believed that X-rays are produced by hydrodynamic shocks occuring somewhere in the stellar winds, thus connected to inhomogeneities, time variability and non-monotonic velocity fields.

The production and effect of X-rays in WN atmospheres has been studied recently in Kiel by means of a simple two-component model (Baum et al., in preparation). The "normal" gas component has the temperature and density as obtained from the standard model calculations (He and 1.5% N in radiative equilibrium). The second component should simulate the shocked, hot gas and is assumed being homogeneously distributed over the whole atmosphere with a specified filling factor (free parameter); one might think of optically thin, hot filaments. The only considered effect of that hot component is the emission and absorption of radiation by free-free processes. The additional emission and absorption coefficients are then fully specified by the electron temperature of that hot component, T_X (free parameter), while the same density as for the normal component is adopted. Special care is taken to include the specific interaction processes between X-radiation and matter (K-shell absorption, Auger effect). The purpose of the calculations is now to describe correctly the transfer of X-rays within the WR atmosphere in order to obtain the emergent X-ray flux, and, secondly, to study the effects of the X-ray field on the ionization and excitation of the "normal" gas component.

Test calculations have been performed for one specific model (T_* = 35 kK, R_* = 12.5 $R_\odot$, $\dot{M}$ = $10^{-4.4}$ $M_\odot$/yr, v_∞ = 2500 km/s). The parameters of the hot component (temperature T_X, and the filling factor) are varied.

The emergent X-ray fluxes, scaled to the unit distance of 10 pc and expressed in EINSTEIN count rates, are compared in Fig. 7 with typical observations. Obviously the observed rates can be reproduced with reasonable values for the filling factor (a few percent) and for the hot-gas temperature ($T_X \approx 5\ 10^6$ K). The latter value corresponds to a thermal velocity of about 200 km/s, which is a plausible shock amplitude.

As a second, important result we find that in models which reproduce the observed emergent X-ray fluxes, the X-ray field within the atmosphere has no significant influence on the He and N

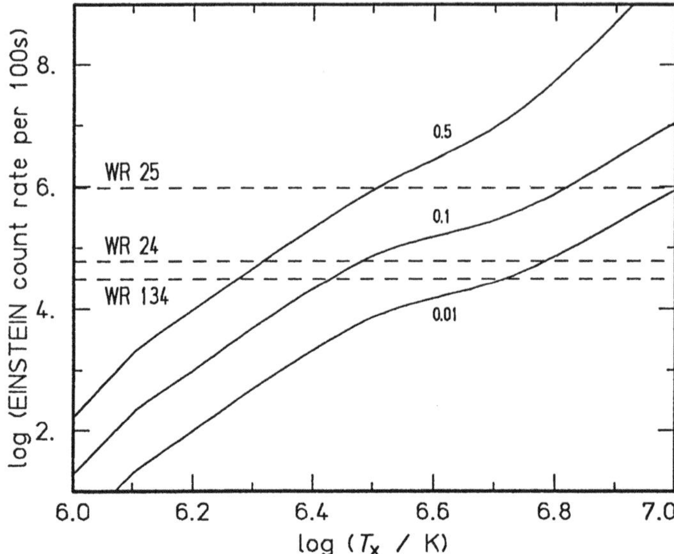

Fig. 7. X-ray emission from WR stars. The observed EINSTEIN count rates for WR 24, WR 25 and WR 134 (scaled to the unit distance of 10 pc, but not corrected for interstellar absorption) are compared to predictions of a simple two-component model. The theoretical count rate depends on the assumed temperature T_X (x axis) and the filling factor (labels) of the hot gas component. (After Baum et al., in preparation)

population numbers, i.e. there is no "superionization" produced by X-rays, obviously due to their relatively small number. This implies that the synthetic spectra obtained with the standard model are not affected (except, of course, in the X-ray range itself). Hence spectral analyses performed with the standard model should not be biased by the neglect of the X-rays.

4.4. Conclusions

With the standard model for WR atmospheres the observed spectra (line and continuum) in the UV and visual range can be reproduced, at least to some degree of accuracy. Moreover, the eclipse light curve of V444 Cygni can be consistently explained on the basis of the standard model. Thus the standard model can be considered as an adequate description of the reality, and some confidence is justified into the results obtained by quantitative spectral analyses based on that model. Limitations of the standard model are indicated by the observed X-ray emission, and by time variabilities.

5. The evolutionary status of WR stars

The analyses of WR spectra reviewed above provide severe empirical constraints for the discussion of the evolutionary status of these stars. Different scenarios have been proposed for the evolution of (single) WR stars: direct evolution from the main sequence ("Conti's scenario" for very massive stars: O → Of → WR), homogeneous evolution due to internal mixing (cf. Maeder 1987), or the post-red-supergiant (post-RSG) scenario which has been elaborated in detailed evolutionary calculations.

In Fig. 8 the results from the spectral analyses (cf. Fig. 1, but now omitting those objects with ambiguous parameters) are compared with the most-recent theoretical tracks for post-RSG evolution by Maeder (1990). For this comparison the "stellar temperature" T_* (related to the radius R_* of the static stellar core) is identified with the effective temperature of the evolutionary models which refers to the hydrostatic equilibrium radius. For WNE-s stars this identification is not mandatory, as in those stars the static layers lie in optically thick regions and are thus unobservable. Their location is only guessed from an inward extrapolation of the ad-hoc adopted velocity law, and therefore the significance of R_* (and hence T_*) is not clear in those cases.

The track for an initial mass of M_i = 25 $M_\odot$ can explain the most-luminous WN stars, especially with regard to the surface abundances. While hydrogen gradually vanishes, the track (dashed) crosses the WNL box. The hydrogen-free products of CNO burning appear at the surface when the track (now full-drawn) passes through the region where the WNE-s stars are found. But the evolutionary tracks fail in explaining the majority of the WN stars which fall below the 25 $M_\odot$-track, suggesting that either less-massive stars than predicted can reach the Wolf-Rayet stage, or the luminosity decreases during evolution.

The discrepancy in luminosities is even more pronounced when those tracks from Maeder (1990) are considered which were calculated for the lower metallicity of the LMC. While the average empirical luminosity of the LMC sample is even less than for the Galactic sample, the theoretical track for 25 M_i never returns to the blue side. This raises the minimum WR luminosity to $10^{5.8}$ $L_\odot$ (track for M_i = 40 $M_\odot$), in conflict to the empirical result (cf. also next section).

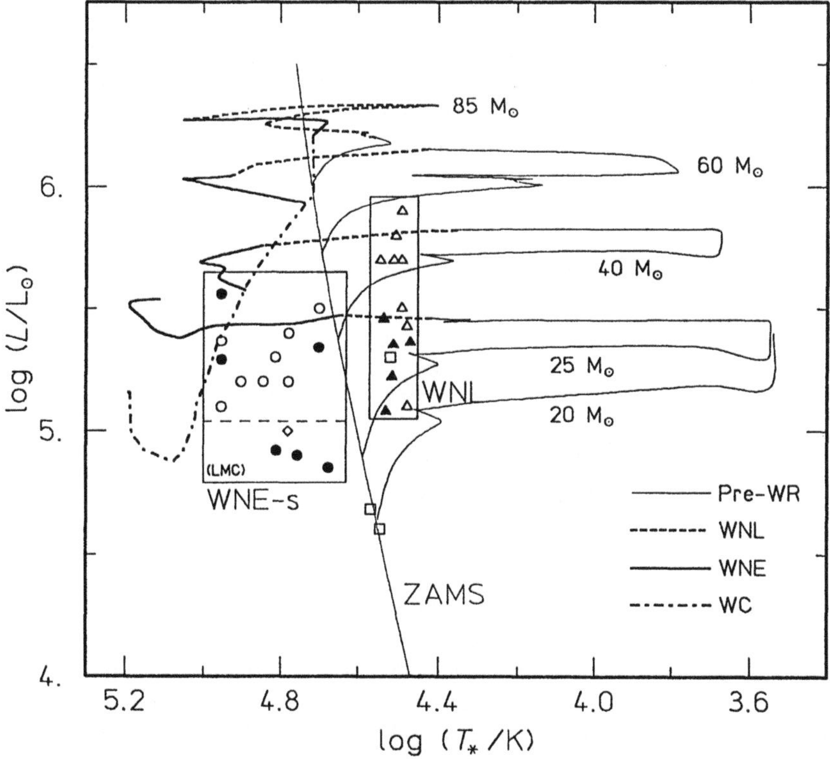

Fig. 8. The empirical location of Wolf-Rayet stars in the HR diagram, compared with evolutionary tracks. The discrete symbols indicate the positions obtained from spectral analyses of Galactic (open symbols) and LMC (filled symbols) members (cf. Fig. 1). The different symbols distinguish between WNL (△), WNE-w (□) and WNE-s (○) subtypes, respectively. The only WC star analyzed so far (WR 111, WC5) is indicated as well (◊). The boxes envelop the positions of the WNL and WNE-s stars, respectively. A dashed line separates the lower part of the WNE-s box which only contains LMC stars, while the upper part of the WNL box is only populated by Galactic objects. Superimposed are evolutionary tracks calculated by Maeder (1990) for solar metallicity, labelled by their initial masses. The evolutionary stages (reflecting surface abundances) along the tracks are indicated by different drawing styles, as explained in the inlet. (After Koesterke et al. 1991)

Abundances as observed in WC stars are predicted by the evolutionary calculations when the products of helium burning show up at the surface. According to Maeder (1990), stars with M_i = 25 $M_\odot$ end in a supernova explosion before reaching the WC stage. All stars with $M_i \geq 40\ M_\odot$ finally reach the same WC track (dash-dotted), but this track misses the location of WR 111 (◊) by far.

We conclude that the post-RSG scenario is a promising attempt to explain the formation of WR stars, but the quantitative agreement of the present theoretical evolutionary tracks with the empirical positions in the HR diagram is still poor. Probably it is the adopted amount of mass-loss during the course of evolution which is to blame.

6. What drives the mass loss of WR stars?

Self-consistent hydrodynamic models for the acceleration of Wolf-Rayet winds are not available. The theory of radiation-driven winds (RDW) is very successful in explaining the comparably weak winds from O and Of stars (cf. Kudritzki et al. 1987), but fails in producing mass-loss as strong as observed in WR stars. This difficulty can be expressed in terms of the so-called momentum problem. When comparing the rate of mechanical momentum contained in the stellar wind ($\dot{M}\ v_\infty$) with the rate of momentum carried with the radiation field (L/c), one obtains empirical ratios which are much greater than unity (cf. Table 1). The RDW theory in its present form, on the other hand, never yields momentum ratios exceeding unity by such large factors. (It must be emphasized, however, that this is not a strict limit, but just a result of numerical models in their present state.)

Table 1. Average momentum ratios for WR subgroups

	$\dot{M}\ v_\infty\ c\ /\ L$ Galaxy	LMC
WNL	5	9
WNE-w	10	≈10
WNE-s	45	49
WC (WR111)	50	

Furthermore, the RDW theory for O stars predicts certain metallicity effects. Are these effects visible in WR stars, when comparing between stars in the Galaxy and the LMC?

A clear correlation between the terminal velocity obtained from the helium lines, v_∞, and WN subtype was found by Koesterke et al. (1991), contrary to previous statements (Abbott & Conti 1987, Willis 1991). The comparison between Galaxy and LMC (Fig. 9) does not reveal any systematic differences. According to the RDW theory, however, LMC stars should have slower winds than their Galactic counterparts (as empirically confirmed for O stars, cf. Garmany & Conti 1985).

As another metallicity effect, the RDW theory predicts that the mass-loss rates scale, roughly, with the square-root of the metallicity, for otherwise identical stars. ("Metallicity" here mainly refers to the abundances of heavy elements such as iron, which are not affected by the chemical evolution of the WR star itself.) Thus $\dot{M}$ should be smaller by 0.3 dex in LMC stars, compared to the Galaxy. However, the average mass-loss rates for the spectral subgroups is exactly the same in both samples analyzed so far (Koesterke et al. 1991), again not in line with the expectation from the RDW theory.

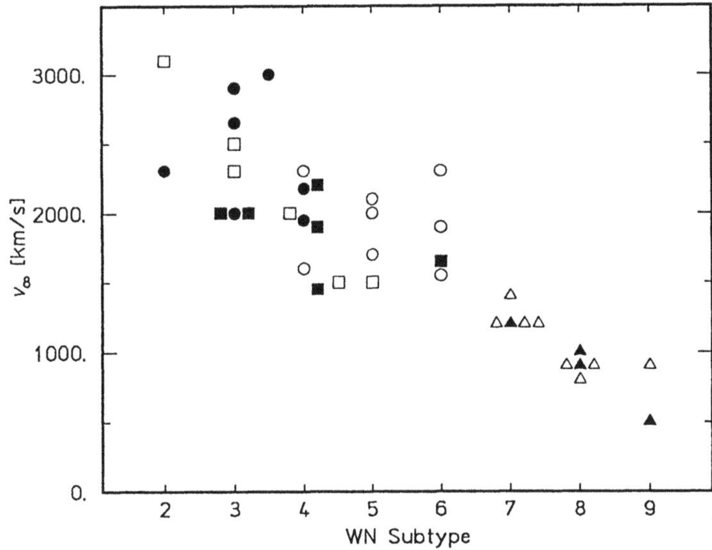

Fig. 9. Terminal velocity v_∞ versus WN subtype. Open and filled symbols denote Galactic or LMC stars, respectively, while the different symbols distinguish between WNL (△), WNE-w (□) and WNE-s (○) subtypes. (From Koesterke et al. 1991)

The metallicity dependence of the mass-loss rate predicted by the RDW theory was also adopted by Maeder (1990) for calculating the evolutionary tracks discussed in the preceding section. This is the reason why he predicts a larger minimum luminosity for WR stars in the LMC than for Galactic stars, contrary to the empirical results (cf. Sect. 5). Hence this gives an indirect evidence that during the relevant evolutionary phases the mass-loss rate does not scale with metallicity.

Radiation pressure is the only candidate for the force which accelerates the WR winds. (At low velocities around the critical point which governs the mass-loss rate, other processes such as stellar vibrations might be important as well.) The hitherto existing RDW calculations, however, cannot explain the large wind momentum observed, and predict metallicity effects which are not observed. The only hope ist that these problems are due to the simplifications and neglects inherent to the available RDW models. One objection is certainly the restriction to stationary solutions with monotonic velocity fields, although these solutions are known to be unstable (cf. Owocki, these proceedings).

7. Outlook

Basic questions about WR stars are yet open; but the quantitative analysis of their spectra, which became possible thanks to the progress in non-LTE modelling, has already provided important empirical information. The hope is justified that a continuation of that work (if financed) will soon improve our understanding of the evolution of WR stars and of the physics of their expanding atmospheres.

Acknowledgements. The contributions of E. Baum, L. Koesterke, U. Leuenhagen, W. Schmutz, E. Schwarz and U. Wessolowski to this research are gratefully acknowledged.

References

Abbott D.C., Conti P.S.: 1987, *Ann. Rev. A&A* **25**, 113
Abbott D.C., Bieging J.H., Churchwell E., Torres A.V.: 1986, *ApJ* **303**, 239
Beals C.S.: 1929, *MNRAS* **90**, 202
Breysacher J.: 1981, *A&A Suppl.* **43**, 203
Castor J.I., Van Blerkom D.: 1970, *ApJ* **161**, 485
Cherepashchuk A.M., Eaton J.A., Khaliullin K.F.: 1984, *ApJ* **281**, 774
Conti P.S., Leep E.M., Perry D.N.: 1983, *A&A* **268**, 228
Garmany C.D., Conti P.S.: 1985, *ApJ* **293**, 407
Hamann W.-R.: 1985, *A&A* **148**, 364
Hamann W.-R.: 1986, *A&A* **160**, 347
Hamann W.-R.: 1987, in *Numerical Radiative Transfer*, W. Kalkofen (ed.), Cambridge University Press, p. 35
Hamann W.-R.: 1990, in *Properties of Hot Luminous Stars*, C.D. Garmany (ed.), ASP Conference Series Vol. 7, p. 259
Hamann W.-R., Wessolowski U.: 1990, *A&A* **227**, 171
Hamann W.-R., Dünnebeil G., Koesterke L., Schmutz W., Wessolowski U.: 1991, *A&A* **249**, 443
Hamann W.-R., Leuenhagen U., Koesterke L., Wessolowski U.: 1992, *A&A* (in press)
Hillier D.J.: 1987a, *ApJ Suppl.* **63**, 947
Hillier D.J.: 1987b, *ApJ Suppl.* **63**, 965
Hillier D.J.: 1988, *ApJ* **327**, 822
Hillier D.J.: 1989, *ApJ* **347**, 392
Hillier D.J.: 1991, *A&A* **247**, 455
van der Hucht K.A.: 1991, in *Wolf-Rayet Stars and Interrelations with Other Massive Stars in Galaxies*, Proceedings of IAU Symp. No. 143, K.A. van der Hucht and B. Hidayat (eds.), Kluver Academic Publishers, Dordrecht, p. 19
van der Hucht K.A., Conti P.S., Lundström I., Stenholm B.: 1981, *Space Sci. Rev.* **28**, 227
Koesterke L., Hamann W.-R., Schmutz W., Wessolowski U.: 1991, *A&A* **248**, 166
Koesterke L., Hamann W.-R., Kosmol P.: 1992, *A&A* (in press)
Kornilov V., Cherepashchuk A.M.: 1979, *Sov. Astron. Letters* **5**, 214
Kudritzki R.P., Pauldrach A., Puls J.: 1987, *A&A* **173**, 293
Maeder A.: 1987, *A&A* **178**, 159
Maeder A.: 1990, *A&A Suppl.* **84**, 139
Mihalas D., Hummer D.G.: 1973, *ApJ* **179**, 827
Nugis T.: 1991, in *Wolf-Rayet Stars and Interrelations with Other Massive Stars in Galaxies*, Proceedings of IAU Symp. No. 143, K.A. van der Hucht and B. Hidayat (eds.), Kluver Academic Publishers, Dordrecht, p. 75
Nussbaumer H., Schmutz W., Smith L.J., Willis A.J.: 1982, *A&A Suppl.* **47**, 257
Pollock A.M.T.: 1987, *ApJ* **320**, 283
Schmutz W., Hamann W.-R., Wessolowski U.: 1989, *A&A* **210**, 236
Smith L.F.: 1968, *MNRAS* **138**, 109
Smith L.J., Willis A.J.: 1982, *MNRAS* **201**, 451
Underhill A.B.: 1991, *ApJ* (in press)
Wessolowski U., Schmutz W., Hamann W.-R.: 1988, *A&A* **194**, 160
Willis A.J.: 1991, in *Wolf-Rayet Stars and Interrelations with Other Massive Stars in Galaxies*, Proceedings of IAU Symp. No. 143, K.A. van der Hucht and B. Hidayat (eds.), Kluver Academic Publishers, Dordrecht, p. 265
Willis A.J., Wilson R.: 1978, *MNRAS* **182**, 559
Wolf C.J.E., Rayet G.A.P.: 1867, *Comptes Rendus* **65**, 292

Spectrophotometry of Wolf-Rayet Stars

Andrzej Niedzielski

Institute of Astronomy, N. Copernicus University, ul.Chopina 12/18, PL-87-100
Toruń, Poland

This work is based on two different sets of **observational data**: optical and IUE, which were gathered for seven galactic WR stars: HD 4004, 16523, 17638, 191765, 192103, 192163 and 193928. The optical medium dispersion spectra (22-28 Å/mm in the blue and 80-86 Å/mm in the red wavelength range) as well as their reduction are described in detail elsewhere (Niedzielski & Nugis 1991). The archival IUE low resolution spectra were collected from IUE database in Vilspa. Altogether we use 42 optical and 37 low resolution IUE spectra for the 7 program stars. Basing on the IUE spectra we were able to identify and measure the strongest lines in both WN and WC stars including HeII (n-3) lines. When compared to Nussbaumer et al. (1981) or Willis & Wilson (1978) our resulting equivalent widths (EW) agree within 0.11 dex (in Log EW) in average.

Having the lines identified and line widths (FWHM) measured it was possible to notice **the stratification of WR envelopes** immediately. As shown in Kuhi (1973) in WN envelopes this effect is easily seen on Ionisation Potential vs. FWHM graphs. We noticed that in the case of WC stars this effect is more masked (by blending) but also seen. Basing on such graphs we were able to estimate the terminal velocity of the wind in WR envelopes. A more detailed analysis of the FWHM data leads to an interesting observation: the widths of Pickering HeII (n-4) lines seem to correlate with n. The higher the upper level n in (n-4) transition the lower the FWHM velocity - Fig.1. This suggest that we are dealing with **stratification in HeII (n-4) lines**. This effect was discussed by Hillier et al. (1983) and Hillier (1987). Such stratification, if confirmed, leads to a conclusion that different He II lines are effectively formed at different radii of WR envelope, where the matter expands with different velocities - the representative point approximation is no longer valid. The Ionisation Potential vs. FWHM graphs, being affected by stratification within one ion lines, can not be used for studying the WR envelopes neither. Moreover, this effect gives the unique opportunity to study the irregularities in WR winds by observing profiles of several He II lines at the same time.

To estimate **the hydrogen content** in the envelopes of the studied stars we used the standard Balmer-Pickernig decrement method as described in Conti et al. (1983). We found that several Pickering HeII lines can not be used for such analysis because of serious blending with lines of other ions. We excluded from this analysis in WN stars lines $\lambda\lambda$ 3888, 4100, 4200 Å. The lines $\lambda\lambda$ 3968 and 4025 were used carefully. In the case of WC stars we found that only the lines $\lambda\lambda$ 3858, 4025, 4100, 4542 (in HD 192103) and 4860 can be used for such study. We used selected lines to obtain the H/He ratio. In each case we used two odd lines to approximate the Pickering decrement shape and one even line. In Table 1 the final results are given. For the WN stars the mean values of the determined H/He ratio are given. For the WC stars we conclude that H/He is close or equel to zero.

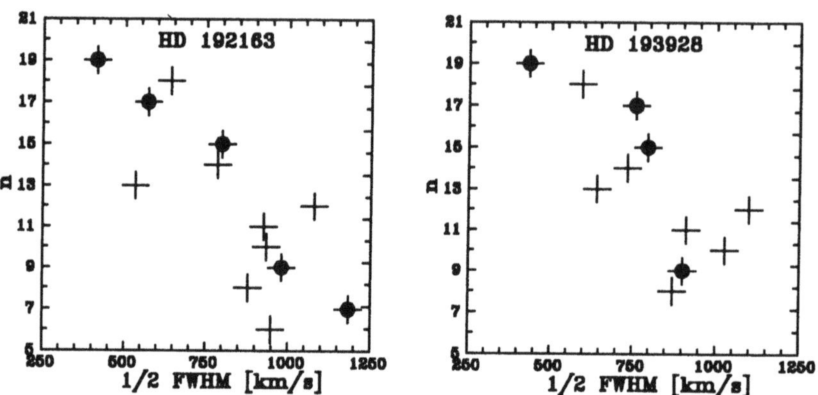

Fig. 1. The stratification in WR envelopes as seen in He II Pickering series lines of HD 192163 and HD 193928. The dark dots denote the least blended lines of this series.

The temperatures of WR stars were estimated in two ways: by fitting to the observed continuum the b-b distribution and by assuming the Zanstra mechanism from line intensities. The Zanstra temperatures were determined by assuming recombination mechanism for HeII lines (as in Nussbaumer et al. 1981) including photoionisation from the second level. The effective recombination coefficients were taken from Hummer & Storey (1987) for $T_e = 3 \times 10^4$ kK and $N_e = 1 \times 10^{11}$ cm^{-3}. We used both (n-3) and (n-4) HeII lines to determine T_Z, we excluded however the even n Pickering lines in WN stars (because of blending with hydrogen). In the case of WC stars we used the least blended lines only.

When fitting a theoretical distribution to the observed spectrum we assumed the Planck function. The fits were done only in the IUE region every 100 Å from 1200 to 3200 Å and the resulting temperatures are given in Table 1. We find this method rather rough and subjective. The reddening correction, ferrum quasi-continuum and strong P Cygni absorptions make the T_{BB} determination very approximate.

The luminosities of the studied WR stars can be estimated from the effective temperature definition after finding the radius through flux integration. The observed, dereddened optical and IUE spectra were integrated together with the unobserved region 0-1200 Å approximated by Planck distribution with the estimated temperatures. The resulting radii of the continuum forming region together with luminosities are shown in Table 1.

The temperatures and luminosities estimated for the studied WR stars allow us to find their **location on the HR diagram** (Fig.2). All of them lay close to the ZAMS, on the right of it, between the position of 25 and 60 $M_{\odot}$ stars. This location is in qualitative agreement with the evolutionary tracks for single, massive stars with strong mass-loss (long dashes) and for more massive components of close binary systems (short dashes). High luminosities and temperatures as well as low hydrogen abundances

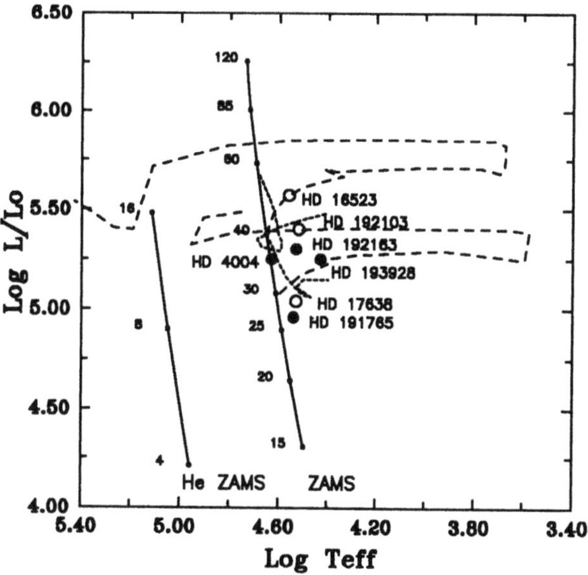

Fig. 2. The estimated position of the program stars on the HR diagram.

Table 1: *Summary of results.*

Star HD	Type WR	V_∞ [km/s]	E_{B-V}	T_{BB} [kK]	T_Z [kK]	$\frac{N(H)}{N(He)}$	$\frac{R}{R_\odot}$	$\mathrm{Log}\,\frac{L}{L_\odot}$	$\frac{M}{M_\odot}$
4004	WN5	3393	0.68	51.7	33.2	0.38	8.8	5.3	35
191765	WN6	2819	0.44	35.4	33.5	0.48	8.5	5.0	22
192163	WN6	2790	0.56	34.0	33.5	0.46	12.9	5.3	35
193928	WN6	2363	1.30	25.0	28.9	0.19	19.5	5.3	37
16523	WC5	2955	0.85	40.2	32.6	(0.0)	15.6	5.6	51
17638	WC6	2613	1.10	35.1	33.0	(0.0)	9.6	5.0	22
192103	WC8	2503	0.50	33.5	32.2	(0.0)	15.5	5.4	36

and high velocities in envelopes confirm the evolutionary status of WR stars as evolved massive O stars which undergone a substantial mass-loss.

References

Conti P.S., Leep E.M., Perry D.N. (1983) *Astrophys. J.* **268**, *228*

Hillier D.J. (1987) *Astrophys. J. Suppl.* **63**, *947*

Hillier D.J., Jones T.J., Hyland A.R. (1983) *Astrophys. J.* **271**, *221*

Hummer D.G., Storey P.J. (1987) *MNRAS* **224**, *801*

Kuhi L.V. (1973) *in: 'Wolf-Rayet and High-Temperature Stars', IAU Symp. 49, (Bappu M.K.V., Sahade J. eds), p. 205*

Niedzielski A., Nugis T. (1991) *Astron. Astrophys. Suppl.* in press.

Nussbaumer H., Schmutz W., Willis A.J. (1981) *Astron. Astrophys. Suppl.* **47**, *257*

Seaton M.J. (1979) *MNRAS* **187**, *73*

Willis A.J., Wilson R. (1978) *MNRAS* **182**, *559*

Near-Infrared Spectroscopy of Galactic Wolf-Rayet Stars

Ian D. Howarth[1] and Werner Schmutz[2]

[1]Department of Physics and Astronomy, University College London, Gower Street, London WC1E 6BT;
[2]Institut für Astronomie, ETH Zentrum, CH-8092 Zürich

Abstract: We present high-quality near-IR spectra of 24 Galactic WR stars, of a broad range of subtypes, selected as having known distances. The data cover the region 0.97μm–1.12μm and include the λ10830-Å He I triplet. Measurements of He I and He II lines, together with the absolute magnitudes, yield temperatures, luminosities, mass-loss rates, and terminal velocities for our sample. We extend that sample by including results for a further 12 stars of known distance, taken from the literature. The spectroscopic mass-loss rates are in excellent agreement with those estimated from radio data, and exceed the 'single-scattering limit' by large factors. Mass-loss rates depend only weakly on mass, but there is a tight correlation between surface mass flux and temperature. Terminal velocities correlate loosely with subtype for both WR sequences. Principally because of revisions to the adopted absolute magnitudes, our luminosities average slightly fainter than found previously; that exacerbates discrepancies with the predictions of evolutionary models, but reconciles results for Galactic and LMC stars. However, comparison with core-helium-burning mass–luminosity tracks suggests that the spectroscopic luminosities may be systematically too faint by ~0.5 dex. We suggest that the WNC stars in our sample may represent an intermediate phase between WNL and WCE subtypes. A full version of this paper is being submitted to *Astronomy & Astrophysics.*

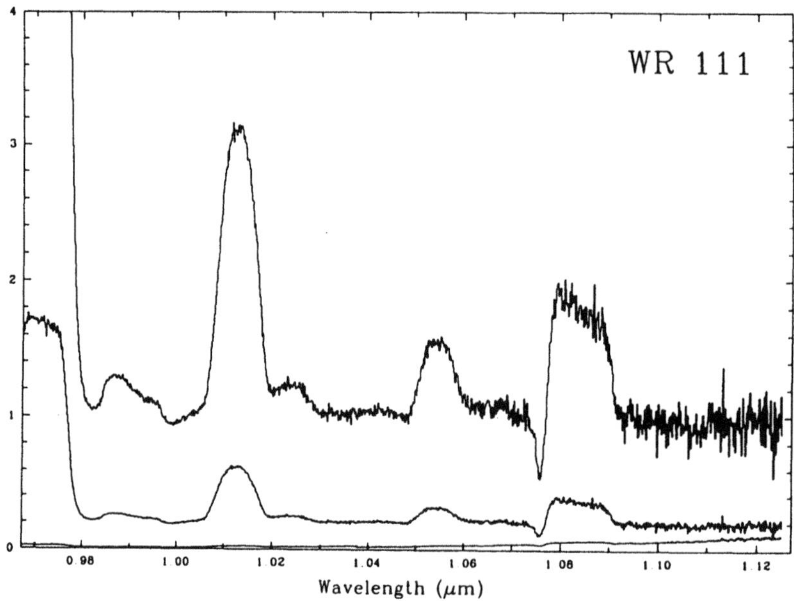

The near-IR spectrum of the WC5 star WR 111 (HD 165763).

Luminous Blue Variables

D. John Hillier

Institut für Astronomie und Astrophysik der Universität München, Scheinerstraße 1, W–8000 München 80, Federal Republic of Germany.

Abstract: Properties of Luminous Blue Variables (LBVs) are reviewed. We focus on the relationship of LBVs to other emission line stars (e.g. Ofpe/WN stars, B[e] stars), noting the overlap in properties that occur between the different classes. The importance of the peculiar Hubble-Sandage 'Var A' in M33 is emphasized. We highlight evidence that suggests that not all LBVs undergo a dramatic increase in mass loss during the LBV outburst. Recent results achieved in modeling of LBV spectra are discussed: Model ionization structures, and model profiles are presented to emphasize the complex radiative transfer effects occurring in LBV stellar winds. Finally, new results concerning the infamous LBV Eta Carinae, and its associated Homunculus, are presented.

1 Introduction

The term Luminous Blue Variable (or LBV) was proposed by Conti (1984) as a generic term for the most massive variable (blue) stars. This group includes the Hubble-Sandage variables, the S Doradus stars, and the peculiar variables η Car and P Cygni. Typically LBVs are of spectral type A, B or O, and have bolometric magnitudes less than -9 (see, for e.g. Humphreys 1989).

Whilst the term LBV has become popular in the literature, it should be noted that the definition is very broad, and not based on spectroscopic classification criteria. The lack of a spectroscopic criterion is unsurprising, since the spectroscopic class changes with time. Because of the loose definition of the LBV class, it is likely that the group contains objects whose evolutionary state and variability are unrelated. Irregular variability is the key phenomenon that distinguishes LBVs.

Over recent years the term LBV has become synonymous with evolved objects near the Humphreys-Davidson Instability Limit (Humphreys and Davidson 1979). The instability limit is a loosely defined luminosity cutoff above which luminous red supergiants do not exist. There are no red counterparts for the most luminous blue stars. Humphreys, on the basis of IRIS data, argues that this cannot be a selection effect. Therefore the variability in LBVs has been linked to the instability which allows the most massive stars to loose sufficient mass to prevent them from becoming red supergiants. The LBV phase is therefore crucial to massive star evolution. Unfortunately the instability mechanism for the major outbursts is not understood, and the mass loss during the LBV phase is treated in an ad-hoc manner.

The following paper is organized as follows. In Sect. 2 we discuss the known LBVs, and the nature of their variability. Section 3 discusses the relationship of LBVs to other (peculiar) emission line stars while polarization observations of LBVs are discussed in Sect. 4. Recent developments in modeling LBV spectra are considered in Sect. 5, and mass loss from LBVs is discussed in Sect. 6. Finally, in Sect. 7, we discuss new spectroscopic observations which provide new insights into the infamous LBV η Car.

Given the enormity of the literature on LBVs we make no attempt to be complete. The aim has been to portray general properties of LBVs, highlighting what we consider to be some of the more significant uncertainties. Up-to-date references for individual sources can readily be obtained from the Simbad database[1], while an extensive insight into LBVs can be gleaned from IAU Colloquium 113 (Davidson, Moffat, and Lamers 1989), which was devoted to LBVs.

2 Known LBVs and their Variability

In our galaxy, there are 4 well recognized LBVs (P Cygni, AG Car, η Car, and HR Car) with two more, HD 160529 (Sterken et al. 1991) and WRA 751 (Hu et al. 1990) added to this list recently. In the LMC the well known LBVs are S Dor (R88, HD 35343), R 71 (HD 269006), R 127 (HD 269858), with an additional object, R110 (HD 269662), recently proposed by Stahl et al. (1990). A further 20 or so LBVs are known in other galaxies. The scarcity of LBVs belies their importance.

Irregular variability, the key phenomenon that characterizes LBVs, has been discussed extensively by Lamers (1987). Lamers groups the variability into 3 classes depending on amplitude. In the first class we have stars such as η Car and P Cygni which underwent large outbursts with a variation in visual magnitude of more than 3 magnitudes. The timescale for such outbursts is unknown, but must be measured in centuries (or longer). In the second class we have objects such as AG Car and S Dor which have undergone moderate variations of 1 to 2 magnitudes, with a recurrence time measured in decades. The spectral type was seen to change from early B (at min) to late A (at max). The third class of variability is microvariability, where erratic variations of 0.1 to 0.2 magnitudes are observed to occur over a time scale of months.

Because of the long variability timescales our knowledge of LBV variability is severely hindered by observational selection effects. For example, in the present century P Cygni has undergone no outbursts, and only shows small amplitude (< 0.2 mags) variability (de Groot 1989). Without observations prior to 1900 (the major activity in P Cygni occurred prior to 1700) P Cygni would not be classified as an LBV! We simply may have missed many LBVs because they are currently in a quiescent state.

[1]The Simbad database is operated at CDS, Strasbourg, France

3 Related Objects

With only a handful of LBV candidates in the galaxy, and a further handful in the LMC, our knowledge of LBVs is necessarily limited. New sources of insight into LBV phenomena need to be sought. One such source is to compare the relationship of LBVs with other evolved luminous stars.

3.1 Ofpe/WN Stars

This class of objects was first identified by Walborn (1977), and has been discussed extensively by Bohannan and Walborn (1989). Their spectra are characterized by the occurrence of both high excitation emission (e.g. N III and He II emission), and low excitation emission (e.g. N II and He I). The Balmer series is also in emission. In some sense they are intermediate in type between extreme O3 stars, and the luminous WN7/WN8 stars.

McGregor et al. (1988b) have obtained IR spectra of a sample of emission line stars in the LMC. Eight objects were found to have He I 2.058 μm emission stronger than Hγ indicating that they are helium rich. Of these 8 objects, 5 belonged to the Ofpe/WN class confirming their advanced evolutionary state.

Until recently, Ofpe/WN stars were considered separately from LBVs however in 1980, R127 (HD29858), a defining member of the Ofpe/WN class, underwent a large outburst (Stahl et al. 1983, Walborn 1984). At maximum, its spectrum closely resembled that of S Dor (Stahl et al. 1983). Subsequently Stahl (1986) noted that AG Car in its recent minimum closely resembled Ofpe/WN stars, and hence could have been classified as such.

An obvious question immediately arises: Are all Ofpe/WN stars dormant LBVs, and if not, what distinguishes R 127 from the others? The question has important evolutionary implications since if all Ofpe/WN stars are LBVs the number of known LBVs in the LMC has almost tripled!

3.2 P Cygni Stars

As a working definition of P Cygni stars we take only those stars of spectral type B which show optical P Cygni profiles on metal lines, as well as on H and He I lines. The obvious example is, of course, P Cygni. P Cygni is classified as an LBV: Are all P Cygni stars LBVs? The timescales involved do not allow us to answer this question directly.

Given the gross similarity of some of these objects to P Cygni we must consider these stars in the same context as LBVs. A good example is HD 316285. While its variability has not been shown (principally from a lack of observations) it is a massive star with P Cygni profiles exhibited by transitions arising from H, He I, Fe II, and Na. Its present mass loss is extreme ($> 10^{-4} M_{\odot}\,\mathrm{yr}^{-1}$), and it is nitrogen rich (Hillier et al. 1988).

A somewhat worrisome development for P Cygni stars (and hence LBVs) concerns R81. This star was called the P Cygni of the LMC by Wolf et al. (1981b). Subsequent long term high precision photometry has shown this star to be an eclipsing binary with a period of 74.6 days (Stahl et al. 1987). It is likely to be a contact system, and shows phase dependent profile variations. How does the binarity influence the spectroscopic appearance of this system, and what influence, in general, does binarity have on LBVs and P Cygni stars (see also Sect. 3.5)?

3.3 B[e] Stars

The B[e] stars are characterized (Zickgraf et al. 1986) by the following:

1. Broad H emission, sometimes with P Cygni profiles.
2. Narrow emission lines of Fe II, [Fe II], and [O I].
3. Hot dust emission as evidenced by a strong infrared excess.

An extensive discussion of 8 of these objects is given by Zickgraf et al. (1985,1986)

A common belief is that these objects are not variable, and are therefore distinct from LBVs. However, as I indicated during the meeting, this lack of variability is probably an observational artifact — our observational timescales are just too short. The general belief is that they have evolved off the main sequence. The dust formation, when it occurred around these objects, was probably a dramatic event. If we were to observe such an event today, what would we call it? Supporting evidence of variability comes from Shore (1991) who found that S22 (discussed by Zickgraf et al. 1986) has varied strongly in the UV.

Rotation is postulated as the key ingredient to explain the spectral appearance of B[e] stars. The narrow emission lines, and the hot dust emission are supposed to originate inside a dense disk, while the broad permitted lines arise from a 'normal' polar CAK wind. McGregor et al. (1988b) observed 6 of the B[e] stars discussed by Zickgraf et al.(1986). All were found to have strong Bγ emission, while three also showed the first overtone CO bandheads in emission at 2.3 μm. The CO emission will also arise in the dense disk.

As these stars contain a disk, their appearance changes depending on the inclination. Inclination effects were used by Zickgraf et al. (1986) to explain some of the observed spectral differences between these objects. As the objects are very non-spherical, one must ask whether an object, viewed at two different angles, could be classified differently.

Finally, with regard to LBVs, I note that η Car possesses many similarities to B[e] stars. It is only distinguished, at least with regard to the definition, by the lack of narrow [O I] emission. Its terminal wind velocity of approximately 500km s^{-1}, is comparable to that shown by S12 (Zickgraf et al. 1986). It shows strong broad Balmer emission together with narrow emission of Fe II and [Fe II]. It also has hot dust, and in addition, a disk has previously been postulated to explain the spectral appearance of η Car in the IR and at optical wavelengths (e.g. Warren-Smith et al. 1979). Further we note that the Homunculus would be less than an arcsecond across if η Car were in the LMC.

Eta Carinae shows evidence of abundance anomalies (Davidson et al. 1986), indicating the presence of CNO processed material at the stellar surface. This may be a characteristic difference between η Car and the B[e] stars, however we note that B[e] stars are very poorly studied. Abundances in these objects are simply not known.

The luminosities of B[e] stars are comparable to the LMC LBVs (Zickgraf et al. 1986).

3.4 CO Emission Line Stars

Among the luminous emission line stars there is a group of objects which show the first overtone band of CO at 2.3 μm in emission (Whitelock et al. 1983b, McGregor et al. 1988a,1988b). Many in this subgroup also show both Bγ and He I 2.058 μm in emission. CO is easily dissociated around hot stars, hence the existence of CO emission is very surprising unless the CO is shielded from the hot photospheric radiation field. As noted previously, 3 of the 6 LMC stars to show CO emission are classified as B[e], and hence have been postulated to contain disks. The CO emission stars are a peculiar group of objects and in at least some, binarity may be playing a role in producing the observed emission.

The association with LBVs arises from the detection of CO in emission in moderate resolution spectra of HR Car (McGregor et al. 1988a, McGregor 1989). The emission, while weak, is definitely present. No CO emission has been seen in spectra of η Car, although the envelope conditions (i.e., strong mass loss, hot dust, presence of disk) are probably comparable to some CO emission line stars. The presence of CO in emission around HR Car exemplifies the wide spectral properties exhibited by LBVs.

It is worth noting that HR Car also exhibits some other properties that can be associated with B[e] stars. The spectrum shows numerous narrow emission components, while at the same time the Hβ profile shows evidence for large outflow velocities (Hutsemèkers and Van Droom 1991). It does not have hot dust, although cool dust is present (McGregor et al. 1988a).

3.5 Hubble-Sandage Variables

Hubble-Sandage Variables in external galaxies (excluding the clouds) provide the means to substantially increase the sample of LBVs that can be studied, and hence allow new insights into LBV evolution. The discovery of these variables dates back to the early 1920's. Extensive light curves for one star in M31, and 4 in M33, were presented by Hubble and Sandage (1953). These stars showed variations of 1 to 2 magnitudes, while their spectra are characterized by a strong UV continuum, and emission lines of H, He I, Fe II, and [Fe II], similar to S Dor (Humphreys 1975,1978).

The most striking enigma to arise from studies of Hubble-Sandage Variables concerns 'Var A' in M33 whose photometric history has been discussed by both Hubble and Sandage (1953), and Rosino and Bianchi (1973). The star showed a slow increase from 17.5 to 16th magnitude between 1920 and 1950 before declining to its present magnitude of about 18.8. Its absorption spectrum at maximum was similar to an early F star, and showed evidence for emission in the Balmer series, possibly with P Cygni profiles.

In 1985/86 spectra obtained by Humphreys et al. (1987) showed 'Var. A' to have the spectrum of an M3 red supergiant! Its present luminosity ($M_{BOL} = -9.5$) is comparable to that at maximum, and is emitted mainly in the IR. In order to explain this behaviour, Humphreys et al. proposed that Variable A was at a critical stage of its evolution where the balance between radiation pressure, turbulence, and gravity is tenuous. The bolometric luminosity of 'Var. A' (which is similar to that of R 71) places it close, possibly below, the Humphreys-Davidson Instability Limit. 'Var. A' is an important object, and MUST not be ignored when discussing LBVs.

A further important result concerning Hubble-Sandage Variables comes from the extensive studies by Gallagher et al. (1981), and Kenyon and Gallagher (1985). They propose that LBVs actually arise from two distinct sources. The more luminous sources (such as η Car, AG Car) probably arise from a single massive star approaching (or near) the Humphreys-Davidson Instability Limit. Such objects are found in regions of massive star formation. On the other hand, the Hubble-Sandage Variable 'AF And' occurs in an interarm region in M31, and shows no obvious nitrogen enhancement. They suggest that such systems are binaries, with the emission line spectrum and variability intrinsically connected with their binarity.

4 Polarization

Currently, polarization observations are becoming available for both P Cygni stars (Taylor et al. 1991) and W-R stars (Schulte-Ladbeck et al. 1990). Polarization offers a sensitive diagnostic of the envelope geometry. With an estimate of the envelope asphericity we can investigate the validity of conclusions obtained from spherical models. The geometry of the envelope is also critical if we are to understand the wind dynamics.

Extensive polarization monitoring of P Cygni has been carried out by Hayes (1985), and more recently by Taylor et al. (1991). P Cygni is found to be variable in polarization with a maximum amplitude of 0.8%. Due to the large uncertainty with the interstellar correction, the mean polarization of P Cygni is difficult to determine. Taylor et al. suggest that there is no preferred axis, and hence that the envelope of P Cygni is basically spherical. The large variations in polarization occur by stochastic ejections of matter at different angles, and indicate that along some sight lines the electron column density varies by an order of magnitude.

Another valuable diagnostic tool is the measurement of polarization across emission lines. This provides additional information on the structure of the emitting envelope, and also provides constraints on the interstellar polarization (Schulte-Ladbeck et al. 1991).

5 Modeling Considerations

The modeling of LBVs is still in its early stages, but the new sophisticated codes of Hillier (e.g. 1990), Hamann, Schmutz and collaborators (e.g. Hamann and Wessolowski 1990), the Munich code (e.g. Gabler et al. 1989), and other codes currently under development, should allow very significant progress to occur within the next few years.

The most detailed model to date is for the related Ofpe/WN star R84 by Schmutz et al. 1991. They were able to obtain a reasonable fit to both the H and He (He I and He II) spectra together with the UV and optical continuua. The derived parameters were $L = 5 \times 10^5 L_\odot$, $T_{eff} = 28500\,K$, $R = 30\,R_\odot$, $V_\infty = 400\,\mathrm{km\,s^{-1}}$, $\dot{M} = 2.5 \times 10^{-5}\,M_\odot\,\mathrm{yr}^{-1}$. The derived H/He abundance of 2.4 (by number) confirms the advanced evolutionary stage of R84.

The only LBV to be modeled in detail is P Cygni. The early work on P Cygni was based on the Sobolev approximation, and typically used the core-halo approximation. This core halo approximation is now known to be invalid. To obtain a reliable model for P Cygni it is essential to treat the continuum and line formation simultaneously.

Drew (1985) used the Sobolev method in conjunction with the 'on the spot approximation' for the continuum transfer to investigate P Cygni. Her work illustrates the importance of metal cooling, and indicates that He is neutral in the radio emitting region. More sophisticated continuum transfer calculations are unlikely to change these conclusions. We would emphasize, however, that the details of the temperature and ionization structure are very sensitive to the treatment of the radiation flow in the Lyman continuum.

At present, no detailed model exists for an LBV during different phases of its variability cycle. It has been commonly assumed that the variation in visual luminosity in LBVs was caused by the ejection of an optically thick shell. This shell redistributes UV flux to the optical, causing the star to appear visually brighter. This is consistent with observations of some LBVs which suggest no change in the Bolometric Luminosity during outburst (Appenzeller and Wolf 1981, Wolf 1989). The consequences of the shell model were developed by Davidson (1987).

Leitherer et al. (1989a) have showed that this simple picture is incorrect. During outburst, the radius of the star must change. A simple scaled up increase in mass loss cannot explain the observed appearance during outburst. Leitherer et al. (1989a) were forced to conclude that the LBV phenomenon is not just one of mass ejection, but involves a fundamental change in the outer envelope of the central star. In support of their arguments we note that evidence has been accumulating that some LBVs have similar mass loss rates at both outburst, and minimum (Sect. 6).

With regards to the modeling of LBVs I wish to discuss several important points regarding models of P Cygni (and related) stars which can profoundly influence interpretation of their spectra. These findings have arisen in experience obtained over the last 4 years in modeling both HD 316285, and to a lesser extent, P Cygni. The work is largely unpublished (but see Hillier, et al. 1988), primarily because of the poor results — I was unable to generate a model to simultaneously fit Hα and the strong Paschen series observed in HD 316285. The difficulty of modeling this object is exacerbated by a poor distance determination, and the large reddening ($E_{B-V} \approx 2$).

5.1 Ionization instability

The first point regards an inherent instability of the atmosphere. This instability can be investigated by two different approaches. Firstly, we can self-consistently determine the mass loss and atmospheric structure for a given (g, R, Teff). Pauldrach and Puls (1990) showed that when this is done, there is a region of instability where changes in the adopted parameters of only a few % can cause a large variation in the mass loss. This instability is due to the influence of the Lyman continuum on the ionization balance of Fe.

In the approach I have adopted the mass loss is taken as a free parameter. The mass loss and luminosity are then adjusted to fit the observation, with new models generated using an old model for the initial population estimates. In this approach it is found that there is an instability strip where small changes in one of the adopted parameters can have a profound influence on the resulting ionization structure. To see this we consider the following models with $R = 60R_{\odot}$, $V_{\infty} = 500\,\mathrm{km\,s^{-1}}$ ($\beta = 1$) and $\dot{M} = 5 \times 10^{-5} M_{\odot}\,\mathrm{yr}^{-1}$.

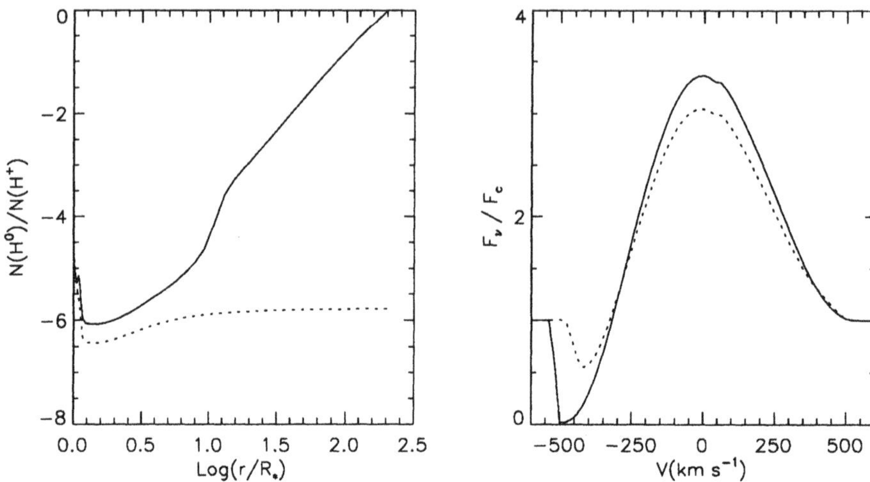

Figure 1: The hydrogen ionization structures, and the resulting Hβ profiles for the L= 5×10^6 (solid lines) and L= 6.5×10^6 models discussed in the text.

In Fig. 1a we show the resulting H ionization structure for two models, one with a luminosity of $5 \times 10^5 L_\odot$, and the other with a luminosity of $6.5 \times 10^5 L_\odot$. The ionization structure is very different — in the first model the hydrogen ionization is low. The profiles of the H lines are also very different (Fig. 1b), particularly for the Balmer series. The P Cygni absorption is considerably stronger in the lower ionization model. This effect is easily explained by the large optical depth of the resonance lines in the low luminosity model which ensures a larger population in the n=2 state. A change in luminosity of only a few percent is sufficient to cause the switch between the two ionization structures. In reality, of course, such large changes in the ionization structure will have a profound influence on the wind dynamics, as noted above.

The observed P Cygni profiles in the extreme P Cygni star HD 316285 cannot be explained by a classical $\beta = 1$ velocity law ($V(r) = V_o + (V_\infty - V_o)(1 - R_*/r)^\beta$) with the high ionization model. It requires a slow (e.g. linear or $\beta = 3$) velocity law, or a low ionization envelope.

5.2 Absorption Structure in the Balmer Series

The second point concerns the observed profiles of the Balmer series. In the extensive literature on P Cygni there are references to the appearance of double (and even triple) absorption components on higher members of the Balmer series (e.g. Adams and Merrill 1957, Kolka 1983). Such effects are usually explained by invoking shell ejection, or another similar phenomena. It is worth noting, however, that multiple absorption profiles can be obtained in certain models purely from radiative transfer effects which give rise to complex variations in the line source function.

An example profile obtained from one such model (not meant to represent P Cygni) is shown in Fig. 2a, together whith the corresponding source function Fig. 2b. The effects vary from line to line, and are not present in all models.

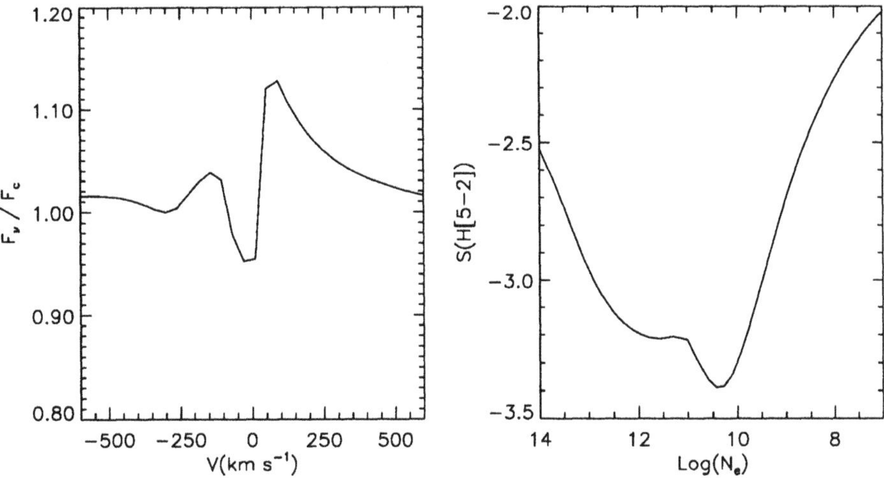

Figure 2: An example of a model $H\gamma$ profile showing two absorption components. In the adjacent figure, the behaviour of the source function as a function of electron density is illustrated.

We would argue that such effects are important in P Cygni stars, and can explain some of the observed profiles without invoking shell ejection. Such effects will also be important for atomic species, such as Fe II, which are strongly coupled to the H ionization structure. The variability in the observed profiles may simply arise from variations in the ionization structure and source functions caused by relatively small variations in mass flux. It is easy to envisage quite complex profile changes if we combine the source function sensitivity with the 'clumping instability' discussed during this Conference (Owocki 1991).

6 Mass Loss From LBVs

As an extensive discussion of mass loss in LBVs is given by Lamers (1989a), this topic needs only a brief discussion here.

It is generally assumed that the mass loss during outburst is much larger than that at minimum. This scenario arises primarily from R71, whose spectral characteristics undergo a dramatic change between minimum and maximum. (Wolf, et al. 1981a). $H\beta$ is observed to be considerably stronger at maximum (in both equivalent width [EW] and flux). Leitherer et al. (1989b) found that the mass loss for R71 increased by a factor of 30 from $6 \times 10^{-7}\, M_\odot\,\mathrm{yr}^{-1}$ at minimum, to $2 \times 10^{-5}\, M_\odot\,\mathrm{yr}^{-1}$ at maximum.

Recent evidence suggests, however, that the mass loss variation exhibited by R71 is not typical of all LBVs.

The first piece of evidence concerns AG Car. At minimum, McGregor et al. (1988a) found a mass loss rate of $4 \times 10^{-5}\, M_\odot\,\mathrm{yr}^{-1}$ (converted to d=6 kpc, Lamers et al. 1989b). On the other hand, Wolf and Stahl (1982) found a mass loss rate of approximately $10^{-4}\, M_\odot\,\mathrm{yr}^{-1}$ near maximum. These values are the same within the errors, and suggest that the mass loss of AG Car varies by at most a factor of 3 between its minimum and maximum states. Supporting evidence comes from the observations of Whitelock et al. (1983a), whose spectra indicate that the flux of the Paschen-β line remained approximately constant, while the continuum flux at $1.25\,\mu$m increased by a factor of 2.5.

More recently, Stahl et al. (1990) have reported observations of the new LBV R 110 (HD 269662) both at and before outburst. They find similar mass loss rates of $1 - 4 \times 10^{-6}\, M_\odot\, \mathrm{yr}^{-1}$, and hence the mass loss in this LBV is also seen not to vary a great deal during its variability cycle. This object also provides a counter example to Lamers (1989a) conjecture that all LBVs show a similar mass loss at maximum.

It is also worth noting that Leitherer et al. (1985) indicate that the Balmer series flux remained constant in S Dor between July 1983 (just after maximum) and August 1984 (near minimum). This observation suggests that the mass loss rate of S Dor also varies by a factor of less than 3.

Inherently related to mass loss variability is the change in velocity law during the outburst, which to date is poorly known for LBVs. In a radiation driven wind one might expect the terminal velocity to change with outburst phase. As a consequence, wind–wind interactions may occur. If there exists a slow wind phase of sufficient duration it is conceivable that a faster wind could compress this material, and perhaps cause shocks. Such material could give rise to emission features normally not associated with a classical stellar wind. This type of scenario has been considered, for example, in relation to the formation of planetary nebulae shells where a fast wind sweeps up gas previously ejected by a slow red-giant wind (e.g. Kahn 1983, Kowk 1983). The timescales and conditions are, of course, completely different and with LBVs the event can, in principle, occur many times.

7 New Observations of η Car

Eta Car is one of the most famous and best studied LBVs. It underwent a period of major activity in the 1840's, showing variations of more than two magnitudes (Walborn and Liller 1977), and reaching a peak visual magnitude of -1^{m}. During the late 1850's and the 1860's it suffered a fairly rapid decline to 7th magnitude, which has been attributed (primarily) to the formation of dust (e.g. Andriesse, et al. 1978) It suffered a second (minor) outburst towards the end of last century, but since then has been fairly stable with only slow, longterm fluctuations in brightness.

Surrounding η Car there is a main condensation (approximately 17″ by 12″) generally referred to as the Homunculus. The Homunculus is believed to have been ejected during the period of violent activity in the 1840's. Outside the Homunculus there are fainter condensations, some of which have very large (i.e., $1000\,\mathrm{km\,s^{-1}}$) proper motions (Walborn et al. 1978). These condensations generally show a spectrum quite distinct from that of the central object or the Homunculus (Walborn 1976, Davidson et al. 1982).

Because of its high luminosity, and because the ejecta are nitrogen rich (Davidson et al. 1986), it is believed to be a massive star undergoing a period of extreme mass loss. Speckle masking observations by Hofmann and Wiegelt (1988) showed a bright component as well as three faint starlike objects approximately 12 times fainter than the main source. Thus η Car may be a multiple system, or alternatively the 'companions' may be due to inhomogeneities in the dust condensation around η Car.

Recently, Space Telescope images of η Car have been published by Hester et al. (1991). They find that the morphology of the Homunculus indicates that it is a thin shell with very well defined edges, and is clumpy on scales of 0.2″ (10^{16} cm). They reject the bipolar model generally proposed for η Car and argue that the data may indicate an oblate shell with polar blowouts.

Recent ground-based observations have also provided new insights into η Car. In 1986, David Allen of the Anglo-Australian Observatory and I performed a detailed spectroscopic investigation of η Car and the Homunculus. Spectral maps were obtained at resolutions of 3Å and 1Å, and at a spatial resolution of 1″. The wavelength range is from 4200Å to 1μm while the accuracy of the fluxes in the final data set is estimated to be 20%. These data strongly constrain any model for η Car and the Homunculus.

A paper detailing the observations, and providing first results has been submitted (Hillier and Allen 1991). A brief summary is presented here.

Our observations confirm earlier work suggesting that the Homunculus is primarily a reflection nebula. The continuum, and the broad H and He I emission lines arise from dust scattering photons into the line of sight. These photons originate in (or near) the central object. In addition, we find low excitation (e.g.[Ni II], [Cr II]) emission (hereafter termed intrinsic emission) arising from the Homunculus gas. The velocities of the intrinsic emission are distinct from those of the dust scattered features.

To interpret the velocities we adopt the simplest model possible by assuming the Homunculus arises through a radial flow emanating from a point source. We define V_r as the radial outflow velocity of some scattering clump, which has a normalized position vector $\vec{r}$ relative to the central source. We take $\vec{O}$ as the unit vector pointing from the source to the observer. Defining V_g as the velocity of the gas emission, as measured in the observer's frame, and V_d the velocity of the dust emission (and adopting the usual convention with positive velocities denoting redshifted emission) we have

$$V_g = -V_r \cos\theta \tag{1}$$

and

$$V_d = V_r(1 - \cos\theta), \tag{2}$$

where

$$\cos\theta = \vec{r} \cdot \vec{O}. \tag{3}$$

In both lobes of the Homunculus the H and He emission is redshifted, consistent with dust scattering. On the other hand, the gas emission is blue shifted in the south-east (SE) lobe (implying material moving towards us), and red shifted in the north-west (NW) lobe.

Both the velocity of the emission features arising from dust scattering, and the velocity of features arising from intrinsic emission in the Homunculus gas, are observed to increase along the south-east major axis, however *their separation remains constant.* This indicates that the velocity variation is due to projection, and *not* due to a change in the velocity of the material giving rise to both the scattered and intrinsic emission lines. The data are consistent with the SE lobe being a partial shell (radius 9″) expanding at a velocity of 650 km s^{-1}.

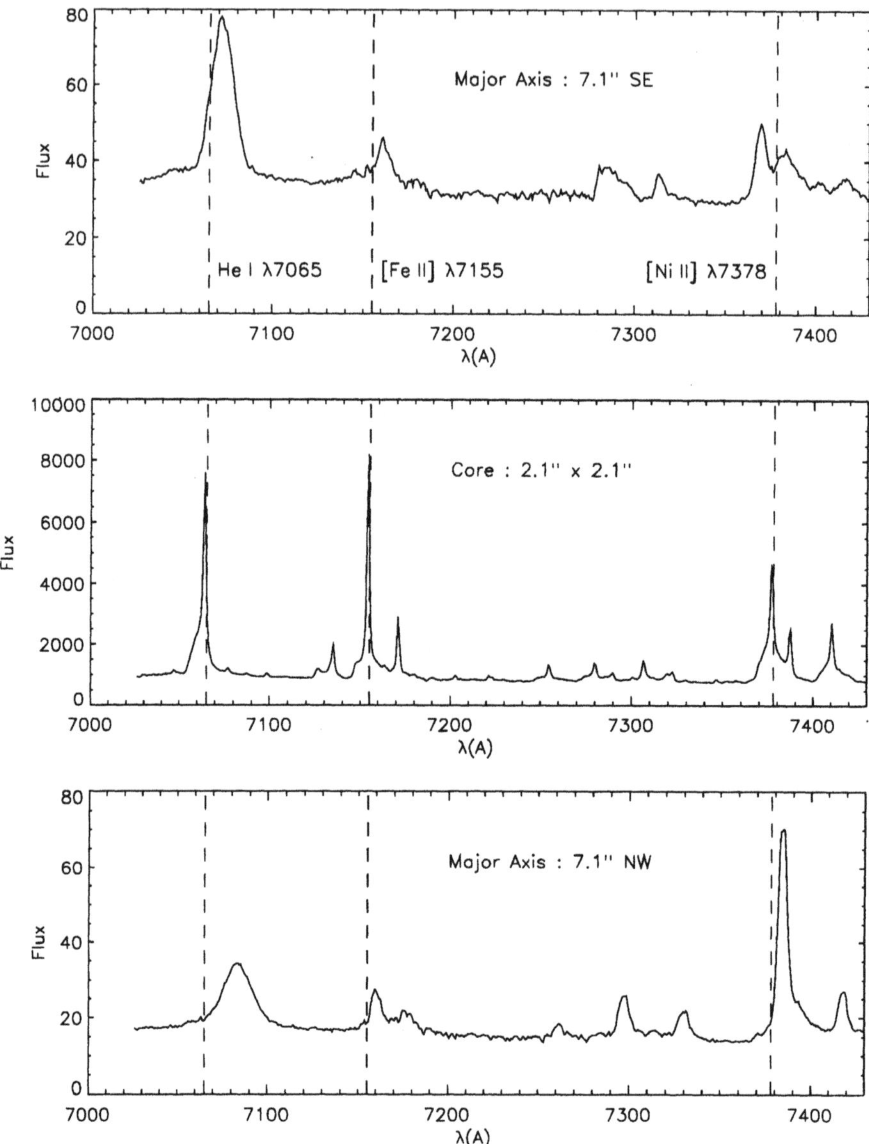

Figure 3: Spectra from the central 2.3" × 2.3" area, and from two locations (7"SE, and 7"NW) along the major axis. The He I emission at 7065Å is redshifted in both lobes, and broader than on the central source. Intrinsic [Ni II] emission is seen in both lobes — In the SE lobe it is blue shifted while in the NW lobe it is redshifted. The redshift exhibited by the forbidden line in the NW lobe is less than that shown by the dust scattered He I lines. Note the dramatic weakening (as compared with the central spectrum) of [Fe II] λ7155 relative to He I, particularly in the SE lobe. In the NW lobe, the [Fe II] λ7155 line also has an intrinsic emission component.

Moderate resolution spectra (7000 – 7450Å) from the core, and two locations along the major axis, are presented in Fig. 3, to highlight the profile and velocity shifts.

While the Homunculus spectrum arises predominantly from scattering, it is distinct from that of the core. In particular, the [Fe II] and Fe II emission lines are much stronger (relative to the continuum, and to the H and He I emission lines) in the core spectrum, than in the lobe spectra (Fig. 3). Why does the He I/[Fe II] line ratio change when both the continuum and He lines are produced by scattering of light from the central source?

We argue that we must be viewing η Car from a special direction. Light from the central object must be heavily absorbed by dust while the [Fe II] lines are formed outside (or coincident with) the dust and hence suffer little extinction. The [Fe II] lines appear strong in the spectrum of the core because of the heavy extinction suffered by light arising from the central object. On the other hand, the dust in the lobes must see light from the central source that has suffered little extinction. While the [Fe II] lines are also scattered they will appear weak.

The lobe spectra show P Cygni profiles on the H and Fe II emission lines which we suggest reflect the spectrum of the central source, and hence provide information on the stellar wind. In addition our spectra show that He II λ4686 is absent, limiting the effective temperature of η Car to less than 30000K.

8 Conclusion

Luminous Blue Variables are a diverse group of objects exhibiting a wide range in spectral appearance and variability behaviour. LBVs occupy a similar location in the H-R diagram to Ofpe/WN stars and B[e] stars, and have properties which overlap objects of these classes. The relationship between the various objects is uncertain, but abundance analysis may provide the necessary key to probe the evolutionary connections. The H/He ratio, and the C/N and O/N ratios are key diagnostics. To date, only limited abundance information is available.

The long variability timescales are a strong impairment to understanding LBVs. We still have no firm theory for the mechanism which causes the LBV phase. While phenomenological discussions have been given (e.g. Appenzeller 1989) we cannot start from the basic stellar parameters (e.g. L, M, rotation rate, composition) and predict the mass loss through the LBV phase from first principles.

LBVs appear to show wide variations in their mass loss variability between maximum and minimum. Codes are now available that can be used to make reliable mass loss determinations as a function of the variability phase. Such information is a prerequisite to understanding the LBV phenomena. The available transfer codes also show that radiative transfer effects may have a profound influence on the spectral appearance of LBVs, and profoundly affect our interpretation.

Recent observations (both ground-based, and from space) have provided new insights into η Car and its Homunculus. These observations reveal how complex geometrical effects can influence the spectral appearance of such objects, and provide a strong reminder of the difficulty of interpreting the spectra of unresolved objects.

References

Adams, W.S., Merrill, P.W., 1957, ApJ 125, 102
Andriesse, C.D., Donn, B.D., Viotti, R., 1978, MNRAS 185, 771
Appenzeller, I., 1989, in: IAU Col. 113, *Physics of Luminous Blue Variables*, eds. K. Davidson, A.F.J. Moffat, H.J.G.L.M. Lamers, Kluwer, Dordrecht, p. 195
Appenzeller, I., Wolf, B., 1981, in: The Most massive Stars, ESO Workshop, eds. S. D'Odorico, D. Baade, and K. Kjär, p131
Bohannan, B., Walborn, N.R, 1989, PASP 101, 520
Conti, P.S., 1984, in: IAU Symp. 105, *Observational Tests of the Stellar Evolution Theory*, eds A. Maeder, A. Renzini, Kluwer, Dordrecht, p.233
Davidson, K., 1987, ApJ 317, 760
Davidson, K, Dufour, R.J., Walborn, N.R., Gull, T.R., 1986, ApJ 305, 867
Davidson, K., Moffat, A.F.J., Lamers, H.J.G.L.M. (eds.) 1989, IAU Col. 113, *Physics of Luminous Blue Variables*, Kluwer, Dordrecht
Davidson, K, Walborn, N.R., Gull, T.R., 1982, ApJ 254, L47
Drew, J.E, 1985, MNRAS 217, 867
Gabler, R., Gabler, A., Kudritzki, R.P., Puls, J., Pauldrach, A., 1989, A&A 226, 162
Gallagher, J.S., Kenyon, S.J., Hege, E.K., 1981, ApJ 249, 83
de Groot, M., 1989, in: IAU Col. 113, *Physics of Luminous Blue Variables*, eds. K. Davidson, A.F.J. Moffat, H.J.G.L.M. Lamers, Kluwer, Dordrecht, p. 257
Hamann. W-.R, Wessolowski, U., 1990, A&A 227, 171
Hayes, D.P., 1985, ApJ 289, 726
Hester, J.J., Light, R.M., Westphal, J.A., Currie, D.G., Groth, E.J., Holtzman, J.A., Lauer, T.R., O'neil, E.J. Jr., 1991, AJ 102, 654
Hillier, D.J., 1990, A&A 231, 116
Hillier, D.J., Allen, D.A., 1991, submitted to A&A.
Hillier D.J., McGregor P.J., Hyland A.R., 1988, Mass Outflows from Stars and Galactic Nuclei, eds. L. Bianchi and R. Gilmozzi, Kluver, Dordrecht, p. 215
Hofmann, K.-H., Weigelt, G., 1988, A&A 203, L21
Hu, J.Y., de Winter, D., Thé. P.S., Pérez, M.R., 1990, A&A 227, L17
Hubble, E., Sandage, A., 1953, ApJ 118, 353
Humphreys, R.M., 1975, ApJ 200, 426
Humphreys, R.M., 1978, ApJ 219, 445
Humphreys, R.M., 1989, in: IAU Col. 113, *Physics of Luminous Blue Variables*, eds. K. Davidson, A.F.J. Moffat, H.J.G.L.M. Lamers, Kluwer, Dordrecht, p. 3
Humphreys, R.M., Davidson, K., 1979, ApJ 232, 409
Humphreys, R.M., Jones, T.J., and Gerhz, R.D., 1987, AJ 94, 315
Hutsemékers, D., Van Droom E.,1991, A&A 248, 141
Kahn, F.D., 1983, in: IAU Symp. 103, *Planetary Nebulae*, ed D.R. Flower, Kluwer, Dordrecht, p. 305
Kenyon, S.J., Gallagher, J.S., 1985, ApJ 290, 542
Kolka I., 1983, Publ. Academy. of Sci. Estonian SSR 32, 51
Kwok, S., 1983, in: IAU Symp. 103, *Planetary Nebulae*, ed D.R. Flower, Kluwer, Dordrecht, p. 293
Lamers, H.J.G.L.M., 1987, in *Instabilities in Luminous Early Type Stars*, eds.

H.J.G.L.M. Lamers and C.W.H. de Loore (Reidel), p 99.
Lamers, H.J.G.L.M., 1989, in: IAU Col. 113, *Physics of Luminous Blue Variables*, eds. K. Davidson, A.F.J. Moffat, H.J.G.L.M. Lamers, Kluwer, Dordrecht, p. 135
Lamers, H.J.G.L.M., Hoekzema, N., Trams, N.R., Cassatella, A., Barylak, M., 1989, in: IAU Col. 113, *Physics of Luminous Blue Variables*, eds. K. Davidson, A.F.J. Moffat, H.J.G.L.M. Lamers, Kluwer, Dordrecht, p. 271
Leitherer, C., Appenzeller, I., Klare, G., Lamers, H.J.G.L.M., Stahl. O., Waters, L.B.F.M., Wolf, B., 1985, A&A 153, 168
Leitherer, C., Schmutz, W., Abbott, D.C:, Hamann W.-R., Wessolowski, U., 1989, ApJ 346, 919
Leitherer, C., Schmutz, W., Abbott, D.C:, Torres-Dodgen, A.V., Hamann W.-R,, Wessolowski, U., 1989, in: IAU Col. 113, *Physics of Luminous Blue Variables*, eds. K. Davidson, A.F.J. Moffat, H.J.G.L.M. Lamers, Kluwer, Dordrecht, p. 285
McGregor, P.J., in: IAU Col. 113, *Physics of Luminous Blue Variables*, eds. K. Davidson, A.F.J. Moffat, H.J.G.L.M. Lamers, Kluwer, Dordrecht, p. 165
McGregor, P.J., Hyland, A.R., Hillier, D.J., 1988, ApJ 324, 1071
McGregor, P.J., Hillier, D.J., Hyland, A.R., 1988, ApJ 334, 639
Owocki, S, 1991, this volume
Pauldrach, A.W.A, Puls, J., 1990, A&A 237, 409
Rosino, L., Bianchini, A, 1973, A&A 22, 453
Schulte-Ladbeck, R.E., Nordsieck, K.H., Nook, M.A., Magalhães, A.M., Taylor, M., Bjorkman, K.S., Anderson, C.M., 1990, ApJ 365, L19
Schulte-Ladbeck, R.E. Nordsieck, K.H., Taylor, M., Bjorkman, K.S., Magalhães, A.M., Wolff, M.J., 1991, ApJ (in press)
Schmutz, W, Leitherer, C., Hubeny, I., Vogel, N., Hamann, W-R, Wessolowski, U., 1991, ApJ 372, 664
Shore, S.N., 1991, in Proceedings of *Variable and Non-Spherically Symmetric Outflows from Stars*, eds. L. Drissen, C. Leitherer, A. Nota. (in press)
Stahl, O., 1986, A&A 164, 321
Stahl, O., Wolf, B., Klare, G., Cassatella, A., Krautter, J., Persi, P., Ferrari-Toniolo, M., 1983, A&A 127, 49
Stahl, O., Wolf, B., Zickgraf, F.J., 1987, A&A 184, 193
Stahl, O., Wolf, B., Klare, G., Jüttner, A., Cassatella, A., 1990 A&A, 228, 379
Sterken, C., Gosset, E., Jütner, A., Stahl, O., Wolf, B., Axer, M., 1991, A&A 247, 383
Taylor, M., Nordsieck, K.H., Schulte-Ladbeck, R.E., Bjorkman, K.S., 1991, AJ 102, 1197
Warren-Smith, R.F., Scarrott, S.M., Murdin, P., Bingham, R.G., 1979, MNRAS 187, 761
Walborn, N.R., 1977, ApJ 215, 53
Walborn, N.R., 1976, ApJ 204, L17
Walborn, N.R., 1984, in: IAU Symp. 108, *Structure and Evolution of the Magellanic Clouds*, eds. S. van den Bergh, K.S. de Boer, Kluwer, Dordrecht, p. 239
Walborn, N.R., Blanco, B.M., and Thackeray, A.D., 1978, ApJ 219, 498
Walborn, N.R., Liller, M.H., 1977, ApJ 211, 181
Whitelock, P.A., Carter, B.S., Roberts, G., Whittet, D.C.B., Baines, D.W.T., 1983, MNRAS 205, 577

Whitelock, P.A., Feast, M.W, Roberts, G., Carter, B.S., Catchpole, R.M., 1983, MNRAS 205, 1207
Wolf, B., 1989, in: IAU Col. 113, *Physics of Luminous Blue Variables*, eds. K. Davidson, A.F.J. Moffat, H.J.G.L.M. Lamers, Kluwer, Dordrecht, p. 91
Wolf, B., Appenzeller, I, Stahl, O., 1981, A&A 103, 94
Wolf, B., Stahl, O., 1982, A&A 112, 111
Wolf, B., Stahl, O., de Groot, M.J.H., Sterken, C., 1981, A&A 99, 351
Zickgraf, F.-J., Wolf, B., Stahl, O., Leitherer, C., Klare, G., 1985, A&A 143, 421
Zickgraf, F.-J., Wolf, B., Stahl, O., Leitherer, C., Appenzeller, I., 1986, A&A 163, 119

OBSERVED EVOLUTIONARY CHANGES IN THE VISUAL MAGNITUDE OF THE LUMINOUS BLUE VARIABLE P CYGNI

Mart de Groot[1] and Henny J.G.L.M. Lamers[2,3]
[1] Armagh Observatory, College Hill, Armagh BT61 9DG, Northern Ireland
[2] SRON Laboratory for Space Research, Sorbonnelaan 2, 3584 CA Utrecht, The Netherlands
[3] Astronomcial Institute, Princetonplein 5, 3584 CC Utrecht, The Netherlands

Abstract
Historical observations of the star P Cygni (B1 Ia$^+$) show that the star has steadily increased its visual brightness between 1700 and 1988 by 0.15 ± 0.02 magn/century. If we assume that the luminosity of the star remained constant (as predicted by evolutionary tracks of massive stars), the increase in brightness must be due to a steady decrease of the Bolometric Correction and hence of T_{eff} at a rate of 6 ± 1 percent per century. This is the first time ever that the evolution of a star has been observed in secular photometric variations.

The redward evolution of P Cyg in the HR-diagram occurs on a Kelvin-Helmholtz time-scale for core-contraction and envelope-expansion, but the observed rate of change is a factor two faster than predicted by the evolutionary tracks. The discrepancy can be explained by either assuming that dynamical effects speed up the post main-sequence evolution of massive stars or that the core mass of the star is smaller than predicted.

FORBIDDEN LINES OF [Fe II] AND [N II] IN P CYGNI'S ENVELOPE

Garik Israelian[1] and Mart de Groot[2]
[1]Buyrakan Observatory, Armenia
[2]Armagh Observatory, Northern Ireland

Abstract
From high-resolution spectra of the Luminous Blue Variable P Cygni we determine electron densities in those parts of its envelope where lines of [Fe II] and [N II] are formed. The results allow us to deduce certain properties of the star's envelope and stongly suggest that the ejection of matter from P Cygni occurs in randomly directed "blobs".

Observations with an echelle spectrograph and CCD detector of the B1 Ia^+ star P Cygni have revealed the presence of forbidden emission lines of [Fe II] and [N II] (Stahl *et al.* 1991). Interestingly, these forbidden lines were also found *a posteriori* upon inspection of high-dispersion photographic spectogrammes used by De Groot (1969) in his analysis of P Cygni. These spectrogrammes go back to 1942, but the forbidden lines were not found at the time of the study undertaken in the sixties.
Three reasons can be given for this failure:

1. The forbidden lines have intensities of a few percent above the continuum. In P Cygni with its numerous absorption and emission lines the decision where to draw the continuum is a difficult one and rather broad lines of low intensity are easily missed;
2. The signal-to-noise ratio of Stahl *et al.'s* spectrogrammes is many times higher than that of the older photographic material;
3. The delivered wisdom on the spectrum of P Cygni at the time of the earlier study was that there were no forbidden lines. Thus, such lines were not looked for and, as a result, not found.

This is an excellent example of how preconceptions can seriously interfere with astrophysical studies to the detriment of progress in the field.

The [Fe II] lines are essentially flat-topped indicating that they originate in a spherically symmetric expanding atmosphere with a constant velocity over the region of formation of these lines. All [Fe II] lines show a width, in velocity units, of about 230 km s^{-1}. The Fe II lines observed in the UV (Lamers et al 1985) show the same velocity, validating the assumption that the two kinds of lines are formed in the same part of the envelope.
For hot stars, the occurrence of forbidden lines becomes possible when collisional excitation dominates, i.e. when the geometrical dilution factor

$W \leq 10^{-13}\ N_e\ T_e^{-1/2}$ (Viotti, 1976)

Lamers *et al.* (1985) have estimated the distance of the shell with constant velocity where the Fe II lines are formed. They find a distance of $300R_*$, giving a geometrical dilution factor

$W = 2.8 \times 10^{-6}$

If, furthermore, we assume $T_e = 10^4$ K (Drew, 1985) we find

$N_e > 2.8 \times 10^9$

Also according to Viotti (1976), the intensity ratio of [Fe II] λ4244 and Fe II λ4233 can be used to estimate the electron density:

$I_{4244}\ /\ I_{4233} = 10^{11}\ N_e^{-1}$

From Stahl *et al's* spectra we estimate $I_{4244}\ /\ I_{4233} \geq 10$. This gives

$N_e < 10^{10}$

The good agreement between the two values of N_e derived above shows that our assumption about the value of the dilution factor and, therefore, of the radius of the shell where the Fe II and [Fe II] lines are formed, is essentially correct. We will assume, therefore, that

$N_e \sim 5 \times 10^9$ at $r = 300\mathrm{R}_*$

The [N II] lines have rounded, roughly parabolic, profiles indicating that they are formed in a part of the envelope where there is a velocity gradient. Since their shapes are similar, we will assume that they are all formed in the same part of the envelope. With the wind velocity in P Cygni's envelope monotonically increasing outward and the width of the [N II] lines, in velocity units, 160 km s^{-1}, these lines must be formed closer to the star than the Fe II lines. For the region where the [N II] lines are formed we can use the intensity ratio between the red and yellow [N II] lines to estimate the electron density: $(I_{6548} + I_{6583}) / I_{5755} = 7.53 \exp(2.5 \times 10^4 T_e^{-1}) / (1 + 2.7 \times 10^{-3} N_e T_e^{-1/2})$ (Osterbrock, 1974).
In P Cygni this intensity ratio equals 1.25. Again, assuming, as above $T_e = 10^4$, we find

$N_e = 1.4 \times 10^6$.

The portion of Stahl *et al.'s* (1991) tracing containing the red [N II] lines near Hα is reproduced in Figure 1.

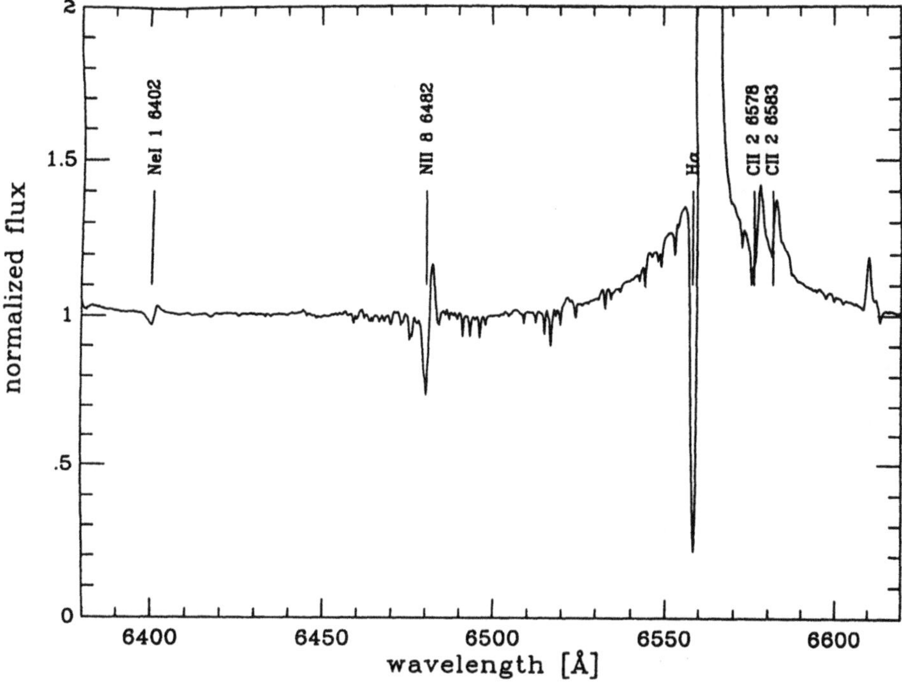

Figure 1.The spectrum of P Cygni showing the [N II] lines at λλ 6548 and 6583. Courtesy of O. Stahl.

Thus, we find that in the outer region at $\sim 300\mathrm{R}_*$, where the [Fe II] lines are formed, the electron density is much higher than in the inner region, at $\leq 200\mathrm{R}_*$, where the [N II] lines are formed. This can be understood when one realises that the [Fe II] lines are formed in a part of the envelope which is still being accelerated and too close to the star to allow the formation of a dense shell.

Also, the ejection of matter from the star seems to occur in randomly-directed "blobs" (Taylor *et al.* 1991). Such blobs will be accelerated as they travel away from the star and replenish the outer shell moving at the terminal velocity. That outer shell probably is a true axi-symmetric shell indicating that the blob ejections, while random in direction, do show an axi-symmetric pattern when integrated over a sufficiently long period.
The observed [N II] lines are formed in these accelerating blobs and the value for N_e deduced from their strengths is an integration over the entire envelope of the star. If at any time there are only a few such blobs, then their true density is higher than the one derived above by a factor $1/f$, where f is the filling factor of the blobs with respect to the total envelope.
With $N_e = 5 \times 10^9$ at $r = 300R_*$, one would expect $N_e = 10^{10}$ at $r = 200R_*$ if the density in the envelope decreases as r^{-2}. Our above figure of $N_e = 1.4 \times 10^6$ for the [N II] line-forming region would then suggest a filling factor $f = 1.4 \times 10^{-4}$. This seems a rather low value in view of the fact that the blobs do influence the degree of polarization of P Cygni's light. Also, the short-term variations in the light curve of P Cygni (Percy *et al.* 1988; de Groot, 1990) seem to suggest that these may be caused by the ejection of blobs containing sufficient matter to influence the total light of the star. A final conclusion on the number, size and density of the blobs must wait until more accurate measurements can be made.
The above-derived values can be compared with the results obtained by Stahl and Wolf (1986) in the case of a number of emission-line supergiants in the LMC. The [N II] lines in their sample of stars have velocities of only 10 to 20 km s^{-1}, whereas in P Cygni the [N II] lines give a velocity of 160 km s^{-1}. Stahl and Wolf (1986) also derive electron densities for the stars in their sample and find values larger than 2×10^5 to 4×10^6. Our value of 1.4×10^6 is in perfect agreement with their results.
So, while in P Cygni the [N II] lines are formed in a region with the same electron density as in other emission-line supergiants, they indicate a much higher velocity in a part of the atmosphere which is still being accelerated. The reason for this difference seems to be that in the LMC supergiants the [N II] lines are formed relatively farther from the star at the interface between the stellar wind and the interstellar medium, while in P Cygni they are formed much closer to the star. This can be interpreted as evidence for the continuing activity of P Cygni.

References

de Groot, M. 1969, Bull. Astron. Inst. Netherlands. 20, 255

de Groot, M. 1990, in Properties of Hot Luminous Stars, C.D. Garmany (ed.). Astron. Soc Pacific Conf. Ser., No. 7, p. 165

Drew, J.E. 1985, Mon. Not. R. astr. Soc., 217, 867

Lamers, H.J.G.L.M., Korevaar, P. and Cassatella, A. 1985, Astron. Astrophys., 149, 29

Osterbrock, D.E. 1974, Astrophsyics of Gaseous Nebulae, Freeman and Co. San Francisco, p. 101

Percy, J.R. *et al.* 1988, Astron. Astrophys., 191, 248

Stahl, O. and Wolf, B. 1986, Astron. Astrophys., 158, 371

Stahl, O., Mandel, H., Szeifert, Th., Wolf, B. and Zhao, F. 1991, Astron Astrophys., 244, 467

Taylor, M., Nordsieck, K.H., Schulte-Ladbeck, R.E. and Bjorkman, K.S. 1991, Astron. J., 102, 1197

Viotti, R. 1976, Astrophys. J., 204, 293

Spectroscopic observational data on P Cygni and comparison with the model of the bistable wind

Indrek Kolka
Tartu Astrophysical Observatory, 202444 Tõravere, Estonia

OBSERVATIONAL DATA

In recent years different variability cycles in the behaviour of P Cygni have been discovered both in photometric and spectroscopic data (Percy et al. 1988, Kolka 1989). The characteristic time–scales range from 50 days to 200 days. Our new photographic spectroscopic observations in 1989–90 (12Å/mm, 3600...4000Å) cover a period of about 200 days confirming the existence of shorter cycles (40 ... 50 days). One can see cyclic variations in the radial velocity of absorption troughs in different H I and He I lines in the observed near – UV wavelength region (Fig. 2b). The most displaced points of minimum residual intensity were observed in H_9 and their pattern suggests in addition a trend of velocity–shifts with longer time–scale ($\geq$ 100 days) against the shorter cycles. Earlier (e.g. Markova, Kolka 1989) we have observed different absorption–like details in one spectral line. On recent spectrograms we could not distinguish clearly the neighbouring components mainly due to lower resolution (12Å/mm against to 9Å/mm) and poorer S/N–ratio (25 ...30). One can check the appearance and disappearance of possible components in Fig. 1 showing the evolution of the profile of H_{14}.

The behaviour of line intensities can be checked by monitoring the variations of equivalent widths. Here we present (Fig. 2a,c) the values of equivalent widths (W_a, W_e) of nonrectified P Cygni–type absorption and emission components. The variability pattern is more complicated than at radial velocities. There is an indication of the existence of the yet two times shorter (20 ...25 days) time–scale in the variability of W_e and W_a.

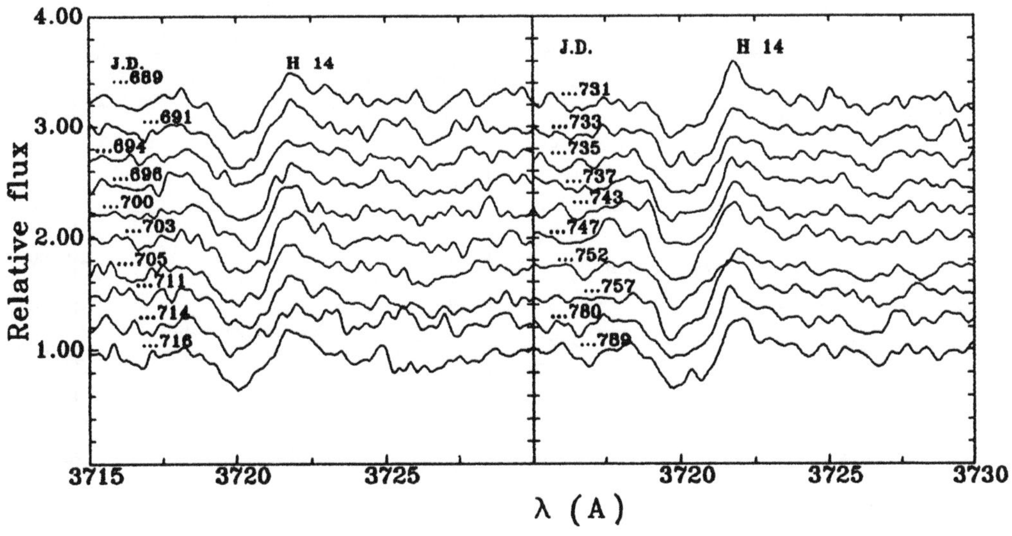

Fig. 1. The evolution of the Balmer–line (H14) profile in 1989.

The weak lines of higher excitation (He I 3927Å, O II 3727Å, S III 3718Å) are probably formed in deeper layers of the stellar wind and have therefore hardly any emission components. The precision of the measurements is not high in this case but the obvious trends in the values of W_a (see Fig. 2d) indicate the differences in their reaction to velocity cycles.

VARIABILITY OF SPECTRAL LINES AND THE MODEL OF THE BISTABLE WIND

The bistability of wind models in the case of P Cygni (Pauldrach, Puls 1990) means that the process of matter outflow can take place in two modes with up to four times different mass loss rates and excitation conditions. According to the model estimates the cycle–length between two switches of the high density mode is 70 ... 80 days. This interval covers about two main observed velocity cycles and it seems that both high and low mass loss periods can initiate similar velocity pattern. Or perhaps the bistability period is over–estimated?

One important parameter of the bistable model is the ionization state: high mass loss rate should coexist with species in the singly ionized stage and low mass loss rate with the doubly ionized species. The model calculations by Pauldrach and Puls (1990) showed that in this sense the most sensible agents are C, Si, S having ionization limits far in the hydrogen Lyman continuum which can vary strongly. We tried to check the role of the Lyman continuum quanta by monitoring the behaviour of the high excitation ($\sim 20eV$) lines mentioned above. The changes in the equivalent widths (W_a) do not correlate obviously with the velocity variations and the possible influence of the variable ionizing radiation on the W_a-s through the varying excitation conditions is probably mixed with the opposite influence of the mass loss rate.

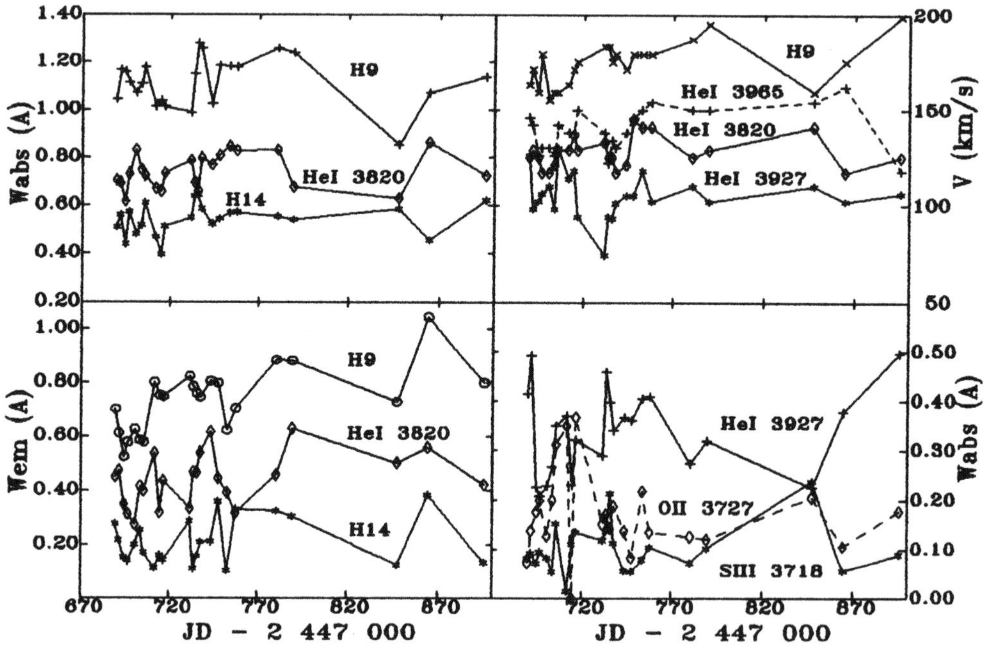

Fig. 2. The variability of spectral line parameters.

However, the "shell–pushing" scenario to explain the variability in spectral lines is probably promising. We compared some quite crude model profiles which take into account the enhanced–density shell at different evolutionary epochs and actual observed profiles of He I 3820Å and H_9 and found interesting qualitative similarity. It turned out that the best way for comparison is to find observed and modeled "difference–spectra" by subtraction of the averaged–over–cycles spectrum from every individual spectrum. The results are presented in Fig. 3a and Fig. 3b. The marked wavelength intervals on the unfortunately noisy plots of the observed spectra should be compared with details on the calculated spectra. The bistable model could fit to this picture when we have more precise data on time–scales and more realistic (may be without spherical symmetry?) model profiles to compare with observed ones having much higher S/N–ratio.

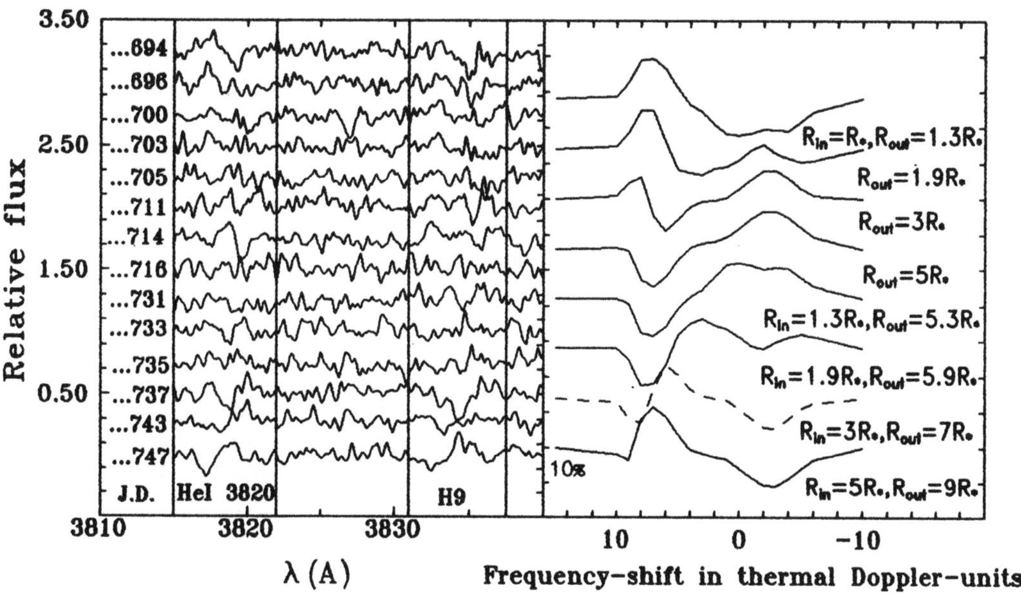

Fig. 3. The observed (left) and calculated (right) difference–spectrum at different epochs.

REFERENCES

Kolka, I. (1989). In: Physics of Luminous Blue Variables (K. Davidson, A.F.J. Moffat, H.J.G.L.M. Lamers, Eds.), Kluwer *Acad. Publ.*, pp. 265–267.

Markova, N., Kolka, I. (1989). An Atlas of Spectral Line Profiles of P Cygni in 1981–83. pp. 3–15, (*Tartu Teated* nr. 103).

Pauldrach, A.W.A., Puls, J. (1990). *Astron. Astrophys.* **237**, 409–424.

Percy, J.R. et al. (1988). *Astron. Astrophys.* **191**, 248–252.

THE SPECTROSCOPY OF UNUSUAL HIGH LUMINOSITY STARS: HD 168607 AND 6 CAS

E.L. Chentsov

Special Astrophysical Observatory of USSR Academy of Sciences

Nizhnij Arkhyz, Stavropol Territory, 357147, USSR

HD 168607, B9.4 Ia-0 and 6 Cas, A2.5 Ia-0 are hypergiants, they belong to the brightest stars of Galaxy (M_v = $-8^m.9$ and $-8^m.5$ respectively). Besides that these two stars answer all requirements of the belonging to P Cyg-type with exception only of the strong photometric variability, pseudophotospherical by nature, and the forbidden emissions. In the spectrum of HD 168607 the role of the latters play shell emissions Fe II λ 6318 and 6384 A, though.

Spectral data consisting of 13 spectrograms of HD 168607 and 74 spectrograms of 6 Cas were accumulated from 1976 to 1988. Spectrograms were recorded with the main stellar spectrograph of the 6-m telescope at the SAO AS USSR, dispersions on the whole 14 and 9 A/mm.

As the equally important for us were line shifts and profile anomalies, plates measurements were made with both oscilloscope comparator and microdensitometer.

Visual spectra of HD 168607 and 6 Cas were described in detail by Chentsov & Luud (1989), Aydin (1979), Sokolov & Chentsov (1984). The spectrum of 6 Cas at first sight is typical for white supergiants. It differs from the spectrum of α Cyg, for example, only in delicate but very important details, such as broad (±1500 km/s) base to the emission peak of H_α profile, the splitting of H_α and H_β absorptions into two components from time to time, the weak variable emissions on the red wings of H_β and the strongest Fe II lines and their variable asymmetry.

HD 168607 spectrum more reminds the spectrum of S Dor at outburst and spectra of similar objects, such as R 55, R 60 which are presented in Stahl's et al. (1985) atlas. All accessible members of the Balmer serie have broad emission bases (±2000 km/s). The splitting into components is observed not only in H_α and H_β but also in the weak Fe II absorptions.

The above-mentioned is illustrated in Figure 1. In particular the sharpness and splitting of H_β absorption and peculiarity of Fe II profiles in HD 168607 spectrum are seen in Fig. Ia. Fe II lines are so weak that their form become distinct only after the summing up of several lines. This summing profile usually has the overfall of residual intensity only several per cent. At the same time this profile is P Cyg-profile with the redshifted emission and blueshifted two-component absorption.

Fig. Ib shows the type of profile asymmetry predominating in the spectrum of 6 Cas. If we divide the profile into two by a vertical straight line passing through the main minimum, we find that the blue part of the profile produces a larger contribution to the equivalent width than does the red one. The main factor that deforms the profiles is not the weak emission on the red flanks but the additional absorption on the blue flanks.

Figure 2 shows the radial velocities and differential shifts of the individual lines and their homogeneous groups as functions of the formations depths ("kinematic sections" of the atmospheres). For the ordinary white supergiants such as β Ori or α Cyg such relations are clear and simple - Figure 3. For 6 Cas (Fig. 2b) the simplicity gets broken. H_α and H_β often give two values of the velocity, but they can be very close for the secondary, blueshifted H_β component and for the main H_α component: in the upper layers Balmer course stops. In the case of HD 168607 it's better to speak not about the "kinematic section", but about the simultaneous existence of several velocity levels presented by H I and Fe II lines' components. Apparently we deal with the time from time appearance of the isolated shells.

The next obvious task is to find out if there is a similarity with ejections of shells in the stellar wind of P Cygni, and maybe even to estimate the ejection time scale.

Unfortunately it is possible for 6 Cas only. The weakness and the small altitude over the horizon didn't allow to obtain for HD 168607 more than one or two time points per a season. But the sets of our observations of 6 Cas are not yet long and filled enough. It is clear we deal with the cycle profiles evolution, but none of our sets describes the full cycle and quite distinctly. Therefore Figure 4a shows the composite picture. It was made up by shifting of the several velocity sequences along the time axis. Of course only sufficiently long sequences were used. By the choice of the shift's variant the depths of red H_β wing in the λ 4862 A were taken into consideration - Fig. 4b. One can see in Fig. 1b the examples of the highest and lowest positions of the mentioned point.

According to our version the profiles' variations in the spectrum of 6 Cas follow the same scheme that we have in P Cyg spectrum: "the absorption components appear near the emission peak and tend to move to the short-wavelength side of the profile, where they disappear" (Markova & Kolka, 1988). But in the P Cyg case it's possible to trace with the help of any line which has moderate emission. In the case of 6 Cas it appears, that H_β is the only one suitable line in the visible part of the spectrum.

As for the characteristic time and the amplitude of velocity variations, they also are very close for 6 Cas and P Cyg. With the help of Fig. 4a it's possible to find for 6 Cas : $\Delta t = 150 \div 170^d$ and $\Delta v = 100 \div 120$ km/s.

Both objects are very perspective for the further spectroscopic and photometric investigations. We would like to draw attention to them at least of two observatories, the northern for 6 Cas and southern for HD 168607. The best variant could be the long-term (at least half a year) monitoring of spectral variations similar to that one which was arranged recently for P Cygni (Stahl et al., 1991).

The special problem is the determination of stars' radial velocities of mass-centers. For HD 168607 this problem is solved by means of gas-dust complex M17 to which it belongs. For 6 Cas is possible the use of association Cas OB5 members and especially its optical companion. Namely 6 Cas B has the spectral class O9 and at the same time it is probably the spectroscopic binary (Barsukova & Chentsov, 1990), but not 6 Cas A as it was supposed by Talavera & Gomez de Castro (1987).

REFERENCES

Aydin, C.: 1979, Ap. Sp. Sci., v.64, 481

Barsukova, E.A., Chentsov, E.L.: 1990, Astrofiz. Issled. (Izv. SAO), v. 29, 101

Chentsov, E.L., Luud, L.: 1989, Astrofisika, v. 31, 5

Markova, N., Kolka, I.: 1988, Ap. Sp. Sci., v. 141, 45

Sokolov, V.V., Chentsov, E.L.: 1984, Astrofiz. Issled. (Izv. SAO),v.18, 8 = 1985, Bull.SAO, Allerton Press, v.18, 5

Stahl, O., Wolf, B., de Groot, M.: 1985, A & Ap Suppl., v.61, 237

Stahl, O., Mandel, H., Szeifert, Th., Zhao, F.: 1991, A & Ap, v.244, 467

Talavera, A., Gomez de Castro, A.I.: 1987, A & Ap, v.181, 300

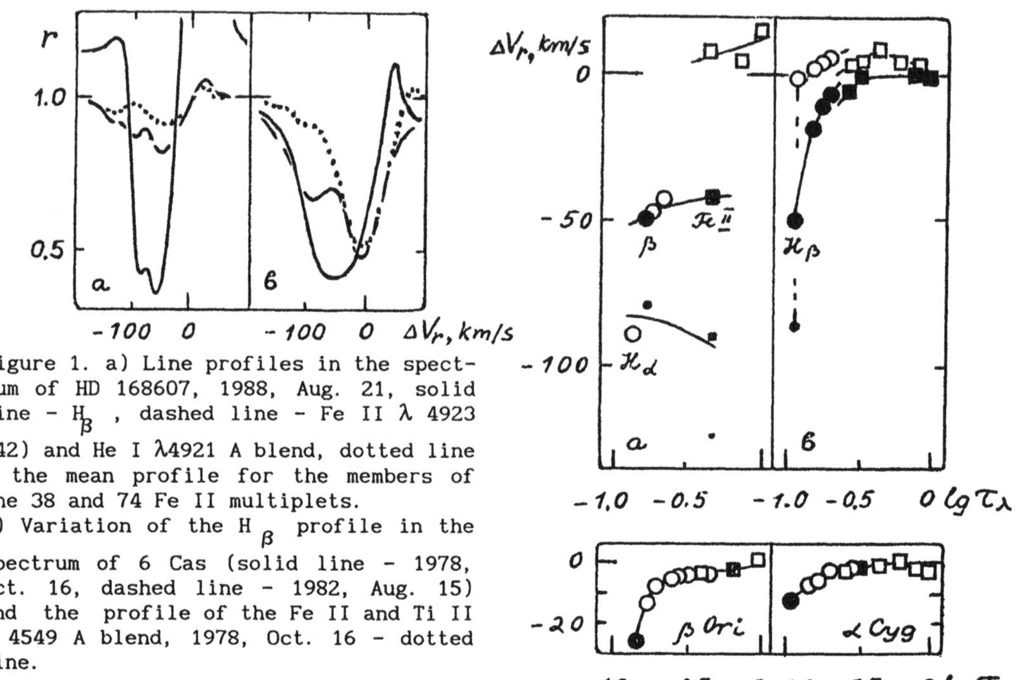

Figure 1. a) Line profiles in the spectrum of HD 168607, 1988, Aug. 21, solid line - H_β , dashed line - Fe II λ 4923 (42) and He I λ4921 A blend, dotted line - the mean profile for the members of the 38 and 74 Fe II multiplets.
b) Variation of the H_β profile in the spectrum of 6 Cas (solid line - 1978, Oct. 16, dashed line - 1982, Aug. 15) and the profile of the Fe II and Ti II λ 4549 A blend, 1978, Oct. 16 - dotted line.

Figure 2. Radial velocities as a functions of the optical depth or "kinematic sections" of the atmospheres of HD 168607 (a) and 6 Cas (b). The zero points of the vertical axes are the supposed radial velocities of the centers of mass of the stars. The dates of the observations are as in Fig. 1: a) 1988, Aug. 21 for HD 168607, b) 1978, Oct. 16 (solid lines and filled symbols) and 1982, Aug. 15 (dashed lines and open symbols) for 6 Cas.

Figure 3. Same as Fig. 2 for the supergiants of closes spectral classes.

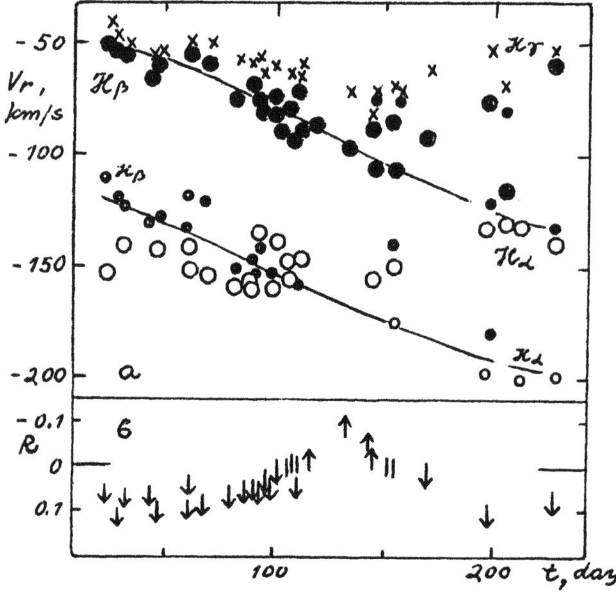

Figure 4. 6 Cas. The composite picture of the variations of the radial velocities of the hydrogen absorption components (a, big symbols correspond to the main components) and the depth of the red wing of H_β (b).

Discrete Absorption Components in O-Type Stars

Ian D. Howarth

Department of Physics and Astronomy, University College London,
Gower Street, London WC1E 6BT

Abstract: The primary observational characteristics of variability in UV P-Cygni lines of O stars are illustrated. 'Narrow components' seen in almost all spectra are just one aspect of discrete absorption components, or DACs. The blueshifted DACs develop at $\sim 0.5v_\infty$ and then migrate to higher velocities; their production is probably related to the line-driven instability.

1 Introduction

Luminous O stars usually exhibit relatively sharp (full-width at half depth $\sim 0.1v_\infty$) violet-displaced absorption enhancements near the terminal velocity in UV P Cygni profiles. These 'narrow components' were recognized in the earliest *Copernicus* spectra of OB stars (*e.g.*, Underhill 1975; Morton 1976; Snow & Morton 1976), and are illustrated by a few examples in Figure 1.

Prinja & Howarth (1986) showed that the narrow components are quite strongly variable, and that the extended P Cygni absorption trough is also variable, though less so. Subsequent time-series *IUE* spectroscopy (Prinja, Howarth, & Henrichs 1987; Prinja & Howarth 1988) has revealed the systematics of this variability. Typically, a rather broad absorption feature first becomes apparent around $\sim 0.5v_\infty$; that 'discrete absorption component', or DAC, then evolves to higher velocity, becoming narrower and stronger in the process (Fig. 2). The narrow components are now seen to be just one aspect of DAC variability, but one that is readily identified in one-off 'snapshot' spectra. This accounts for the observation that the narrow components often show structure and can sometimes be resolved into two or more subcomponents.

In this brief overview I shall summarize some observational aspects of discrete absorption components. A more extensive review of recent observations is given by Prinja (1992), and their interpretation is discussed in Howarth (1992).

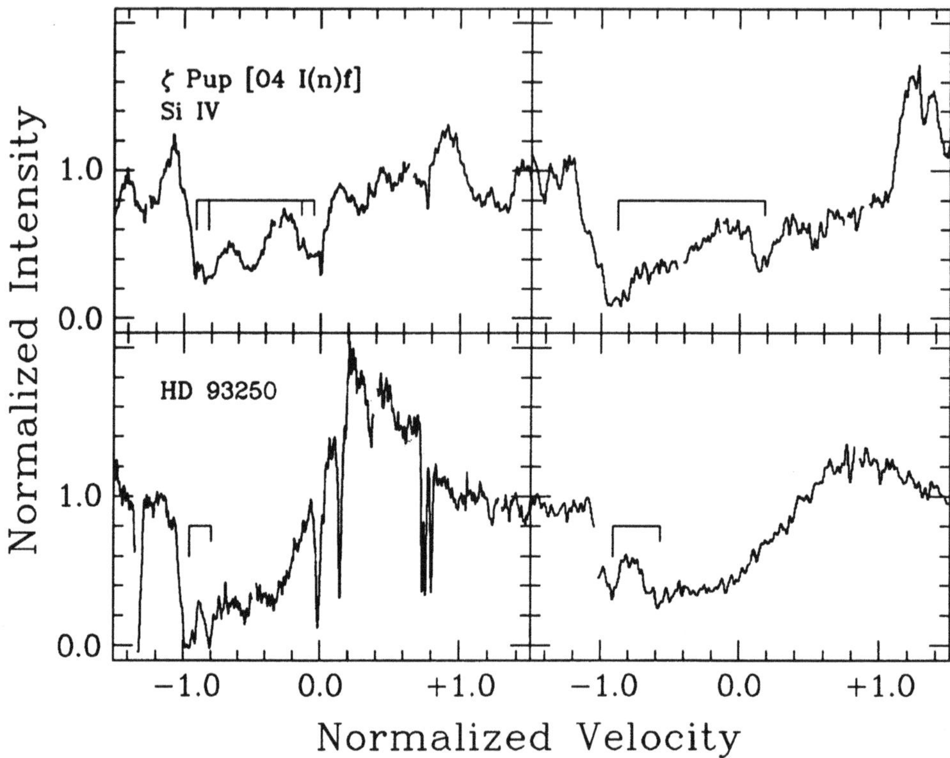

Fig. 1. Narrow components are present in O stars of all spectral types: early, late, supergiant, and main-sequence. In this figure velocities (with respect to the rest position of the short-wavelength doublet components) are normalized to terminal velocities, and intensities to local continuum levels.

2 Observational Characteristics

2.1 Distribution

In luminous OB stars the narrow components are usually seen in the same UV resonance lines of abundant ions that show P Cygni profiles. (Exceptions include the B0 Ia star HD 152667, for which Howarth (1984a) reports narrow components in excited Fe III lines; and ζ Pup, where Prinja *et al.* (1992) have recently demonstrated DAC behaviour in the subordinate N IV λ1718 line.)

Extensive surveys of narrow components carried out by Lamers, Gathier & Snow (1982), Prinja & Howarth (1986), and Howarth & Prinja (1989) have shown that these features are present in essentially all O stars (Fig. 1); similar features are seen in B supergiants. By inference, DACs are also present in all O stars and B supergiants.

Be stars show complex, narrow absorptions in UV resonance lines (*e.g.,* Grady, Bjorkman & Snow 1987), but their characteristics are not identical to those of the DACs, and it is not clear that the same mechanisms are responsible. The same is true of low-gravity hot subdwarfs with stellar winds (*e.g.,* Howarth 1987). Narrow components

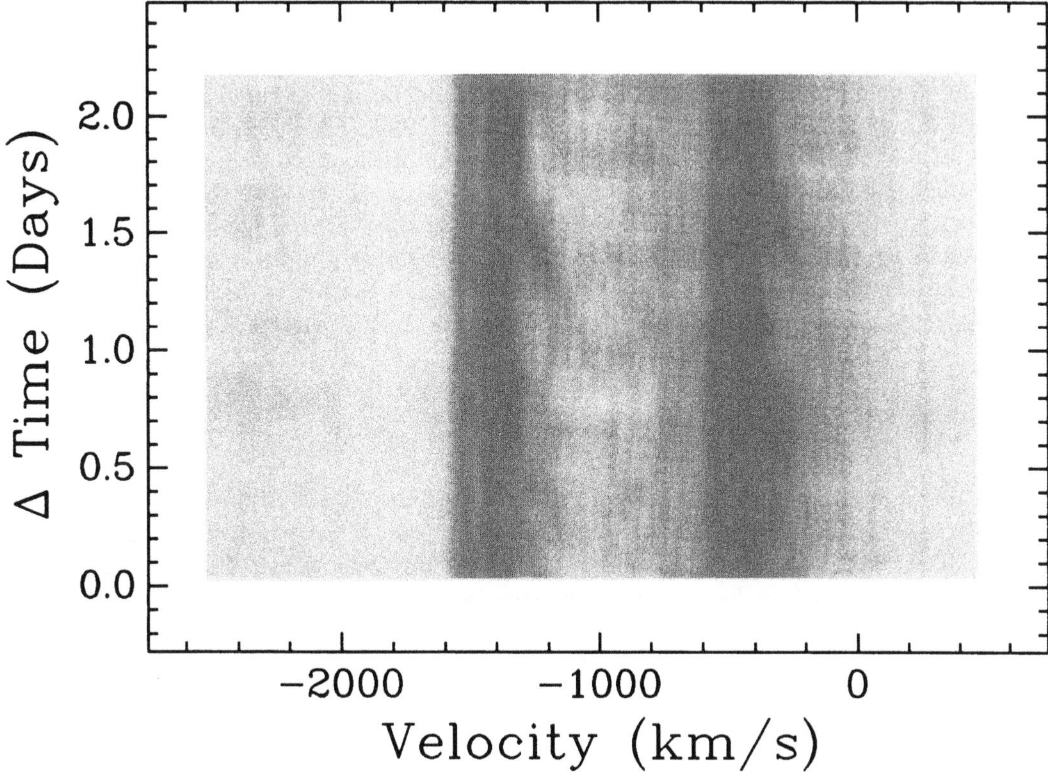

Fig. 2. Greyscale representation of time series *IUE* observations of the N V doublet in HD 149757 (ζ Oph). One strong DAC can be seen to form at intermediate velocities, accelerate, and merge into the narrow components (additional DACs can be seen in high-contrast plots of the data). The doublet separation is 970 km/s and the terminal velocity is 1500 km/s.

do not seem to be a widespread characteristic of Wolf-Rayet winds, but that may be in part observational selection because of the large optical depths in their resonance lines; WR stars certainly show line-profile variability (*e.g.* Willis *et al.* 1989).

2.2 Column densities

Narrow components have column densities, N^N, which are typically of order 14.5 dex cm^{-2} in C^{3+}, N^{4+}, and Si^{3+}. The columns loosely correlate with the column density in the 'underlying' P Cygni profile, N^P (Howarth & Prinja 1989); $N^N \simeq 0.1 N^P$. The limited range in observed columns is in part a selection effect—in weak P Cygni profiles the narrow components are undetectable, while much stronger profiles are saturated.

In some (though not many) cases, the narrow components go to zero intensity (*e.g.* HD 93250; Fig. 1). When they do not, doublet ratios are consistent with optically-thin line-forming regions which cover the projected face of the star, and not with optically-thick 'clumps' which do not (Howarth 1984a).

2.3 Effects of Rotation

Prinja (1988) discovered a correlation between $v_e \sin(i)$ and the timescale for the recurrence and development of DACs. That correlation has been confirmed and extended by Henrichs, Kaper & Zwarthoed (1988) and by Prinja (1990). Figure 3 (kindly supplied by Raman Prinja) illustrates how discrete components appear more frequently and evolve more rapidly in stars with greater $v_e \sin(i)$. It is important to note that this relationship is with *projected* rotation velocity; if it were with equatorial rotation velocity the relationship would be unlikely to be evident in the available sample of a half-dozen or so stars. There is also a weak indication that low column densities in narrow components are restricted to stars with small $v_e \sin(i)$ (Howarth & Prinja 1989).

3 Discussion

The rather sudden appearance of DACs at intermediate velocities shows that they are basically controlled by a mechanism which is *intrinsic* to the wind. That mechanism is most likely the line-driven instability discussed by Owocki (1991a, and references therein). All winds driven by radiation pressure in the lines are prone to the instability, and all O stars show DACs. The scattered radiation field inhibits the growth of the instability near to the star (Lucy 1984, Owocki 1991b), explaining why the low-velocity variability is so much less than at higher velocities (*e.g.,* Fig. 3 of Howarth 1984b).

References

Grady, C.A., Bjorkman, K.S., and Snow, T.P, 1987. *Ap. J.,* **320**, 376.

Henrichs, H.F., Kaper, L., and Zwarthoed, G.A.A., 1988. In *A Decade of UV Astronomy with the IUE Satellite,* ESA SP-281, **2**, 145.

Howarth, I.D., 1984a. *M.N.R.A.S.,* **206**, 625.

Howarth, I.D., 1984b. *M.N.R.A.S.,* **211**, 167.

Howarth, I.D., 1987. *M.N.R.A.S.,* **225**, 33P.

Howarth, I.D., 1992. In *Variable and Non-Spherically Symmetric Outflows from Stars,* ed. L. Drissen, C. Leitherer, and A. Nota, A.S.P. Conference Series, in press.

Howarth, I.D., and Prinja, R.K., 1989. *Ap. J. Suppl.,* **69**, 527.

Lamers, H.J.G.L.M., Gathier, R., and Snow, T.P., 1982. *Ap. J.,* **258**, 186.

Lucy, L.B., 1984. *Ap. J.,* **284**, 351.

Morton, D.C., 1976. *Ap. J.,* **203**, 386.

Owocki, S.P., 1991a. In *Wolf-Rayet Stars and Other Massive Stars in Galaxies,* ed. K.A. van der Hucht & B. Hidayat, Kluwer, Dordrecht, p. 155.

Owocki, S.P., 1991b. In *Stellar Atmospheres: Beyond Classical Models,* ed. L. Crivellari, I. Hubeny, & D.G. Hummer, Kluwer, Dordrecht, p. 235.

Prinja, R.K., 1988. *M.N.R.A.S.,* **231**, 21P.

Prinja, R.K., 1990. In *Angular Momentum and Mass Loss for Hot Stars,* ed. L.A. Willson & R. Stalio, Kluwer, Dordrecht, p. 223.

Prinja, R.K., 1992. In *Variable and Non-Spherically Symmetric Outflows from Stars,* ed. L. Drissen, C. Leitherer, and A. Nota, A.S.P. Conference Series, in press.

Prinja, R.K., and Howarth, I.D., 1986. *Ap. J. Suppl.,* **61**, 357.

Analysis of Circumstellar Spectra of S Dor

Krzysztof Gesicki

Copernicus Astronomical Center, ul.Chopina 12/18, 87-100 Toruń, Poland

A new method of solution of radiative transfer for spectral line formed in an expanding circumstellar shell I have presented somewhere else (Gesicki, 1991). This method, together with the so called equivalent two level atom method, allowed me for non-LTE calculations of BaII lines in circumstellar envelope of Rho Cas. If the BaII circumstellar lines were observed in other objects, it would be possible to perform similar analysis. Such possibility exists in some of Luminous Blue Variables, since during their maximum phase they can reach such a spectral type, where one can expect to see the BaII lines. Especially interesting can be the star R 110, which has been classified as F0 Ia supergiant (Stahl et al., 1989). Unfortunately there are no published observations of BaII lines for R 110 and for other LBVs.

The computer code prepared for mentioned above analysis can be used for calculation of the shape of one spectral line formed in a two level atom. The radiative transfer is solved exactly, in the observers frame. In the shell a stellar disc can be placed, with emitting flux arbitrarily depending on frequency and distance from the disc center. The advantage of the applied method is that it can be used for an arbitrary velocity field across the circumstellar envelope. The comoving frame methods and the Sobolev approximation do not allow this.

The Figure 1 presents an example of how the spectral line shape can depend on the assumed velocity field in a circumstellar shell. The flux on the stellar surface is assumed constant, i.e. there is no spectral line in photospheric continuum an the presented profile originates entirely in the circumstellar envelope. The shell thickness is four stellar radii, the range of velocities is the same in both cases. The distribution of matter in the envelope, i.e. the density structure, is determined from a boundary density and a given velocity field, by the continuity equation.

Interesting observations indicating variable velocity in circumstellar matter I found in paper of Wolf and Stahl (1990), part of it I present here in Figure 2. Except of the small variability in line shapes there is presented a clear evidence of collapsing matter (inverse P-Cyg type profiles for date Dec.16,1989).

I have tried to reproduce the presented above observed profiles with my code. The calculated profiles can be compared to the profiles observed, both shown in Figure 2. In my calculations I have assumed, as before, no underlying spectral line of photospheric origin. I have varied the outer radius of the envelope and the velocity field in it. The assumed runs of velocity for profiles from Figure 2 are shown in Figure 3. The line is

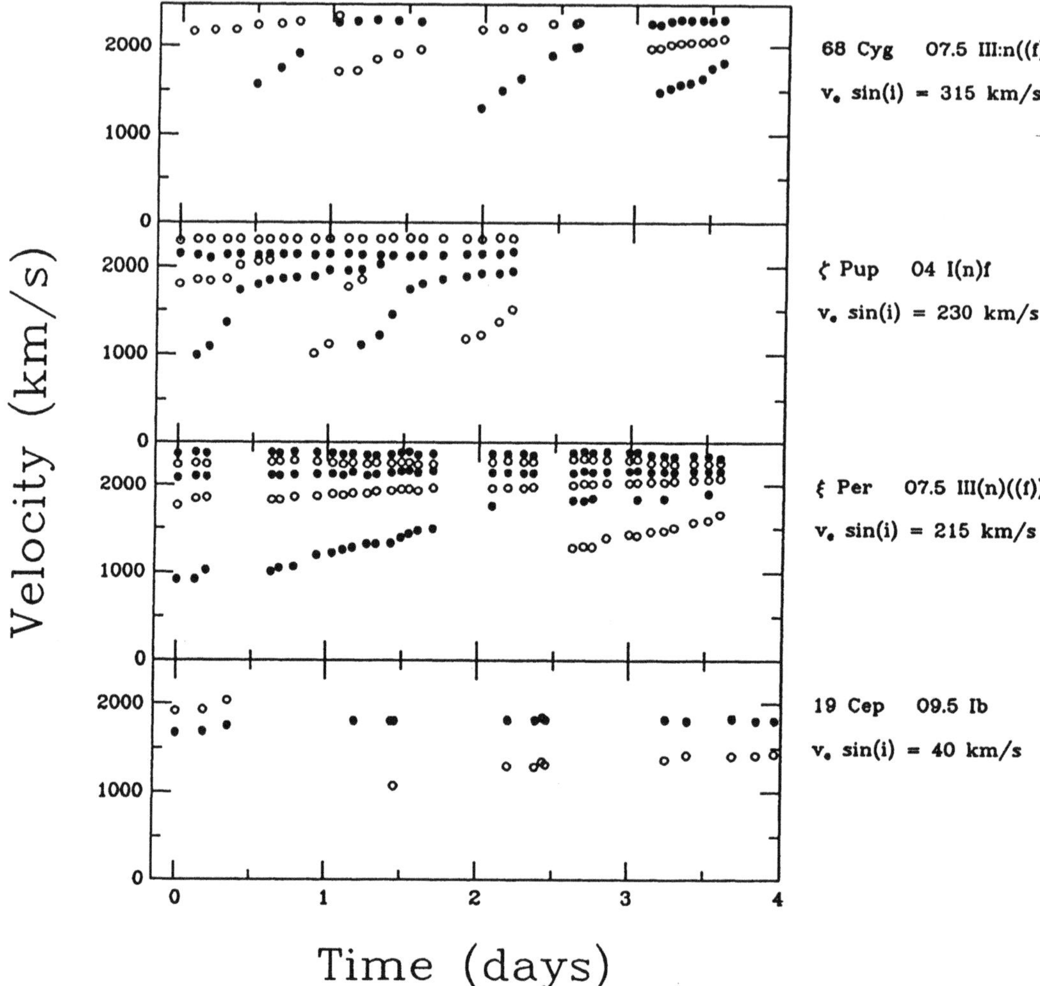

Fig. 3. Central velocities of DACs as a function of time in stars of different $v_e \sin(i)$. Faster projected rotation velocities are associated with faster development and more-frequent recurrence.

Prinja, R.K., and Howarth, I.D., 1988. *M.N.R.A.S.*, **233**, 123.
Prinja, R.K., Howarth, I.D., and Henrichs, H.F., 1987. *Ap. J.*, **317**, 389.
Prinja, R.K., *et al.*, 1992. *Ap. J.*, in press.
Snow, T.P., and Morton, D.C., 1976. *Ap. J. Suppl.*, **32**, 429.
Underhill, A.B., 1975. *Ap. J.*, **199**, 691.
Willis, A.J., *et al.*, 1989. *Astr. Ap. Suppl.*, **77**, 269.

formed by pure scattering in a two level atom. The Doppler broadening is assumed constant, with value of 10km/s. After performing calculations with many different possible structures of the envelope I can draw a few conclusions:

1) thick or thin envelope can produce similar line profiles. The Figure 2 presents line shapes for shells of thickness 1 R_* and 4 R_* and this conclusion can be clearly seen.

2) the small changes in the main absorption component, between first (Nov.6,1987) and second (Jan.23,1989) observation, can be interpreted as a change in the velocity gradient in the shell (compare the upper and middle curves in Figure 3). The acceleration in internal layers should be weaker in the second case, what is in agreement with the next (third) observation, showing that these layers are collapsing.

3) the third (Dec.16,1989) profile is much more complicated. The presence of the redshifted absorption component can easily be explained with the assumption that some matter in the shell is collapsing. I have tried some other possibilities that can have a physical sense, but I failed to reproduce this feature. The strong emission between the two absorptions cannot be reproduced without additional source of population of the upper atomic level. This fact can most naturally be interpreted as the collisional excitation resulting from heating of the infalling matter. But the heating should be in layers with velocity close to zero, because the emission peak indicates this value. Therefore I have assumed that the infall of matter is decelerated close to the stellar surface. In these internal layers I have artificially enlarged the population of the upper atomic level, to mimic the collisional excitation. The assumed velocity field and the resulting profiles are presented in the lowest parts of Figures 2 and 3. It can be seen that calculated profiles resemble quite nice the observed ones.

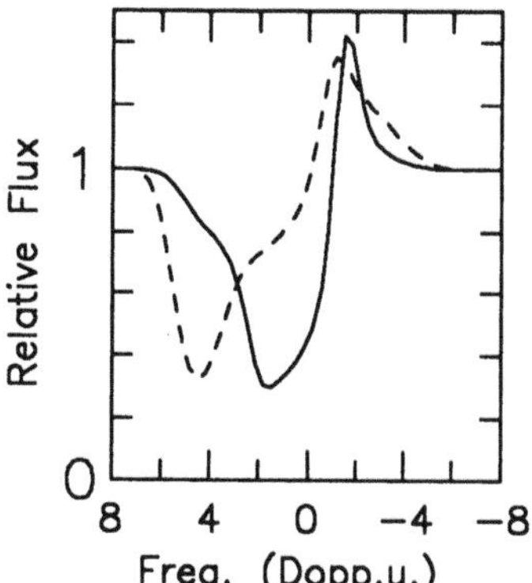

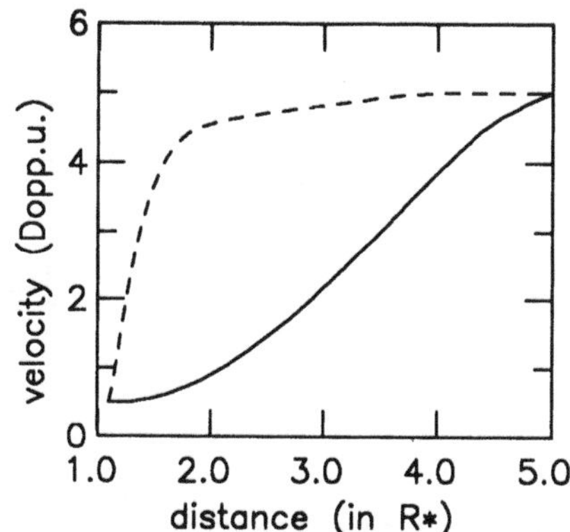

Fig. 1. Two circumstellar line profiles (left part of the figure) formed in envelopes with different velocity structure (shown in right part of the figure). The velocity as well as the frequency are expressed in the Doppler width units.

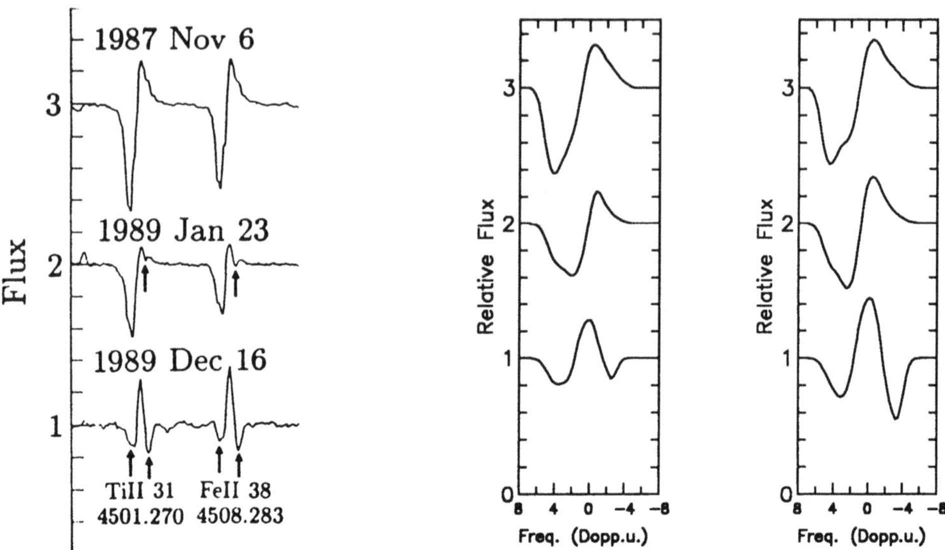

Fig. 2. The left part presents an example of the observed circumstellar profiles in star S Dor. This is a copy of Fig.3 from paper of Wolf and Stahl (1990). The middle and right parts present the calculated circumstellar line profiles for models with shell thickness of 1 R_* and of 4 R_* respectively. The sequence of theoretical line profiles corresponds to the sequence of the observed ones

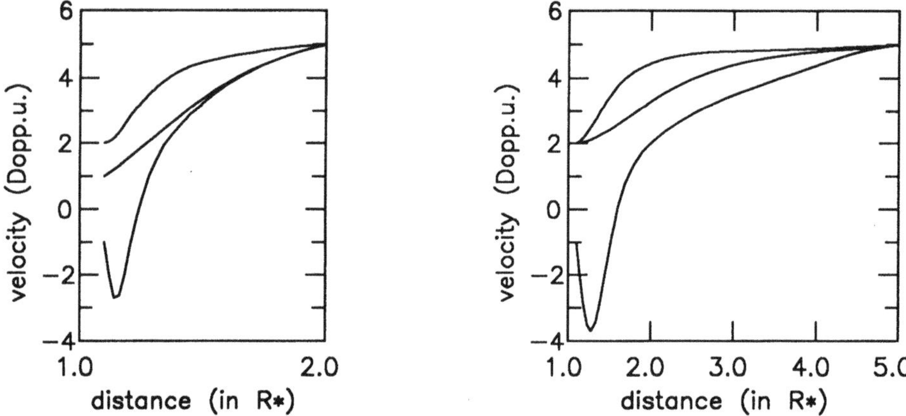

Fig. 3. The velocity field across the envelope for profiles shown in Fig.2. The left part presents models with shell thickness of 1 R_*, the right part with thickness of 4 R_*. The sequence of lines from the upper to the lowest corresponds to the sequence of profiles from top to bottom, shown in Fig.2

References

Gesicki,K. (1991): *Astron.Astrophys.*, in press

Stahl,O., Wolf,B., Klare,G., Juttner,A., Cassatella,A. (1989): VILSPA IUE Preprint No.45

Wolf,B., Stahl,O. (1990): *Astron.Astrophys.*, **235**, 340-344

COMOVING FRAME CALCULATIONS FOR λ-CEPHEI

Nadine Rons, Mark Runacres
Astrofysisch Instituut, Vrije Universiteit Brussel
Pleinlaan 2, B-1050 Brussel, Belgium,
Ronny Blomme
Koninklijke Sterrenwacht België
Ringlaan 3, B-1180 Brussel, Belgium

Abstract

Many stars continuously eject material and thus surround themselves with a stellar wind. We can observe this wind indirectly through its characteristic effects on the stellar spectrum. In particular, this mass flow influences the line profiles.

These so called P-Cygni profiles can be observed in the visual part of the spectrum of early-type stars and Wolf-Rayet stars, as well as - and even more clearly - in the ultraviolet part. In order to study the stellar wind structure, a computer programme was written which calculates theoretical P-Cygni profiles using the Comoving Frame Method. As input, this code can use the results obtained from other programmes, such as a non-LTE code. Through parameter adjustments a fit to an observed profile can be made.

The Comoving Frame Method

The major advantage of this method is the angle independance of opacity and emissivity, as seen in the frame moving along with the wind. The model is constructed semi-empirically with pre-specified velocity and opacity laws. Several laws have been introduced. We currently use the laws from Groenewegen et al. (1989).

Following Mihalas et al. (1975) the calculation of a P-Cygni profile takes place in three steps:

First the source function is calculated as a function of frequency and depth. In the Comoving Frame Method this is done by solving the equation of radiative transfer in a frame which instantaneously moves along with the wind material.

Secondly the emergent intensity is integrated along a set of lines through the stellar wind. These are parameterized by p and lie parallel to the line of sight. This integration is executed for an appropriate set of observer frame frequencies. To do this, the source function S and the optical depth τ, calculated in the first step, are needed.

$$I(p, \nu_{OBS}) = \int S(r, \nu_{CMF})\, e^{-\tau}\, d\tau + I_{surface\ star}(p, \nu_{OBS})\, e^{-\tau_{tot}}$$

The second term contains the intensity emitted at the stellar surface and is added only for those lines which cross the star.

Finally, in the third step, the profile is obtained after an integration of the emergent intensities over the disk surface of the stellar wind.

$$I(\nu_{OBS}) = \int_{\sigma\ disk} I(p, \nu_{OBS})\, d\sigma$$

Resulting profiles

Figure 1 shows an example of such an observed profile and two theoretical ones. A better fit is obtained for the emission part than for the absorption part. Since the code was not yet able to calculate a doublet, this discrepancy can be explained by the fact that we fitted a theoretical singlet to an observed doublet, while the doublet character mainly appears in the absorption trough.

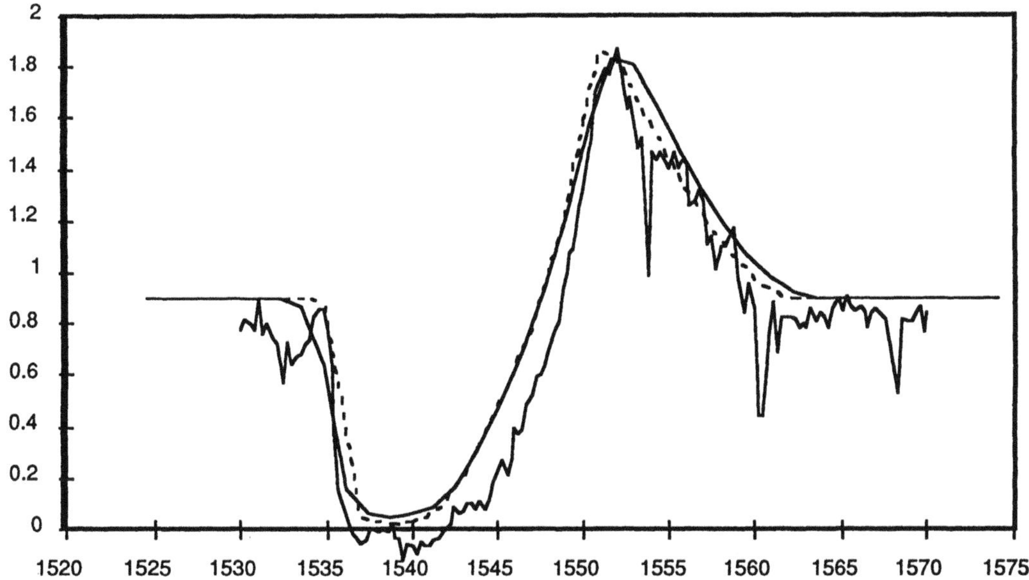

Fig. 1: The C IV doublet (λλ 1548, 1551) of λ Cep (O6 I (n)fp) : normalized emergent intensity versus wavelength in Å.

Future work

A strong interaction exists between the radiation field and the level populations of the different ionization stages of the chemical elements. These populations are strongly related to the optical depth, for which, up to now, we have used a parameterized law. Our work on a non-LTE code will enable us in the future to calculate the number densities corresponding to the levels that are considered. An opacity law can then be derived and inserted in the comoving frame code.

References

Groenewegen, M.A.T., Lamers, H.J.G.L.M., 1989, Astron. Astrophys. Suppl. Ser. 79, 359.

Mihalas, D., Kunasz, P.B., Hummer, D.G., 1975, Astrophys. J. 202, 465.

IV. Non-classical Phenomena in Early-Type Stars

Nonradial Pulsations of O- and B-Stars

Dietrich Baade

European Southern Observatory
Karl-Schwarzschild-Str. 2, D-W-8046 Garching, Germany

Abstract: The available observational facts are reviewed which show that in the hot, high-luminosity corner of the HR-diagram nonradially pulsating stars are very common. For individual stars a proof of nonradial pulsation is not always easy to furnish, but the global evidence is very strong. With the possible exception of supergiants, the OB-star instability region is separated from other nonradially pulsating stars such as the δ Scuti stars by a zone starting around B7 where even low-amplitude pulsators suddenly become rare. OB-stars are also theoretically much less elusive than has sometimes been suggested in the past since the recent upward revision of stellar opacities now enables driving of the pulsations also by the classical κ-mechanism possible. Within the OB-domain, sub-patterns of pulsational properties begin to show up in the observational data. As an example with other interesting repercussions, some attention is given to the double-dichotomy between Be- and Bn-stars. Both are rapid rotators and among B-stars earlier than B7 roughly equally abundant. But the incidence of low-order nonradial pulsation modes is much higher in Be-stars, and only Be-stars go through outburst-like phases of strongly enhanced mass loss rates. Interdependencies between (a) rapid rotation and chemical structure, (b) abundances and pulsation, and (c) pulsation and rotation might in some combination lead to these two different B-star sub-populations. Observational possibilities to discriminate between various alternatives are briefly discussed.

1 Introduction

Nonradial pulsations have been observed or suspected (or both) in stars all across the HR diagram. Except for a probable 'supergiant bridge', the variable O- and B-stars appear to be separated from the next cooler group of variables, the radially and nonradially pulsating δ Scuti stars, by a broad gap between about B7 (e.g., Smith and Penrod 1984; Baade 1989a,b; Balona et al. 1992) and A2 (Breger 1979). Within this gap, variable stars are difficult to find, and reported amplitudes barely exceed the limit of detectability (McNamara 1987).

The blue boundary of the instability zone has yet to be explored. The hottest probable nonradial pulsators known to date are ζ Pup (Baade 1991b) and λ Cep (Henrichs 1991); their respective spectral types are O4 If and O6 I(n)fp, i.e., both are supergiants. On or close to the main sequence, the earliest examples are 10 Lac (O8 III; Smith 1977),

υ Ori (O9.5 V; Smith 1981a), and the emission-line star ζ Oph (O9.5 Vn; Vogt and Penrod 1983; Kambe et al. 1990 and references therein). It would be premature, however, to infer just from this enumeration that pulsational variability extends to earlier spectral types in the supergiant sequence. Only variability in the wind is, as one would expect, clearly more pronounced among the more luminous stars (Fullerton 1990, 1991).

For a long time, pulsating OB-stars were considered a somewhat esoteric subject because those known were very few in number, and there was no straightforward way to make models of OB-stars pulsate. Both aspects have changed quite substantially over the past decade so that OB-stars may finally find a rather improminent place within the menagerie of pulsating stars.

However, there are three questions which do not in this form exist for other nonradially pulsating stars:

a) Are g-modes excited in OB-stars? – Not considering white dwarfs no other group of stars is presently known to display so many candidate g-modes. This is of substantial interest because g-modes penetrate a star more deeply than do p-modes and, therefore, reflect the not otherwise observable conditions in the stellar core (e.g., Cox et al. 1992). However, the high rotation rates of many OB-stars require a more differentiated classification scheme of nonradial pulsation modes (e.g., Lee and Saio 1986).
b) Does nonradial pulsation contribute to mass loss?
c) Can nonradial pulsation affect a star's evolution?

It is well possible that already in a few years' time, we will know that these questions are irrelevant or even wrong. However, at present, they form part of the attraction of the subject of nonradial pulsations in O- and B-stars. The main stages of the evolution of the field are documented in a number of review articles (e.g., Smith 1981b, 1986; Baade 1986, 1987a,b,c, 1988; Cuypers 1991; Fullerton 1991; Sterken and Jerzykiewicz 1991; Walker 1991). A comprehensive overview of the present status is given in the proceedings of a recent, dedicated workshop (Baade 1991a). This paper can, therefore, be restricted to highlighting the main issues and elaborating on a few new results or considerations.

2 Is it nonradial pulsation?

This question would probably have been pursued less insistingly without the extensive photometry that has been done of numerous Be stars. In most Be-stars, the amplitudes are variable and relatively small; but in virtually all Be-stars are they measurable and associated with a stable period. The observed periods appear correlated with independently inferred rotational periods (Balona 1990) to an extent which is not generally expected if the variability is due to nonradial pulsation alone. If, furthermore, the lengths of the (roughly sinusoidal) cycles and subtle differences between odd and even cycles, which suggest that the true period is twice the cycle length (Harmanec 1984), are taken into account, these periods become statistically indistinguishable from the computed rotational periods. The simplest conclusion is, then, that the variability is due to some co-rotating surface features ('spots').

As a circumstellar alternative to spots, a structure of co-rotating radial spokes has been suggested (Harmanec 1991). However, this does not appear a very realistic option because at best it would only account for the variability of emission lines stars. Even then would the observed acceleration of sub-features of line profiles cause serious physical contradictions with what else is known about circumstellar disks of early-type stars. Furthermore, it is not clear whether magnetic fields which would probably be required to keep the spokes in place could have escaped direct measurements. For these reasons, the discussion below will not pursue this model much further. At any rate, many of the comments made below on the spot model apply analogously to the spoke model.

For low amplitudes, the eigenmodes of slowly rotating stars are described by spherical harmonics (Ledoux 1951). However, the latter form a complete set. Therefore, the significance of the approximation of observed line profile variations by spherical harmonics has been questioned because anything on a sphere can be expanded into a series of spherical harmonics. So, why should one, against the apparently strong photometric evidence, consider nonradial pulsation at all? Actually, there are many reasons as becomes clear if also the evidence which a) spectroscopy in general and b) observations of OB-stars other than Be-stars in particular have supplied is admitted to the discussion. This, however, is very essential:

- Most broad-lined OB-stars with low-order line-profile variability also display regularly-spaced smaller structures which periodically travel across the Doppler profile. In very rapidly rotating stars, the phase velocity is roughly inversely proportional to the spatial wavelength, $1/|m|$, of the pattern. Nonradial pulsation theory predicts this for modes with long periods in the co-rotating frame (e.g., Saio and Lee 1991). Two big and $|m|$ small co-rotating star spots could have a similar effect. But no effort has been made so far to make it appear at least plausible that groups of spots form so distinct patterns each of which is extremely regular and restricted to the equator.
- There are multi-mode pulsators such as the B0.7 III-star ϵ Persei (Gies and Kullavanijaya 1988) whose periods are not m-commensurate. The suggestion of commensurability has been made (Harmanec 1987), though, but it ignores that from the acceleration of the features at the line center also the mode orders could be determined with sufficient accuracy.
- Traveling sub-features have also been observed in supergiants which in spite of their broad lines have physical rotation periods of several days whereas the inferred period of revolution of the observed patterns (sometimes called super-period: $P_{super} = |m| \times P_m$ where m is the nonradial mode order) does not exceed 1-2 days (Baade et al. 1990, Baade 1991b).
- Substantial contrast variations of individual subfeatures between line center and wings (e.g., Baade 1984) are hardly compatible with a scalar field such as spots. Conversely, this behavior is the necessary consequence of the horizontal velocity component which dominates g-modes.
- In stars seen at small inclination angles, i, large horizontal velocity amplitudes to and from pole and equator will be observable at all phases of a wave propagating about the star's rotation axis. During the revolution, the line profile modulation associated with such a wave will vary strongly in contrast. Most prominent, however, is the reversal of their direction of propagation when the wave moves from one

hemisphere (in stellar azimuth) to the other (Kambe and Osaki 1988). This has, in fact, also been observed (Baade 1987, see also Baade 1984).

A similar effect cannot be expected from star spots. In high-S/N observations, traveling bumps can usually be followed from one line wing to the other. Only at the extreme limits of the profile is the contrast of the pattern too low for detection. This is sufficient to infer that putative spots must be closely confined to equatorial regions. Obscuration by the stellar disk and limb darkening then make it impossible for such spots to be strongly noticeable throughout their entire path about the stellar rotational axis.

This list is to be compared with the capabilities of single-channel photometry which basically yields only three quantities (plus their variations): period, amplitude, and shape of the light curve. It does not surprise, then, that observers using different observing techniques arrive at different conclusions.

The spectroscopic arguments presented above do not prove that nonradial pulsation is a standard property of OB-stars. However, for a fair number of individual stars their identification as nonradial pulsators is inescapable. The apparent relative uniformity of many phenomena over a wide range of various parameters such as mass, luminosity, rotation rate, etc. clearly suggests a generalization of this conclusion.

In more debatable cases simultaneous spectroscopy and multi-channel photometry will enable the ambiguity between a velocity field and the surface distribution of the line-to-continuous absorption coefficient ratio to be significantly reduced. Because photometry cannot exploit the Doppler imaging effect, such an analysis will be limited to low-degree modes ($\ell \leq 5$). When periods are determined independently and temperature variations are properly accounted for beforehand, the modeling of any residual line profile variations will yield solutions with a much higher level of confidence (cf. Lee and Saio 1990a, Lee et al. 1991).

The vector nature of the velocity fields of nonradial pulsations will in most cases still permit an inversion of the observations only by adopting an explicit model. In order to test the nonradial pulsation hypothesis this will inevitably involve spherical harmonics. However, if for every period one single stellar eigenfunction is sufficient to achieve a good fit (as much of the previous work has suggested already), it will be difficult to deny the significance of such a result.

3 The classical β Cephei stars

Historically, the spectacular line profile variability of some β Cephei stars gave the first observational impetus to study nonradial oscillations of stars (Ledoux 1951) because it was thought that this would be a way to explain the observed line doubling. Only high-definition profiles measured after the photographic era showed (Smith 1981 and references therein; Crowe and Gillet 1989) that the line doubling probably is due to atmospheric shocks caused by the highly supersonic pulsation velocities of a *radial* mode. But nonradial modes are also excited in β Cephei stars (e.g., Smith 1981).

Although new candidates have been nominated from time to time, only about two dozen stars have during the course of several decades been promoted to this exclusive class (Rountree Lesh and Aizenman 1978). The apparent narrowness, $\Delta T_{\text{eff}}/\overline{T}_{\text{eff}}$

≤ 0.1, of the instability strip between B0.5 and B2 occupied by them served as a convenient formal, but not physically understood classification criterion. The discovery of a quasi-continuum of lower-amplitude variables all around that strip has shattered this convention and obviously calls for a revision of the definition of the class of β Cephei stars.

Two suggestions have been put forward to this effect. Smith (1980) has proposed that the term β Cephei star be restricted to OB-stars which have at least one radial mode excited. Balona (e.g., 1991; see also Sterken and Jerzykiewicz 1990), on the other hand, calls an OB-star β Cephei if its observed (super-) period is too short ($\leq$0.3 day) to be due to rotational modulation. The former suggestion appears conceptually stronger because it is impossible that groups of stars overlap which do and which do not pulsate radially. Accordingly, the classification will always be unique (although, perhaps, time dependent). By contrast, it is not obvious that a star's break-up rotation period represents a similarly unambiguous limit which cannot be undercut by NRP periods. Balona's argument for his definition is that only for so short periods a confusion of spots and pulsation is exlcuded. But this ignores that there are many more discriminating criteria (see Sect. 2) and therefore unnecessarily limits the scope of the research that needs to be done in this area.

Until recently, the radial-mode criterion had the additional virtue of maintaining the narrowness of the β Cephei instability strip. However, Fullerton et al. (1991) meanwhile announced the detection of radial velocity variations in the O7 II star HD 34656 which are incompatible with the line profile-variations expected for nonradial pulsations but with 8.2 hours have a period which is too short to be orbital. Rather, this suggests pulsation in the radial fundamental mode. If confirmed, this result is remarkable as it would defy all the numerous previous searches for β Cephei stars outside their classical reserve even though very few of these surveys, if any, explored domains with so different effective temperatures.

Waelkens et al. (1992) have suggested that with increasing metallicity the blue edge of the β Cephei instability strip moves blueward. But, before speaking of an extension of the β Cephei strip or even a new instability strip, it will in the case of HD 34656 be necessary to consider the 'horizontal' alternative because of the high luminosity class, II, assigned to this star: In early-type supergiants, radial as well as nonradial pulsations are generally expected (e.g., Lovy et al. 1984) so that HD 34656 could rather fall into this category. On the other hand, an apparent single-mode pulsator like HD 34656 has never before been found in an OB-supergiant (cf. van Genderen 1991).

4 Candidate driving mechanisms

The narrowness of the classical β Cephei instability strip has been a serious problem for any attempt to identify the driving mechanism. For example, Osaki's (1974) suggestion that an oscillatory mode of the convective core resonantly couples with a global eigenmode of the star had to be dropped for just this reason: since all non-degenerate, hot stars have a convective core, this instability, if existent, should affect stars over a much wider range in effective temperature. Now, many years later, observations have in fact uncovered such a broad instability zone, and the work by Lee and Saio (1986,

1990b) has revived and substantially refined Osaki's original idea. An important component of this model is the modification of the properties of low-frequency oscillations by rapid rotation. This introduces modes with negative energy which do not exist in non-rotating stars but in fast rotators can pump energy into a positive mode so that a coupled overstable mode appears.

More recently, another driving mechanism was revived, namely the classical κ-mechanism. This mechanism had, of course, been exhaustively investigated. But all model calculations failed to excite pulsations because there was not enough driving. Only by means of an at its time rather artificial manipulation of the opacity, namely the introduction of a small 'opacity bump' in the outer layers (Stellingwerf 1978), could the κ-mechanism be forced to function also in OB-stars. However, now it has been found that at temperatures above 10^5 K numerous not previously considered iron lines increase the opacity such as to roughly provide this extra opacity. Accordingly, the κ-mechanism now is a very viable contender for the driving of nonradial pulsations in normal OB-stars (Cox et al. 1992). Although there is not (yet) an explanation of the apparently more strongly biased distribution of radial pulsators in the HR diagram, clearly the previously used argument against nonradial pulsations in OB-stars, namely that we cannot be sure whether such pulsations can be driven, is no longer valid.

5 A connection to mass loss?

In slowly rotating main-sequence B-stars, nonradial pulsation does not seem to lead to mass loss. This ought to be expected because not even the very highly supersonic amplitudes which in some β Cephei stars are associated with a radial mode (Crowe and Gillet 1989) seem to have a very significant effect of this kind (Blomme and Hensberge 1985).

With the discovery of nonradial pulsations also in Be-stars (Baade 1982, Vogt and Penrod 1983) the available negative evidence for a pulsationally driven mass loss in OB-stars was ignored for a while; partly so because of the desire to finally, after more than a century of intensive observations, understand the strongly episodic mass loss from Be-stars (e.g., Baade et al. 1988). For a while, it was believed that in λ Eri (B2 IVe) there was a correlation between pulsation amplitude and the strength of the Hα emission (Bolton 1982, Smith and Penrod 1984, Penrod 1986), and a generalization was attempted (e.g., Baade 1987a, Osaki 1986). Yet, later work (Smith 1989, Bolton and Štefl 1990) could not confirm the behavior reported for λ Eri.

It is undeniable, though, that in many Be-stars (cf. Baade 1982, 1984, 1991) and some broad-lined supergiants (Baade 1988, Baade et al. 1990) the ratio, V/R, of the violet and the red component of emission lines undergoes regular variations. Only in one star, 28 CMa (Baade 1982), has this variability been shown to be periodic with the same period as the stellar absorption profiles. In μ Cen, the period, if any, seems to be different from the stellar period. Several nights worth of high-S/N profiles of the He I 667.8 line unexpectedly suggest that the V/R variability is due to some extra emission which is very unlikely to be circumstellar in origin (Baade 1991c). This would invalidate previous notions (Smith and Penrod 1984, Baade 1987a) that the V/R variability somehow is the response of the circumstellar disk to, e.g., the local temperature variations of the underlying star.

The existence of V/R variations also in supergiants is not without interest because supergiants are not known to have circumstellar disks. On the other hand, the observed line profiles show unambiguously that the variability takes place close to the stellar equatorial plane.

In any case, the rapid V/R variability of emission lines requires a concentrated observational effort in order to permit at least a correct description of the symptoms to be given. If in rapidly rotating Be-stars and supergiants there is additional line emssion which is due to, e.g., photospheric shocks, the idea of a causal connection between pulsation and mass loss might be revived for a second time.

Another connection (it does not compete with the former) presently deserves to be taken more seriously. Because in line-driven winds the force increases with any additional outward velocity, the wind responds with extreme instabilities to any such perturbation (Owocki 1992). It seems well possible that the photospheric variability due to nonradial pulsation is sufficient to initiate such instabilities which, then, could e-fold several dozens of times (Owocki 1992 and references therein). An observational manifestation of this effect could be the discrete components observed in UV resonance lines of virtually all luminous OB-stars (Henrichs 1991).

6. Position dependence of pulsation properties in the Hertzsprung-Russell Diagram

Generally, pulsations are of significant interest only if they can be used as an independent diagnostic of the structure and evolutionary state of the stars concerned. For individual stars, this so-called asteroseismology normally requires the identification of the radial overtone of the observed modes. With the exception of the radial modes of β Cephei stars (e.g., Waelkens 1981, Shobbrook 1985; see also Baade 1986), no serious effort of this kind has been undertaken for OB-stars. Furthermore, many of the observed low-order modes have very long periods. If this identifies them as g-modes, exact determination of the radial overtone may be practically impossible as the g-mode spectrum becomes dense towards long periods (e.g., Unno et al. 1989). However, in ζ Oph (O9.5 Vne), Kambe et al. (1990 and references therein) identify also shorter-period variations which have $|m|$-values between 4 and 7 with g-modes but of low order, n. As stated in the Introduction and Section 7, g-modes are seismologically particularly rich so that the study of pulsating OB-stars bears some promise.

Because of broad, low-contrast lines, differential atmospheric expansion, other intrinsic variability, etc., present-day analyses of OB-stars occasionally suffer more strongly from general uncertainties than does the analysis of other types of stars. Therefore, any additional, pulsational diagnostic would be most welcome. For instance, for single OB-stars there is the irritating ambiguity in the direction of their evolution in the HR diagram. A first step forward could be made already if just groups of stars could be distinguished on the basis of their pulsational behavior. A very simple example of the practical value of such information is the pulsational identification of a high-Galactic latitude B-star as a little to moderately evolved object, namely a β Cephei star, rather than a post-AGB star (Waelkens and Rufener 1988). Accordingly, star formation several kiloparsec above the Galactic disk still continues at a finite level.

The following descriptions form an attempt to demarcate groups of OB-stars according to their pulsation properties. The differentiation made is only preliminary and occasionally not necessarily much more than a call for verification by more extensive observations. This applies especially to the O-stars where processes in the wind further enhance the complexity of the variability of many lines even in the visible part of the spectrum (Fullerton 1990, Fullerton et al. 1992). But as a whole, the emerging pulsational cartography of this part of the HR diagram should after some modest further iteration begin to be diagnostically useful.

- **Supergiants** generally only show semi-regular or even irregular light and radial velocity curves (van Genderen 1991 and references therein). The dominant time scales appear consistent with the radial fundamental or also nonradial modes (Lovy et al. 1984). Fourier decomposition into a number (15) of very-low amplitude components has been attempted only for the A2 Iae star α Cyg (Lucy 1976). In view of the interesting results of this analysis, it is surprising that the length and density of the series of observations of α Cygni are apparently still unsurpassed by other normal early-type supergiants.
- **Broad-lined supergiants** ($v \sin i \geq 100\,\mathrm{km\,s^{-1}}$) may show the typical signature of nonradial modes of intermediate order ($|m| \approx 8$) in their profiles (Baade and Ferlet 1984; Baade 1988, 1991b; Baade et al. 1990). The super-period, $|m| \times P_m$, is of the order of 1-2 days and, therefore, generally shorter than the presumable rotation period. Only few stars of this class have been studied in some detail. It comprises several of the hottest candidate nonradial pulsators known to date (cf. Introduction).
- **Near-main sequence stars** with narrow lines and spectral types between O8 and B5 were the first OB-stars in which low-order variability of their profiles was discovered (Smith 1977, 1981b). At least the cooler members of this spectroscopically defined class of variables seem to have their photometric counterpart in what Waelkens (1992) after "pulsating mid-B stars" now calls "slowly pulsating B stars". The most important properties of these stars are their multi-period variability and the length of their periods which in most cases are between 1 and 3 days. This combination excludes any explanation other than low-degree (low ℓ), high-order (high-n) nonradial g-mode pulsation.
- **β Cephei stars**, if defined as radial pulsators, appear to be confined to much more specific domains in T_{eff} than are other pulsating OB-stars. When β Cephei stars pulsate also nonradially, the periods of these modes are rather similar to the radial modes (4-6 hours) and, therefore, significantly shorter than the periods of low-order nonradial modes in most other OB-stars.
- **Be-stars** usually show both one low- ($|m| \approx 2$) and one or more higher-order ($|m| \approx 8$) modes. Especially the former undergo substantial amplitude variations on time scales of a year (e.g., Balona et al. 1992 and references therein). The periods of these modes are around 0.5-2 days.
- **Bn-stars** are in the Bright Star Catalog equally abundant as Be-stars except at late sub-classes where they rapidly outnumber the latter. They also exhibit a similar distribution of $v \sin i$ values but, by definition, have never been observed to display

emission lines. Pulsationally, they are distinct from Be-stars in that they show low-order pulsation modes with detectable amplitudes only rarely.
This was first found spectroscopically (Smith and Penrod 1984; Baade unpublished) but tentative confirmation may perhaps be inferred also from photometry. Cuypers et al. (1989) and Balona et al. (1992) observed a total of 30 Be-stars with tabulated spectral types earlier than B7. Of these, 27 were variable (many periodic), one was variable but possibly a peculiar case, one was perhaps variable, and only three appeared constant. Among the comparison stars, there were seven stars with spectral type B6 or earlier and a tabulated $v \sin i$ of around 200 km s^{-1} or higher. Since reports about line emission do not seem to exist, these stars ought to be Bn-stars. Of these 7 objects, five apparently were constant and only 2 were variable. The difference in the incidence of photometric variability in Bn- and Be-stars thus seems large; but the two samples are not directly comparable because of their respective nearly opposite selection biases. In any event, virtually all Bn-stars appear to undergo higher-order (higher $|m|$) nonradial mode pulsations (which are not photometrically detectable) in very much the same way as do Be- and other broad-lined stars (Smith and Penrod 1984; Baade 1987a and unpublished).

- `Maia variables` may exist (McNamara 1987) in the transition region beyond ~B7, towards the δ Scuti stars. At the suggested time scales of 0.5-2 days, multiple amplitudes of only a few mmag are very difficult to detect.
- `Pre-main sequence stars` have not been surveyed. However, in the Herbig Ae-star HD 163296 low-order line profile variations were observed which appear consistent with nonradial pulsation (Baade and Stahl 1989); the observations were insufficient for a thorough time series analysis.
- `Wolf-Rayet stars`, although not usually included among OB-stars, shall nevertheless briefly be mentioned. Radial velocity variations of emission line complexes have been reported and attributed to nonradial pulsation (Vreux 1985). It is doubtful, however, whether this can be reconciled with theory (Maeder 1985, 1986) and how the coupling between star and wind could be arranged in order to have the reported effect. In optical absorption troughs of some WNL stars, discrete components have been observed (Stahl et al. 1988) which clearly arise in the wind.
- δ `Scuti stars` provide for an enlightening comparison because they are indisputed as pulsating variables (Breger 1979, Smith 1981, Yang 1991) and in their behavior share many of the properties described above for OB-stars.

A curious conclusion from the above list is that the presently known (super-)periods of nonradial pulsations seem to depend relatively little on temperature or luminosity or other parameters and mostly fall into the range of 1 to 3 days. Perhaps the most deviating group are the β Cephei stars where also the periods of nonradial modes are in the domain of radial modes.

7 The Be/Bn-star double-dichotomy

The differences and similarities between Be- and Bn-stars have survived every and all attempts of their explanation. It is still unknown whether Be- and Bn-stars represent alternative or sequential stages in the evolution of rapidly rotating B-stars. The more recent discovery also of pulsational differences (see Sect. 6) has not so far made this task any easier. But the Be/Bn problem can now serve as an illustration of how different pulsation properties of groups of stars might be used to improve the understanding of the other differentiations between the groups. Since it has proven difficult to deduce the solution analytically, observations need to supply more facts. The planning of such observations should of course be guided by theoretically inspired guesses. One set of such considerations is outlined in this section.

In series of quasi-static stellar models, rapid rotation only plays a relatively minor role. However, the evolutionary path changes significantly if the rotationally induced enhanced mixing is included (Maeder 1987). The stronger mixing and quasi-homogeneous evolution are the result of a baroclinic instability which develops because in a rapidly rotating star surfaces of constant pressure and entropy intersect each other at a small but finite angle. The additional mixing permits a star to burn a larger fraction of its original hydrogen supply. This has led to the suggestion (Maeder 1987) that blue stragglers are the descendants of initially very rapidly rotating stars. CNO surface abundance anomalies are predicted by this model and seem to have found tentative confirmation from the atmospheric analysis of some broad-lined stars (Schönberner et al. 1988). If more firmly established, this would constitute a connection between rapid rotation and chemical structure.

An alternative explanation of the statistical identity of observed periods and inferred surface rotation rates in Be-stars (Balona 1990) is that the variability is due to g-modes. In Saio and Lee's (1991) model, the observed periods of g-modes reflect the rotation rate of the stellar core. The available observations would, therefore, imply that core and surface rotation rates are the same, i.e. Be-stars rotate as solid bodies. As Balona (1990) points out, it is not obvious how rigid rotation can be maintained against the contrary effect of non-homologous evolution. More important in the present context is that rigid-body rotation would disqualify Be-stars as objects which currently undergo homogeneous evolution as in Maeder's (1987) calculations where it results from the ABCD instability which requires differential rotation. However, at an earlier stage, the precursors of Be-stars may, of course, have had a different rotational profile (see also below).

Contrary to Lee and Saio's model, the κ-mechanism would drive nonradial pulsations regardless of whether or not the periods match the time scales of convection or rotation of the core. Since Cox et al. (1992) find the periods of κ-mechanism driven g-modes to depend primarily on the structure, i.e. mainly the hydrogen abundance gradient, of the semiconvection zone, the observed periods may be carriers of even more important information. (However, the problem to explain the apparent correlation between rotation and pulsation periods would regain its full size.) Accordingly, Be-stars could actually be in a phase of homogeneous evolution. They do not rotate quite as rapidly as assumed in Maeder's calculations; but if this is a problem, homogeneous evolution will hardly be realized in any actual star.

Besides this effect of rotation on evolution, there are connections also between non-radial pulsation and both chemical abundances and rotation. Pulsation and abundances are, of course, linked by the opacity (e.g., Cox et al. 1992). Concerning rotation, it has been shown (Ando 1982) that nonaxisymmetric nonradial pulsation modes transport angular momentum between different layers of the star. In a rapidly rotating star, a change in the rotation profile might affect internal mixing processes and thereby, as seen above, alter the evolutionary track.

Putting these three factors and their mutual connections together, a number of conjectures can be formulated. The compilation given below of such considerations is *not* intended to form a logical, or even physical, sequence. Rather, from these and perhaps other building blocks hypotheses can be assembled which are not theoretically unreasonable but can be subjected to observational tests:

- A rapidly rotating star initially evolves homogeneously and during this phase enriches its envelope with CNO-processed material.
- For different envelope abundances, different nonradial modes are driven. (In κ-mechanism models such as the one of Cox et al. [1992], the driving is provided by a narrow zone not far below the surface.)
- Homogeneous evolution terminates because mass loss brakes the rapid rotation.
- Homogeneous evolution ends/resumes when nonradial pulsation has distorted/restored the rotational profile.
- In the presence of rapid rotation, nonradial pulsation enables episodes of strongly enhanced mass loss: an Oe/Be-star forms.

Depending on how these statements are combined to a working hypothesis, Be-stars can be the ancestors or the descendants of Bn-stars or the two form parallel branches in the evolution of OB-stars. In order to observationally discriminate between alternatives, observations should explore also new territory. One such area which in rapidly rotating stars has been exploited only to a very modest extent is the atmospheric abundance analysis. However, the technical quality of the observations and the capabilities of the atmospheric analysis techniques have considerably matured. A strictly differential abundance analysis of Be- and Bn-stars can therefore be undertaken which is accurate enough to strongly constrain evolutionary tracks.

8 Conclusions

Spectroscopic and photometric variability is so widespread among OB-stars that in some regions in the HR diagram, e.g. around B1-B3, non-variable stars may be difficult to find. Often the variations follow complex but common patterns which are most easily understood as nonradial pulsation. In fact, there is only one group of intrinsically variable OB-stars, namely the helium variable stars (Landstreet 1992 and references therein), where it is clear that nonradial pulsation cannot account for the observations. In O- and other very luminous stars, the absorption spectrum forms in an expanding atmosphere. In many spectral lines the signature imposed by outward moving wind inhomogeneities (Fullerton 1990, Fullerton et al. 1992) will mask possible other variations.

An important group which still has to be checked are the Bp stars. Their chemical abundance pecularities are a surface phenomenon which is probably due to diffusion. But the slow velocity of diffusion processes may not easily compete with much larger pulsation amplitudes, as is also suggested by the generally low rotation velocities of Bp-stars. Therefore, one would not expect to detect long-period pulsations in Bp-stars. But in view of the narrow spectral lines an observational test for low-order nonradial pulsation can easily be done. A negative result would give rise to studies why Bp-stars could avoid pulsation. Slow rotation, magnetic fields, and late spectral types could be among the properties that may quench pulsation.

Models designed to compete with nonradial pulsation will have to address more facets of the observations than merely the period. Conversely, the alleged adaptiveness of the nonradial pulsation model remains a prejudice until it has been demonstrated that the variability associated with one period requires more than one stellar eigenfunction (plus atmospheric response, etc.) for its explanation.

The attractive potential of nonradial pulsations lies in their coupling with evolution, composition, structure, rotation, mass loss, etc. However, the reality of such connections still needs to be scrutinized, at least observationally. It is important that this is done in a broad, comparative approach which pays equal attention to groups of 'normal' as well as of more special objects (which may become normal, too, after they are better understood). More accurate abundance analyses should form part of this effort.

References

Ando, H. (1982): *Astron. Astrophys.* **108**, 7

Ando, H. (1986): *Astron. Astrophys.* **163**, 97

Baade, D. (1982): *Astron. Astrophys.* **105**, 65

Baade, D. (1984): *Astron. Astrophys.* **134**, 105

Baade, D. (1986): In *Highlights of Astronomy*, Vol. 7, ed. J.P. Swings, D. Reidel, Dordrecht, p. 255

Baade, D. (1987a): In Proceedings IAU Coll. No. 92 *Physics of Be Stars*, eds. A. Slettebak and T.P. Snow, D. Reidel, Dordrecht, p. 361

Baade, D. (1987b): In Proceedings IAU Symp. No. 132 *The Impact of very High S/N on Stell. Physics*, eds. G. Cayrel de Strobel and M. Spite, D. Reidel, Dordrecht, p. 193

Baade, D. (1987c): In Proceedings IAU Symp. No. 132 *The Impact of very High S/N on Stellar Physics*, eds. G. Cayrel de Strobel and M. Spite, D. Reidel, Dordrecht, p. 217

Baade, D. (1987d): *IAU Inform. Bull. Var. Stars* No. 3124

Baade, D. (1988): In *O, Of and Wolf-Rayet Stars*, eds. P.S. Conti and A.B. Underhill, NASA/-CNRS *Monograph Series on Nonthermal Phenomena in Stellar Atmospheres*, NASA SP-497, p. 137

Baade, D. (1989a): *Astron. Astrophys.* **79**, 423

Baade, D. (1989b): *Astron. Astrophys.* **222**, 200

Baade, D. (ed., 1991a): *Rapid Variability of OB-Stars: Nature and Diagnostic Value*, ESO Conference and Workshop Proceedings No. 36, European Southern Observatory, Garching

Baade, D. (1991b): In *Rapid Variability of OB Stars: Nature and Diagnostic Value*, ed. D. Baade, ESO Conference and Workshop Proceedings No. 36, ESO, Garching, p. 21

Baade, D. (1991c): In *Rapid Variability of OB Stars: Nature and Diagnostic Value*, ed. D. Baade, ESO Conference and Workshop Proceedings No. 36, ESO, Garching, p. 217.

Baade, D., Dachs, J., v.d. Weygaert, R., Steeman, F. (1988): *Astron. Astrophys.* **198**, 211
Baade, D., Ferlet, R. (1984): *Astron. Astrophys.* **140**, 72
Baade, D., Schmutz, W., v. Kerkwijk, M. (1990): *Astron. Astrophys.* **240**, 105
Baade, D., Stahl, O. (1989): *Astron. Astrophys.* **209**, 268
Balona, L.A. (1990): *Mon. Not. R. astr. Soc.* **245**, 92
Balona, L.A. (1991): In *Rapid Variab. of OB Stars: Nature and Diagnostic Value*, ed. D. Baade, ESO Conference and Workshop Proceedings No. 36, ESO, Garching, p. 249
Balona, L.A., Cuypers, J., Marang, F. (1992): *Astron. Astrophys. Suppl. Ser.*, in press
Blomme, R., Hensberge, H. (1985): *Astron. Astrophys.* **148**, 97
Bolton, C.T. (1982): In IAU Symp. No. 98 *Be Stars*, eds. M. Jaschek and H.-G. Groth, D. Reidel, Dordrecht, p. 181
Bolton, C.T., Štefl, S. (1990): In *Angular Momentum and Mass Loss for Hot Stars*, eds. L.A. Willson and R. Stalio, Kluwer, Dordrecht, p. 185
Breger, M. (1979): *Publ. Astron. Soc. Pacific* **91**, 5
Cox, A.N., Morgan, S.B., Rogers, F.J., Iglesias, C.A. (1992): *Astrophys. J.*, in press
Crowe, R., Gillet, D. (1989): *Astron. Astrophys.* **211**, 365
Cuypers, J., Balona, L.A., Marang, F. (1989): *Astron. Astrophys. Suppl. Ser.* **81**, 151
Cuypers, J. (1991): In *Rapid Variability of OB Stars: Nature and Diagnostic Value*, ed. D. Baade, ESO Conference and Workshop Proceedings No. 36, ESO, Garching, p. 83
Fullerton, A.W. (1990): Ph.D. thesis, Univ. of Toronto, Toronto
Fullerton, A.W. (1991): In *Rapid Variab. of OB Stars: Nature and Diagnostic Value*, ed. D. Baade, ESO Conference and Workshop Proceedings No. 36, ESO, Garching, p. 3
Fullerton, A.W., Gies, D.R., Bolton, C.T. (1991): *Astrophys. J. Letters* **368**, L35
Fullerton, A.W., Gies, D.R., Bolton, C.T. (1992): *Astrophys. J..*, submitted
van Genderen, A.M. (1991): In *Rapid Variability of OB Stars: Nature and Diagnostic Value*, ed. D. Baade, ESO Conference and Workshop Proceedings No. 36, ESO, Garching, p. 117
Gies, D.R., Kullavanijaya, A. (1988): *Astrophys. J.* **326**, 813
Harmanec, P. (1984): *Bull. Astron. Inst. Czechoslov.* **35**, 193
Harmanec, P. (1987): *IAU Inf. Bull. Var. Stars* No. 3097
Harmanec, P. (1991): In *Rapid Variability of OB Stars: Nature and Diagnostic Value*, ed. D. Baade, ESO Conference and Workshop Proceedings No. 36, ESO, Garching, p. 265
Henrichs, H.F. (1991): In *Rapid Variability of OB Stars: Nature and Diagnostic Value*, ed. D. Baade, ESO Conference and Workshop Proceedings No. 36, ESO, Garching, p. 199
Kambe, E., Osaki, Y. (1988): *Publ. Astron. Soc. Japan* **40**, 313
Kambe, E., Ando, H., Hirata, R. (1990): In *Progress of Seismology of the Sun and Stars*, eds. Y. Osaki and H. Shibahashi, Springer, Berlin, p. 449
Landstreet, J.D. (1992): these proceedings
Ledoux, P. (1951): *Astrophys. J.* **114**, 373
Lee, U., Jeffery, C.S., Saio, H. (1991): In *Rapid Variability of OB Stars: Nature and Diagnostic Value*, ed. D. Baade, ESO Conference and Workshop Proceedings No. 36, ESO, Garching, p. 245
Lee, U., Saio, H. (1986): *Mon. Not. R. astr. Soc.* **221**, 365
Lee, U., Saio, H. (1990a): *Astrophys. J.* **349**, 570
Lee, U., Saio, H. (1990b): *Astrophys. J.* **360**, 590
Lovy, D., Maeder, A., Noels, A., Gabriel, M. (1984): *Astron. Astrophys.* **133**, 307
Lucy, L.B. (1976): *Astrophys. J.* **206**, 499
Maeder, A. (1985): *Astron. Astrophys.* **147**, 300
McNamara, B.J. (1987): *Astrophys. J.* **312**, 778
Maeder, A. (1986): In *Highlights of Astronomy*, Vol. 7, ed. J.P. Swings, D. Reidel, Dordrecht, p. 273

Maeder, A. (1987): *Astron. Astrophys.* **178**, 159
Owocki, S. (1992): these proceedings
Osaki, Y. (1974): *Astrophys. J.* **189**, 469
Osaki, Y. (1986): *Publ. Astron. Soc. Pacific* **98**, 31
Penrod, G.D. (1986): *Publ. Astron. Soc. Pacific* **98**, 35
Rountree Lesh, J., Aizenman, M. (1978): *Ann. Rev. Astron. Astrophys.* Vol. 16, p. 215
Saio, H., Lee, U. (1990): In *Progress of Seismology of the Sun and Stars*, eds. Y. Osaki and H. Shibahashi, Springer, Berlin, p. 453
Saio, H., Lee, U. (1991): In *Rapid Variability of OB Stars: Nature and Diagnostic Value*, ed. D. Baade, ESO Conference and Workshop Proceedings No. 36, ESO, Garching, p. 293
Schönberner, D., Herrero, A., Becker, S., Eber, F., Butler, K., Kudritzki, R.-P., Simon, K.-P. (1988): *Astron. Astrophys.* **197**, 209
Shobbrook, R.R, (1985): *Mon. Not. R. astr. Soc.* **214**, 33
Smith, M.A. (1977): *Astrophys. J.* **215**, 574
Smith, M.A. (1980): *Astrophys. J.* **240**, 149
Smith, M.A. (1981a): *Astrophys. J.* **248**, 214
Smith, M.A. (1981b): In *Workshop on Pulsating B Stars*, ed. G.E.V.O.N. and C. Sterken, Obs. de Nice, Nice, p. 317
Smith, M.A. (1986): In *Hydrodynamic and Magnetohydrodynamic Problems in the Sun and Stars*, ed. Y. Osaki, Univ. of Tokyo, Tokyo, p. 145
Smith, M.A. (1989): *Astrophys. J. Suppl. Ser.* **71**, 357
Smith, M.A., Penrod, G.D. (1984): In *Relation Between Chromospheric-coronal Heating and Mass Loss*, eds. R. Stalio and J.B. Zirker, Trieste Observ., Trieste, p. 394
Stahl, O., Vreux, J.-M., Baade, D., Magain, P. (1988): unpublished
Stellingwerf, R.F. (1978): *Astron. J.* **83**, 1184
Sterken, C., Jerzykiewicz, M. (1990): In *Confrontation Between Stellar Pulsation and Evolution*, eds. C. Cacciari and G. Clementini, Astron. Soc. Pacific Conference Series Vol. 11, p. 236
Sterken, C., Jerzykiewicz, M. (1991): In *Rapid Variability of OB Stars: Nature and Diagnostic Val.*, ed. D. Baade, ESO Conference Proceedings No. 36, ESO, Garching, p. 105
Unno, W., Osaki, Y., Ando, H., Saio, H., Shibahashi, H. (1989): *Nonradial Oscillations of Stars*, Univ. of Tokyo Press, Tokyo
Vogt, S.S., Penrod, G.D. (1983): *Astrophys. J.* **275**, 661
Vreux, J.-M. (1985): *Publ. Astron. Soc. Pacific* **97**, 274
Waelkens, C. (1981): *Astron. Astrophys.* **97**, 274
Waelkens, C. (1992): *Astron. Astrophys.*, in press
Waelkens, C., Van den Abele, K., Van Winckel, H. (1991): *Astron. Astrophys.*, in press
Waelkens, C., Rufener, F. (1988): *Astron. Astrophys.* **201**, L5.
Walker, G.A.H. (1991): In *Rapid Variability of OB Stars: Nature and Diagnostic Value*, ed. D. Baade, ESO Conference and Workshop Proceedings No. 36, ESO, Garching, p. 27
Yang, S.S. (1991): In *Rapid Variability of OB Stars: Nature and Diagnostic Value*, ed. D. Baade, ESO Conference and Workshop Proceedings No. 36, ESO, Garching, p. 129

Analysis of line profile variations in pulsating B stars.

C.Waelkens and C.Aerts
Astronomisch Instituut,Celestijnenlaan 200 B, 3001 Leuven, Belgium

1 Introduction

The advent of modern high-resolution spectrographs equipped with sensitive detectors has revealed an extremely rich pattern of variability of the line profiles in early-type stars. A recent review of these observations is given in the 1990 ESO workshop on rapid variability in OB stars (Baade 1990). Surprisingly enough, however, it appears that no agreement exists on even the basic interpretation scheme of this variability in several stars. Competing models involve nonradial pulsation (NRP), rotating spots, corotating circumstellar features, binarity.

Part of the confusion is certainly due to the intrinsic complexity of the observed variations and the practical difficulties for getting extensive data sets. A second problem is the mathematical richness of the most popular model, the NRP-hypothesis, which may involve so many parameters that it virtually does not have any predictive power, especially if multiple modes and variable amplitudes are involved.

Fairly few studies so far have succeeded in expressing the richness of line profile variations of nonradially pulsating stars in a quantitative way, a notable exception being the study of ε Per by Gies and Kullavanijaya (1988). Below we describe how a quantitative determination of pulsation mode parameters can be achieved by combining the analysis of line profile variations with Balona's moment method and that of photometric amplitudes. We discuss how the method yields unambiguous results for some β Cephei variables.

2 Balona's moment method

A line profile $f(v)$, expressed as a function of the velocity in the direction of the observer, is characterized by its moments defined as

$$< v^n > \equiv \frac{\int_{-\infty}^{+\infty} v^n f(v)\, dv}{\int_{-\infty}^{+\infty} f(v)\, dv}. \tag{1}$$

A complete characterization of course involves all moments, but one may hope that a reasonably complete mode identification is possible from the first three moments. For a star pulsating with a single pulsation frequency ω, the following theoretical expressions are obtained:

$$\left\{ \begin{array}{lcl} < v > & = & v_p A(l,m,i) \sin(\omega t) \\ < v^2 > & = & v_p^2 B(l,m,i) \sin(2\omega t) + v_p v_\Omega D(l,m,i) \sin(\omega t) + v_p^2 I(l,m,i) + b_2 v_\Omega^2 + \sigma^2 \\ < v^3 > & = & v_p^3 C(l,m,i) \sin(3\omega t) + v_p^2 v_\Omega E(l,m,i) \sin(2\omega t) + \\ & & [v_p^3 F(l,m,i) + v_p v_\Omega^2 G(l,m,i) + v_p \sigma^2 H(l,m,i)] \sin(\omega t). \end{array} \right. \tag{2}$$

The full expressions for the coefficients are given by Aerts (1992, this volume).

The various coefficients can then be compared with the observed ones. In practice, one determines the pulsation frequency ω from the variations of the first moment, and then determines multifrequency fits to the three moments. A quantitative criterion for identifying the parameters (l, m, i) is then defined by the requirement that best agreement must be found between the observations and the theoretical expressions. This is achieved through the construction of a discriminant. Various alternatives of this discriminant can be devised (Aerts, this volume). We consider here the discriminant $\gamma_{lm}(i)$, which involves only the leading terms in the expressions (2) for the moments.

Having determined the pulsation parameters and the inclination, one can then estimate the rotation velocity v_Ω and the width σ of the intrinsic profile from the other terms in the expressions (2) for the moments.

The method is easily generalized for multiperiodic stars. The frequencies can then again be determined from the variations of the first moment. However, the complexity of the expressions for the higher moments increases very fast with increasing number of frequencies, since, the velocity being a vectorial quantity, cross terms between the various components are important.

3 Application to selected β Cephei stars

The β Cephei stars are slightly evolved early-B stars that since long are known to be regular pulsators. Monoperiodic as well as multiperiodic pulsation occurs, the latter necessarily in at least one nonradial mode. These stars then constitute ideal test cases for methods for mode identification. Extensive photometric studies of the β Cephei stars exist: we rely on these studies as well for the period determination as for the independant estimate of the pulsation parameter l from comparison of the amplitudes in different colors (Heynderickx 1991).

We show in Figure 1 the variation of the three moments and the discriminant $\gamma_{lm}(i)$ for the monoperiodic variable δ Ceti. A minimum is found for a radial mode, in agreement with photometric observations. We point out, however, that already for this simplest case the agreement between theory and observation is not perfect. Theoretical expectation is that the second moment for a radial mode only involves a term in 2ω, while it can be seen on the figure that also a small contribution with ω occurs.

An interesting case is the star KK Velorum, the photometry of which suggests monoperiodic pulsations with a period of 0.216 days. The absence of any detectable color variations in this star suggests a mode with $l = 4$ or more (Heynderickx 1991). The minimum for the discriminant $\gamma_{lm}(i)$ is found for $(l, m) = (4, 0)$. However, as is apparent from Figure 2, the first moment of the line profiles is dominated by a term in 2ω, which can only occur if a second pulsation mode with this frequency occurs. A careful analysis of the photometry reveals that the double frequency is also significant there, with an appreciable color variation. A multiperiodic solution is then possible where 2ω corresponds to a radial or an $l = 1$ mode, and ω to a mode with $(l, m) = (4, 2)$.

Finally, we have applied the method to the multiperiodic β Cephei star ν Eridani. A difficulty arises with this and most multiperiodic β Cephei stars in the sense that long beat periods are involved, so that a proper deconvolution of all modes is impossible with a limited dataset, which is the rule rather than the exception for line profile observations. Furthermore, the Fourier decomposition of the higher moments clearly reveals that much power is contained in the interaction terms, which render the constructed discriminant fairly inappropriate. It is nevertheless possible to assign a radial mode to the principal oscillation, in agreement with the photometry, and to attribute the remaining variations to $l = 1$ modes. However, several almost equivalent spectroscopic solutions are possible, and the combination of both line profiles and photometry is necessary for the final mode identification.

4 Evaluation

This study of the line profile variations of some β Cephei stars shows that a quantitative determination of the pulsation parameters of early-type stars from such observations is possible, but also that it is by no means trivial. Already for the simplest case, a monoperiodic radial pulsator, the agreement between theory and observation is not perfect. For multiperiodic stars, the cross-terms in the higher moments imply that many observations are needed, well distributed over the various beat periods. Such beat periods tend to be long in β Cephei stars, as they probably are in most pulsating B stars, and so the practical possibility of unambiguous mode identification from spectroscopic observations is most often remote. An interesting exception is the β Cephei star HD 147985, for which three fairly-large-amplitude modes with short beat periods occur (Waelkens and Cuypers 1985).

It seems then that real progress can only be made if all available information is fully exploited. We therefore strongly plead for a combined photometric and spectroscopic approach. The method outlined here may also be applied for the 53-Persei stars or "slowly pulsating B stars" (Waelkens 1991), for which the photometric observations yield the periods and suggest that $l = 2$ modes are involved (Watson 1988; Heynderickx 1991). However, the very long beat periods in these stars practically exclude full spectroscopic coverage.

The situation is virtually hopeless, however, for rapid rotators such as broad-lined variables and Be stars. A major problem is of course that the theoretical assignment of a single value of l to pulsation modes in rapid rotators breaks down (e.g.Lee and Saio 1987). Moreover, both photometry and the moment method are not well suited for treating modes with high $l-$values as are often claimed for such stars. If, however, the claims that an $l = 2$ mode occurs in Be stars (Penrod 1986, Baade 1988) make sense, the appreciable color variations that should be associated with such modes should be observable.

5 References

Baade, D., 1988, *IAU Symposium 132: The Impact of Very High S/N Spectroscopy on Stellar Physics*,G.Cayrel de Strobel and M.Spite,edrs., 217

Baade, D., 1990, edr., *ESO Workshop on Rapid variability of OB-stars: nature and diagnostic value*

Gies, D.R., Kullavanijaya, A., 1988, *Astrophys. J.* **326**, 813

Heynderickx, D., 1991, Ph D Thesis KU Leuven

Lee, U., Saio, H., 1987, *Mon. Not. R. astr. Soc.* **224**, 513

Penrod, G.D., 1986, *Publ. Astron. Soc. Pacific* **98**, 35

Waelkens, C., 1991, *Astron. Astrophys.* **246**, 453

Waelkens, C., Cuypers, J., 1985, *Astron. Astrophys.* **152**, 15

Watson, R.D., 1988, *Astrophys. Space Sci.* **140**, 255

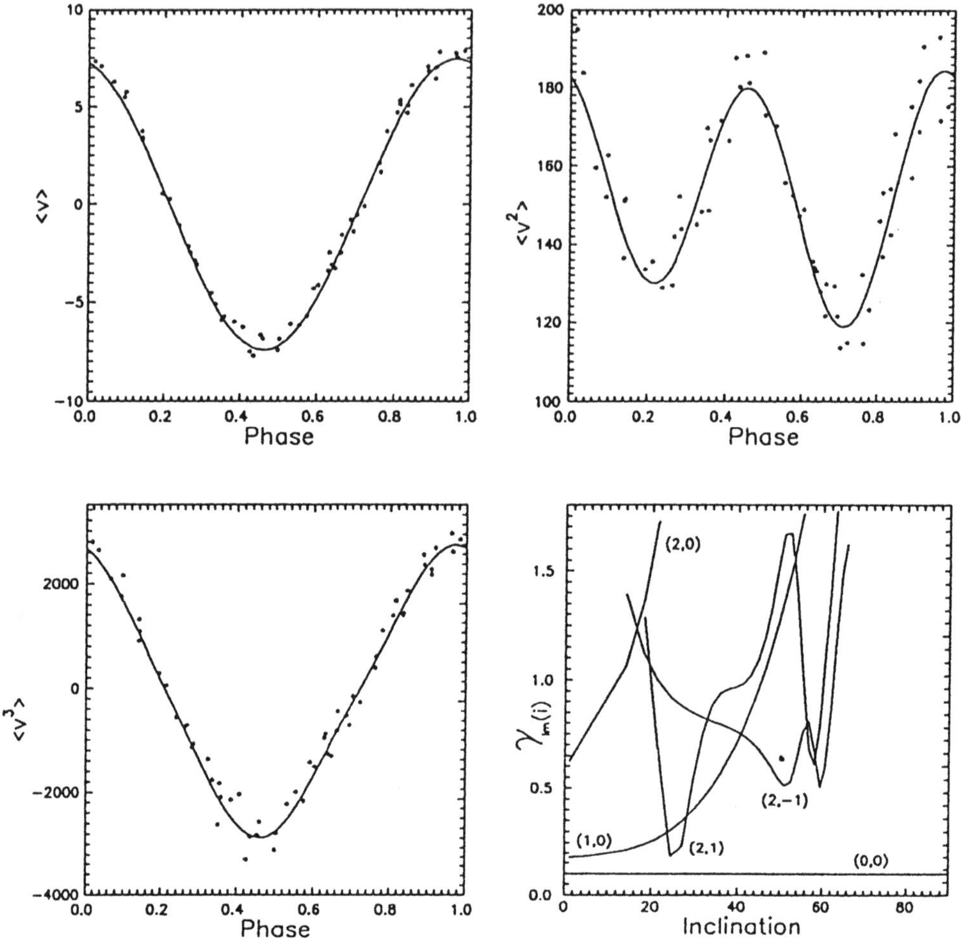

Figure 1: $< v >, < v^2 >, < v^3 >$ and the discriminant for the β Cephei star δ Ceti

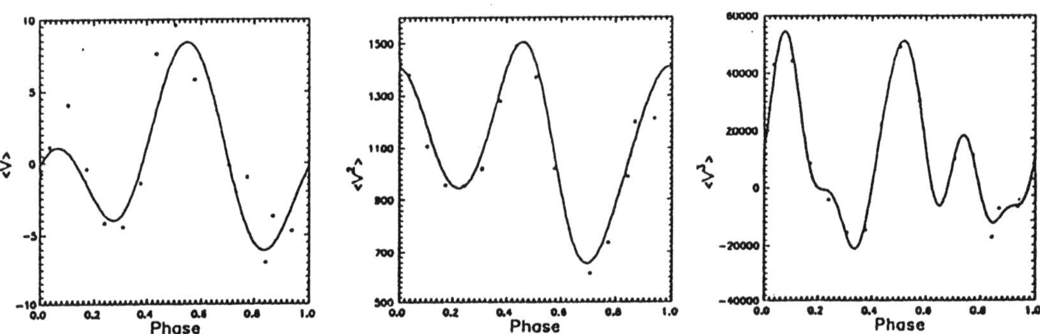

Figure 2: $< v >, < v^2 >, < v^3 >$ for the β Cephei star KK Velorum

Mode identification of pulsating stars from line profile variations with the moment method.

C.Aerts
Astronomisch Instituut,Celestijnenlaan 200 B, 3001 Leuven, Belgium

1 Introduction

The velocity field generated in the photosphere of the star by a nonradial pulsation leads to periodic variations in the profiles of spectral lines. Thus, high resolution digital spectroscopy makes a study of line profiles eminently suitable as a means of mode identification. Usually, one relies on trial and error fitting techniques to model observed profiles, a procedure of which it is not clear that it leads to a unique solution. Moreover, there is no simple way of taking into account the temperature variations which can significantly affect the profile. To overcome these problems, Balona developed a method which is based on the time variations of the moments of line profiles. We extend this method and discuss some different possibilities for the construction of a discriminant. Waelkens and Aerts(1992, this volume) discuss some results of the method for mono- and multiperiodic β Cephei stars.

2 Definition and calculation of the moments

We characterize a line profile $(f * g)(v)$ by its first few $(n = 1, 2, 3)$ moments

$$< v^n > \equiv \frac{\int_{-\infty}^{+\infty} v^n f(v) * g(v)\, dv}{\int_{-\infty}^{+\infty} f(v) * g(v)\, dv}, \tag{1}$$

where v is the velocity at a point on the surface of the star in the direction of the observer, $f(v)$ is the velocity field and $g(v)$ is the intrinsic profile (assumed to be a Gaussian). For a star pulsating with a single frequency ω, the $n^{\rm th}$ moment will vary with frequencies as high as $n \cdot \omega$:

$$\left\{ \begin{array}{lcl} < v > &=& v_p A(l,m,i) \sin(\omega t) \\ < v^2 > &=& v_p^2 B(l,m,i) \sin(2\omega t) + v_p v_\Omega D(l,m,i) \sin(\omega t) + v_p^2 I(l,m,i) + b_2 v_\Omega^2 + \sigma^2 \\ < v^3 > &=& v_p^3 C(l,m,i) \sin(3\omega t) + v_p^2 v_\Omega E(l,m,i) \sin(2\omega t) + \\ & & \left[v_p^3 F(l,m,i) + v_p v_\Omega^2 G(l,m,i) + v_p \sigma^2 H(l,m,i) \right] \sin(\omega t) \end{array} \right. \tag{2}$$

where l and m are the degree and the azimuthal number of the pulsation mode
v_p is the amplitude of the pulsation velocity
$v_\Omega \equiv \Omega R \sin i$ is the projected rotational velocity
i is the inclination angle of the star, positively oriented from the

direction of the observer to the rotation axis $\vec{\Omega}(i \in [0°, 360°[)$

σ is the width of the intrinsic profile

b_2 is a constant depending on the limb-darkening coefficient

and $A, B, C, D, E, F, G, H, I$ can be obtained as a function of l, m, i by the integrations in (1) giving:

$$
\left\{
\begin{aligned}
A(l,m,i) &\equiv \frac{2N_l^m}{1+\frac{2}{3}\beta} a_{l,m,0}(J_{l,0}^0 + KH_{l,0}^0) \\
B(l,m,i) &\equiv \frac{2N_l^m}{1+\frac{2}{3}\beta} (a_{l,m,1} + a_{l,m,-1})(J_{l,1}^1 + KH_{l,1}^1) \\
D(l,m,i) &\equiv \frac{(N_l^m)^2}{1+\frac{2}{3}\beta} \sum_{k=-l}^{l} (-1)^k a_{l,m,k} a_{l,m,-k} \left[(1+K(l+1))^2 \right. \\
&\quad \times (B_{l,l}^{3,k} + \beta B_{l,l}^{4,k}) + (K(l+1))^2 (B_{l+1,l+1}^{1,k} + \beta B_{l+1,l+1}^{2,k}) \\
&\quad \left. -2K(l+1)(1+K(l+1))(B_{l,l+1}^{2,k} + \beta B_{l,l+1}^{3,k}) \right] \\
I(l,m,i) &\equiv \frac{(N_l^m)^2}{1+\frac{2}{3}\beta} \sum_{k=-l}^{l} a_{l,m,k}^2 \left[(1+K(l+1))^2 (B_{l,l}^{3,k} + \beta B_{l,l}^{4,k}) \right. \\
&\quad -2K(l+1)(1+K(l+1))(B_{l,l+1}^{2,k} + \beta B_{l,l+1}^{3,k}) \\
&\quad \left. +(K(l+1))^2 (B_{l+1,l+1}^{1,k} + \beta B_{l+1,l+1}^{2,k}) \right] \\
C(l,m,i) &\equiv \frac{(N_l^m)^3}{1+\frac{2}{3}\beta} \sum_{k=-l}^{l} \sum_{r=-l}^{l} (-1)^{k+r+1} a_{l,m,k} a_{l,m,r} a_{l,m,-k-r} I_{l,l,l}^{k,r,-k-r} \\
E(l,m,i) &\equiv \frac{(N_l^m)^2}{1+\frac{2}{3}\beta} \sum_{k=-l}^{l} a_{l,m,k} \left(a_{l,m,-k-1} L_{l,l}^{k,-k-1} + a_{l,m,-k+1} L_{l,l}^{k,-k+1} \right) \\
F(l,m,i) &\equiv \frac{(N_l^m)^3}{1+\frac{2}{3}\beta} \sum_{k=-l}^{l} \sum_{r=-l}^{l} a_{l,m,k} a_{l,m,r} \left(a_{l,m,k+r} I_{l,l,l}^{k,r,k+r} \right. \\
&\quad \left. + a_{l,m,r-k} I_{l,l,l}^{k,r,r-k} + a_{l,m,k-r} I_{l,l,l}^{k,r,k-r} \right) \\
G(l,m,i) &\equiv \frac{3N_l^m}{1+\frac{2}{3}\beta} \left[a_{l,m,0} \left(J_{l,0}^2 + KH_{l,0}^2 \right) \right. \\
&\quad \left. -\frac{1}{2} (a_{l,m,2} + a_{l,m,-2}) \left(J_{l,2}^2 + KH_{l,2}^2 \right) \right] \\
H(l,m,i) &\equiv \frac{2N_l^m}{1+\frac{2}{3}\beta} a_{l,m,0} \left(J_{l,0}^0 + KH_{l,0}^0 \right),
\end{aligned}
\right.
\tag{3}
$$

with

$$J_{l,k}^n \equiv \int_0^1 (\mu^2 + \beta\mu^3)(1-\mu^2)^{n/2} P_l^k d\mu \tag{4}$$

and

$$H_{l,k}^n \equiv (l+1) \int_0^1 (1-\mu^2)^{n/2} [(\mu^2 + \beta\mu^3) P_l^k - (\mu + \beta\mu^2) P_{l+1}^k] d\mu \tag{5}$$

(Balona 1986b),

$$B_{l,n}^{k,m} \equiv \int_0^1 \mu^n P_l^k P_n^m d\mu \tag{6}$$

(Balona 1989),

$$L_{l,n}^{k,m} \equiv \int_0^1 (\mu+\beta\mu^2)(1-\mu^2)^{1/2}[\mu P_l^k + K(l+1)(\mu P_l^k - P_{l+1}^k)] \times[\mu P_n^m + K(n+1)(\mu P_n^m - P_{n+1}^m)]d\mu \tag{7}$$

and

$$I_{l,n,q}^{k,m,p} \equiv \int_0^1 (\mu+\beta\mu^2)[\mu P_l^k + K(l+1)(\mu P_l^k - P_{l+1}^k)] \times[\mu P_n^m + K(n+1)(\mu P_n^m - P_{n+1}^m)][\mu P_q^p + K(q+1)(\mu P_q^p - P_{q+1}^p)]d\mu \tag{8}$$

(Aerts, in preparation).

3 Construction of some possible discriminants

We compare the Fourier components derived from the measurements with the theoretically calculated Fourier components of the first, second and third moments. It follows from (2) that the terms varying with $\omega, 2\omega, 3\omega$ in $< v >, < v^2 >, < v^3 >$ respectively depend only on the pulsation velocity.

3.1 The two dimensional discriminant

We solve $v_p^{(1)}, v_p^{(2)}, v_p^{(3)}$ from the three terms discussed above and compute the associated average value $< v_p >$. We then define the discriminant as the function

$$\gamma_{lm}(i) \equiv \sqrt{\frac{1}{6}\left[(v_p^{(1)} - < v_p >)^2 + (v_p^{(2)} - < v_p >)^2 + (v_p^{(3)} - < v_p >)^2\right]}. \tag{9}$$

The discriminant then gives us the most probable values for (l, m, i) as those for which a minimum is found for $\gamma_{lm}(i)$. Examples are shown in Waelkens and Aerts(1992, this volume). The disadvantage of such a discriminant is that we assume an average of $< v_p >$ which is not necessarily correct, because the equations (2) are not perfectly fulfilled by the observations.

3.2 The three dimensional discriminant

To overcome the disadvantage of γ,we construct a more general discriminant in the following way: let AA, BB, CC be the observed amplitudes of the three terms discussed. The best set of (l, m) is the one which satisfies the equations

$$\left\{\begin{array}{l} AA - v_p A(l,m,i) = 0 \\ BB - v_p^2 B(l,m,i) = 0 \\ CC - v_p^3 C(l,m,i) = 0 \end{array}\right. \tag{10}$$

most accurately. We thus minimize the function

$$\Gamma_{lm}(v_p, i) \equiv \sqrt{|AA - v_p A(l,m,i)|^2 + |BB - v_p^2 B(l,m,i)| + |CC - v_p^3 C(l,m,i)|^{2/3}} \tag{11}$$

for each set of values (l, m), giving us a most probable v_p and i for this choosen set. Note that the discriminant has the dimensions of a velocity. Figure 1 shows an example for an $l = 2$, $m = 0$ mode where $\Gamma_{lm}(v_p, i)$ is minimal for $v_p = 5.2\,\mathrm{km/s}$ and $i = 285°$. The set of values (l, m) for which $\Gamma_{lm}(v_p, i)$ is minimal is considered as the one defining the pulsation mode. We thus consider a whole possible range of values for v_p and i. Figure 1 shows that, by fixing a wrong average $< v_p >$, we might miss some deeper minima in the discriminant, leading to a wrong mode identification.

4 Conclusions

The moment method seems to be suitable for determining the pulsation parameters of early-type stars but line profile variations are by no means staightforward to interpret (Waelkens and Aerts, this volume). It is clear that we have used only part of the information contained in the moments. A future development will be the construction of a discriminant which includes all significant information given by the moments and does not rely only upon a few leading terms.

5 References

Aerts, C., Waelkens, C., Balona, L.A., 1990, *Confrontation between Stellar Pulsation and Evolution, Astronomical Society of the Pacific Conference Series* **11**, 290

Balona, L.A., 1986a, *Mon. Not. R. astr. Soc.* **219**, 111

Balona, L.A., 1986b, *Mon. Not. R. astr. Soc.* **220**, 647

Balona, L.A., 1987, *Mon. Not. R. astr. Soc.* **224**, 41

Balona, L.A., 1989, *Progress of Seismology of the Sun and Stars, Lecture Notes in Physics* 367

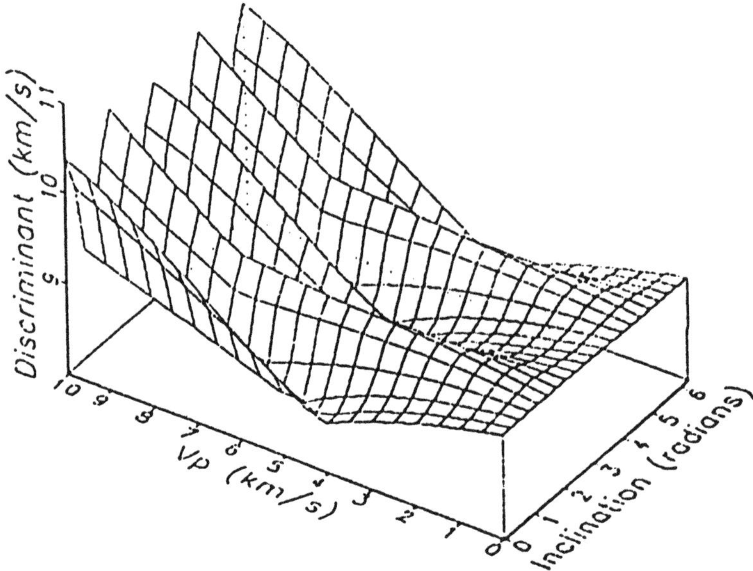

Figure 1: An example of a discriminant $\Gamma_{20}(v_p, i)$, showing a minimum at $v_p = 5.2\,km/s$ and $i = 285°$.

Mode Identification in Beta Cephei Stars from UV Observations

H. Cugier, D.A. Boratyn

Astronomical Institute of the Wrocław University, PL-51-622
Wrocław, Kopernika 11, Poland

1 Introduction

The well known β Cephei variables are a group of stars of spectral type B, whose light and radial velocity periods lie between about two and seven hours. Ledoux (1951) showed that nonradial oscillations could explain some unusual properties observed for these objects.

Usefulness of UV obsevations for mode discrimination in β Cephei stars has been pointed out by Stamford and Watson (1979). In the present paper we discuss further implications of both radial and nonradial oscillations for the observed light variations using an analytic approach, first formulated by Dziembowski (1977) and subsequently developed by Watson (1988) and others.

2 Basic parameters and predicted light curves

We assume that local perturbations of the stellar radius and effective temperature are given by

$$\Delta R = \varepsilon R P_{lm}(\cos\Theta)\cos(\omega t + m\Phi) \tag{1}$$

and

$$\Delta T_{\text{eff}} = B\varepsilon T_{\text{eff}} P_{lm}(\cos\Theta)\cos(\omega t + m\Phi + \psi_T)\,. \tag{2}$$

r,Θ,Φ is the spherical coordinate system whose polar axis ($\Theta = 0$) coincides with the rotation axis. $r = R$ represents the equilibrium radius of the star. ε is a small parameter andP_{lm} is the associated Legendre polynomial. The pulsation frequency ω can be estimated from either photometry or spectroscopy. The integration, over a disk, of the corresponding specific intensity variations leads to magnitude variations, $\Delta m(\lambda, t)$, seen by a distant observer.

For a given mode l of oscillation several parameters must be specified. We adopted the phase shift $\psi_T = \pi$ and the pulsation constant $Q = 0\overset{d}{.}027$ as representative values for β Cephei stars. B (or equivalently $R_{\rm ad} = \frac{B}{B_{\rm ad}}$, where $B_{\rm ad}$ corresponds to the adiabatic case) is a free parameter determined during the best-fit procedure.

Figure 1 shows an example of the predicted light ranges $\Delta m(\lambda)$, for different values of $R_{\rm ad}$.

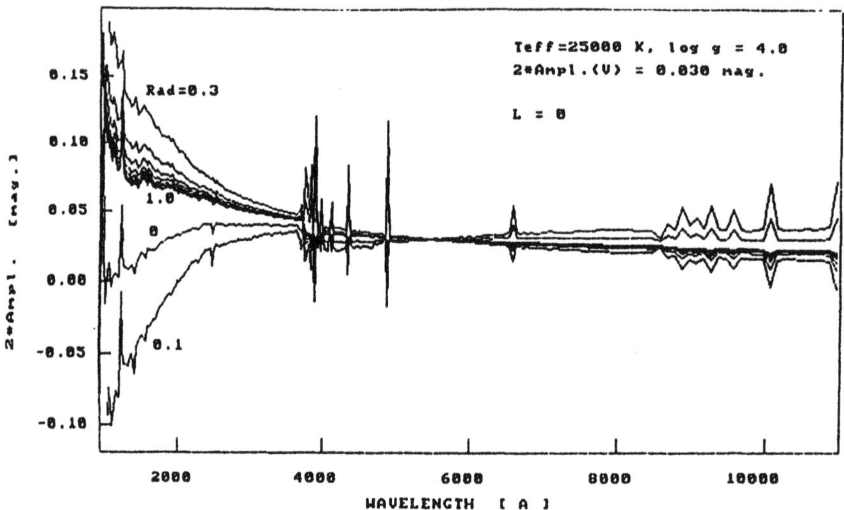

Fig. 1. Monochromatic light ranges for $l = 0$ normalized to $0\overset{m}{.}030$ at $\lambda =$ 547.5 nm.

An important point is that the shape of $\Delta m(\lambda)$ do not depend on the aspect angle Θ_0 , which together with Φ_0 determine direction to the observer.

3 Individual stars

Light curves from UV photometers and spectrophotometers on board of OAO-2, ANS, TD-1A, "Copernicus" and Voyager 2 are available for a number of β Cephei stars. Most of the singly-periodic stars examined had monochromatic amplitudes consistent with radial oscillations, $l = 0$ (cf. Fig. 2). However, in some cases we cannot ruled out another mode identifications, e.g., $l = 1$ for α Lup.

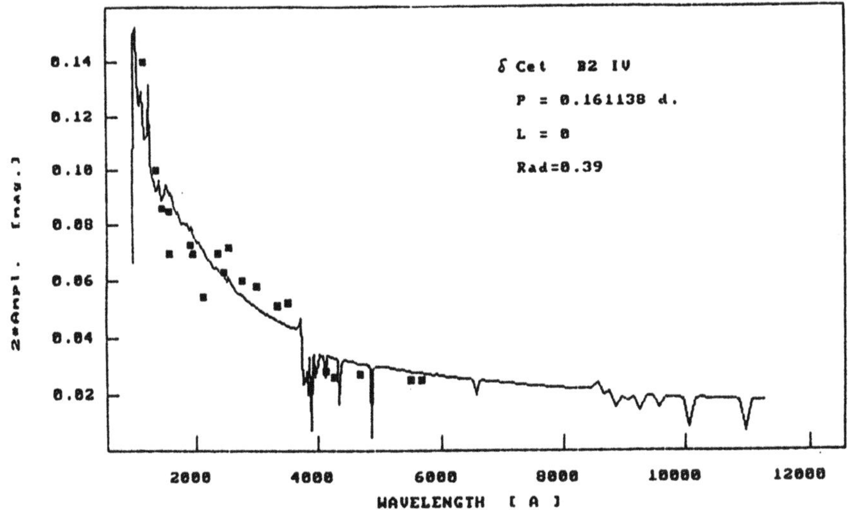

Fig. 2. The best-fit solution for the single-periodic star, δ Cet.

The light variations of multiperiodic stars are more complicated. In the case of σ Sco we found $l = 0$ for $P_2 = 0\overset{d}{.}246836$ and $l = 1$ for $P_1 = 0\overset{d}{.}23967$. However, for β CMa we obtain $l = 2$ for $P_2 = 0\overset{d}{.}251300$, $l = 1$ for $P_1 = 0\overset{d}{.}25003$ and $l = 0$ for $P_3 = 0\overset{d}{.}23904$.

In our approach, the azimuthal spherical harmonic number m cannot be determined. An analysis of line profiles (cf. Ledoux 1951) or rotationally spliting of frequencies (cf. Jerzykiewicz 1978) is necesssary to identify m.

References

Dziembowski, W.: 1977, Acta Astron. **27** 203

Jerzykiewicz, M.: 1978, Acta Astron. **28** 465

Ledoux, P.: 1951, Astrophys. J. **114** 373

Stamford, P.A., Watson, R.D.: 1979, Proc. IAU Coll. **46**, Univ. of Waiato, New Zaeland

Watson, R.D.: 1988, Astrophys. Space Sci. **140** 255

Magnetic fields in hot stars

J. D. Landstreet [1] [2]

[1]Department of Astronomy, University of Western Ontario, London, Ontario, Canada
[2]Observatoire Midi-Pyrenees, Toulouse, France

1 Introduction

The stars of the upper main sequence show a remarkable variety of chemical peculiarities. This variety of surface chemical abundance patterns is particularly clear when comparison is made with lower main stars, almost all of which have chemical abundances quite similar to that of the sun, except for one or two individual elements such as Li. The great variety of abundance peculiarities observed in upper main sequence stars clearly shows the existence of powerful chemical segregation processes acting in and near the surface of many upper main sequence stars, since it has not proven possible to understand the observed abundances as the result of nuclear processing or mixing with the deep interior.

These chemical segregation processes almost certainly act in *all* upper main sequence stars to a greater or lesser extent. It is essential to understand the segregation mechanisms in order to correctly interpret abundance studies of stellar spectra, and to know when stellar atmospheric abundances provide information about the material from which the stars formed or about internal nuclear processing, and when this information is masked or distorted by processes intrinsic to the atmospheres and envelopes of the stars observed. In understanding the physics of the segregation processes, the obviously chemically peculiar stars will play an important role, as it is in these stars that the segregation mechanisms are most clearly at work.

The chemically peculiar stars of the upper main sequence show peculiarities which have some tendency to be systematic with effective temperature, although at any given effective temperature there is a large dispersion of abundances. Furthermore, the chemical peculiars divide into two sets, in one of which large magnetic fields are known to occur, while in the other set the stars seem to have no fields or at least rather weak fields. The classes of abundance anomalies widely used are as follows:

Range of T_e	Magnetic stars	Non-magnetic stars
7-10,000 K	Ap SrCrEu	Am, λ Boo
10-14,000 K	Ap Si	Ap HgMn
13-18,000 K	He weak (Si, SrTi)	He weak (PGa)
18-22,000 K	He strong	

(Note that weak fields have recently been reported in at least one Am star, rasising the question of whether the division above is a clear as now supposed; cf. Mathys and Lanz 1990)

The most outstandingly peculiar of all the chemically peculiar stars of the upper main sequence are the magnetic stars, in which abundance anomalies of two dex or even more relative to solar abundances are found for some elements. These stars will almost certainly play an especially important role in furnishing observational constraints on the physics of the segregation mechanisms. In this paper, I review the observations of such stars, the immediate inferences about their nature that may be extracted more or less directly from observations, and current progress in modelling the atmospheres. I shall describe some ideas about how the chemical peculiarities may have arisen in these stars, and how the physics may be affected by the presence of strong magnetic fields. I shall also include a brief survey of magnetic fields found in white dwarfs, as models of these stars have a number of features in common with our ideas about magnetic upper main sequence stars.

2 Field measurements in main sequence stars

Magnetic fields in upper main sequence stars are detected by using some aspect of the Zeeman effect. It is worthwhile to review briefly what kinds of information about a magnetic field are furnished by the Zeeman effect, so that it will be clearer how the observations are interpreted and modelled.

When an atom is placed in a magnetic field, each individual atomic level splits into a number of discrete states, whose energy separation is proportional to the field strength of the applied field (see for example Condon and Shortley 1951). Absorption and emission of light can generally only occur between states whose magnetic quantum numbers m differ by 0 or 1. Transitions with $\Delta m = 0$ give rise to one or several lines symetrically spread around the wavelength λ_o of the single line that arises between these levels when no field is present. These lines are known as π components. Transitions with $\Delta m = +1$ give rise to a second group of lines displaced from λ_o; transitions with $\Delta m = -1$ are displaced symmetrically to the other side of λ_o. These two groups of lines are known as σ components. The displacement of either group of σ components from λ_o is directly proportional to the strength of the applied magnetic field.

The π and σ components of the split atomic line also have distinctive polarization properties. In a magnetic field aligned along the line of sight, the π components (in absorption) are linearly polarized normal to the field direction, while the σ components are linearly polarized parallel to the field. If the field is parallel to the line of sight, the π components are not visible, and the two groups of σ components have opposite circular polarizations.

The splitting and polarization properties of spectral lines in the presence of a magnetic field make possible several methods of detecting and measuring a stellar magnetic field. If the field is sufficiently large that the Zeeman splitting of the line components is larger than any other line broadening mechanism (usually the dominant broadening is provided by stellar rotation), the observed spectral line may be resolved into two or three discrete components, whose separation is proportional to the field strength. Because the splitting depends on the magnitude of the field and not on its direction, a measurement of the separation provides a measure of the mean field modulus $<|\mathbf{B}|>$ averaged over the visible hemisphere of the star, usually called the mean surface field B_s. This kind of splitting is fairly rare to observe, as it requires field strengths well above the typical field of one or two kilogauss found in most magnetic stars, and also rather low projected rotational velocity. The effect is only observable when vsini (in km s^{-1}) is numerically not much larger than B_s (in kG). Examples of the splitting observed in a number of lines of one star may be found in Landstreet et al (1989), and of splitting of a few lines in a larger sample of stars in Mathys (1990).

A more robust method of detecting the field is possible using the polarization properties of the Zeeman effect. When the field has a significant mean component along the line of sight, the opposite circular polarizations of the two groups of σ components permits detection of the field even when the splitting of the line by the field is small compared to other line widening. In this case, the mean wavelength of the spectral line is slightly different when observed in one sense of circular polarization than when observed in the other, because of the systematic separation of the two oppositely polarized groups of σ components. This difference in mean wavelength in opposite senses of circular polarization may be detected even when it is a small fraction of the width of the observed line. An equivalent way of detecting this effect is to measure the circular polarization in the wings of one or many spectral lines; the small shift between the right and left circularly polarized components leads to enhanced absorption of left circularly polarized light in one line wing, and of right circularly polarized light in the other wing. Because this effect is measured differentially, and because circular polarization is normally produced only by a magnetic field, a field as low as some tens of gauss may be measured, even using the Balmer lines of H. A measurement of circular polarization in line wings, or of a wavelength change when the same line is observed in right and left circularly polarized light, provides a measure of the mean component of the field along the line of sight averaged over the visible hemisphere, $<B_z>$, also known as the mean longitudinal field B_ℓ or the effective field B_e. Further discussion of the measurement and interpretation of this quantity may be found in Landstreet (1980, 1982) and Mathys (1989).

In the same way, a transverse field leads to non-zero linear polarization in the line profile, which provides information about the field components $<B_x>$ and $<B_y>$ of the field. As yet few measurements with high resolution are available, although Borra and Vaughan (1976;1977) succeeded some years ago in detecting this effect in β CrB. The linear polarization in the lines also leads to a net linear polarization in broadband light, produced by the non-zero net polarization present in all individual saturated lines. Detection of this effect has been reported by Kemp and Wolstencroft (1974), and recently a programme to study the effect systematically has been initiated by Leroy.

Almost all of the fields detected in upper main sequence stars have been discovered from measurements of $<B_z>$. Fields are detected in a significant fraction of all A and B stars. Roughly 10% of the stars near A0V are detectably magnetic. The stars in which fields are detected are invariably chemically peculiar when compared to the sun, often showing overabundances of one dex or more of such elements as Si, Ca, Ti, Cr, Sr, and many rare earths. He is underabundant in all but the hottest (He-strong) magnetic stars, where it constitutes the main overabundant element. In addition, the magnetic stars usually have unusually small values of projected rotational velocity, and are often variable in brightness, spectral appearance, and magnetic field strength. A summary of the observed properties of magnetic A and B stars may be found in the monograph by Wolff (1983). Recent compilations of magnetic observations are presented by Borra, Landstreet and Thompson (1983), Thompson, Brown and Landstreet (1987), Bohlender et al (1987), and Mathys (1991). A rather complete list of older observations is found in the catalogue of Didelon (1983).

The observed variations have a very simple character. Variations are always strictly periodic, and all variable properties vary with the same period, maintaining a fixed phase relationship among the variables. The variations in light are generally small, of the order of 0.1 magnitude or less. Spectrum variations can be very great; lines of one or more elements such as Si, Ti, Cr, or some of the rare earths may vary from near invisibility to great strength, although lines of other elements such as H and Fe are usually almost constant. The magnetic field variations often (but not always) result in the reversal of the sign of the observed field $<B_z>$. The range of observed periods is very great, from about 0.5 days to around 10^4 days. (The periods of some 300 magnetic Ap and Bp stars are tabulated in the bibliographic catalogues of Catalano and Renson [1984;1988] and Catalano, Renson, and Leone [1991]). A particularly important characteristic of the period of a given star is that it is always long enough that the observed values of vsini could be caused by rotation with the period of the variations.

3 The rigid rotator model and direct inferences from observations

The stability of the periods, the large range of periods observed, the strongly variable character of some spectral lines while other remain almost constant, and the correlation of period with $(\mathrm{vsini})^{-1}$ all strongly suggest that variation of the magnetic stars is produced simply by rotation. That is, we imagine these stars to have surfaces which do not change with time intrinsically on a time scale short enough to observe, but which have large variations in chemical abundances over their surfaces which we observe as the star rotates. The abundances of some elements on the surface are characterised by strong patchiness. The magnetic field varies because it has a large-scale structure that is not symmetric about the rotation axis, which we see from different directions as the star rotates. Stellar brightness varies slightly in particular wavelength bands because the chemical composition of the atmosphere on the visible hemisphere changes as the star rotates, causing small changes in the emergent flux. This model is known as the rigid or oblique rotator model.

A number of direct inferences about the geometrical structure of the field of a magnetic A or B star, and about its evolution, may be extracted more or less directly from the observations using the rigid rotator model as a basis. Perhaps the most important of these inferences is from the fact that a non-zero value of $<B_z>$ is observed at all, which clearly implies that the fields of the magnetic A and B stars are more or less simple and coherent in structure. (In contrast, the complex fields of lower main sequence stars, which resemble the field of the sun, with numerous regions of opposite polarity visible on one stellar hemisphere at a given time, do not produce a measureably non-zero value of $<B_z>$, although selective magnetic line broadening of atomic lines of large π-σ separation per unit field strength in a number of stars reveals fairly clearly the presence of a field of a few kG.) When the observed value of $<B_z>$ varies, as it usually does, it almost invariably varies sinusoidally. This is what we would expect if the field were dipolar in nature, and this observation strongly suggests that the fields of magnetic A and B stars have very important dipole components. This observation does not, however, establish that the field geometry of a magnetic A or B star is nearly dipolar; more complex structure superimposed on a dipole would alter the observed variation of $<B_z>$ only a little, unless the complex structure is comparable in field strength to the dipolar component. Examples of stars showing clear evidence of complex field structure have been reported by Thompson and Landstreet (1985) and Landstreet (1990).

Another important observational characteristic of the fields of magnetic A and B stars confirms the relative simplicity of the field geometries: when Zeeman splitting is observed in spectral lines that split into only two components with no central (undisplaced) line component, as is the case for the Fe II line at 6149.2 Å, the observed components are each relatively narrow, and the two components separate cleanly, with little or no residual absorption between them (see Mathys 1991 for a number of examples). This clearly indicates that the field strength does not vary by a large amount over the visible hemisphere of these (large-field) stars, but is relatively simple. In particular, there are no important regions of zero field: these would lead to absorption at the undisplaced wavelength of the line. Magnetic A and B stars do not have the kind of bimodal field distribution (regions of zero field mixed in with regions of kilogauss field) that is found on the sun (see for example Stenflo 1989).

Over the sample of magnetic A and B stars observed, which is now more than 100 stars, the median field $<B_z>$ observed is about 250 G. The high field tail of the distribution extends to about 20,000 G, while the low field limit is set by the great observational difficulties of measuring a field below about 100 G. The weakest longitudinal field reliably measured is about 100 G (Bohlender and Landstreet 1990a; Donati et al 1990). Assuming that the field geometries of most of these stars are approximately dipolar, the observed median field corresponds roughly to a surface field $<|\mathbf{B}|>$ of the order of one kG. Such a small field is not detectable directly through line spitting; the typical separation of the σ components for such a field would only be of order 0.05 Å at 5000

Å, smaller even than the usual instrumental resolution of a typical stellar spectrograph, and much smaller than the typical rotational broadening of some tens of km s^{-1}, which corresponds to a widening of a few tenths of an Å around 5000 Å. Thus most magnetic A and B stars do not have measureable values of $<|\mathbf{B}|>$. Among the stars that do, the observed field strengths $<|\mathbf{B}|>$ extend up to 3.5×10^4 G (Babcock 1960), but are almost always considerably smaller (Preston 1971; Mathys 1990).

If we consider a large and randomly chosen sample of magnetic Ap stars, the fraction of the sample that have longitudinal fields $<B_z>$ whose sign reverses when it is measured through a full rotation period provides information about the distribution of the inclination angle β of the magnetic axis to the rotation axis over that sample of stars (Preston 1967). If most magnetic stars had small β angles, $<B_z>$ would usually only vary a little through the rotation cycle, regardless of the inclination i of the rotation axis to the line of sight, as we would always observe the star from an almost constant (magnetic) latitude. On the other hand, if all the dipole axes made an angle of $\beta = 90°$ with respect to the rotation axes, we would see both magnetic hemispheres for any inclination i significantly different from zero, and any detectable longitudinal field would always reverse sign as the star turned. The observations that most but not all magnetic stars reverse the sign of their observed fields $<B_z>$, and that the negative and positive extrema do not usually reach the same field strength, turns out to imply that in most stars the angle β is fairly large (say around 70 or 80°), but a significant fraction of the sample have small magnetic inclinations (β of 20 or 30°, or even less). A discussion of the available data is found in Borra and Landstreet (1980).

The observed field strengths are not correlated with rotation. Large fields are found indifferently in both rapid and slow rotators. This probably means that the fields are not being produced by some kind of dynamo currently operating inside the star, since we would expect such a dynamo to be fairly sensitive to rotation, as indeed seems to be the case for solar-type stars. Instead, the field may be a relic (a fossil field) left from the era when the star condensed from the interstellar medium. The electrical conductivity inside a star is so large that a field may remain frozen inside a star for an exceedingly long time; the time scale for ohmic decay is of the order of 10^{10} years (Cowling 1976). The expected long lifetime of a fossil field is supported by detailed theoretical models of the interiors of upper main sequence stars which have been constructed to explore how stellar fields might evolve under the influence of ohmic decay, meridional circulation, and stellar evolution (Moss 1989). The result of such models is that a fossil field in an upper main sequence star is expected to change by only a modest factor during the main sequence lifetime. This result is consistent with observational studies that have tried to detect the evolution of magnetic fields over the time span of the main sequence lifetime by comparing magnetic fields measured in samples of young stars with those observed in old stars; little change in average field properties is apparent (Thompson, Brown, and Landstreet 1987).

A very exiting discovery was made by Kurtz and Wegner (1979) when they detected short period light variability in one of the coolest magnetic Ap stars. About a dozen similar short-period variables have since been found, all among the coolest magnetic Ap stars. These stars are observed to vary periodically with one or several stable periods of between four and 15 minutes, and amplitudes of typically a few millimagnitudes.

The variations are now recognized as pulsations. These pulsations vary in amplitude as the star rotates, becoming undetectable as the observed field $<B_z>$ passes through zero (when the magnetic equator crosses the centre of the visible hemisphere). The pulsations thus seem to take the magnetic axis as an axis of symmetry. In addition, in some of the stars, individual modes are excited and die away on longer time scales of days or weeks. These variations have been identified as non-radial p-mode pulsations of high order (many radial nodes, hence short periods) but low degree (which means that the surface structure is fairly simple, so that the pulsations may be seen in integrated light). However, it has not yet been possible to identify all of the observed modes in any star, nor is it clear what the driving mechanism is. When these pulsations are better understood, they will provide an extremely powerful probe of conditions inside the pulsating stars, in the same way as information is available from helioseismology. A recent review of this field is by Kurtz (1990).

Another very interesting phenomenon is found among the very hottest magnetic B stars, some of which show clear evidence for the presence of circumstellar matter, and perhaps for mass loss. This phenomenon is sometimes found in hot He-weak stars, but is mostly observed among the He-strong stars, the hottest of the known magnetic stars. These stars have excess helium in their atmospheres, unlike all other types of magnetic A or B star. Their ultraviolet spectra often show strong C IV resonance lines in absorption or emission, sometimes with a strong asymmetry towards shorter wavelengths that clearly implies mass outflow. In at least two or three of these objects the wind lines are modulated by the rotation of the star, as though the wind is not the same in all directions from the star, but is controlled or channeled by the magnetic field (Barker et al 1982; Shore et al 1990). In one He-strong star, σ Ori E, the circumstellar material is apparently trapped in a magnetosphere which is dense enough to scatter several percent of the outflowing light when a magnetospheric cloud passes in front of the star, causing weak eclipses (Groote and Hunger 1976; Landstreet and Borra 1978; Hunger 1986; see also Hunger's paper in this volume).

4 Magnetic white dwarfs

Magnetic fields reminiscent of those observed in the magnetic A and B stars, but far more intense, are also observed in white dwarfs. In some stars these fields are detected via the Zeeman effect in spectral lines, just as they may be detected in the main sequence stars with unusually large fields if those stars rotate slowly. At the enormous field strengths found in white dwarfs, which are found to exceed 10^6 G (1 MG), the Zeeman effect is considerably more complex than in main sequence stars, because with the larger fields the atomic perturbations produced by the magnetic field are comparable to the separations between levels of different principal quantum numbers. This greatly complicates the calculation of energy and wavelength shifts, expecially above 50 MG, but enough calculations are now available that it is possible in a number of cases to model some lines and estimate the magnetic field strength in the stellar atmosphere.

Another observational effect is also available to use in detecting fields in white dwarfs, which is particularly important in those stars that show no spectral lines in accessible spectral regions (DC white dwarfs). This effect is a general circular polarization of the continuum radiation from a magnetized star that occurs at a detectable level for fields of about 10^6 G or more. Such polarization is due to the fact that in a medium permeated by a field, all the electrons tend to spiral about field lines in the same sense. This leads to an asymmetry in the absorption and emission of radiation of photons of opposite circular polarizations, and to a net polarization of the emerging radiation. For a field of 10^6 G, the expected circular polarization at 5000 Å is about 0.2%, an easily detectable amount, and this polarization increases with increasing B. With larger fields, linear polarization may also be produced.

Fields have now been detected in more than two dozen single white dwarfs, ranging in field strength from about 10^6 G up to 5×10^8 G (for a review, see Schmidt 1987; some recent continuum polarization observations and discussion of their interpretation may be found in West 1989). Unlike the fields of main sequence A stars, for which the distribution of field strengths shows a peak at small fields (around a few hundred gauss) with a tail extending up to 10^4 G or more, the white dwarf fields seem to have roughly constant frequency per decade of field strength. Only about 2% of all white dwarfs have such fields; the great majority do not have detectable fields at all, which means they certainly have fields of less than 10^6 G, and in most cases probably less than 10^5 G (Angel, Borra, and Landstreet 1981).

The observed fields are mostly found to be constant in time, but about one quarter vary periodically with periods of between 1.5 hours and 3 days. As in the magnetic upper main sequence stars, the variations are strictly periodic, and the sign of the mean longitudinal component of the field, as measured either by the sign of continuum circular polarization or by Zeeman polarization in the lines, frequently is found to reverse sign. The large range of observed periods and the reversal of the field sign both strongly suggest that the magnetic white dwarfs, like the magnetic A and B stars, are rigid rotators. The non-variable stars are presumably either stars in which the rotation is extremely (!) slow, or in which the axis of the magnetic field is aligned fairly closely with the rotation axis of the star (Schmidt and Norsworthy 1991).

It is not yet obvious how the fields of magnetic white dwarfs evolve with time (or equivalently, since white dwarfs are simply cooling slowly, with decreasing effective temperature). Both weak and strong fields occur for hot white dwarfs ($T_e > 15000$ K) as well as for cool stars ($T_e < 8000$ K). This lack of obvious field evolution suggests that the observed fields may again be fossil fields, anchored deep in the star, that are supported by electrical currents generated at an earlier stage of the star's evolution. Estimates of the time scale for ohmic decay support this possibility. In fact, it is entirely possible that the observed fields in white dwarfs may be the descendants of the large fields found in magnetic A and B stars. This view is supported both by the observed frequency of magnetic white dwarfs, and by the size of the observed fields, which are roughly consistent with flux conservation from main sequence fields (Angel, Borra, and Landstreet 1981).

5 Modelling the atmospheres of magnetic stars

The observed fields in magnetic Ap and Bp stars and in magnetic white dwarfs are large enough that the energy density $B^2/8\pi$ (in Gaussian units) is larger than the thermal energy density $3nkT/2 = 3p/2$ in the gas down to optical depths well below the radiating photosphere. In principle, this makes it possible for the field to have important hydrostatic and dynamic effects on the structure of the atmosphere and outer envelope. In particular, it is quite likely that the magnetic field suppresses or at least greatly modifies the photospheric convection layer produced by the H I-II ionization zone in stars of effective temperatures below about 12000 K. If this convection occured in the normal fashion, it would drive mixing that would enforce surface chemical homogeneity in the photosphere and prevent cool Ap stars from developing horizontally inhomogeneous abundance distributions (Michaud 1980).

Other kinds of effects are also possible, but they depend on whether important electrical currents flow in the stellar atmosphere or not. If significant currents flow, the hydrostatic structure (the pressure - optical depth relationship) of the atmosphere may be substantially altered. Currents could be produced by such effects as ohmic decay of the interior magnetic field, field structure changes due to meridional circulation flows inside the star, or changes in stellar structure resulting from evolution, ambipolar diffusion of protons within the H I-II ionization zone, or even dynamical evolution resulting from the distortion of the star by the interior field into a figure not axisymmetric about the rotation axis (Mestel and Takhar 1972). Only exploratory discussions of these possible effects on the atmospheric structure have been published so far; the interested reader might consult Peterson and Theys (1981), Landstreet (1987), and Babel and Michaud (1991c).

A very active area of current study is efforts to map the distributions of various chemical elements over the stellar surface, and to determine the magnetic field structure in which the distributions have developed, for specific individual stars for which adequate observations are available. Because of the great complexity of the physics that may be involved in the chemical separation of elements in stellar atmospheres, it is extremely important to obtain good maps of abundance distributions and the associated field structure for a number of stars, to provide constraints on and tests of theoretical work. Currently a number of groups are working on this problem.

Fortunately, the basic theory of line formation in a stellar atmosphere (at least in LTE) is rather well understood, thanks to the efforts of a series of theoreticians working mainly on interpreting observations of the solar atmosphere (Unno 1956, Rachkovsky 1962, Beckers 1969, and Stenflo 1971, for example). (On the other hand, the non-LTE theory of line formation in a magnetic field is still being worked out; the situation is reviewed by Rees 1987.)

One important area of activity is the derivation of maps for stars in which the spectral lines are moderately broadened by rotation (say, $20 < v\sin i < 50$ km s^{-1}), and in which the local magnetic fields are sufficiently weak (say, $\langle|\mathbf{B}|\rangle < 1$ kG) that the local line profile emitted by a particular spot on the disk is hardly affected (especially, hardly widened) by the field. In this case, observations of one or several suitable lines throughout the rotation period provides enough information to extract a fairly detailed two-dimensional map. This is possible because abundance patches at different stellar latitudes (measured from the rotation axis) have different velocity shifts as they cross the visible hemisphere of the star, and also are visible for different fractions of the rotation period. Consider a star with the rotation axis inclined to the line of sight by about 45°. A spot near the visible rotation pole will be detectable throughout the rotation period, and because of its nearness to the axis, will have fairly small radial velocity excursions. A spot near the stellar equator will be visible for about half of the rotation period, and as it moves across the visible hemisphere, it will be moved by the doppler shift from one wing of the line profile to the other. A spot perhaps 30° below the equator in the direction of the invisible rotation pole will be visible for only a short part of the rotation period, but will have moderate doppler shifts. Thus it is clear that both longitude and latitude information is contained in high resolution, high signal-to-noise ratio observations of a line profile through the rotation cycle. From such data, surprisingly detailed maps may be derived for elements for which fairly strong unblended lines may be observed with high resolution and signal-to-noise ratio. Recent examples may be found in Khokhlova, Rice and Wehlau (1986), Hatzes, Penrod and Vogt (1989), and Rice and Wehlau (1990; 1991). The method has been reviewed by Khokhlova (1986) and by Vogt (1988), among others.

The most important limitation of this class of models up to now is that maps of abundance distributions are derived for just those stars for which in general only minimal constraints on the magnetic field are available. Usually only observations of the mean longitudinal field $\langle B_z \rangle$ are available for these (relatively) broad-lined stars. Thus the abundance maps are not easily related to the magnetic field structure that underlies them. To overcome this problem, several approaches are possible. One may attempt to map stars for which the field is large enough to substantially modify the line profiles, in which case much stronger constraints are available on the field structure. This is the approach taken by me and my collaborators. Alternatively, one may obtain spectropolarimetry of metallic lines along with measurements of line profiles, and use the full information content of the polarimetry to constrain the magnetic field structure. This approach is now being developed by several individuals, such as Donati, Mathys, Piskunov, and Stift.

Any attempt to model the magnetic field geometry and the abundance distributions simultaneously, as is necessary when one uses the information content of the line profiles to constrain both these structures, has two main problems. First, information about the field distribution usually degrades (or at least complicates) information about the abundance distributions, as the simple correspondence between doppler shift in a spectral line and distance from the rotation axis on the stellar disk is confused by local magnetic line broadening. Secondly, the calculation of local line profiles including the effects of magnetic splitting by a field of arbitrary orientation and strength is a lot more costly in computing time than the calculation of a line profile without magnetic effects. One way of dealing with these difficulties is to simplify the maps drastically, for example by modelling the field structure as a simple axisymmetric superposition of a few multipoles, and taking the abundance distribution to be axisymmetric about the magnetic axis. This greatly simplifies the search for successful maps, compensating for the added complexity of calculating the line profiles for comparison with observation at each step of the calculation. This approach has been used to obtain simultaneous models of magnetic field structure and of distributions of several elements for the cool Ap star 53 Cam (Landstreet 1988), the extremely magnetic mid-B star HD 215441 = Babcock's star (Landstreet et al 1989), and the He-strong star HD 64740 (Bohlender and Landstreet 1990b).

Several features emerge quite clearly from the maps and models obtained so far. First, even common elements such as Si, Ti, and Cr may be enhanced in average abundance over the solar abundances by large factors of the order of 10^2 or even 10^3. In HD 215441, for example, about 1% of all the atoms in a large region near one magnetic pole are Si atoms, although this element only contributes 0.003% of the atoms in the sun (Landstreet et al 1989). Secondly, large abundance contrasts are present for a number of common elements, sometimes up to a factor of 10^2 difference in abundance between areas of high abundance and areas of low abundance, while other elements are distributed in an apparently much more uniform fashion. The highly non-uniform elements are not always the same from star to star. Thus, in the Ap star 53 Cam (T_e = 8500 K), almost all the photospheric Ca and Ti atoms are located in large patches, the Ti patch at one magnetic pole and the Ca patch at the other, while Cr, Fe, and the rare earths are more uniformly distributed (Landstreet 1988). On the slightly hotter Ap star HD 125248 (T_e = 9500 K), it is Cr and the rare earths that show very large contrasts, while Ti and Fe are more nearly uniform. Finally, the magnetic fields are apparently not very accurately represented by a simple centred dipole, but instead usually have one pole substantially stronger than the other (or equivalently, a strong quadrupolar component), and sometimes still further departures from dipolar structure are apparent. This difference between the field strength and/or structure at the two magnetic poles is presumably responsible for the very large abundance differences which may occur between the two poles.

Modelling of the field structures of magnetic white dwarfs is simplified in comparison with modelling of magnetic fields in Ap and Bp stars by the lack of obvious surface chemical inhomogenity. However, the mapping problem is made more difficult by the fact that accurate calculations of line shifts and strengths in fields of 50 MG or more are only now becoming available even for H, and are still not available at all for He. The calculation of continuum polarization is in an even less satisfactory state. Furthermore, spectra of white dwarfs are usually only obtained with low resolution and signal-to-noise ratio. Because of these difficulties, determination of field structure has generally proceeded by fitting simple models rather than by more sophisticated methods. Modelling results have been reviewed by Wickramasinghe (1987); some recent results are presented by Achilleos and Wickramasinghe (1989). The available models suggest that the field structure may not in general be terribly far from dipolar, although at least one star is known with a strong localized field (a magnetic spot) superimposed on the global dipole (Schmidt et al 1986).

6 Chemical separation mechanisms

Much work has recently been done on the problem of the mechanism(s) of chemical separation in the presence of a magnetic field, and also to test various possible mechanisms against observational constraints. It is generally believed that the observed abundance peculiarities in the magnetic Ap and Bp stars have been produced by selective diffusion of trace elements in the background medium of H, as suggested some years ago by Michaud (1970; see also his paper in this volume). These trace ions are subjected to the competing influences of gravity downward, and radiation pressure support through spectral lines upward, throughout much of the upper envelope as well as in the atmosphere and even above it. As a result, some elements (for example, He) will sink out of sight. Others (typically elements of low cosmic abundance with complex spectra, for example many rare earth elements) will be driven upward from reservoirs whose depth depends strongly on the atomic structure of the element considered, and may either collect in the atmosphere or be pushed out into the interstellar medium. The efficiency of this process depends strongly on the uncertain hydrodynamic state of the stellar envelope (it will be affected by convection or turbulence as well as by large-scale circulation currents, if they move quickly enough), as well as on possible instabilities, which have not yet been very extensively explored. Judging from the patchiness of the observed magnetic Ap stars, diffusion is also influenced quite strongly on the magnetic field. It is because of the sensitivity of diffusion to competing processes that mapping may ultimately enable us to study the envelopes of the magnetic stars.

The problem, then, assuming that diffusion is the underlying mechanism responsible for the observed abundance distributions, is to evaluate, for a given stellar mass and magnetic field structure, the diffusion velocities of trace elements relative to the general hydrogen gas. These must be calculated for horizontal as well as vertical motions, as a function of the changing chemical composition profile throughout the outer envelope of the star, so that the surface abundances may be computed (starting from some reasonable initial conditions) as a function of time through the (main sequence) life of the star. This general programme has yet to be carried out, although the underlying physics is coming to be well enough understood that it may be feasible in the not-to-distant future.

An ambitious project to compare theory with observation has recently been undertaken by Babel and Michaud (1991a) and Babel (1992). Babel and Michaud attempt to consider all relevant processes needed to explain the distributions of several elements observed for 53 Cam. They start by calculating vertical radiative accelerations for a number of elements (Ca, Sc, Ti, Cr, Mn, Sr) as a function of depth and local abundance, ignoring possible effects of the magnetic field. Where the radiative acceleration exceeds gravity, an element is expected to be driven upward. According to their calculations, Cr and Mn are expected to be pushed upward into the atmosphere but (because the radiative acceleration drops high in the atmosphere) not be lost to interstellar space. Overabundances of these two elements are therefore expected; this is observed for Cr but not for Mn. Sc and Ti have large radiative accelerations everywhere and may leave the star. Thus these elements are expected to have roughly solar abundances, which is observed for Sc, but conflicts with the large abundance of Ti observed in a spot around the stronger magnetic pole. Ca is pushed up into the atmosphere from only a shallow reservoir, and so should not be greatly overabundant, contrary to the large abundance observed in a spot around the weaker magnetic pole of 53 Cam. Sr is also pushed up from a shallow reservoir, and is calculated to be somewhat overabundant, but not as much as is observed.

Babel and Michaud then consider ways in which the magnetic field could modify the situation, either by altering the vertical flows or by generating significant horizontal velocities. Several potentially important processes must be examined. (1) Horizontal flows will result from the fact that ions tend to follow field lines, rather than simply moving vertically, at small optical depths. This leads to a tendency for some elements to concentrate high in the atmosphere where field lines are horizontal (Michaud, Megessier, and Charland 1981). (2) The difficulty ions have in moving across field lines may also result in extra support in regions of horizontal field lines for ions that are supported in the neutral state but tend to fall as first ions. The downward drift of the ions is inhibited by the field, so that the net upward diffusion velocity is increased, and so larger abundances may be supported where field lines are horizontal than where they are vertical. For an element that falls as a neutral but rises an an ion, the effect is reversed. This effect may be particularly important for Si, which is observed to be quite overabundant in the atmospheres of most late B magnetic stars, contrary to the results of simple diffusion models (Vauclair, Hardorp and Peterson 1979; Alecian and Vauclair 1981). (3) An element that rises as a neutral but falls as an ion will rise vertically but fall along the field lines. Where the field is oblique, this may drive the element horizontally towards regions of vertical field. An element that falls as a neutral but rises as an ion will tend to accumulate where the field is horizontal (Michaud, Megessier, and Charland 1981; Megessier 1984). (4) The splitting of spectral lines by the magnetic field leads to some desaturation, with a consequent increase in vertical radiative acceleration (Michaud, Megessier and Charland 1981). (5) The anisotropy of Zeeman splitting as seen by photons propagating at various angles with respect to the magnetic field leads to horizontal radiative acceleration in regions of oblique field lines (Babel and Michaud 1991b). (6) If the magnetic field is strong enough to suppress the convection zone due to the enhanced opacity in the layers where H is partially ionized, the temperature gradient drives ambipolar diffusion of neutral H with respect to the protons in the ionization zone. Because the ion-proton collision cross section is much higher than the ion-H atom cross section, the upward diffusion of protons lifts most ions through this regions (Babel and Michaud 1991c). (7) Finally, if a stellar wind is present, it will superimpose the general outflow velocity on the diffusion velocities, and will gradually lift into the atmosphere deeper layers of the envelope whose chemical composition has been previously altered by diffusion. With an appropriate mass loss rate (of the order of 10^{-12} to 10^{-15} $M_{\odot}$ per year), this effect may bring into the atmosphere elements such as He that would otherwise sink out of sight (Vauclair 1975; Vauclair, Dolez, and Gough 1991). Furthermore, if the wind is controlled or modulated by the magnetic field, the results will be different in different regions of the star.

Babel and Michaud argue that if no mass loss takes place in 53 Cam, even the combined action of all the other effects discussed is unable to explain the observed abundance distributions of Ca and Ti. Ca, which is supported only in a thin layer, is not expected to be better supported where the magnetic field is weak and vertical, so presumably horizontal motions must be important. But the only process to go in the right direction is radiative acceleration, and this cannot explain the large contrast between the two poles. In contrast, Ti is lifted from a deep reservoir. Consequently, horizontal motions are not likely to be important, as the relevant length scale (the stellar radius or circumference) is so much larger than the vertical scale over which diffusion is effective. Ti is expected to be pushed into the interstellar medium near the poles, but perhaps retained near the magnetic equator. The model does not explain the occurrence of a large overabundant spot near one pole. These results lead Babel and Michaud to conclude that very possibly a magnetically controlled stellar wind does operate. The wind needed is sufficiently weak that it would be almost impossible to detect directly; the main evidence for its existence is the apparent impossibility of explaining the observed abundance distributions on a magnetic Ap star without invoking a wind. It remains, of course, for detailed calculations to show that a postulated wind could explain the observed abundances. Models calculated by Babel (1992) for a very weak wind improve the agreement with observations, especially for Ca, but still do not explain the distribution of Ti.

The potential importance of a weak stellar wind is also discussed by Vauclair, Dolez, and Gough (1991). Their calculations show that the He overabundance observed in the hottest Bp stars, the He-strong ones, may be explained by such a wind, as originally shown by Vauclair (1975). Effectively, for the right mass loss rate, the upward movement of gas in the stellar envelope goes faster than downward diffusion of He, so He is swept towards the surface. At the surface, He becomes mainly neutral, so that its mean free path becomes much larger, and the downward diffusion velocity becomes large enough for the He to slip through the general outward movement of the wind. It thus accumulates in the atmosphere, where it is observed. If the wind is modulated by the magnetic field (for example, suppressed in regions where the magnetic field is approximately horizontal), an inhomogenous distribution of He is expected over the stellar surface, as is usually observed.

A second very interesting result of this paper concerns the internal distribution of He in cooler Ap stars. In the absence of a wind, He (which has a very simple spectrum and a large cosmic abundance) is expected to diffuse downward in all Ap stars, and to gradually become dramatically depleted throughout a deep envelope below the photosphere. When this happens, the convection zone driven by the He II-III ionization zone disappears, as does the possiblility of κ-mechanism pulsations driven by this zone of high continuous opacity. Vauclair, Dolez and Gough show that mass loss rates of a few times 10^{-14} $M_{\odot}$ per year will have the effect of lifting He up as far as the region where it becomes mainly neutral, a level in cooler Ap stars which is well below the observable photosphere. The importance of such a sub-photospheric layer of enhanced He abundance is that it could potentially drive the pulsations observed in some of the coolest Ap stars.

Thus, we are beginning to see a convergence of theory and observation which shows great promise of helping to identify and delimit physical effects occuring not only within the visible atmospheres of many A and B stars, but also below and above these atmospheres. This is a very exciting development, and one which is sure to stimulate much interest among astrophysicists in the next few years.

References

Achilleos, N. and Wickramasinghe, D. T. 1989, Ap. J., 346, 444.
Alecian, G. and Vauclair, S. 1981, Astr. Ap.. 101, 16.
Angel, J. R. P., Borra, E. F., and Landstreet, J. D. 1981, Ap. J. Suppl., 45, 457.
Babcock, H. W. 1960, Ap. J., 132, 521.
Babel, J. and Michaud, G. 1991a, Ap. J., 366, 560.
Babel, J. and Michaud, G. 1991b, Astr. Ap., 241, 493.
Babel, J. and Michaud, G. 1991c, Astr. Ap., 248, 155.
Barker, P. K., Brown, D. N., Bolton, C. T., and Landstreet, J. D. 1982, in *Advances in Ultraviolet Astronomy: Four Years of IUE Research*, ed. Y Kondo, J. M. Mead, and R. D. Chapman (NASA CP-2238), p.589.
Beckers, J. M. 1969, Solar Phys., 9, 372.
Bohlender, D. A., Brown, D. N., Landstreet, J. D., and Thompson, I. B. 1987, Ap. J., 323, 325.
Bohlender, D. A. and Landstreet, J. D. 1990a, Ap J. (Letters), 358, L25.
Bohlander, D. A. and Landstreet, J. D. 1990b, Ap. J., 358, 274.
Borra, E. F. and Landstreet, J. D. 1980, Ap. J. Suppl., 42, 421.
Borra, E. F., Landstreet, J. D., and Thompson, I. B. 1983, Ap. J. Suppl., 53, 151.
Borra, E. F. and Vaughan, A. H. 1976, Ap. J. (Letters), 210, L145.
Borra, E. F. and Vaughan, A. H. 1977, Ap. J., 220, 924.
Catalano, F. A. and Renson, P. 1984, Astr. Ap. Suppl., 55, 371.
Catalano, F. A. and Renson, P. 1988, Astr. Ap. Suppl., 72, 1.
Catalano, F. A., Renson, P., and Leone, F. 1991, Astr. Ap. Suppl., 87, 59.
Condon, E. U. and Shortley, G. H. 1951, *The Theory of Atomic Spectra* (Cambridge: Cambridge University Press).
Cowling, T. G. 1976, *Magnetohydrodynamics* (Bristol: Adam Hilger).
Didelon, P. 1983, Astr. Ap. Suppl., 53, 119.
Donati, J.-F. et al 1990, Astr. Ap., 232, L1.
Groote, D. and Hunger, K. 1976, Astr. Ap., 52, 303.
Hatzes, A. P., Penrod, G. D., and Vogt, S. S. 1989, Ap. J., 341, 456.
Hunger, K. 1986, in *Upper Main Sequence Stars with Anomalous Abundances*, ed. C. R. Cowley, M. M. Dworetsky, and C. Megessier (Dordrecht: Reidel), p. 257.
Kemp, J. C. and Wolstencroft, R. A. 1974, Mon. Not. R. Astr. Soc., 166, 1.
Khokhlova, V. L., Rice, J. B., and Wehlau, W. H. 1986, Ap. J., 307, 786.
Khokhlova, V. L. 1985, Sov. Sci. Rev. E, Ap. Space Phys., 4, 99.
Kurtz, D. W. 1990, Ann. Rev. Astr. Ap., 28, 607.
Kurtz, D. W. and Wegner, G. 1979, Ap. J., 232, 510.
Landstreet, J. D. 1980, Astron. J., 85, 611.
Landstreet, J. D. 1982, Ap. J., 258, 639.
Landstreet, J. D. 1987, Mon. Not. R. Astr. Soc., 225, 437.
Landstreet, J. D. 1988, Ap. J., 326, 967.
Landstreet, J. D. 1990, Ap. J. (Letters), 352, L5.
Landstreet, J. D., Barker, P. K., Bohlender, D. A., and Jewison, M. S. 1989, Ap. J., 344, 876.
Landstreet, J. D. and Borra, E. F. 1978, Ap. J. (Letters), 224, L5.
Mathys, G. 1989, Fund. Cos. Phys., 13, 143.
Mathys, G. 1990, Ast. Ap. 232, 151.
Mathys, G. 1991, Astr. Ap. Suppl., 89, 121.
Mathys, G. and Lanz, T. 1990, Astr. Ap., 230, L21.
Megessier, C. 1984, Astr. Ap., 138, 267.

Mestel, L. and Takhar, H. S. 1972, Mon. Not. R. Astr. Soc., 156, 419.
Michaud, G. 1970, Ap. J. 160, 641.
Michaud, G., Megessier C., and Charland, Y. 1981, Astr. Ap., 103, 244.
Moss, D. 1989, Mon. Not. R. Astr. Soc., 236, 629.
Peterson, D. M. and Theys, J. C. 1981, Ap. J., 244, 947.
Preston, G. W. 1967, Ap. J. 150, 547.
Preston, G. W. 1971, Ap. J., 164, 309.
Rachkovsky, D. N. 1962, Izv. Krim. Ap. Obs., 27, 148.
Rees, D. E. 1987, in *Numerical Radiative Transfer*, ed. W. Kalkofen (Cambridge: Cambridge University Press), p. 213.
Rice, J. B. and Wehlau, W. H. 1990, Astr. Ap., 233, 503.
Schmidt, G. D. 1987, in *The Second Conference on Faint Blue Stars*, ed A. G. D. Philip, D. S. Hayes, and J. W. Liebert (Schenectady: L. Davis Press), p. 377.
Schmidt, G. D. et al 1986, Ap. J., 309, 218.
Schmidt, G. D. and Norsworthy, J. E. 1991, Ap. J., 366, 270.
Shore, S. N., Brown, D. N., Sonneborn, G., Landstreet, J. D., and Bohlender, D. A. 1990, Ap. J., 348, 242.
Stenflo, J. O. 1971, in Solar Magnetic Fields, ed. R. Howard (Dordrecht: Reidel), p. 101.
Stenflo, J. O. 1989, Astr. Ap. Rev., 1, 3.
Thompson, I. B., Brown, D. N., and Landstreet, J. D. 1987, Ap. J. Suppl., 64, 219.
Thompson, I. B. and Landstreet, J. D. 1985, Ap. J. (Letters), 289, L9.
Unno, W. 1956, Publ. Astr. Soc. Japan, 8, 108.
Vauclair, S., Hardorp, J., and Peterson, D. M. 1979, Ap. J., 227, 526.
Vauclair, S., Dolez, N., and Gough, D. O. 1991, Astr. Ap., in press.
Vogt, S. S. 1988, in The Impact of Very High S/N Spectroscopy on Stellar Physics, ed G. Cayrel de Strobel and M. Spite (Dordrecht: Kluwer Academic Publishers), p. 253.
West, S. C. 1989, Ap. J., 345, 511.
Wickramasinghe, D. T. 1987, in *The Second Conference on Faint Blue Stars*, ed A. G. D. Philip, D. S. Hayes, and J. W. Liebert (Schenectady: L. Davis Press), p. 389.
Wolff, S. C. 1983, *The A-Type Stars: Problems and Perspectives* (Washington D. C.: NASA SP-463).
ds

The Structure of Ap Star Magnetic Fields

Gautier Mathys

Geneva Observatory, chemin des Maillettes 51, CH-1290 Sauverny, Switzerland

Abstract: Constraints on the structure of the magnetic field of Ap stars are derived from observations of spectral line profiles recorded in circularly polarized light. In particular, an original, quantitative analysis of the crossover effect, through the moment technique, is presented. Departures of the field structure from a centred dipole are found for several stars.

1 Introduction

Medium-high resolution ($\lambda/\Delta\lambda \approx 1.6\ 10^4$) spectra of Ap stars have been recorded in right (RCP) and left (LCP) circularly polarized light at ESO with the Cassegrain Echelle Spectrograph (CASPEC) fed by the 3.6 m telescope. They are used to derive constraints on the spatially unresolved structure of the magnetic field of these stars.

2 The Mean Longitudinal Field

Wavelength shifts $\lambda_R - \lambda_L$ of lines between RCP and LCP spectra can in a weak-line approximation be interpreted in terms of a mean longitudinal magnetic field $\langle H_z \rangle$. $\langle H_z \rangle$ is the line-intensity weighted average of the line-of-sight component of the magnetic field over the visible stellar disk.

$\langle H_z \rangle$ varies through the stellar rotation cycle due to the change of aspect of the visible stellar hemisphere. The shape of the variation curve of $\langle H_z \rangle$ provides information on the geometry of the stellar magnetic field. For most stars of the present program, $\langle H_z \rangle$ has been found to vary sinusoidally (Mathys 1991), in agreement with the behaviour found by Balmer line photopolarimetry (e.g., Borra & Landstreet 1980, Thompson et al. 1987). This is consistent with the view that *the magnetic field of Ap stars has a dominant dipolar component.*

For two stars, HD 119419 and HD 175362, the variations of $\langle H_z \rangle$ are not sinusoidal. The shape of the curve of $\langle H_z \rangle$ variations for these two stars is well approximated by the superposition of a sinusoid with the rotation frequency of the star and of another sinusoid with twice that frequency. These curves are asymmetric, indicating that *the magnetic field of HD 119419 and of HD 175362 does not have the property of cylindrical symmetry about an axis going through the centre of the star.*

3 The Crossover Effect

Due to the combination of rotational Doppler effect and of magnetic field distribution over the stellar surface, the widths of spectral lines in RCP and LCP light can be different: this is the crossover effect, first evidenced by Babcock (1956). This effect is largest close to the phases when the mean longitudinal field reverses its polarity, and it is null at mean longitudinal field extrema.

A quantitative characterization of the crossover effect is given by the second-order moment $R_V^{(2)}(\lambda_0)$ of the line profile about its centre in the Stokes parameter V. In the weak-line approximation, $R_V^{(2)}(\lambda_0)$ is proportional to the *crossover magnetic field* $\langle x\, H_z \rangle$ (Mathys 1989), which is *the first-order moment of the line-of-sight component of the magnetic field with respect to the plane defined by the line of sight and the stellar rotation axis*, weighted by the local line intensity.

Note that the quantity that is directly determined from the observations is the product of the projected stellar equatorial velocity and of the crossover magnetic field, $v \sin i \, \langle x\, H_z \rangle$.

4 Comparison with the Centred Dipole Model

If the field structure is that of a dipole located at the centre of the star, under approximations consistent with those made to derive these quantities, the mean longitudinal magnetic field $\langle H_z \rangle$ and the crossover magnetic field $\langle x\, H_z \rangle$ are predicted to vary as follows:

$$\langle H_z \rangle = \frac{2}{5} H_d (\sin i \, \sin\beta \, \cos \Omega t + \cos i \, \cos\beta), \tag{1}$$

$$\langle x\, H_z \rangle = \frac{3}{16} H_d \, \sin\beta \, \sin \Omega t. \tag{2}$$

H_d is the field strength at the magnetic pole. β is the angle between the stellar rotation axis and the dipole axis. Ω is the angular rotation frequency of the star. Ωt is the angle between the plane defined by the stellar rotation axis and the dipole axis, on the one hand, and the plane defined by the line of sight and the stellar rotation axis, on the other hand, at time t (time origin is chosen in such a way that the dipole axis lies in the latter plane at $t = 0$).

One notes that, according to this model, both $\langle H_z \rangle$ and $\langle x\, H_z \rangle$ are expected to vary sinusoidally. The variations of $\langle x\, H_z \rangle$ should occur in phase quadrature with respect to those of $\langle H_z \rangle$ (corresponding to the well known fact that crossover effect is maximum when the mean longitudinal field reverses its polarity). $\langle x\, H_z \rangle$ is furthermore predicted to reverse symmetrically about 0.

Combining (1) and (2), one derives the following relation:

$$v \sin i \, \frac{\langle x\, H_z \rangle^+ - \langle x\, H_z \rangle^-}{\langle H_z \rangle^+ - \langle H_z \rangle^-} = \frac{15}{32} v. \tag{3}$$

In other words, assuming that the stellar magnetic field is a simple dipole, one can derive the stellar equatorial velocity from a suitable combination of the maximum $\langle H_z \rangle^+$ and minimum $\langle H_z \rangle^-$ of the mean longitudinal field, and of $v \sin i \langle x H_z \rangle^\pm$ (where $\langle x H_z \rangle^\pm$ are the extrema of the crossover field), that is of four quantities directly determined by simple measurements. From v, the stellar radius R can be derived. The consideration of the obtained value of R is a powerful test of the adequacy of the centred dipole model to represent the magnetic field of the considered star.

Within this frame, a study of the crossover effect has been completed for the 4 stars HD 83368, HD 125248, HD 137909, and HD 153882. The curve of variation of $\langle H_z \rangle$ for these 4 stars is sinusoidal. $\langle x H_z \rangle$ also varies sinusoidally. (By contrast, in HD 175362, where the variation of $\langle H_z \rangle$ is non-sinusoidal, the variation of $\langle x H_z \rangle$ also departs from a sinusoid.) The variation of $\langle x H_z \rangle$ occurs symmetrically about 0, except maybe for HD 83368, where a departure from this symmetry is found at the $2.4\,\sigma$ level. The phase lag $\Delta\phi$ between the variations of $\langle H_z \rangle$ and of $\langle x H_z \rangle$ does not significantly differ from 1/4 rotation cycle in either of the 4 stars. All this is consistent with the idea that the magnetic field geometry of the 4 considered stars is close to a centred dipole. However, deriving the stellar radii as described above, one gets reasonable values for 3 of the stars, but one that is unacceptably small for HD 125248. This shows that the magnetic field of this star probably departs significantly from a centred dipole. Alternatively, a more reasonable value of the radius could be obtained by adopting a somewhat different limb-darkening law (no longer fully consistent with the approximations made to derive $\langle H_z \rangle$ and $\langle x H_z \rangle$). Nevertheless, it would then seem reasonable to apply a similar limb-darkening law to the other 3 stars under consideration, which would yield unreasonable radii for them (in particular, for HD 137909).

5 Conclusion

Though they are still very preliminary and partial, the present results illustrate the power of the use of good quality spectropolarimetric observations, even of rather moderate resolution, to set constraints on the spatially unresolved structure of Ap star magnetic fields. It would be untimely to propose a model of the magnetic field of the considered stars at the present early stage of their study, but it is already confirmed that a significant fraction of Ap stars have magnetic fields definitely departing from a simple centred dipole geometry. Further details will be given in forthcoming papers.

References

Babcock H.W., 1956, ApJ 124, 489
Borra E.F., Landstreet J.D., 1980, ApJS 42, 421
Mathys G., 1989, Fundam. Cosmic Phys. 13, 143
Mathys G., 1991, A&AS 89, 121
Thompson I.B., Brown D.N., Landstreet, J.D., 1987, ApJS 64, 219

DIFFUSION, MASS LOSS AND ACCRETION IN STARS

Georges Michaud
Département de physique, Université de Montréal
Montréal, QC, CANADA H3C 3J7

I- Astrophysical Context

Atomic diffusion is a basic physical process and its effects can be absent in stars only if a more effective transport process wipes them out. Self consistent model atmospheres and evolutionary models must include either atomic diffusion or an alternate process, such as turbulence, efficient enough to wipe out gravitational settling and other effects of atomic diffusion. Using this approach, the observed abundance of Li can constrain the age of globular clusters as well as the hydrodynamical process that limits the effect of diffusion within stars (Proffitt and Michaud 1991). The explanation of the presence or absence of abundance anomalies becomes an essential part of our understanding of the structure of stars.

Models assuming that diffusion dominates all other particle transport processes explain a large fraction of the abundance anomalies observed on AmFm, HgMn, magnetic Ap, horizontal branch, white dwarf and He stars. In rotating stars it is possible to calculate the effect of meridional circulation using the model of Tassoul and Tassoul (1982) without introducing any arbitrary parameter. In A and B stars, meridional circulation competes with diffusion: in those objects where rotation is small enough that diffusion is a more efficient transport process than meridional circulation, abundance peculiarities appear and the objects become AmFm or HgMn stars. However, once meridional circulation is slow enough not to interfere with the settling of He, it is also slow enough not to have any effect on the overabundances of heavy elements (Charbonneau and Michaud 1991). For metals, it could only play a role in those A stars where He settling is not sufficient to allow the disappearance of the He convection zone and which are usually called normal A stars.

The observed anomalies are generally smaller than predicted by models that include only those hydrodynamical processes that we know how to calculate without arbitrary parameters: diffusion, meridional circulation and convection (if α is calibrated using the Sun). A detailed agreement between the observed abundance anomalies and the expected ones is not possible without the introduction of some other transport process. Turbulence has been investigated. It was shown by Vauclair *et al.* (1978) that it could not be the main factor in reducing abundance anomalies caused by diffusion in AmFm stars. Charbonneau and Michaud (1991) reinforced that conclusion by adding the simultaneous effect of meridional circulation on the particle transport. It does not appear likely that turbulence could be the dominant competing process in most objects. It is then specially interesting to investigate the potential effects of mass loss and its counterpart mass accretion on the observed abundance anomalies. Accretion has not yet been confirmed except in binary systems and in white dwarfs (Fontaine and Wesemael 1991, §4); it is interesting to look for additional signatures of its effects. Mass loss is observed to be present in all stars where the sensitivity allows to measure it and it is important to understand its effect on abundance anomalies. The presence of chromospheric activity among the F and late A stars suggests the presence of solar type winds (Simon and Landsman 1991).

The solar wind is the only weak stellar wind directly observable: it corresponds to a mass-loss rate of 10^{-14} $M_\odot$ yr^{-1}. We know that the solar wind is chemically differentiated: the elements that are neutral at $T = 10^4$ K are underabundant by factors $\sim$ 2 - 3 in the wind (time averaged in situ particle measurements) and in the corona (Meyer 1985). There is a controversy as to where the chemical separation occurs, the correlation with ionization potentials suggesting that it occurs where $T = 10^4$ K. This can be interpreted as implying that above the photosphere, for mass-loss rates of 10^{-14} $M_\odot$ yr^{-1}, the only separation possible is that occurring where elements are neutral.

The detailed effect of turbulence on meridional circulation is perhaps complicated by horizontal turbulence (Charbonneau and Michaud 1991, Chaboyer and Zahn 1991, Charbonneau 1991b). It was first shown by Charbonneau and Michaud (1991) that horizontal turbulence could wipe out the effect of meridional circulation in limiting gravitational settling. Chaboyer and Zahn (1991) considerably extended that result by showing that the effect of meridional circulation could be modeled by an equivalent vertical turbulence if horizontal turbulence is large enough. Charbonneau (1991b) showed that this was possible only for a limited range of parameters, so that for this effect to dominate requires implicit fine tuning and surprisingly large values of horizontal turbulence. The potential link between stellar evolution, turbulence and chemical abundances is being investigated in detail by the Yale group. See, for instance, Pinsonneault *et al.* (1990).

II- Competition between Mass Loss, Accretion and Diffusion

There are currently no satisfactory models of the origin of mass loss in the regime where it interferes with the appearance of abundance anomalies. Such models are needed. Similarly there are only very rough models of the evaluation of accretion rates (Proffitt and Michaud 1989). We will concentrate on the effect of the assumed mass-loss and accretion rates on chemical anomalies.

The mass-loss rate, dm/dt, is assumed small enough to have a negligible effect on stellar structure. It merely introduces, throughout the static stellar envelope, a global outward velocity of matter, v_w:

$$\frac{dm}{dt} = -4\pi R^2 \rho v_W \quad . \tag{1}$$

This equation expresses flux conservation for the main constituent. For the small accretion rates of interest to abundance anomalies, the signs of both dm/dt and v_W are merely reversed.

A trace element diffusing in the presence of mass loss must satisfy the conservation equation:

$$-N_H \frac{\delta c}{\delta t} = \frac{1}{r^2} \frac{\delta}{\delta r} \left(r^2 c N_H (v_W + v_D) \right) \tag{2}$$

where c is the concentration ($c = n(A)/N_H$) and v_D is the diffusion velocity of an ionized trace element of atomic mass A and charge Z in ionized hydrogen:

$$v_D = -D_{12}\{ \frac{\delta \ln c}{\delta r} + [(A - \frac{Z}{2} - \frac{1}{2})g - Ag_R] \frac{m_p}{kT} - k_T \frac{\delta \ln T}{\delta r} \}. \quad (3)$$

The first term on the right hand side is the "classical" diffusion term. It is 0.0 at the beginning of stellar evolution and seldom becomes important in stars since the equilibrium abundance gradients are rarely approached. The second term is the gravitational settling term modified for the presence of the electric field. It is followed by radiative acceleration and thermal diffusion. An equation valid for any binary mixture and in presence of degeneracy is given in Pelletier et al. (1986). A form for a ternary mixture, useful when two species have large abundances, is given by Proffitt and Michaud (1991; see also Loeb, Bahcall and Milgrom 1989). Effects of partial ionization are discussed in Michaud and Babel (1991) and Babel and Michaud (1991c).

The diffusion coefficients can be found in Paquette et al. (1986; see also Macdonald 1991). Contrary to those of Chapman and Cowling (1970), they take into account that stellar plasmas are not very dilute plasmas. The corrections are important in the interior of main sequence stars where the Coulomb logarithm appearing in the diffusion coefficient is not very large (Noerdlinger 1977) so that the logarithmic approximation made by Chapman and Cowling (1970) and Bahcall and Loeb (1990) is not accurate enough. In particular the energy dependence of the cutoff length is important (see §III of Paquette *et al.* 1986). While the coefficients of Paquette et al. (1986) are to be preferred to those of Chapman and Cowling, there remain uncertainties specially for the thermal diffusion coefficient and these pose interesting challenges to statistical physicists (see Michaud 1991). If diffusion coefficients are arbitrarily reduced, the constraints that diffusion places on hydrodynamics are relaxed. For Li in F stars, the thermal diffusion term is at most 20% of the gravitational settling term.

If the diffusing element is neutral, thermal diffusion becomes much smaller but the diffusion coefficient becomes about 100 times larger (Michaud Martel Ratel 1978). This increases by a factor of 100 the mass loss needed to compete with abundance anomalies at $T_5 = 0.1$ (T_5 is the temperature in units of 10^5 K). Approximate expressions for the critical mass-loss rates are obtained by equating the diffusion and mass loss velocities:

$$\frac{dM}{dt} = \frac{1.6\ 10^{-15} M\ A\ T_5^{1.5}}{Z^2} \quad \text{ionized}$$
$$= 10^{-14}\ M\ A \quad \text{neutral} \qquad (4)$$

where M is the mass of the star in solar units and A the atomic number of the chemical element. For more accurate work one should include the neutral diffusion coefficient explicitly (Michaud *et al.* 1978) instead of the average value used here. For neutral elements, the value given is for neutral elements diffusing in a background of protons; atomic diffusion of a neutral element in neutral hydrogen would be increased by another factor of approximately 2.

The equation for the diffusion velocity must be modified if hydrogen is partly neutral. There then appears an apparent diffusion velocity for both neutral hydrogen and ionized hydrogen which leads to no net transport of hydrogen but, in the ionization region, hydrogen is moving predominantly downwards in its neutral state and upwards in its ionized state. Since ionized elements interact much more with protons than with neutral hydrogen (by a factor of 100 approximately) this ambipolar diffusion leads to a drag velocity that can dominate all diffusion velocities (Babel and Michaud 1991c). To the diffusion velocity, v_D, given by equation (3) must then be added a drag term, giving approximately (see Babel and Michaud 1991 for more details and more exact equations):

$$v_t = v_D + \frac{v_1 D_{t0}}{D_{t1} + D_{t0}}$$

$$v_1 = \left(\frac{n_0}{n_0+n_1}\right)\left(\frac{kT}{m}\right)\left(\frac{n_1}{n_1\nu_{10}+n_0\gamma_{01}}\right)\left[-2\frac{d\ln p_1}{dr} + \frac{d\ln p_t}{dr}\left(\frac{2n_1+n_0}{n_1+n_0}\right) + \alpha\frac{d\ln T}{dr}\right] \tag{5}$$

with $p_t = (2n_1 + n_0)\,kT$.

Except when ambipolar diffusion and mass loss dominate, the competition between radiative acceleration (g_R) and gravity determines whether diffusion leads to over- or underabundances. Accurate radiative accelerations have only been calculated for a few elements. General formulae are available (Michaud *et al.* 1976; Alecian and Artru 1990) but are accurate to at most a factor of 3 (see Michaud 1987 for a discussion). Alecian and Artru (1990) have redetermined radiative accelerations for stellar interiors treating atomic configurations in more detail than Michaud *et al.* (1976). Calculations are needed using more accurate data banks. The required atomic data is becoming available from the Opacity project and from the large compilations of Kurucz (1991). The compilations of Kurucz have the advantage of greater accuracy in the position of levels since he uses measured positions. This allows taking the effect of blends on g_R into account. Detailed calculations are under way for C, N, O and Fe.

It has recently been suggested (see in particular Atukov and Shalagin 1988 and Popov *et al.* 1989) that Light Induced Drift (LID) could dominate particle transport in stars. LID occurs in the presence of a frequency dependence of the radiation flux in the vicinity of resonance lines of an element of interest. To simplify the argument, assume the resonance line is broadened only by the Doppler effect. When it is approaching the star, the element sees a different flux from the one it sees when going away from the star. It is consequently excited preferentially in one of the directions. If the collision cross-section of the excited state is twice that of the ground state, the mean free path in the excited state is half that in the ground state. The differential excitation in the two directions then leads to a drift velocity which could be of the order of the thermal velocity in favourable conditions. The difficulty in applying this effect to calculate abundance anomalies in stars comes from the fact that most chemical species are ionized in stellar atmospheres so that the interaction involves the Coulomb force that is not influenced by the excitation of an interior electron (Leblanc 1989, Leblanc and Michaud in preparation). The effect is then probably much smaller than suggested by Atukov and Shalagin (1988) though more detailed calculations are needed. It could play a role for some elements, specially in the cooler Ap stars where hydrogen and some metals are partly neutral. No detailed calculations have yet been published.

III- AmFm Stars, λ Bootis Stars and the Li Gap.

The parameter free model that has been developed for AmFm stars is in reasonable agreement with observations in so far as the elements that are underabundant have $g_R < g$ just below the H convection zone and settle from it while those that are overabundant have $g_R > g$ and are supported by g_R immediately below the convection zone (Watson 1971, Michaud *et al.* 1976). However more detailed agreement requires the introduction of mass loss (Michaud *et al.* 1983; Michaud and Charland 1986) so that overabundances produced by atomic diffusion do not become much larger than those observed. Furthermore the observations of the Li gap in F stars (Boesgaard and Tripicco 1986) led Michaud (1986) to suggest a link with AmFm stars (see also Michaud and Charbonneau 1991). Mass loss would presumably also play a role in those objects.

At the age of the Hyades, the abundance of Li at the surface of AmFm stars of clusters is affected by evolutionary effects (Richer 1991). The stellar evolution was calculated in detail (Fig. 1) taking all evolutionary effects into account as well as the effect of the settling of He on the structure of the models. The opacities used are from Huebner *et al.* (1977) for the interior opacities and from Kurucz for the atmospheric ones. The settling of He leads to a reduction of the T_{eff} by about 100 K at a given age (see the bottom part of the figure). On the middle part of the figure, different tones of darkness are coded to the Li abundance as a function of time and position in a 1.50 $M_\odot$ star. The dashed line shows the position of the bottom of the superficial convection zone. It is strongly affected by the settling of He.

The surface abundance of Li (upper part of the figure) is the same as in the convection zone, which is assumed completely mixed. Close to the age of the Hyades ($8\ 10^8$ yr) there occurs a dredge up of the Li that had accumulated just below the convection zone during the preceding evolution, when the bottom of the convection zone was slightly above $\Delta M/M = 3\ 10^{-9}$. During the preceding evolution, Li is pushed upwards by $g_R(Li)$ over the small mass interval $10^{-8} < \Delta M/M < 10^{-6}$ where $g_R(Li) > g$ (Fig. 2). This dredge up phase lasts about $3\ 10^8$ yr and is followed by a phase during which matter without any Li is dredged up. It occurs when the convection zone crosses the area where, during the past evolution, $g_R(Li) < g$ (for $\Delta M/M > 10^{-7}$ on Fig. 2). When the convection zone extends deeper in the envelope, the area where gravitational settling did not have time to modify the Li abundance is reached and another increase in surface Li abundance occurs.

The corresponding T_{eff} evolution can be followed from the lower part of the figure. The small abundances do occur both in an early evolutionary phase in which the star could have been an AmFm star in a young cluster and in a later evolutionary phase ($t \sim 10^9$ yr) when it would have had approximately the T_{eff} (=6700) of the Li gap. At the age of the Hyades (taken here as $8\ 10^8$ yr), this star would have (see the upper part of the figure) an overabundance of Li by a factor of about 4 for a mass-loss rate of $10^{-15}\ M_\odot\ yr^{-1}$ but would have an overabundance by a factor of about 30 in the absence of mass loss.

This complex behaviour is expected for all elements that accumulate below the convection zone of AmFm stars. To carry out such calculations requires accurate radiative accelerations throughout the envelope for all elements of interest.

It is premature to say if detailed models including mass loss for individual stars will be able to explain all observed abundance anomalies but it is clear that detailed agreement with observations requires a large amount of modelling since, for all observed elements, it will require calculations similar to those made here for Li including dredge up phases. Similar evolutionary effects probably also play a role for HgMn and magnetic stars. They are probably larger for the Fm stars since the convection zone is deeper than in the hotter Am or HgMn stars but this needs to be confirmed by detailed modelling.

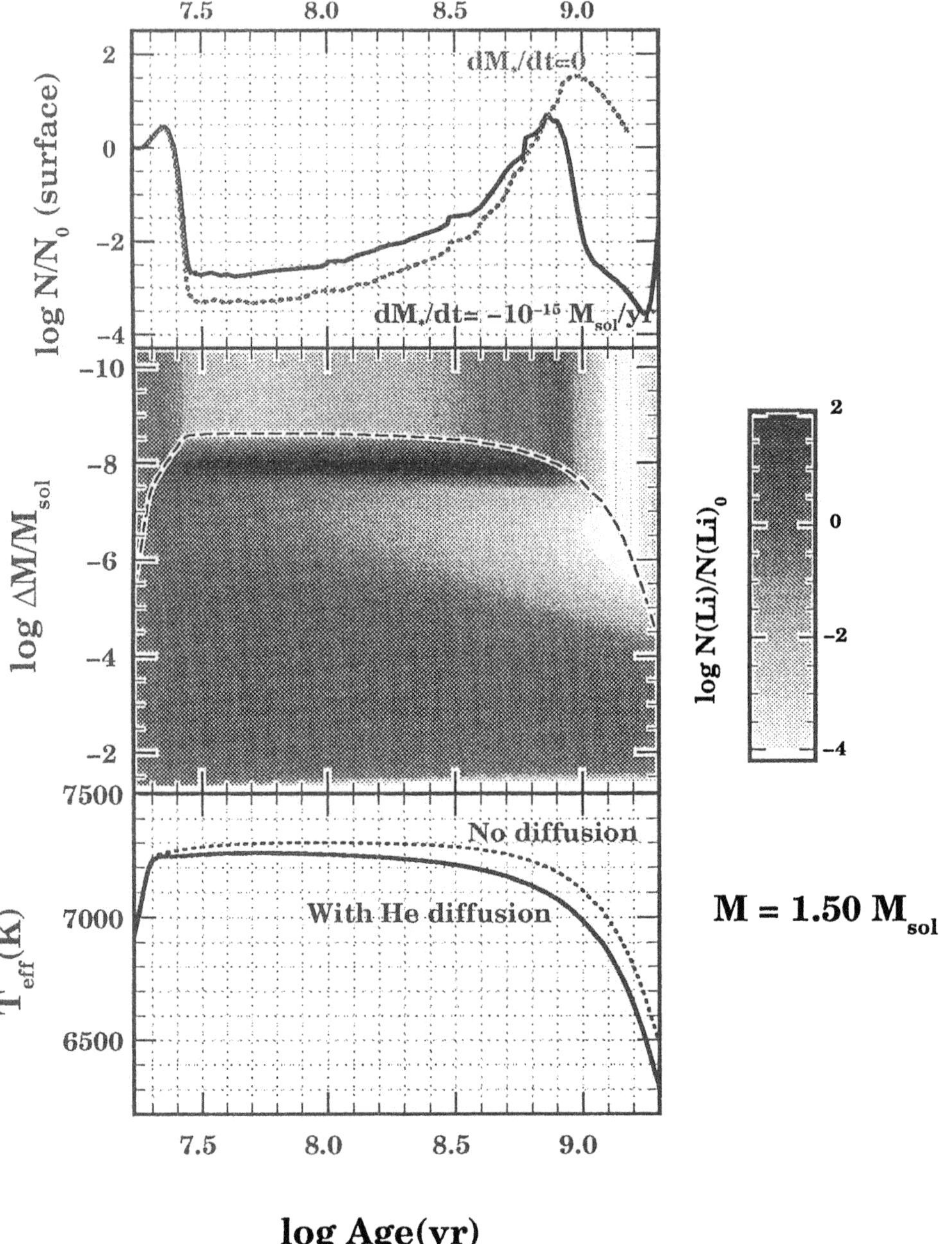

Figure 1 The lower part of the figure shows the T_{eff} evolution as a function of the age of a 1.50 $M_{\odot}$ star when gravitational settling is included (full line) and when it is neglected. The middle part shows the Li abundance coded by tones of grey as a function of age and of the position in the star. The position of the bottom of the convection zone is indicated by the dashed line. The upper part of the figure shows the time variation of the Li abundance in the convection zone and photosphere both in the absence of mass loss and for a mass-loss rate of 10^{-15} $M_{\odot}$ yr^{-1}. See Richer (1991) for other examples.

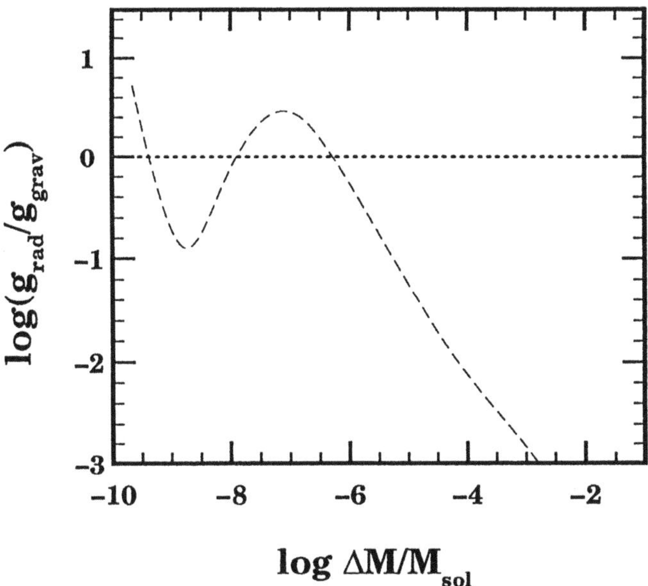

Figure 2 Radiative acceleration as a function of mass in the same star as on Figure 1 at an age of 10^8 yr.

a) λ Bootis stars and mass loss

Mass-loss rates larger than 10^{-12} $M_\odot$ yr^{-1} drag all ionized chemical species from the atmosphere to the interstellar medium (this limit varies with the state of ionization and the temperature but no case has been found when it was larger than this value by orders of magnitude). The abundance anomalies become progressively smaller as the mass-loss rate is increased and elements are carried more effectively into the interstellar matter. In the limit of very large mass-loss rates, there is no anomaly since there is no possibility for separation during the short interval required to go through the stellar envelope (see Michaud and Charland 1986).

It has been suggested by Michaud and Charland (1986) that the underabundances observed in the λ Bootis stars could be explained by gravitational settling in the presence of a mass-loss rate of about 10^{-13} $M_\odot$ yr^{-1}. For such a mass-loss rate, the separation would only occur in the deep envelope. This model can lead to underabundances by factors of 3-5 but no larger. While such anomalies are compatible with most analyses of λ Bootis stars (see Baschek, this volume) they are not with the recent results of Venn and Lambert (1990). Charbonneau (1991a) has further obtained by a detailed solution of gravitational settling in presence of meridional circulation and mass loss that no underabundances would be produced for the suggested mass-loss rate in stars rotating as fast as the λ Bootis stars (equatorial rotation velocities larger than 100 km s^{-1}). This assumed the meridional circulation model of Tassoul and Tassoul (1982) that is successful in explaining the observational limit of the AmFm and HgMn phenomena (see above).

There is however one way in which the calculations of Charbonneau (1991a) could be reconciled with an explanation of the λ Bootis stars by the diffusion model and with the meridional circulation model of Tassoul and Tassoul. It was shown by Michaud and Charbonneau (1991) (see also Chaboyer and Zahn 1991 and Charbonneau 1991b) that horizontal turbulence could eliminate the effect of meridional circulation on gravitational settling. The required horizonal turbulence is large and it is not clear that it is generated in radiative zones. However, it is tempting to suggest that it would be negligible in the radiative zones of the slowly rotating AmFm stars and would progressively increase with rotational velocity becoming large for the λ Bootis stars that have equatorial rotation velocities exceeding 100 km s^{-1}. This has the disadvantage of requiring some fine tuning and adding an implicit assumption, on horizontal turbulence, for the λ Bootis model of Michaud and Charland to apply. It might be worth investigating the required rates quantitatively to see how much fine tuning is required.

An alternate model has been suggested for λ Bootis stars by Venn and Lambert (1990). It was studied quantitatively by Charbonneau (1991a) who obtained that an *accretion* rate of 10^{-13} $M_{\odot}$ yr^{-1} could explain some of the observed underabundances *if* the matter that is accreted is differentiated as observed in interstellar matter. This requires the assumption both of the appropriate accretion rate and of the appropriate chemical separation of the interstellar mater while the model based on diffusion for λ Bootis stars requires the assumption of the mass-loss rate and of horizontal turbulence.

IV- Magnetic Stars

While it now appears very likely that atomic diffusion is important in magnetic stars, there is no complete model of these objects. The underlying hydrodynamics are poorly known but could greatly modify the abundance anomalies that appear at the surface. The number of potential hydrodynamic processes and the various ways in which they could interact throughout the evolutionary life of stars may explain the diversity of anomalies observed at the surface of magnetic Ap stars.

Meridional circulation, turbulence, mass loss, and the presence or absence of convection may play a role. If the results of Charbonneau and Michaud (1991) apply to magnetic stars, they imply that meridional circulation has a negligible effect on the abundance of metals once the equatorial rotation velocity is smaller than 100 km s^{-1}, which is the case for all ApBp stars except for some young He rich stars. Magnetic fields may suppress meridional circulation though there remain uncertainties on the interaction between magnetic fields, meridional circulation and differential rotation (Mestel and Moss 1977; Tassoul and Tassoul 1986; Charbonneau and MacGregor 1991).

Finally the interaction of convection and magnetic fields may lead to the disappearance of convection. However, this is not well established. We will consider both models where it is assumed that convection is suppressed and others where convection is not affected by magnetic fields. Magnetic fields must at least suppress any overshooting above the hydrogen convection zone since, otherwise, the observed horizontal abundance inhomogeneities would not survive (see Michaud 1980). Furthermore there are theoretical arguments suggesting that magnetic fields, even weak ones, suppress turbulence (see for instance Cattaneo and Vainshtein 1991). Since including turbulence would necessarily involve arbitrary constants, turbulence will be neglected in what follows.

a) No convection zones: ambipolar diffusion

Assume first that the convection zone is eliminated by the magnetic field. This is not firmly established but probable. Then ambipolar diffusion becomes an important process.

The ambipolar diffusion of hydrogen occurs because of the ionization gradient. In the ionization region this may dominate all other diffusion velocities. It leads to no real particle transport of hydrogen since all upwards diffusion of protons is cancelled by downwards diffusion of neutral hydrogen. All ionized species (for instance Ca II) interact at least 100 times more strongly with ionized hydrogen than with neutral hydrogen. They are consequently dragged upwards by ambipolar diffusion. The potential effect of ambipolar diffusion on the diffusion velocity of Ca is shown on Fig. 2 of Babel and Michaud (1991c).

Ambipolar diffusion is important in dragging trace species only in the interval where hydrogen ionizes. The T interval over which this occurs is shown as a function of T_{eff} on Fig. 4 of Babel and Michaud (1991c). Ambipolar diffusion may be neglected in AmFm stars as well as in stars with $T_{eff} > 12000$ K. There is a small T_{eff} interval ($11000 < T_{eff} < 12000$ K) where it may play a role in non-magnetic stars but this depends rather sensitively on the extent of convection in those objects.

If the convection zone is suppressed by the magnetic field, ambipolar diffusion dominates in the zone formerly occupied by the convection zone (in what follows we call it the ionization zone, implicitly of hydrogen). All chemical species that are mainly ionized in the ionization zone are dragged by the upwards going protons even if the radiative acceleration is negligible for them. Elements that arrive at the bottom of the ionization zone are carried very rapidly to the top. To stop the dragging by ambipolar diffusion, very large gradients of the dragged element are required. In practice, the flux of protons carries upwards all elements arriving at the bottom of the ionization zone. It may be compared to a convection zone: in a convection zone turbulence is believed to be very large so that any abundance gradient is rapidly homogenized (Schatzman 1969). The convection zone homogenizes abundances between its bottom and its top. Ambipolar diffusion however creates large gradients of heavy element concentration where there would have been a convection zone.

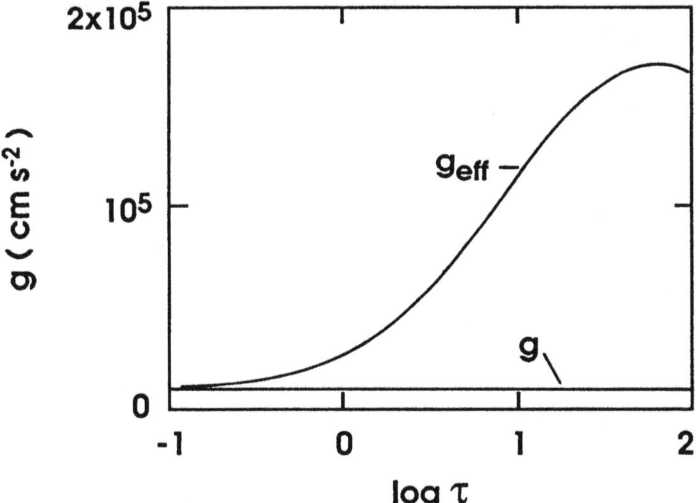

Figure 3 In the presence of an extremely large horizontal magnetic field, ambipolar diffusion leads to an apparent increase of gravity in the line forming region by more than a factor of 10.

There is no detailed model applying ambipolar diffusion yet. The first step is to obtain atmosphere models that take the effect of ambipolar diffusion on the structure of the atmosphere into account. The changes that ambipolar diffusion can make to the structure are potentially very large. This may be seen on Fig. 3 where is shown the maximum structural effect of ambipolar diffusion. It was calculated by assuming that the magnetic field was large enough to eliminate all diffusion of H II. Because ionized hydrogen is completely frozen by the magnetic field, its pressure gradient is modified by the magnetic pressure. However the pressure gradient of neutral hydrogen is not affected by the magnetic field. Using the equations from Burgers (1969), Babel and Michaud (1991c) obtain for neutral hydrogen:

$$\frac{\delta p_0}{\delta r} + n_0 mg = -mn_0[\nu_{01}(v_0 - v_1) - \nu_{0e}(v_0 - v_e)] \qquad (6)$$

where ν_{01} is the collision frequency between protons and neutral hydrogen while ν_{0e} is the collision frequency between protons and electrons. In the presence of a strong magnetic field, the diffusion velocities of protons and electrons must go to zero, which implies that the diffusion velocity of neutral hydrogen must also be zero at equilibrium. The right hand side of the equation must be zero which forces hydrostatic equilibrium for neutral hydrogen alone:

$$\frac{\delta p_0}{\delta r} = -\frac{p_0 gm}{kt} \qquad (7)$$

The proton pressure is then determined by the Saha equation. One then determines an effective gravity using the hydrostatic equation for the total pressure. This is the effective gravity shown in Fig. 3 for the grey case in an atmosphere with $T_{eff} = 8500$ K and log g = 4. The effective gravity could be up to fifteen times larger than gravity in the optical depth interval which covers the photosphere (see Fig. 3). It decreases back to g for optical depths larger than 10^4 (not shown). A comparison with the effect in a Kurucz atmosphere for realistic values of the magnetic field and taking in detail the effect of collisions (see Fig. 7 of Babel and Michaud 1991 c) shows that the maximal effects would only be approached for magnetic fields of 10^5 G or more but the effective gravity can be 2 to 10 times larger than gravity in the line forming regions of some Ap stars. These latter calculations did not include the effect of increased effective gravity on radiative transfer.

b) Effect of mass loss

Mass loss can have a drastically different effect depending on how v_w (see eq. [1]) compares to diffusion velocities. Numerous situations can arise in peculiar stars; only a few have been calculated and we will discuss only two of those. The effect of mass loss on non magnetic stars was discussed above.

i) Helium rich stars

Overabundances of He can materialize through separation in the atmosphere of a T_{eff}= 18000 K star for a mass-loss rate of 10^{-12} $M_\odot$ yr^{-1} but in the atmosphere of a T_{eff} = 25000 K star for mass-loss rate of 3 10^{-13} $M_\odot$ yr^{-1} (see Fig. 7 of Michaud *et al.* 1987; see also Vauclair 1975). It does not appear possible to explain the overabundances that are observed if larger mass-loss rates are present. The decrease of the required mass-loss rate as T_{eff} increases is surprising and may well require that the mass loss be modulated by the magnetic field. It was also concluded by Michaud *et al.* (1987) that the required rate implied that there could be no CNO abundance anomalies where the overabundances of He appeared. This may lead to an observational test. An alternate model would assume that the separation occurs in the wind itself (see Michaud *et al.* 1987).

Models of radiation driven winds for such stars are needed. Magnetic fields probably play a role in those winds.

ii) 53 Cam and Ca II K lines

Recently Babel (1991a, 1991b) investigated the effect of a magnetic field controlled mass loss on the abundances of 53 Cam. He assumed that the magnetic field does not suppress the H convection zone. Consider moderate mass-loss rates ($< 10^{-14}$ $M_\odot$ yr^{-1}), for which the gravitational settling velocity in the atmosphere is larger than the velocity caused by mass loss (see eq. [4]; hydrogen is mainly neutral in the atmosphere of a T_{eff} = 8500 K star). Chemical species that are not supported in the uppermost atmosphere by radiative acceleration accumulate in the atmosphere if they are pushed there by radiative acceleration or dragged there by mass loss. Large abundance anomalies are then possible. Mass loss has also been suggested to be related to the driving mechanism of the pulsations observed on some magnetic Ap stars (Dolez Vauclair and Gough 1991; see also Babel 1991a).

The Ca II K lines offer a good test of the effect of mass loss. In 53 Cam, Landstreet (1988) found it impossible to fit the profile of this line using rings of different abundances related to the magnetic geometry. However, these resonant lines are very sensitive to the Ca distribution in the photosphere. As seen in Figure 4, while the core of the line hardly varies, the wings become much stronger as the mass-loss rate increases from 10^{-15} to 10^{-14} $M_\odot$ yr^{-1}. As can be seen in Figure 3 of Babel (1991b), the Ca abundance in the convection zone and lower atmosphere ($\tau > .01$) increases by more than a factor of 10 when the mass-loss rate increases from 2 to 4 10^{-15} $M_\odot$ yr^{-1} (see also Babel 1991a). This behaviour may be traced to the presence of a large reservoir of Ca slightly below the H convection zone and the small value of the radiative acceleration for the small mass interval between the Ca reservoir and the H convection zone. Evolutionary effects as discussed above for Li have not yet been carried out but could be important. Above the convection zone, Ca is not supported and its abundance gradient is sensitive to the mass-loss rate. The line profile then becomes a sensitive function of the mass-loss rate. The ratio of the wings to the core varies strongly with mass-loss rate. There are three parameters that appear in this model: the size of the two polar caps where mass loss is possible, and the value of the mass-loss rate, assumed to be the same in both of them. The sizes of the two polar caps are only partly independent of each other in that the polar caps were assumed to extend only until the magnetic field made an angle to the vertical chosen to be the same for both of them. There are then two parameters.

It is also assumed in this model that Ca is not dragged where the temperature is very large, in the coronal part of the solution. Using eq. (4) for T_5 = 10, A=40, M = 1.7 and Z= 5 gives a critical mass-loss rate of 10^{-13} $M_\odot$ yr^{-1}, so that it is possible that Ca would not be dragged in such a case for the mass-loss rates considered. This critical velocity could however overestimate the separation in the wind (Michaud *et al.* 1987).

The fit that is obtained is shown on Figure 5 for the time variation of the Ca II K lines. It is quite acceptable. Furthermore, Babel (1991b) also obtained that the agreement was then very good for both the average abundances and the time dependent abundance variations of the lines of Ca I in addition to Fe, Cr, and Sr. There is however no explanation for the geographic distribution of Ti nor the absence of Mn lines. While this model is not the final one for 53 Cam, it suggests that stratification, probably related to mass loss, will play a role in the final model. The progressive decrease of the Ca abundance from the top of the convection zone through the line forming region leaves a signature in the K lines of Ca II.

V- The Abundance Connection

Once it becomes possible to determine line profiles with accuracy, the use of abundance anomalies as constraints to stellar evolution and to stellar hydrodynamics becomes a reality. The observation of AmFm stars in clusters makes it possible to study the matter that lied deeper in the star during the preceding evolution and to test it for potential effects of diffusion and other particle transport processes (see § III). Observations and modelling have been done for Li but many other anomalies are available in AmFm stars and remain to be exploited. In magnetic stars with $T_{eff} > 18000$ K, while the presence of He overabundance seems to imply the presence of mass loss, the abundance of CNO imposes additional constraints and the observed anomalies constrain the possible mass loss models and the link to magnetic fields (§IV b i). Similarly, the time dependence of the profile of the Ca II K lines suggests the presence of a Ca abundance gradient in the atmosphere of this star. It appears to be related to the presence of the magnetic field and probably to mass loss. The other observed abundance anomalies constrain the model; abundance determinations for other chemical elements would further constrain the model (§IV b ii).

If 53 Cam has been so useful to test the diffusion model for magnetic stars, it is because most of its properties have been determined observationally. The magnetic field geometry has been fixed and the abundance cartography has been precisely determined. The age of the star is missing; if available, it would allow to determine the role played by the preceding evolution as was done for Li in AmFm stars in §III. This requires that some cluster Ap stars be studied as carefully as 53 Cam has been studied. While the faintness of cluster Ap stars makes this difficult the availability of larger telescopes and sensitive detectors makes it possible.

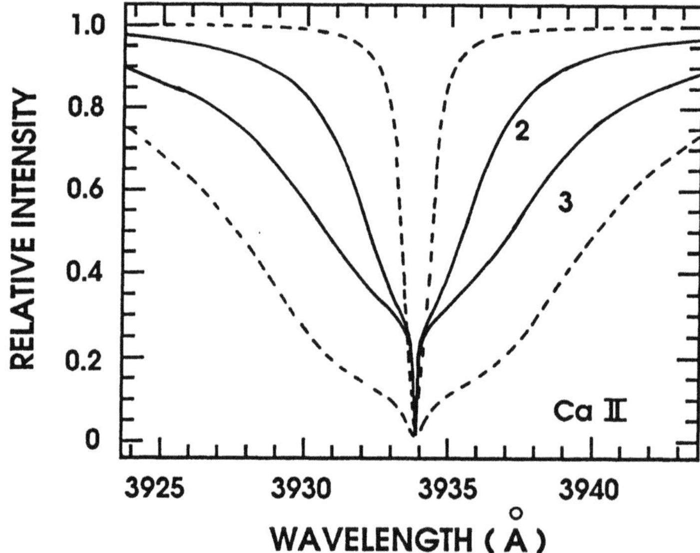

Figure 4 The dashed lines show the profile of the Ca II K line for two different homogeneous abundances while the full lines show the profile for mass-loss rates indicated, in units of 10^{-15} $M_\odot$ yr^{-1}. This figure was adapted from Babel (1991a, 1991b)

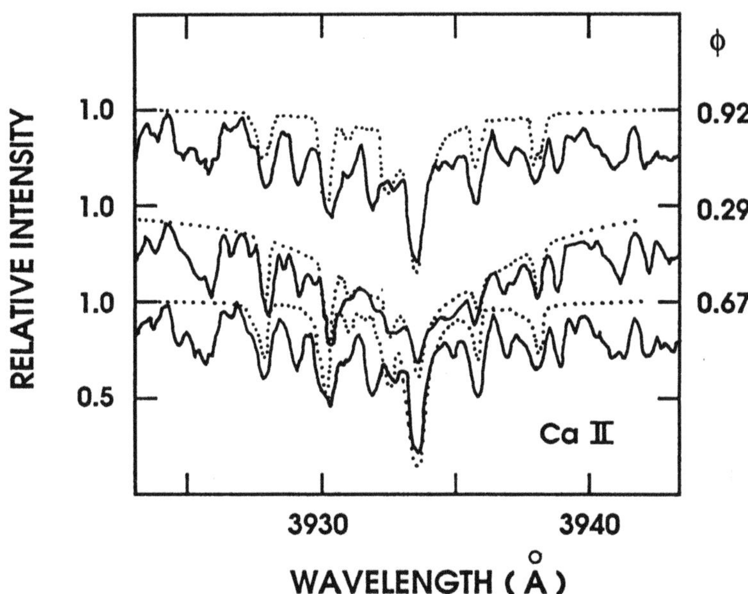

Figure 5 Comparison of the observed and synthesized spectra of the Ca II K line at three different phases. A mass-loss rate of 3 10^{-15} $M_{\odot}$ yr^{-1} was assumed. This is a subset of the phase coverage shown in Babel (1991a, 1991b).

REFERENCES

Alecian, G., and Artru, M.-C. 1990, A&A, 234, 323.

Atukov, S. N., and Shalagin, A. M. 1988, Sov. Astron. Lett. 14, 284.

Babel, J. 1991a, Thèse No. 2494, Université de Genève, Genève, Suisse.

Babel, J. 1991b, A&A, submitted for publication.

Babel, J., and Michaud, G. 1991a, ApJ, 366, 560.

Babel, J., and Michaud, G. 1991b, A&A, 241, 493.

Babel, J., and Michaud, G. 1991c, A&A, in press.

Boesgaard, A. M., and Tripicco, M. J. 1986, ApJL, 302, L49.

Burgers, J. M. 1969, Flow Equations for Composite Gases, (New York: Academic Press).

Cattaneo, F., and Vainshtein, S. I. 1991, ApJL, 376, L21.

Chaboyer, B., and Zahn, J.-P. 1991, preprint.

Chapman, S., and Cowling, T. G. 1970, The Mathematical Theory of non-uniform Gases (3d ed.; Cambridge: Cambridge University Press).

Charbonneau, P. 1991a, ApJL, 372, L33.

Charbonneau, P. 1991b, preprint.

Charbonneau, P., and MacGregor, K. B. 1991, ApJ, in press

Charbonneau, P., and Michaud, G. 1991, ApJ, 370, 693.

Dolez, N., Vauclair, S., and Gough, D. O. 1991, *preprint*.

Fontaine, G., and Wesemael, F. 1991, in IAU Symposium 145, **Evolution of Stars: the**

Photospheric Abundance Connection, Golden Sands, Bulgaria, 25-31 August, ed. G. Michaud and A. Tutukov (Kluwer), pp. 421-434.
Huebner, W. F., Merts, A. L., Magee, N. H., and Argo, M. F. 1977, Los Alamos Sci. Lab. Rep., LA-6760-M.
Kurucz, R. L. 1991, private communication.
Landstreet, J. D. 1988, ApJ, 326, 967.
Landstreet, J. D., Barker, P. K., Bohlender, D. A., Jewison, M. S. 1989, ApJ, 344, 876.
Leblanc, F. 1989, Université de Montréal, Mémoire de maîtrise.
Loeb, A., Bahcall, J. N., and Milgrom, M. 1989, ApJ, 341, 1108.
MacDonald, J. 1991, ApJS, 76, 369.
Mestel, L., and Moss, D. L. 1977, MNRAS, 178, 27.
Meyer, J. P. 1985, ApJS, 57, 151.
Michaud, G. 1980, AJ, 85, 589.
Michaud, G. 1986, ApJ, 302, 650.
Michaud, G. 1987, Physica Scripta, 36, 112.
Michaud, G. 1991, Annales de Physique, in press.
Michaud, G., and Babel, J. 1991, in Stellar Atmospheres: Beyond Classical Models, NATO Advanced Research Workshop, NATO ASI Series, Ed. L. Crivellari, I. Hubeny and D. G. Hummer (Doldrecht: Kluwer) 375-386.
Michaud, G., and Charbonneau, P. 1991, Space Science Reviews, 57, 1.
Michaud, G., and Charland, Y. 1986, ApJ, 311, 326.
Michaud, G., Charland, Y., Vauclair, S., and Vauclair, G. 1976, ApJ, 210, 447.
Michaud, G., Dupuis, J., Fontaine, G., and Montmerle, T. 1987, ApJ, 322, 302.
Michaud, G., Martel, A., and Ratel, A. 1978, ApJ, 226, 483.
Michaud, G., Tarasick, D., Charland, Y., and Pelletier, C. 1983, ApJ, 269, 239.
Noerdlinger, P. D. 1977, A&A, 57, 407.
Paquette, C., Pelletier, C., Fontaine, G., and Michaud, G. 1986, ApJS, 61, 177.
Pelletier, C., Fontaine, G., Wesemael, F., Michaud, G., and Wegner, G. 1986, ApJ, 307, 242.
Pinsonneault, M. H., Kawaler, S. D., and Demarque, P. 1990, ApJS, 74, 501.
Popov, A. K., Shalagin, A. M., Streater, A. D., and Woerdman, J. P. 1989, Phys. Rev., 40, 867.
Proffitt, C. R., and Michaud, G. 1989, ApJ, 345, 998.
Proffitt, C. R., and Michaud, G. 1991, ApJ, 371, 584.
Richer, J. 1991, Thèse de doctorat, Université de Montréal, in preparation.
Schatzman, E. 1969, A&A, 3, 331.
Simon, T., and Landsman, W. 1991, ApJ, 380, 200.
Tassoul, J.-L., and Tassoul, M. 1982, ApJS, 49, 317.
Tassoul, J.-L., and Tassoul, M. 1986, ApJ, 310, 786.
Vauclair, S. 1975, A&A, 45, 233.
Vauclair, G., Vauclair, S., and Michaud, G. 1978, ApJ, 223, 920.
Venn, K. A., and Lambert, D. L. 1990, ApJ, 363, 234.
Watson, W. D. 1971, A&A, 13, 263.

PHOTOSPHERIC ABUNDANCES IN late-A AND Am-Fm STARS

Claude Burkhart
Observatoire de Lyon
69561 St-Genis-Laval Cedex France

Marie-France Coupry
Observatoire de Paris
61 Avenue de l'Observatoire
75014 Paris France

The Main Sequence A stars with intermediate mass provide us a laboratory to observe and study non-thermal processes such as miscroscopic diffusive separation of elements, meridional circulation, turbulent particle transport because these hydrodynamical processes are not masked by mass flows (more massive MS stars) or convection (less massive MS stars). In particular, the Am stars with thin convection zones and slow rotation are a right place to observe diffusion.

In that prospect, 25 late-A and Am-Fm stars, mostly field stars, were observed with high spectral resolution and high signal-to-noise ratio at the ESO CAT and the CFH telescopes. Photospheric abundances were determined for Li, Al, Si, and Fe from a model atmosphere abundance analysis; they are discussed in Burkhart & Coupry (1991). In that region of the H-R diagram where are met SB2 Am stars and δ Scuti pulsating stars, more than one third of the spectra display noteworthy peculiarities, principally in their line profiles. The microturbulent velocity fitting parameter, ξ_t, is a by-product of the abundance analysis. The results of these two last items are summarized in this contribution.

1. Curious-looking line profiles

Line profiles carry, of course, information about the distribution of velocities across the stellar disk, but in that region of the H-R diagram many stars (Fig. 1) emphasize the danger of interpreting them straightforwardly. Beware of interpreting (variable) asymmetrical line profiles as the signature of δ Scuti pulsations without paying attention to any possible occurence of multiple spectroscopic binarity. Abundance results may have been spoiled by unrecognised multiple spectroscopic binarity. When using projected rotational velocities of catalogues, a part of them corresponds to observations of unresolved lines due to the grouping of 2 (or more) systems of stellar lines...

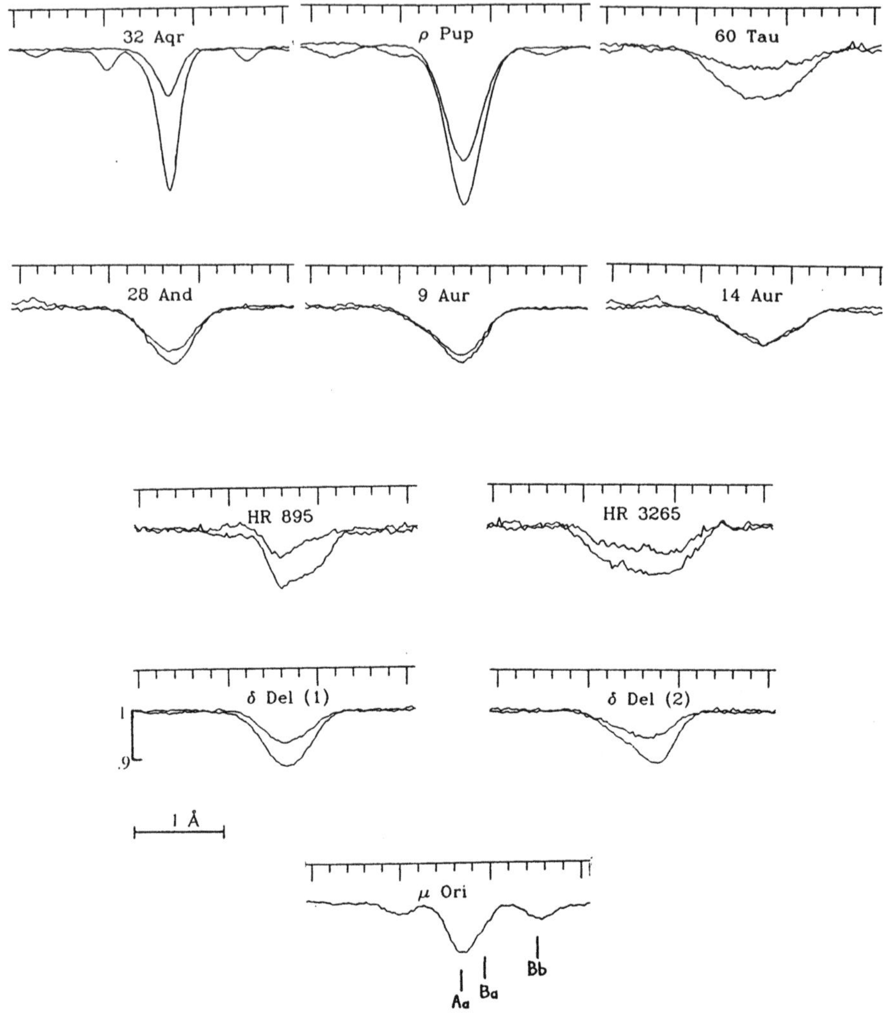

Fig. 1 Line profiles

Line-set : FeI-6678 (the strongest) and Ca-6717 lines. The ratio, Ca/Fe, of the strength of both lines sorts out the Am character, i.e., a weak calcium, when Ca/Fe $\ll 1$

First line : symmetrical line sets of reference arranged in order of increasing projected rotational velocity

Second line : 3 Ca normal stars; 28 And and 9 Aur with extended blue wing profiles and 14 Aur with isosceles triangle-shape profiles

Third line : 2 Ca weak stars; HR 895 with strongly asymmetrical profile with blue side steeper than red side, HR 3265 with rotational broadened profile and weak asymmetry (red side steeper than blue side)

Fourth line : this Ca weak star gives an example of observed variable line shape if δ Scuti pulsation coexists with multiple spectroscopic binarity (SB2)

Fifth line : FeI-6678 line of μ Ori, a SB3 star. The line of the brighter component is asymmetrical with blue side steeper than red side : the third component is detected in the red wing

EFFECTIVE TEMPERATURE AND MODELS FOR EARLY A STARS: APPLICATION TO 78VIR (A2P)

R. MONIER
ESA IUE Observatory , Apartado 50727, 28 MADRID, SPAIN

ABSTRACT. The energy distribution (ED) of the cool Ap star 78Vir is presented from 1200Å to 22000Å at phase 0.0 of its 3.7220 dys rotation period. The ED is used to calculate the integrated flux and the Infrared Flux Method (IRFM,Blackwell and Shallis, 1977) is applied to derive the effective temperature and the angular diameter of 78Vir. A grid of models was then computed for the found effective temperature within its uncertainty, surface gravities around 4.0 and three different chemical compositions. A chisquare technique is used to sort out the best fit(s) to the observed ED.

1. Outline of the method

To construct the energy distribution (ED) of 78Vir, I have merged ultraviolet spectra covering the range from 1200Å to 3200Å with published optical spectrophotometry from 3300Å up to 8000Å (Pyper and Adelman,1983) and infrared photometry in the J and K bands from Groote and Kaufmann (1983). The optical spectrophotometry has been transformed into absolute fluxes using Hayes and Latham's (1975) calibration of Vega. The infrared magnitudes were converted into absolute fluxes using Hayes' (1979) infrared calibration of Vega. The ultraviolet, optical and infrared absolute fluxes were then merged together to yield the energy distribution from 1200Å to 22000Å. Once the ED was available, the IRFM was then used to derive the effective temperature.
The IFRM (Blackwell and Shallis,1977) allows a direct determination of the effective temperature from the integrated flux, F, which defines it. The later was measured by fitting cubic splines throughout the entire ED and integrating under the splines. The infrared fluxes were then used together with predicted fluxes to derive the angular diameter using the relation:

$$f_\lambda = 0.25\theta^2 F_\lambda \ (1)$$

f_λ is the observed infrared flux and F_λ is the model flux at the infrared wavelength computed for a first guess effective temperature. Once θ is derived, the effective temperature is found from its definition:

$$F = 0.25\theta^2 \sigma T_{eff}^4 \ (3)$$

A grid of models was then computed varying the effective temperature inside its allowed range $T_{eff} \pm \delta T_{eff}$ as found with the IRFM (δT_{eff} ,uncertainty on T_{eff}) , the surface gravity in the range $logg \pm 0.25$, and for 3 different abundance sets (solar, 3 times solar and 10 times solar). The observed and model fluxes have been ratioed to their values at 5000Å to allow for comparison. The quality of the fit between the model and the observations

2. Microturbulence

We confirm the "low ξ_t scale" which will produce a "high abundance scale" for the elements studied from only strong lines.

The microturbulence parameter, ξ_t, exhibits as a function of temperature a maximum for both late-A and Am-Fm stars on or near the Main Sequence (Fig. 2). This maximum occurs for a temperature which corresponds
- with the intersection of the Instability Strip and the Main Sequence (physical motions linked with pulsational instabilities in the atmosphere ?),
- with the occurence of very shallow convective zones (velocities linked with "shallow convection" ?).

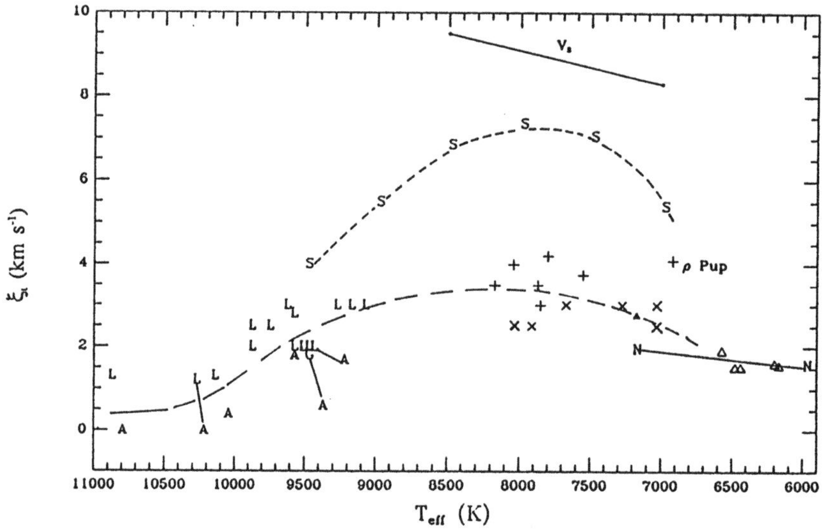

Fig. 2 Microtubulence versus T_{eff}
+ Am, deficient Ca × normal star ▴ unknown Ca △ late-F (this work)
L hot A and Am stars (Lemke, 1989)
A hot A and Am stars (Adelman, 1991)
G Vega (Gigas, 1986)
NN F stars (Nissen, 1981)
S curve for A and Am stars; V_s sonic velocity in the atmosphere (Smith, 1971)
Short lines connect results obtained by different authors for the same star

References

Adelman S.J., 1991, MNRAS 252, 116
Burkhart C., Coupry M.F., 1991, A&A 249, 205
Gigas D., 1986, A&A 165, 170
Lemke M;, 1989, A&A 225, 125
Nissen P.E., 1981, A&A 97,145
Smith M.A., 1971, A&A 11, 325

was evaluated by calculating the chisquare (χ^2) of the observed energy distribution versus the model; the model(s) which best fit having the smallest chisquare. The ultraviolet, optical and infrared fluxes fluxes have been weighted by their respective uncertainties σ_i : $\pm$ 5% in the ultraviolet , $\pm$ 1.5% in the optical and $\pm$ 3.5% in the infrared . This procedure lowers the weight of the ultraviolet and infrared fluxes in the calculation of the chisquare. This is justified as the absolute calibration of the infrared and ultraviolet data is less accurate by a factor of 3 than that of the optical fluxes.

2. Results for 78Vir

Adopting a mean V= 4.92 for all phases of 78 Vir, the integrated flux is equal to 2.6810^{-7} at phase 0.0. The uncertainty on the integrated flux is estimated to be 10 to 11%. A first guess of the effective temperature equal to 9500K was adopted to start the IRFM. Convergence was achieved after three iteration yielding $\theta = 0.343 milliarcsec$ and T_{eff}= 9200K. The uncertainty of the effective temperature arising from those on the integrated flux and on the angular diameter (about 7%) is about $\pm$ 6.3% that is $\pm$ 290K. A grid of 43 models of parameters Teff in the range 9200K$\pm$300K , *logg* in the range 4.00 to 4.50 and for the three chemical compositions [M/H]= $\odot, 3 * \odot and 10 * \odot$ has been computed and compared to the ED at phases 0.0. Clearly , models of solar and 3 times solar composition all have chisquares larger than the models of 10 solar composition of same effective temperatures and surface gravities. The contours of constant chisquares for models of 10 solar composition are displayed in figure 1 . The cross locates the minimum chisquare corresponding to the model of effective temperature 9200K, *logg*= 4.00. The ED at phase 0.0 is compared to this model in figure 2. The model reproduces well the observed Balmer Jump and the Paschen continuum. The fit to the mid ultraviolet spectrum shows that the model lacks line blanketing there. A model with overabundances larger than 10 solar could possibly reproduce the mid ultraviolet spectrum, although , another process, Zeeman splitting of the lines, also would enhance the line blanketing. From 1500Å to 2000Å the model has too much flux also indicating it lacks line blanketing or continuous opacity poorly known in this region.

3. Conclusion

The combined usage of the IRFM and model atmospheres is likely to tie down the fundamental parameters of early A stars allowing to bracket their effective temperature and chemical composition. This method is not very sensitive to surface gravity which should be determined by fitting Balmer lines. In the case of 78Vir, the effective temperature appears to be 9200K$\pm$100K, *logg*=4.0$\pm$0.25 and the abundances close or higher than 10 times solar.

References

Blackwell D.E. and Shallis M.J., **1977**, *Monthly Notices of the Royal Astronomical Society*,**174**,489

Groote and Kaufmann,**1983**,*A and A Sup*,**53**,91

Hayes D.S., **1979**,*in Problems of calibration of Multicolor photometric systems*,**Dudley Observatory Report, 14**,29

Hayes D.S. and Latham M.R.,**1975**,*ApJ*,**197**,593

Pyper D.M. and Adelman S.J.,**1983**,*A and A Sup*,**51**,365

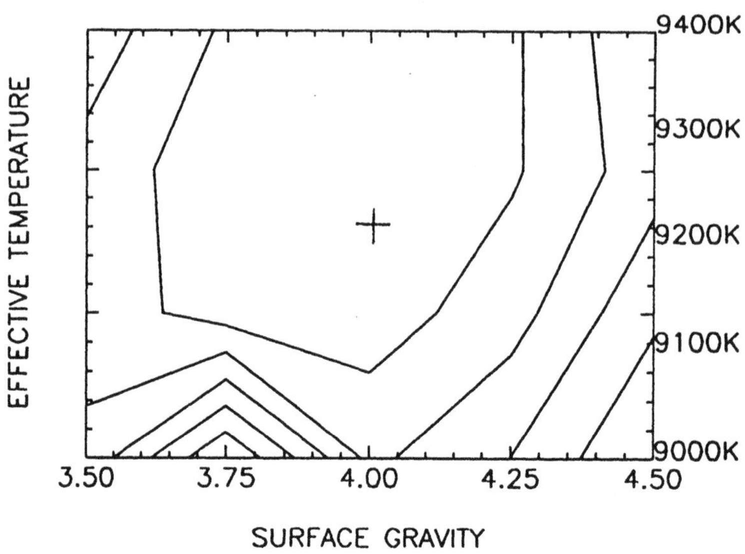

Figure 1: Contours of constant chisquare

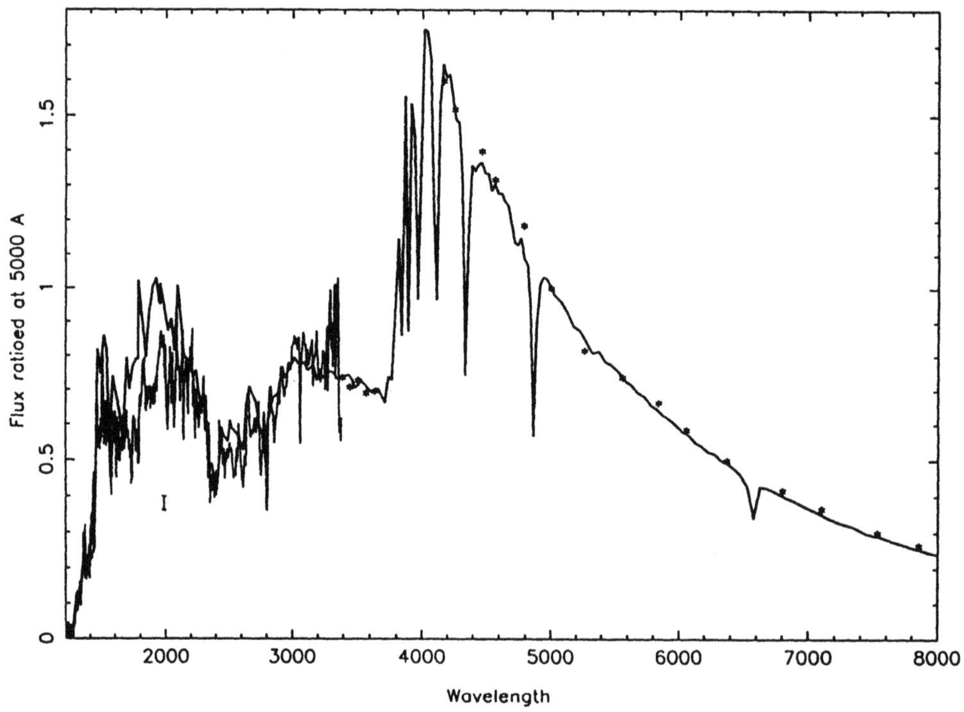

Figure 2: Comparison of the ED of 78Vir to the model of effective temperature 9200K, *logg*= 4.00 and 10 solar metallicity

Silicon Abundances in Normal and Mercury-Manganese Type Late-B Stars

K.C. Smith

Department of Physics & Astronomy, University College London, Gower Street, London WC1E 6BT, England

Abstract: Atmospheric silicon abundances have been obtained for a selection of main-sequence normal and mercury-manganese (HgMn) type late-B stars based on an analysis of co-added archival *IUE* spectra (*cf.* Smith and Dworetsky 1990a, 1990b). Abundances were derived from the resonance doublet Si II λ1525, 1533 by fitting synthetic spectra computed in LTE using Kurucz (1979) model atmospheres. The inferred silicon abundances for the normal stars are in excellent agreement with the solar abundance from Anders & Grevesse (1989), with the notable exception of α Lyr which is distinctly Si-deficient. The derived silicon abundances for the HgMn stars are also approximately solar, but with a scatter somewhat larger than expected on the basis of internal errors alone. Two stars in the HgMn sample are found to have pronounced underabundances of silicon and four exhibit overabundances which are mild but sufficiently large to be significant. These results are broadly consistent with the prediction of diffusion theory that the radiation force in non-magnetic stellar atmospheres can support an abundance of silicon within a factor of two of the normal value (Vauclair, Hardorp and Peterson 1979).

Full details of this work will be published elsewhere.

References

Anders, E., Grevesse, N. (1989): Geochim. Cosmochim. Acta **53** 197

Kurucz, R. L. (1979): Astrophys. J. Suppl. **40** 1

Smith, K. C., Dworetsky, M. M. (1990a): In *Evolution in Astrophysics: IUE Astronomy in the Era of New Space Missions, ESA SP-310,* ed. by E.J. Rolfe (ESA Publ., Noordwijk) pp. 279–282

Smith, K. C., Dworetsky, M. M. (1990b): In *Evolution of Stars: The Photospheric Abundance Connection (Poster Papers),* ed. by G. Michaud, A. Tutukov and M. Bergevin (Université de Montréal, Montréal) pp. 81–82

Vauclair, S., Hardorp, J., Peterson, D.M. (1979): Astrophys. J. **227** 526

RADIATIVE ACCELERATIONS ON GA AND AL IONS IN STABLE ATMOSPHERES OF CP STARS

J. Budaj, M. Žboril, J. Zverko

Astronomical Institute Slovak Academy of Sciences
CS-059 60 Tatranská Lomnica, Czechoslovakia

Abstract: Radiative accelerations of ions (Ga I, II and Al I-V) are computed in the atmospheres of CP stars.

1 Introduction

Diffusion processes in stable atmospheres of CP stars suggested by Michaud (1970) seem to be able to explain a number of features of these stars. However, there are still some predictions in the theory which deserve confirmation by observations. In this contribution we give some preliminary results of our study on radiative accelerations of several chosen ions.

2 Radiative acceleration

We computed radiative accelerations of the ions Ga I and Ga II, and Al I through Al V due to the bound - bound transitions. We used the following formula:

$$a(\tau) = \sum_{l,u} \frac{h\nu}{mc}(1 - e^{\frac{-h\nu}{kT(\tau)}})B_{lu}\frac{N_l(\tau)}{N(\tau)}\int_0^\infty F(\nu,\tau)\varphi(\nu,\tau)d\nu \qquad (1)$$

where hν, m, c, τ, B_{lu} are the energy of transitions from the lower level -l to the upper level $-u$, mass of ion, speed of light, Rosseland optical depth and transition probabilty or Einstein coefficient (it is given by $B_{lu} = 1.336\ 10^6 \lambda f$, λ in Angström and f- is the oscillatory strength), respectively. $N_l(\tau)$, $N(\tau)$, $\varphi(\nu,\tau)$, $T(\tau)$, $F(\nu,\tau)$ are number of ions in l-state, total number of ions, normalized Voigt profile, temperature and radiative flux, respectively. $e^{\frac{-h\nu}{kT(\tau)}}$ is the correction for negative absorbtion. Radiative fluxes were computed using a modified code SYNSPEC developed by Hubený (1987). All computations were based assuming LTE. CGS units were used.

3 Results

3.1 Ga I, Ga II

For Ga I and Ga II see Fig.1 and Fig.2. Our results are only qualitative since we used the set of data by Wiese and Martin (1980) instead of the data used by Alecian (1987). The model of the atmosphere with $T_{eff} = 12000K$ and $log\ g = 4.0$ has been chosen (Kurucz 1979). In computing fluxes in Ga lines all important blending lines (Kurucz and Peytremann 1975) except Si II autoionizing lines have been taken into account. These Si II lines can reduce the flux up to 1/3 in 1414 Ga II line which is dominant in the spectrum (Artru 1986).

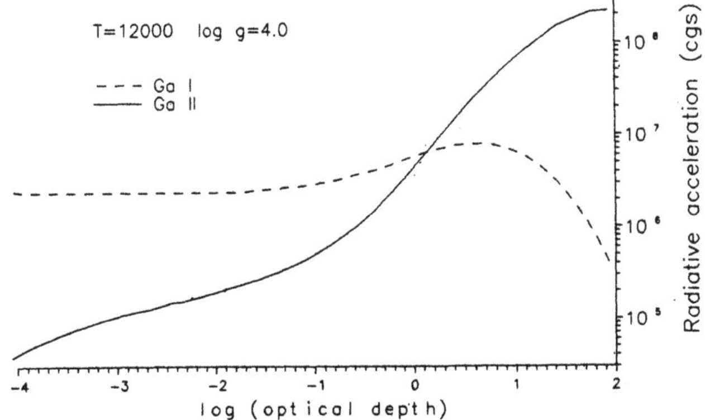

Fig. 1. Radiative accelerations of Ga I and Ga II ions.

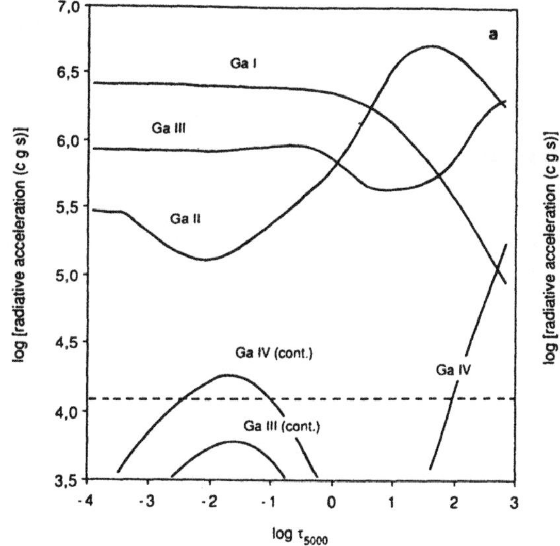

Fig. 2. Radiative acceleratios of Ga ions obtained by Alecian (1987).

3.2 Al I - V

Computations were done using again a particular Kurucz model $T_{eff} = 8500K$ $log\ g = 4.0$ (cgs). Line list (Kurucz and Peytremann 1975) contains several thousands Al lines. Thus we had to select only strongly absorbing lines, gathering them subsequently. Radiative acceleration of Al IV ion has been found as negligible because of the high energy of the lower levels of all well placed lines. There are no lines of Al V ion whose wavelengths are greater than 290 Angstrőms. As a consequence, the acceleration of Al V is by many orders of magnitude less than the gravitational one. As to Al I-III (see Fig.3) we have not at our disposal several lines from a longward part of spectra (mainly up to 8200 Angstrőm) so far.

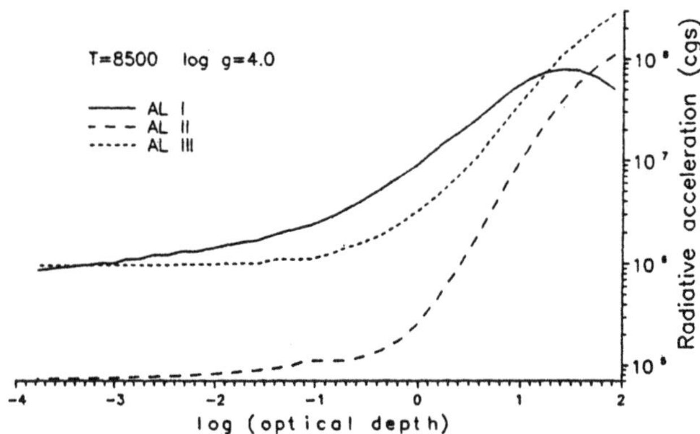

Fig. 3. Radiative accelerations of Al I-III ions.

4 Acknowledgement

We would like to thank Drs. J. Klačka, M. Saniga and R. Komžík for useful critical comments on this paper.

References

Alecian, G. (1987): *Astron. Astrophys.* **60**, 153
Artru, M. C. (1986): *Astron. Astrophys.* **168**, L5
Hubený, I. (1987): *Scient. and Techn. Report Astron. Inst. Czechoslov. Acad. Sci.Ondřejov*
Kurucz, R. L. (1979): *Center for Astrophys. Reprint Series* No. 1050.
Kurucz, R. L., Peytremann E. (1975): *Table of Semiempirical gf Values* (Cambridge: Smithsonian Astrophysical Observatory), *Special Report* No. 362
Michaud, G. (1970): *Astrophys. J.* **160**, 641
Wiese, W. L., Martin, G. A. (1980): *Wavelength and Transition Probability for Atoms and Atomic Ions, NSRDS-NBS 68*

Si II AUTOIONIZATION LINES in STRATIFIED ATMOSPHERE of Bp STAR

M. Zboril and J. Budaj

Astronomical Institute, Slovak Academy of Sciences
CS-059 60 Tatranská Lomnica, Czecho-Slovakia

Abstract: Synthetic spectra were computed under the assumption of atmosphere stratification for Bp type star. At λ5200 depression both Si normal and autoionization lines strengthen at λ5187 and λ5204 and may contribute to λ5175 peaked contributor of λ5200 depression.

1 Introduction

As a consequence of effort a decade ago to shed a light into the absorption features in Ap spectra a considerable attention was payed to the investigation the role of autoionization phenomena in Si II ion. Artru et al (1981a) paper contains much useful line-list of 809 lines of Si II covering the range 900-8000Å. If silicon abundance is enhanced to 100 times the solar value Artru et al (1981b) were able to interpret the absorption observed in TD1 S2/68 spectra at λ1400 and partly λ4200 absorption but not the continuum depressions observed at λ5200 and λ6300 in Ap stars. Spectrophotometrically the λ5200 feature extends from λ4950 up to λ5840. The origin of the features is not explained satisfactionary and line-blocking, bound-free discontinuities and autoionization features are candidates for their explanation. In this contribution we attempt to check in a simple procedure the relation of Si II autoionization lines and λ5200 depression. The depression will be dealt with in a manner that a semi-artificial stratification of the atmosphere will be applied and the synthetic spectra computed under this assumption at wavelengths corresponding to λ5200 depression.

2 Si II autoionization lines - facts

In Artru et al (1981b) line-list of Si II autoionization lines a correlation between the strongest ones (characterized by large enough gf-values and sufficient numbers both of lower and upper energy levels) and the three λ4200, 5200 and 6300 depressions is detected. As a result of a typical silicon overabundance the silicon becomes the most decisive element in the atmosphere of Ap stars except for hydrogen and helium. The strength of Si II autoionization lines is maximal at effective temperature temperatures around 12000 K where line blocking fails to explain the depressions (Maitzen and Muthsam 1980). There is also a direct dependence between the depression $\lambda 1400$ and silicon abundance (Artru et al. 1981b).

3 The λ5200 feature

The depression is known to become more pronounced with with increasing temperature (e.g. Cowley 1981). Maitzen and Muthsam (1980), Maitzen and Seggewiss (1980) found there were two separate contributors to the depression, the narrow one peaking at higher temperatures at $\lambda 5175$ while the broad shallow component remains relatively constant. Using model atmospheres with Ap enhanced elemental abundances Maitzen and Muthsam (1980) found out good agreement between synthetic and observed spectra at $\lambda 5200$ for low temperatures ($\sim 8000K$). For hotter temperatures when $\lambda 5200$ feature strengthens model depressions decreases. To engage in this subject we computed synthetic spectra for Kurucz (1979) line-blanketed model atmosphere with $T_{eff} = 14000K$ and $\log g = 4.0$ with solar composition in first run, while the atmosphere stratification was introduced in second run. Artru's et al. (1981a) set of autoionization lines was included in computations. The stratified atmosphere should be a consequence of an accumulation of atoms which could originate as a result of diffusive processes controlled by the radiative acceleration of atoms (e.g. Michaud 1970). The synthetic spectra were computed to cover the whole range of $\lambda 5200$ depression, i.e. from $\lambda 4950$ up to $\lambda 5840$. In first run, the peaked region wavelengths λ5150-5250, was covered almost singly ionized lines of Si (both types) and Fe,Ti,Al and C. If iron-peak elements are overabundant, the strength of depression may increase. Before the second run in code SYNSPEC (Hubeny 1987) we inserted the behaviour of abundance of any element considered with the optical depth in the atmosphere. We chose the model atmosphere with $T_{eff} = 14000K$ for our study, hot Si, HgMn and some SrCrEu stars were observed to have roughly effective temperatures of this value. For element abundances of our interest (Si,Ca,Mn,Ni and Fe) we adopted values used in model denoted "AB" in Muthsam (1970) paper. Abundances were stratified but the resultant abundance of every element was equal to the adopted value by Muthsam. From Michaud's (1970) temperature-dependent radiative acceleration on Si atoms we put 1.9 dex overabundance in $\log \tau$ ($\lambda = 5000$) from -4.0 up to 0.0 while 2.3 overabundance in $\log \tau$ larger than 0.0. Concerning calcium we applied Borsenberger et al. (1981) results and in $\log \tau$ from -4.0 up to -3.5 1.0 dex overabundance was included but in log depths from -3.5 up to 0.5 -1.0 dex value adopted. The solar values are for rest of depths. Manganese abundance stratification was estimated by Alecian and Michaud (1981), at log depths from -4.0 up to -2.5 1.0 dex overabundance, at log depth interval (-2.5,0.0) -0.045 dex underabundance and for rest depths 2.0 dex overabundance values were implemented. Ferrum and nickel abundances were modified in a way that for ferrum at log depths from -4.0 up to 0.01 dex overabundance and for log depths larger than 0.0 1.3 dex overabundance values were adopted. The same is for nickel but for log depths larger than 0.01 dex overabundance was considered. In the wavelength interval $\lambda 5150 - 5240$ the substantial increase of strength occured almost at wavelenths centered at $\lambda 5187$ and $\lambda 5204$ due to normal and autoionization Si II lines (λ5187) and mainly autoionization lines in the case of λ5204. Si II autoionization lines thus may contribute to peaked λ5175 contributor of the λ5200 depression.

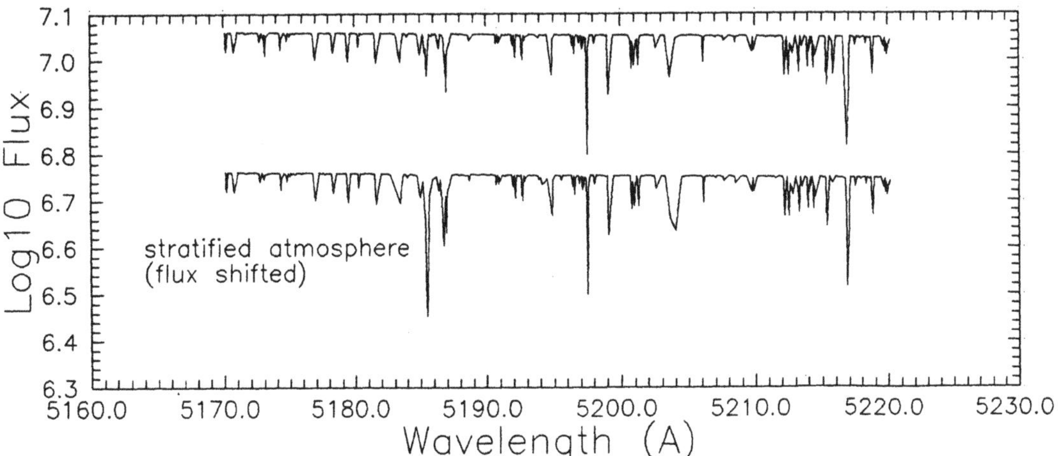

Fig. 1. Synthetic spectra at λ5200 depression.

4 Conclusion

Multiplets of Si II autoionization may be only partly explanation of the features and are more likely in UV part of the spectrum. Nevertheless, effect of increased strength of Si lines due to atmosphere stratification and enhanced Si abundance may influence peaked region at λ5175 of λ5200 depression. At least the following inevitable steps should be made to improve the procedure applied: 1. fully self-consistent computations 2. exact calculations dealing with radiative accelerations on selected atoms 3. to use more complete and temporary Fe line-list 4. direct comparison with observations

References

Alecian G., Michaud G. (1981): *Astrophys. J.* **245**, 226
Artru M. C., Jamar C., Petrini D., Praderie F. (1981a): *Astron. Astrophys. Suppl. Ser.* **44**, 171
Artru M. C., Jamar C., Petrini D., Praderie F. (1981b): *Astron. Astrophys.* **96**, 380
Borsenberger J., Michaud G., Praderie F. (1981): *Astrophys. J.* **243**, 533
Buchholz M., Maitzen H. M. (1979): *Astron. Astrophys.* **73**, 222
Cowley C. R. (1981): *Astrophys. J.* **246**, 238
Hubeny I. (1987): *Scient. and Techn. Report Astron. Inst. Czechoslov. Acad. Sci. Ondřejov*
Kurucz R. L. (1979): *Center for Astrophysics Reprint Series* No.**1050**
Maitzen H. M., Seggewiss W. (1980): *Astron. Astrophys.* **83**, 328
Maitzen H. M., Muthsam H. (1980): *Astron. Astrophys.* **83**,334
Michaud G. (1970): *Astrophys. J.* **160**, 641
Muthsam H. (1979): *Astron. Astrophys.* **73**, 159

He - Variables

K. HUNGER, M. FARTHMANN, U. HEBER
Institut für Theoretische Physik und Sternwarte der Universität Kiel
Olshausenstr. 40, W-2300 Kiel, Germany

He - variables are supposed to be oblique rotators whose spotty surfaces are caused by diffusion. Diffusion generally requires special conditions which in most cases are only met in rather shallow zones of the atmosphere. The question arises whether there exist stars with chemically (vertically) stratified photospheres besides the known (horizontally) inhomogeneous surfaces. Candidates that have been proposed are 3 Sco (Norris and Strittmatter, 1975), a Cen (Norris and Baschek, 1972), and some He- poor variables like HD 28843 = HR 1441 and HD 49333 = HR 2509. The latter have been analyzed on the basis of 20Å/mm spectrograms (ESO 1.5m + Coudé) by Groote et al. (see Hunger, 1986), with the result that HD 49333 indeed is considered a bona fide candidate. The low signal to noise ratio (IIaO - plates), however, did not allow to draw quantitative conclusions.

Therefore, an enlarged sample of 6 stars classified as He - variables, including the above named objects, were analyzed on the basis of high SNR ($\geq$ 100) spectrograms. (The spectra were taken with the same telescope, but with the Echellec spectrograph and a CCD detector.) A minimum of 10 frames with integration time $\approx$ 1 h and a resolution of 0.2Å were obtained by S. Dreizler in 1989. Inspection of the spectrograms immediately revealed that HR 1347 and HR 4089 are binaries, whereas HR 1100 does not exhibit any He lines (may be due to incomplete phase coverage ?) and HR 3448 is a fast rotator with non variable He lines.

The remaining 2 candidates, HD 28843 and HD 49333, were analyzed by means of Kurucz ATLAS 6 model atmospheres which are adapted to the non-solar He content, and a modified version of an (LTE) line formation program (Schönberner and Wolf, 1974). The atmospheric parameters are obtaind from the H_γ- profile, the β - and c_1 - index :

Table 1:

star	$T_{\rm eff}$(K)	log g	period (days)	v sin i (km/sec)
HD 28843	15000	4.0	1.37	100
HD 49333	16000	3.9	2.18	60

(The rotation periods are from Waelkens (1985) (HD 28843) and Pedersen (1979) (HD 49333). For HD 49333, a variable magnetic field has been observed, ranging from - 1.0 to + 1.2 kG (Boehlender et al, 1991), and having the same period.

For the horizontal distribution (no stratification) of helium, 2 He-rich caps of equal radius and He content, but not necessarily symmetric orientation are assumed. This means that a total of 7 parameters have to be adjusted to match the observed phase variation of $W_{\lambda 4471}$. (sin i is determined from radius and rotation period. The fit procedure is described by Groote, 1991, private communication). The resulting geometry is listed in Table 2.

In Fig. 1, the profiles of HeI 4471Å are shown for various phases. Accordingly, HD 28843 can be fully reproduced by an (vertically) unstratified atmosphere. However, for HD 49333 the

Table 2:

star		α^o	δ^o	R	ϵ_{He-cap}	$\epsilon_{He-disk}$
HD 28843	Cap 1	65	100	0.41	0.18	0.0001
	Cap 2	180	80	0.41		
HD 49333	Cap 1	340	50	0.50	0.091	0.002
	Cap 2	160	135	0.50		

excessively broad wings of HeI λ 4471 cannot be explained unless one assumes a drastic increase of He with depth.

Stratification means the introduction of additional 3 parameters : the optical depth τ_0 (λ 4000Å) of transition from the He-poor top layer to the He-rich bottom layer, and the relevant abundances, $\epsilon_{He-poor}$ and $\epsilon_{He-rich}$. In the present case, the parameters can be determined more or less independently for the disk and cap, as it happens that at phase $\Phi = 0.9$, cap 1 is nearly centered on the line of sight, while at $\Phi = 0.45$ almost no cap is visible.

We first demonstrate the fit procedure for the cap ($\Phi = 0.9$). In Fig. 2 (left panel) the profile of 4471 is shown for successive τ_0, while $\epsilon_{He-poor}$ and $\epsilon_{He-rich}$ are kept constant. The best fit is obtained with $\tau_0 = 0.1$. In Fig. 2 (right panel), $\epsilon_{He-rich}$ and τ_0 are constant, while $\epsilon_{He-poor}$ is increased. The best fit yields $\epsilon_{He-poor} \approx 5\%$. In Fig. 3 (left panel), finally, $\epsilon_{He-poor}$ and τ_0 are constant , while $\epsilon_{He-rich}$ is varied. The best match is obtained for $\epsilon_{He-rich} \approx 50\%$.

Examination of phase $\Phi = 0.45$ (Fig. 3 (right panel)) reveals that the disk profile exhibits similar discrepancies as observed in the caps. Hence one is led to the conclusion that also the disk is stratified. With the same procedure as applied to the cap, the following result is obtained: $\epsilon_{He-poor} \approx 0.13\%$, $\epsilon_{He-rich} \approx 3\%$ and $\tau_0 \approx 1.5$. The latter result is somewhat suprising as one expects that beyond $\tau_0 = 1$, no reliable information should be obtainable. But it must be born in mind that at some depth $\tau_0 \geq 1$, diffusion stops and $\epsilon_{He-poor} \approx 9\%$ must eventually be reached. Fig. 4 shows the phase variation of the HeI λ 4471Å profile, if stratification both of the cap and disk are assumed. The latter is illustrated in Fig. 5. (The transition region is handdrawn. The detailed shape is determined by diffusion. It has, however, little influence on the quoted results.)

In Fig. 6, the phase variation of MgII 4481 is shown. The profile variations can be reproduced if one assumes that Mg is enriched in caps which have the same geometry as those of He. ($\epsilon_{Mg-cap} = 10^{-4}$, and $\epsilon_{Mg-disk} = 5\ 10^{-6}$.) The same corelation is observed in HD 28843 (not shown here). This is in contrast to what is observed in the He rich variables (σ Ori E) whose metals vary in antiphase with helium.

ACKNOWLEDGEMENT. We thank Stefan Dreizler (Kiel), who obtained the ESO spectra for us.

REFERENCES

Bohlender, D.A., Landstreet, J.D., Thompson, I.B.: 1991, preprint
Hunger, K.: 1986, IAU coll. No.87, Reidel, po 261
Norris, J., Baschek, B.: 1972, A&A 21, 385
Norris, J., Strittmatter, P.A.: 1975, ApJ 196, 515
Pedersen, H.: 1979, A&AS 35, 313
Waelkens, C.: 1985, A&AS 61, 127
Schönberner, D., Wolf, R.E.A., : 1974, A&A 37, 87

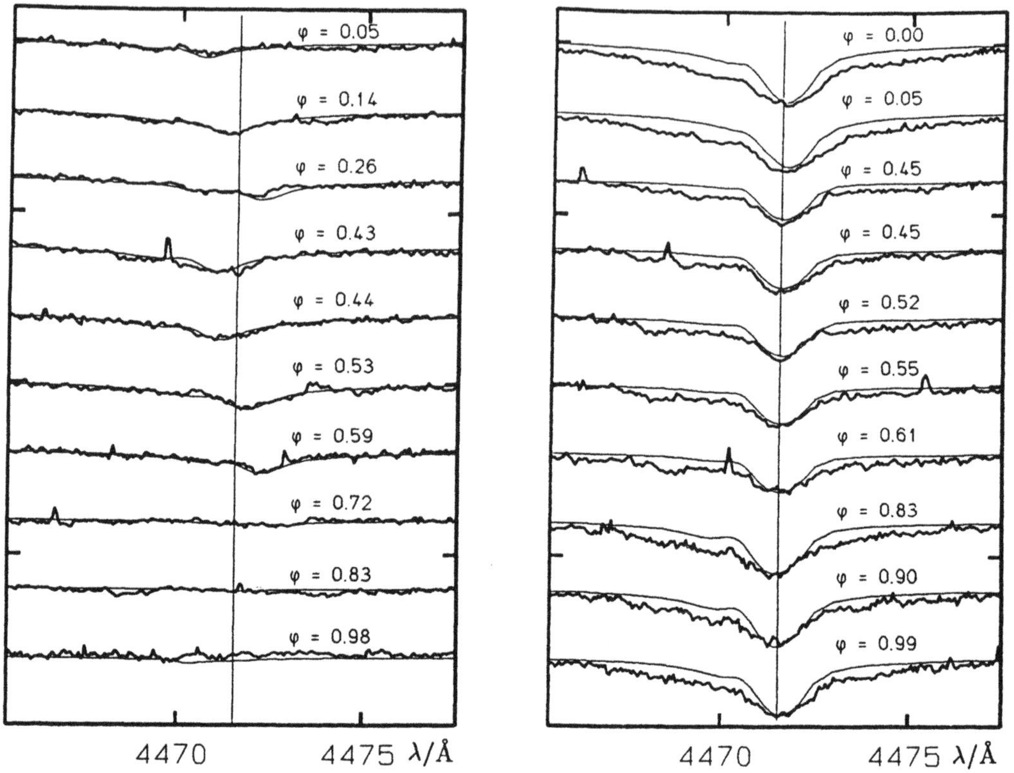

Figure 1: The profile of HeI 4471Å for various phases. For the theory, no (vertical) stratification is assumed. Left HD 28843, right HD 49333.

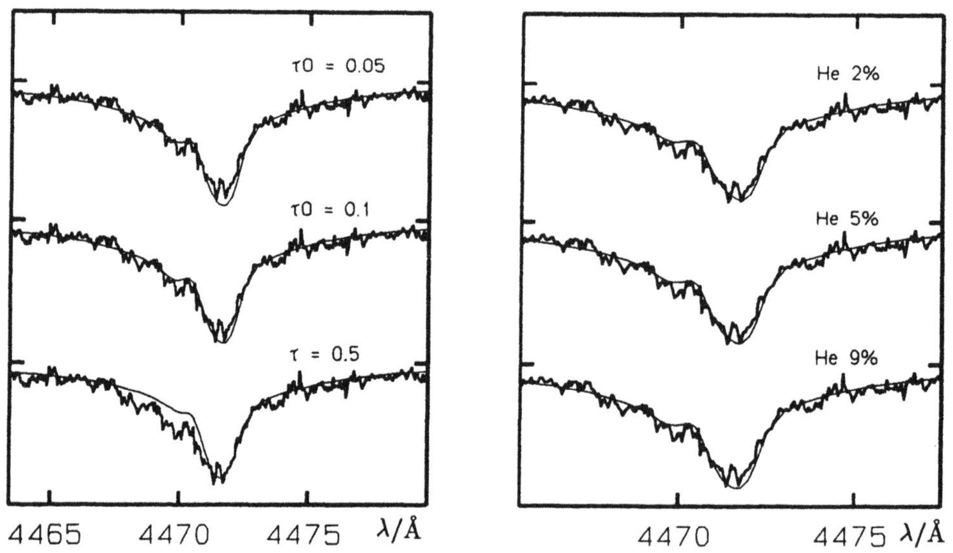

Figure 2: HeI 4471Å profile (HD 49333) for various transition depths τ_0, (left panel), and for various top layer He abundances $\epsilon_{HE-poor}$ (right panel), for $\Phi = 0.9$.

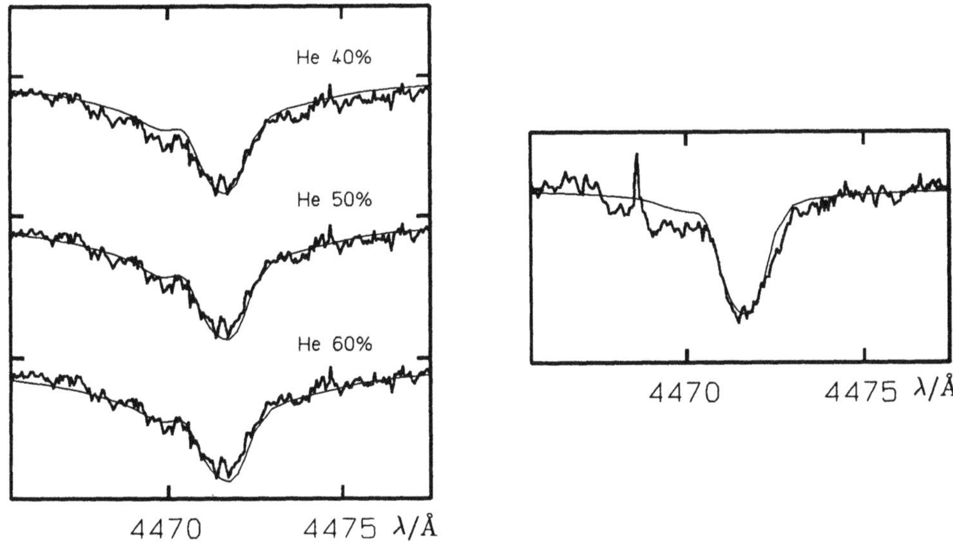

Figure 3: HeI 4471Å profile for various bottom layer He abundances $\epsilon_{He-rich}$, for $\Phi = 0.9$ (left panel). HeI 4471Å profile for $\Phi = 0.45$ (almost no cap visible) with practically the same geometry as is used for Fig. 1 (right panel).

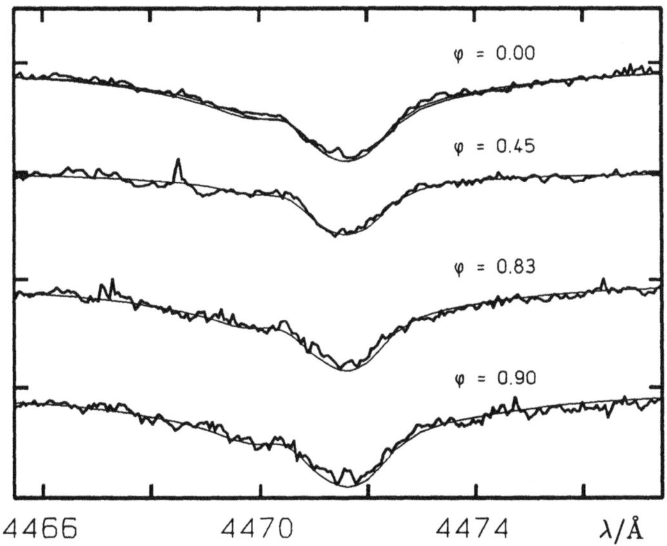

Figure 4: Phase variation of HeI 4471Å, if both, cap and disk, are stratified.

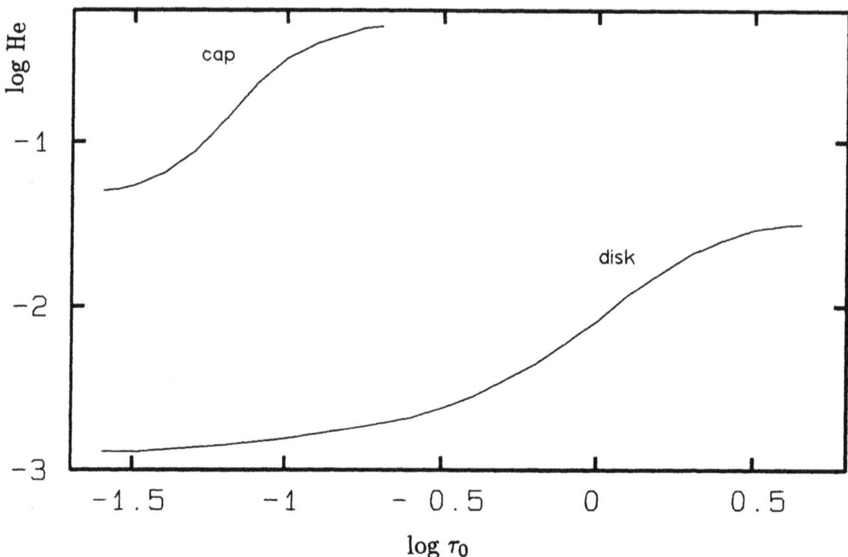

Figure 5: Approximate depths variation of the He abundances. (The transition region is handdrawn.)

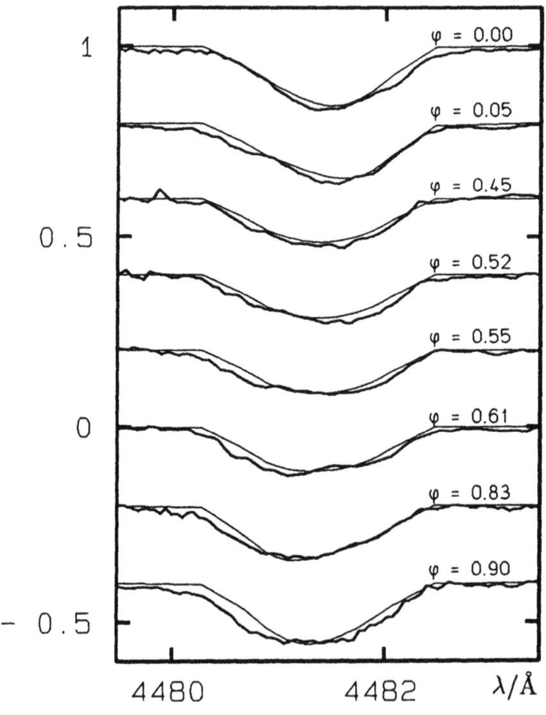

Figure 6: Phase variation of MgII 4481Å. For the theory, the unstratified He-geometry is assumed. For the He-cap $\epsilon_{Mg-cap}=10^{-4}$, and for the disk $\epsilon_{Mg-disk}=5\ 10^{-6}$ is used.

Variable Hα Emission in the Helium-Strong Star δ Orionis C

D.A. Bohlender [1], C.T. Bolton [2], G.A.H. Walker [1]

[1]Department of Geophysics and Astronomy, University of British Columbia, Vancouver, B.C., Canada
[2]David Dunlap Observatory, Richmond Hill, ON, Canada

Abstract: Recent Hα spectroscopy of the helium-strong star δ Ori C (= HD 36485) has revealed the presence of double-peaked Hα emission in this sharp-lined and strongly magnetic B3V star. Changes in the emission are evident over a 3.5 hr time span, and the profile variations are periodic with a period on the order of $1^{\mathrm{d}}.5$

1 Introduction

The helium-strong stars are the hottest known examples of peculiar magnetic Ap-Bp stars. Also known as intermediate helium stars, these objects typically have $N_{\mathrm{He}}/N_{\mathrm{H}}$ ≈ 1 in their atmospheres. Approximately two dozen have been identified (Drilling and Hill 1986) and they range in temperature from 19000 to 25000K, corresponding to spectral types of B2 or B3.

Like the more extensively studied Ap stars, the helium-strong stars are often photometric, spectroscopic, and magnetic variables with periods on the order of a few days. These variations are generally interpreted in terms of the oblique rotator model. A globally ordered magnetic field (usually predominantly dipolar) interacts with diffusive and radiative forces and leads to variations in the atmospheric structure and abundances over the surface of the star. Spectral and photometric variations arise when different regions of the patchy stellar atmosphere are presented to the observer as the star rotates. Several helium-strong stars have the additional complications of variable Hα emission (Bolton *et al.* 1986) and rotationally modulated winds (Barker *et al.* 1982; Shore and Brown 1990), and a few are also firmly established as non-thermal radio sources (Drake *et al.* 1987). The radio emission is thought to arise from gyrosyncrotron emission from mildly relativistic electrons in the stellar magnetospheres.

δ Ori C (HD 36485) is a particularly interesting cool ($T_{\mathrm{eff}} = 19000$K) member of the helium-strong class. This sharp-lined ($v \sin i = 32$ km s^{-1}) B3 star has been discussed in detail by Bohlender *et al.* (1987) and Bohlender (1989). It's constant -3.4 kG effective field is the strongest of the entire class of known magnetic helium-strong stars, and it is also a non-thermal radio emitter. Prior to the observations we discuss below, δ Ori C was not known to be a spectrum variable, although Bohlender's (1989) analysis, based on the peculiar He I profiles of the star, does suggest a non-uniform photospheric helium abundance.

2 Observations

We have obtained 10Å/mm spectra centered on Hα of δ Ori C with the DAO 1.8-m telescope, Cassegrain spectrograph and Ford Aerospace 512 $\times$ 512 CCD, as well as 10.8Å/mm spectra with the DDO 1.8-m telescope, Cassegrain spectrograph and Thomson 1024 $\times$ 1024 CCD as part of a larger project to investigate the photospheres and magnetospheres of the helium-strong stars. Typical exposure times were 20 minutes.

The first night's spectrum immediately indicated that δ Ori C has double-peaked Hα emission, and succeeding spectra clearly showed that the emission is variable. Some of the spectra are illustrated in Fig 1. Figure 2 shows a sequence of spectra obtained over one 3.5 hr time span and indicates that the emission is changing even on this short time scale. The profile variations appear to be periodic with a period on the order of 1.5 days, but further observations are needed to permit a better period determination.

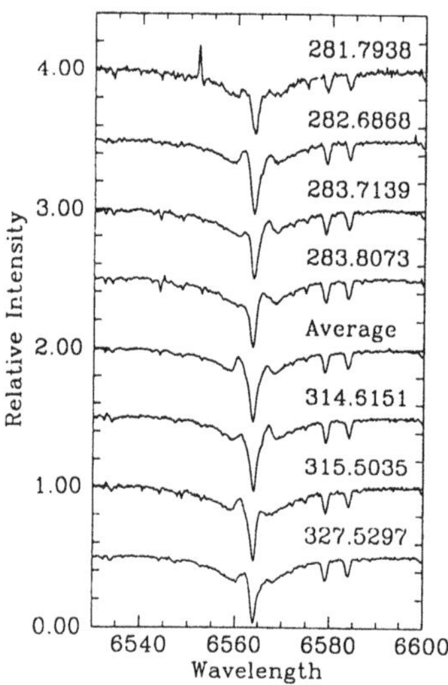

Fig. 1. A selection of Hα profiles of the helium-strong star δ Ori C. Labels for each spectrum give the JD (+2448000) of the midpoint of each observation; the spectrum labelled *average* is the average of the seven spectra shown in Fig 2.

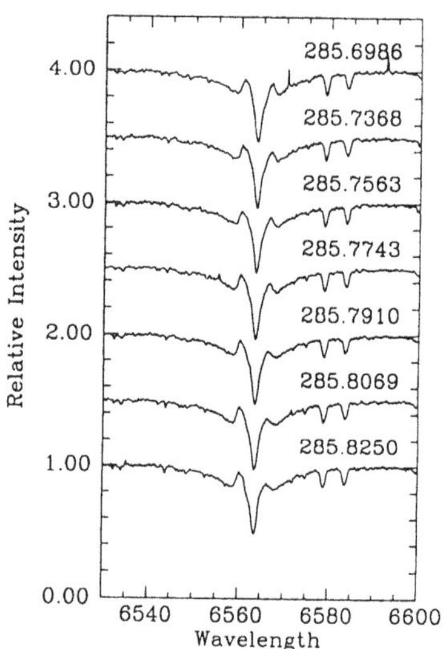

Fig. 2. Short-term variations in the Hα emission of δ Ori C. These spectra span a time of approximately 3.5 hours. Labels for each spectrum give the JD (+2448000) of the midpoint of each observation.

3 Discussion

While considerably weaker, the Hα variability of δ Ori C is quite similar in nature to that of the prototypical helium-strong star σ Ori E, if consideration is given to the fact that the latter star has v sin $i \approx 150$ km s^{-1}. The spectacular variability of σ Ori E's Hα emission is thought to be due to the rigid rotation of two distinct clouds of circumstellar gas that are trapped in its magnetosphere near the intersection points of the rotational and magnetic equators (Bolton *et al.* 1986).

When strongest, δ Ori C has emission peaks with intensities of approximately 15% of the continuum level at about 140 km s^{-1} on either side of line center. Emission is apparent to ± 200 km s^{-1} and variability in the core of the line indicates that the emission extends through the line core. If we assume that the strong magnetic field forces the circumstellar material into rigid rotation about the star then the peak emission occurs at 3.4 R_* and extends to 5.25 R_* above the photosphere. Bolton *et al.* (1986) have derived very similar numbers for σ Ori E.

If the Hα variability of δ Ori C is due to rotation, the rotation axis of the star must have $i \approx 10°$ in order to account for the short period and small v sin i . This would also be consistent with the constant longitudinal magnetic field as well as with the estimated surface magnetic field strength (Bohlender 1989) and the observed behavior of the UV resonance lines (Shore and Brown 1990). In contrast, σ Ori E has $i \gtrsim 45°$ (Hunger *et al.* 1989) and possibly $\approx 90°$ (Bolton *et al.* 1986). δ Ori C therefore appears to provide us with an opportunity to investigate the cool components of the winds and magnetospheres of the helium-strong stars from a different vantage point.

DAB is grateful to the Director of the Dominion Astrophysical Observatory for generous allocations of 1.8-m telescope time. This research was partially supported by the Natural Sciences and Engineering Research Council of Canada.

References

Barker, P.K., Brown, D.N., Bolton, C.T., and Landstreet, J.D. 1982, in *Advances in Ultraviolet Astronomy: Four Years of IUE Research*, ed. Y. Kondo, J.M. Mead, and R.D. Chapman (NASA CP-2238), p. 589.

Bohlender, D.A. 1989, *Ap.J.*, **346**, 459.

Bohlender, D.A., Brown, D.N., Landstreet, J.D., and Thompson, I.B. 1987, *Ap.J.*, **323**, 325.

Bolton, C.T., Fullerton, A.W., Bohlender, D.A., Landstreet, J.D., and Gies, D.R. 1986, in *IAU Colloquium 92, The Physics of Be Stars*, ed. A. Slettebak and T.P. Snow (Cambridge: Cambridge University Press), p. 82.

Drake, S.A., Abbott, D.C., Bastian, T.S., Bieging, J.H., Churchwell, E., Dulk, G., and Linsky, J.L. 1987, *Ap.J.*, **322**, 902.

Drilling, J.S., and Hill, P.W. 1986, in *Hydrogen Deficient Stars and Related Objects*, eds. K. Hunger *et al.* (Dordrecht: Reidel) p. 499.

Hunger, K., Heber, U., and Groote, D. 1989, *Astron. Astrophys.*, **224**, 57.

Shore, S.N., and Brown, D.N. 1990, *Ap.J.*, **365**, 665.

Lambda Bootis Stars

Bodo Baschek

Institut für Theoretische Astrophysik der Universität Heidelberg
Im Neuenheimer Feld 561
D-6900 Heidelberg 1, Germany

Abstract. A brief discussion of the properties, in particular of the abundance pattern, and of the origin of the λ Bootis stars is given.

Definition and Properties of the Group

The λ Bootis stars are a small group of peculiar early A type stars, characterized by normal hydrogen lines, a weak Ca II K line, and extremely weak metal lines (e.g., Mg II λ 4481 Å). The prototype of this class, λ Boo = HD 125 162, was discovered by Morgan et al. (1943). There have been many investigations to search for members and to establish the characteristic properties of the group. Recently Gray (1988) proposed the following morphological *working definition* based on *optical* spectra: The λ Bootis stars are a distinct set of apparently metal-poor A type stars with broad hydrogen-line wings and weak Mg II λ 4481 Å lines.

Utilizing low-resolution spectra obtained with the IUE satellite, Baschek et al. (1984) showed that the λ Bootis stars can be discriminated from normal stars also in the *ultraviolet* region by their strong absorption features at $\lambda\lambda$ 1600, 1657 (C I), 1931 (C I), and 3040 Å. The ultraviolet metal lines on the whole, however, are weaker in the λ Bootis stars . Furthermore, the steep decrease of the continuum below 1600 Å occurs at shorter wavelengths.

On the one hand the λ Bootis stars have the same properties as *main-sequence* A stars of population I, i.e. their effective temperatures, absolute magnitudes or luminosities, spatial velocities, and projected rotational velocities ($v \sin i \simeq 100\,\mathrm{km\,s^{-1}}$) are the same. On the other hand, their *surface* compositions are *peculiar* and resemble to some extent those of mild population II stars such as the horizontal-branch A stars. The λ Bootis stars can be distinguished from the horizontal-branch stars by their broader hydrogen-line wings (higher gravity), by their space velocities, and by details of their abundance patterns. The surface compositions of the λ Bootis stars will be discussed in the next section.

According to Bohlender and Landstreet (1990) *no magnetic fields* are observed in the λ Bootis stars with an error of about 250 G. If magnetic fields exist at all, they are significantly smaller than those in the Ap stars.

Recently evidence for the presence of *circumstellar dust/gas* around the λ Bootis stars has accumulated from observations of infrared excesses and narrow components in the Ca II K and Na I D lines (Sadakane and Nishida, 1986; Gray, 1988; Venn and Lambert, 1990; Holweger and Stürenburg, 1991).

Chemical Compositions – Problems of Analysis

Here we discuss only the abundance pattern of λ *Boo itself* in some detail; the peculiarities of π^1 Ori, 29 Cyg and the other λ Bootis stars are less pronounced. In their pioneering work, Burbidge and Burbidge (1956) obtained an underabundance of 1/30 for Ca and 1/10 for Fe relative to the solar values. Kodaira (1967) found $\leq 1/5$ for Mg, and a normal composition for O from infrared spectra.

The curve-of-growth analysis of optical lines by Baschek and Searle (1969) yielded the following pattern: O normal, Mg, Ca $\leq 1/5$, and the Fe group elements about 1/3 solar. From IUE spectra Baschek et al. (1984) estimated that C and possibly N may be normal or even enhanced. The analysis by Baschek and Slettebak (1988) of ultraviolet lines showed that C, N, O are slightly overabundant to normal, whereas Mg, Al, Si, S, Mn, Fe, Ni are about 1/3 solar. This is in agreement with the results obtained by Baschek and Searle.

Recently, however, Venn and Lambert (1990) derived, from high-quality optical spectra, considerably larger deficiencies for Mg, Ti, Fe, Sr (about 1/100 in λ Boo), while C, N, O, S turned out to be solar. In contrast, first results of an LTE analysis by Stürenburg (1991) indicate less extreme underabundances (for Mg, Si, Ti, Fe, Sr, Ba).

Thus there is agreement in so far as the λ Bootis stars have approximately normal C, N, O and a deficiency of the metals. The degree of the deficiency, however, remains controversial. Unfortunately Venn and Lambert restricted their analysis to the optical region. The ultraviolet flux, in particular the intensity decrease below about 1600 Å and the bulk line blocking, is very sensitive to the metal abundances (cf. Baschek et al., 1984, fig. 4).

The analyses of the spectra of the λ Bootis stars pose several *problems*:

(a) It is difficult to measure the equivalent widths or profiles for the faint and broad lines accurately.

(b) The role of non-LTE effects has yet to be explored. The radiation field around $\lambda\lambda$ 1200–1600 Å, where important photoionisation edges of neutral elements are located, is different for the metal-poor λ Bootis stars and the standard stars so that even relative abundance determinations by a differential analysis may be affected.

(c) The absorption features at λ 1600 and λ 3040 Å are yet unidentified. The λ 1600 Å feature is about 80 Å wide and shows no structure on IUE high-resolution spectra (Baschek et al., 1984). It is disturbing that this by far strongest ultraviolet absorption remains unexplained. One might speculate that an autoionisation transition, or molecular or dust absorption involving H or C, N, O are its source (the underabundant metals need probably not be taken into consideration). An attractive suggestion for the λ 1600 Å feature, which is observed also in white dwarf stars, is the quasi-molecular absorption of H_2 which corresponds to the minimum energy separation of the potential curves of the X $^1\Sigma_g^+$ and B $^1\Sigma_u^+$ states. However, before this identification can be accepted, one should understand why normal A stars which have essentially the same temperature–pressure stratification as the λ Bootis stars do not show this feature. And what about the λ 3040 Å feature (which is not seen in white dwarf spectra)? Furthermore, the occurrence of the λ 1600 and λ 3040 Å features in some horizontal-branch A stars, i.e. in low-metallicity stars of lower surface gravities (Jaschek et al., 1985), has to be explained.

(d) Finally, the choice of standard stars for comparison requires care since many (or all?) "normal" A type stars, such as e.g. Vega, exhibit peculiar abundance patterns.

Nature and Origin

Before interpreting the abundance pattern of the λ Bootis stars we point out its striking similarity with that of the Vega-like stars. One may regard λ Boo as "rotating Vega" or, conversely, Vega as mild "non-rotating λ Bootis star" (Baschek and Slettebak, 1988). Clearly all these abundance peculiarities in the A stars are a *surface* phenomenon only.

The interpretation of the chemical composition of the λ Bootis stars as "birth composition", i.e. by star formation from a low-abundance fluctuation of the interstellar matter, is contradicted by the high abundances of C, N, and O (as compared with population II stars). Also a nuclear origin of the abundances, i.e. "contamination" of the surface by processed matter, is ruled out by the observed CNO pattern.

Michaud et al. (1983) and Michaud and Charland (1986) have suggested that the λ Bootis stars are related to the metallic-line stars and proposed a *diffusion* model characterized by a *mass loss* rate of $\dot{M} \simeq 10^{-13}\,M_\odot\,yr^{-1}$.

In this way moderate underabundances of the order of 1/3 solar can be achieved after about 10^9 yr. It was pointed out by Baschek and Slettebak (1988) that meridional circulation induced in the rapidly rotating λ Bootis stars may present difficulties, and by Gray (1988) that the time scale for development of the abundance pattern may be too long. Subsequent calculations by Charbonneau (1991) including mass loss and meridional circulation did not produce the characteristic abundances of the λ Bootis stars.

The similarity between the abundance pattern in the λ Bootis stars and in the interstellar gas in which the metals are depleted by grain formation, led Venn and Lambert (1990) to suggest the following model: the λ Bootis stars are formed when the star accretes circumstellar gas which is separated from the grains. The dust grains remain in a disk or cicumstellar cloud and can be observed by infrared excess radiation. Detailed calculations of this *accretion/diffusion model* by Charbonneau (1991) showed that an accretion rate of the order of $-\dot{M} \simeq 10^{-13}\,M_\odot\,yr^{-1}$ indeed results in the development of the characteristic abundance pattern of the λ Bootis stars for the right range of effective temperatures. Clearly the grain separation and the accretion processes themselves have yet to be modelled. Nevertheless, diffusion combined with accretion is a promising mechanism for the origin of the λ Bootis phenomenon.

References

Baschek, B., Searle, L.: 1969, Astrophys. J. **155**, 537
Baschek, B., Slettebak, A.: 1988, Astron. Astrophys. **207**, 112
Baschek, B., Heck, A., Jaschek, C., Jaschek, M., Köppen, J., Scholz, M., Wehrse, R.: 1984, Astron. Astrophys. **131**, 378
Bohlender, D.A., Landstreet, J.D.: 1990, Mon. Not. R. astr. Soc. **247**, 606
Burbidge, E.M., Burbidge, G.R.: 1956, Astrophys. J. **124**, 116
Charbonneau, P.: 1991, Astrophys. J. **372**, L33
Gray, R.O.: 1988, Astron. J. **95**, 220
Holweger, H., Stürenburg, S.: 1991, Astron. Astrophys., in press
Jaschek, M., Baschek, B., Jaschek, C., Heck, A.: 1985, Astron. Astrophys. **152**, 439
Kodaira, K.: 1967, Publ. Astron. Soc. Japan **19**, 556
Michaud, G., Charland, Y.: 1986, Astrophys. J. **311**, 326
Michaud, G., Tarasick, D., Charland, Y., Pelletier, C.: 1983, Astrophys. J. **269**, 239
Morgan, W.W., Keenan, P.C., Kellman, E.: 1943, *An Atlas of Stellar Spectra*, University of Chicago Press, Chicago
Sadakane, K., Nishida, M.: 1986, Publ. Astron. Soc. Pacific **98**, 685
Stürenburg, S.: 1991, Poster IV.5, this Workshop
Venn, K.A., Lambert, D.L.: 1990, Astrophys. J. **363**, 234

This article was processed by the author using the TEX Macropackage from Springer-Verlag.

A spectroscopic abundance analysis of Lambda Bootis stars
First results in LTE

S. Stürenburg
Institut für Theoretische Physik und Sternwarte der Universität Kiel, Olshausenstr. 40,
W-2300 Kiel, Germany

Abstract
An abundance analysis has been carried out for some members of the class of rapidly rotating metal-poor Population I A-type stars, called the λ Bootis stars .

In this poster LTE abundances for several elements are presented. Furthermore, some of the problems encountered in an investigation of λ Boo stars are illustrated.

Introduction
In 1943 Morgan, Keenan, and Kellman have taken a classification spectrum of the star λ Boo (HR 5351). Remarkably, the spectrum showed only the hydrogen lines (classifying λ Boo as A0 star) and a very weak CaII K line. Other lines were not detectable, not even the strongest line in A-type stars, MgII 4481Å. Due to this discovery a new class of stars was introduced, namely, the metal poor A-type λ Boo stars .

First abundance analyses for some λ Boo stars were carried out by Baschek & Searle (1969) which later were confirmed by the work of Baschek & Slettebak (1988) based on UV spectra. The results was a slight overabundance for C, N, and O, a moderate underabundance of about -0.5 dex for the heavier elements (Si, S, Mn, Fe, Ni) and a somewhat stronger underabundance for Mg, Al, and Ca. The recently published work by Venn & Lambert (1990), however, quoted much more severe underabundances reaching -2.0 dex.

The aim of this work is to provide further elemental abundances for a member of λ Boo stars from high resolution optical spectra. With these additional informations we hope to bring more light into the λ Boo phenomenon.

Observations
The Table 1 of Gray (1988) contains 16 A-type stars which he has classified as λ Boo stars, including λ Boo itself. For 13 of these stars high resolution spectra were recorded during two observing periods in October 1989 and June 1990 at La Silla, Chile, with the ESO Coudé Echelle Spectrometer, the Short Camera, and the CCD #9 (1024×640 pixels) and during one observing period on November 1989 at Calar Alto, Spain, with a Coudé Spectrometer and a CCD. A first overview of the spectra is given by Holweger & Stürenburg (1991) for eight λ Boo stars and one normal A-type star in three wavelength bands.

Analysis
The main problem in the analysis of these stars is the high rotational velocity (the most "normal" A-type stars have high rotational velocities too, which is the reason why almost no published analyses exist (one exception is the poster by M. Lemke (this conference)). Due to this problem the continuum of the spectra could not be fixed correctly during the reduction phase. As turned out later, the continuum reaches the 100% level for higher $v \cdot \sin i$ only in a few points.

Furthermore, the high rotation causes a strong blending of the spectral lines making equivalent widths useless for deriving abundances. The only alternative is to calculate synthetic spectra and try to fit them to the observed ones. For the synthesis as many lines as possible are needed. Each line contributes to the equivalent width of the strong and broad features in the spectra. For most of our wavelength bands more than 40 lines were taken into account. But it is not clear whether all important lines have been included.

Solar oscillator strengths were determined by calculating synthetic spectra with the Holweger-Müller model atmosphere and fitting them to the Liegé Atlas. Only for a few well known lines — like CaII K or MgII 4481Å — were the values taken from the literature.

Rotation
For comparison with the observed spectra the synthesis must consider the rotational velocity of the stars. A simple but reliable method is the convolution of the synthetic spectra with a rotational profile. We used the approximation due to Unsöld (1968). Rotational broadening of blend lines results in an unexpected behavior. A distinct "ghost" line appears between the two spectral lines after convolution. The reason

for that is the finite range of the rotational profile. Figure 1 shows as an example a synthetic spectrum before and after the convolution with a rotational profile and a corresponding observed spectrum.

The abundance of Mg and the rotational velocity was determined from the MgII 4481Å feature. For each star synthetic spectra with different Mg abundances were calculated and each was broadened with several $v \cdot \sin i$. If the velocities differ not too much the profiles will roughly intersect each other in two points. The wavelength separation of these two points is a measure for the abundance. After having determined the correct Mg abundance the rotational velocity can be obtained.

Circumstellar matter

A further interesting but somewhat troublesome fact is the existence of circumstellar matter around several — perhaps all — of our λ Boo stars . As Holweger & Stürenburg (1991) have shown there is one star (HR 4881) with a strong circumstellar absorption peak in CaII K comparable to that of β Pic (HR 2020). The latter one is the prototype of the β Pictoris stars which have both circumstellar gas (causing sharp absorption features) and dust (causing an infrared excess, seen by IRAS). Lagrange-Henri et al. (1990) have reported a search of β Pic stars. In a relatively large sample among well known shell stars and IRAS objects they have found only very few candidates of these rare type of stars.

Furthermore, they published the CaII K profile of the shell star HR 7731 (as an example) which shows a peculiar profile with a triangular shape. Remarkably, one star (HR 1989) of our sample shows very similar triangular line profiles. We don't have a spectrum of CaII K but many of the FeII and TiII lines exhibit such a profile. Because no theory for the formation of such triangular line profiles was found in the literature the following tentative explanation is proposed.

Due to the high rotation (or any other effect) stellar material flows outward to form a spherical envelope or an equatorial ring. The rotational velocity decreases as r^{-1} with the radius to conserve angular momentum. It can easily be shown that then the maximum line-of-sight component at the circumstellar shell visible in absorption at the stellar limb decreases as r^{-2}. This means that the width of the circumstellar absorption features decreases rapidly with decreasing optical depth in the envelope. As one can see in Figure 2a the result is a nearly triangular profile. If the envelope of the star is concentrated far away from the stellar surface the circumstellar absortpion will be seen as a sharp line peak (Fig. 2b). If the density of the shell decreases faster than in Fig. 2a the triangular profile will be cut (Fig. 2c). In such a case it is hard to identify the star as shell star. What happens if a star with a shell like in Fig. 2a has a lower rotational velocity (less than 100 km/s) ? Figure 2d shows that the resulting profile looks very normal and even here the identification as shell star is difficult. Therefore, it must be kept in mind that some of the spectral lines could be contaminated by circumstellar absorption. We tried to take this fact into account in the abundance analysis presented here.

Results

The program stars were listed in Table 1 with the adopted parameter T_{eff}, $\log g$, $v \cdot \sin i$. The obtained preliminary LTE-abundances — with respect to the solar abundances — were shown in Fig. 3 as a function of $v \cdot \sin i$. The solid line marks the solar abundance and the dashed line marks an underabundance of a factor of 10. The result for the sample of our λ Boo stars is a slight underabundance of 0.3 dex for carbon, a moderate underabundance of about 0.5 dex for Si, Ti, Fe, Sr, and Ba, and a somewhat stronger underabundance for Mg. But it can be seen also that the values scatter over a wide range and in some cases even overabundances were found to occur.

References

Baschek B., Searle L., 1969, ApJ 155, 537
Baschek B., Slettebak A., 1988, A&A 207, 112
Gray R.O., 1988, AJ 95, 220
Holweger H., Stürenburg S., 1991, A&A , in press
Lagrange-Henri A.M., Ferlet R., Vidal-Madjar A., Beust H., Gry C., Lallement R., 1990, A&AS 85, 1089
Morgan W.W., Keenan P.C., Kellman E., 1943, An Atlas of Stellar Spectra, University of Chicago
Unsöld A., 1968, Physik der Sternatmosphären, 2. Aufl., Springer Verlag
Venn K.A., Lambert D.L., 1990, ApJ 363, 234

Table 1: The program stars.

HR	HD	name	T_{eff}	$\log g$	$v \cdot \sin i$ km/s
12	319		8140	3.80	60
541	11413		7950	3.83	125
1570	31295	π^1 Ori	8970	4.25	120
1989	38545	131 Tau	8970	3.60	200
4881	111786		7450	3.93	140
7400	183324	35 Aql	9260	4.22	90
7736	192640	29 Cyg	7990	3.99	80
7764 A	193281		8080	3.58	90
7764 C	193256		7860	3.74	240
7959	198160		7970	3.98	190
8203	204041		8100	4.03	70
8437	210111		7450	3.75	60
8947	221756	15 And	9020	3.91	100

Fig. 1

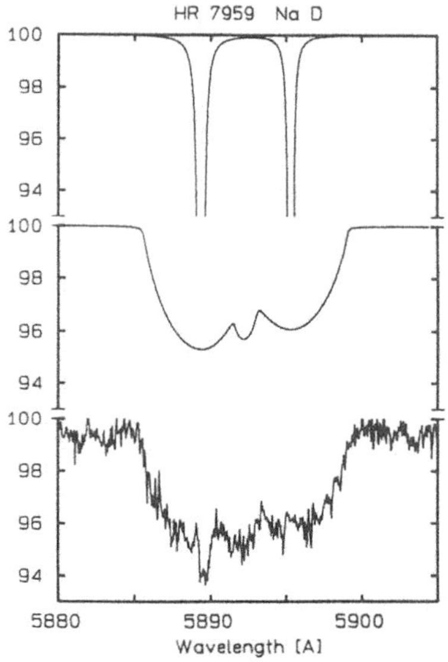

HR 7959 Na D

Wavelength [A]

Fig. 2a

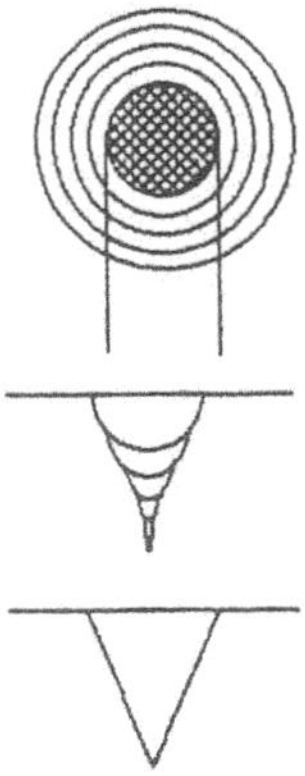

Fig. 2b

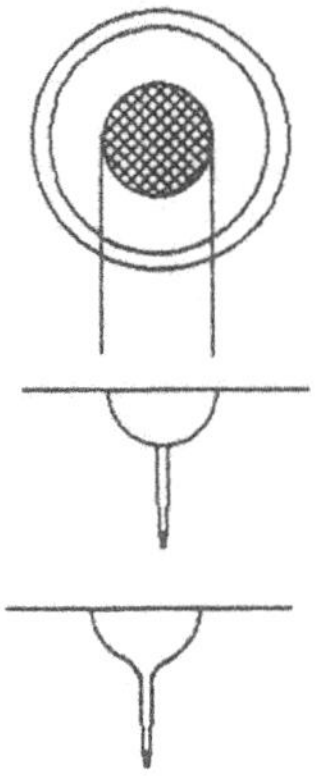

Fig. 2c

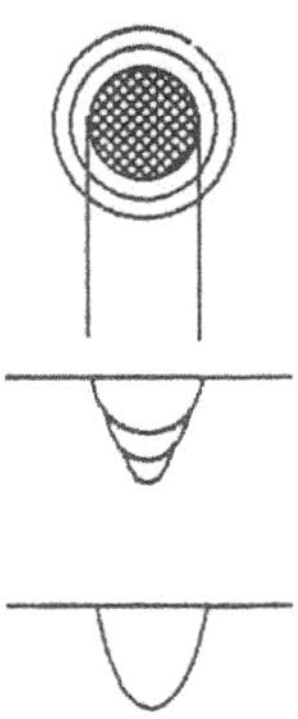

Fig. 2d

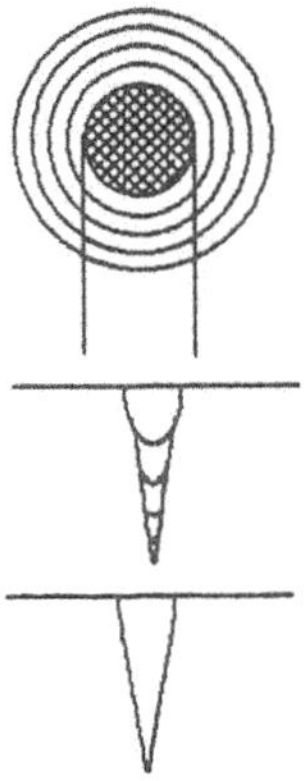

Fig. 3

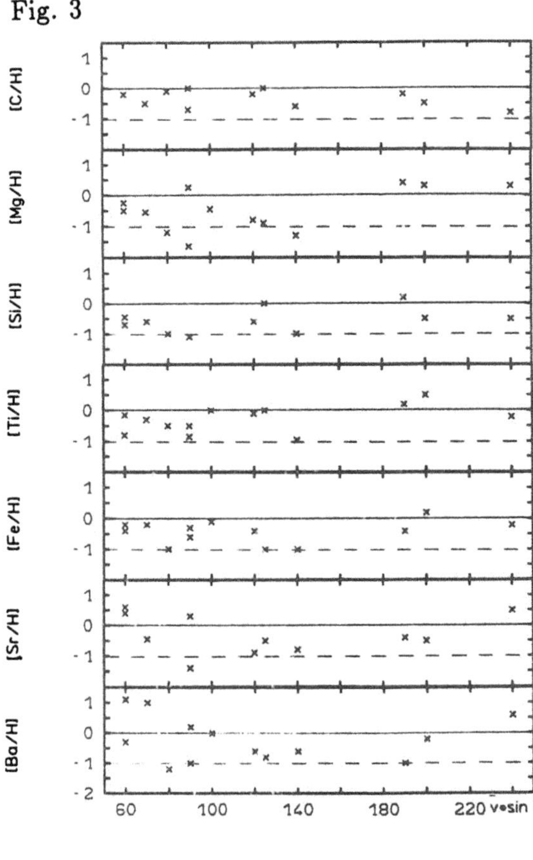

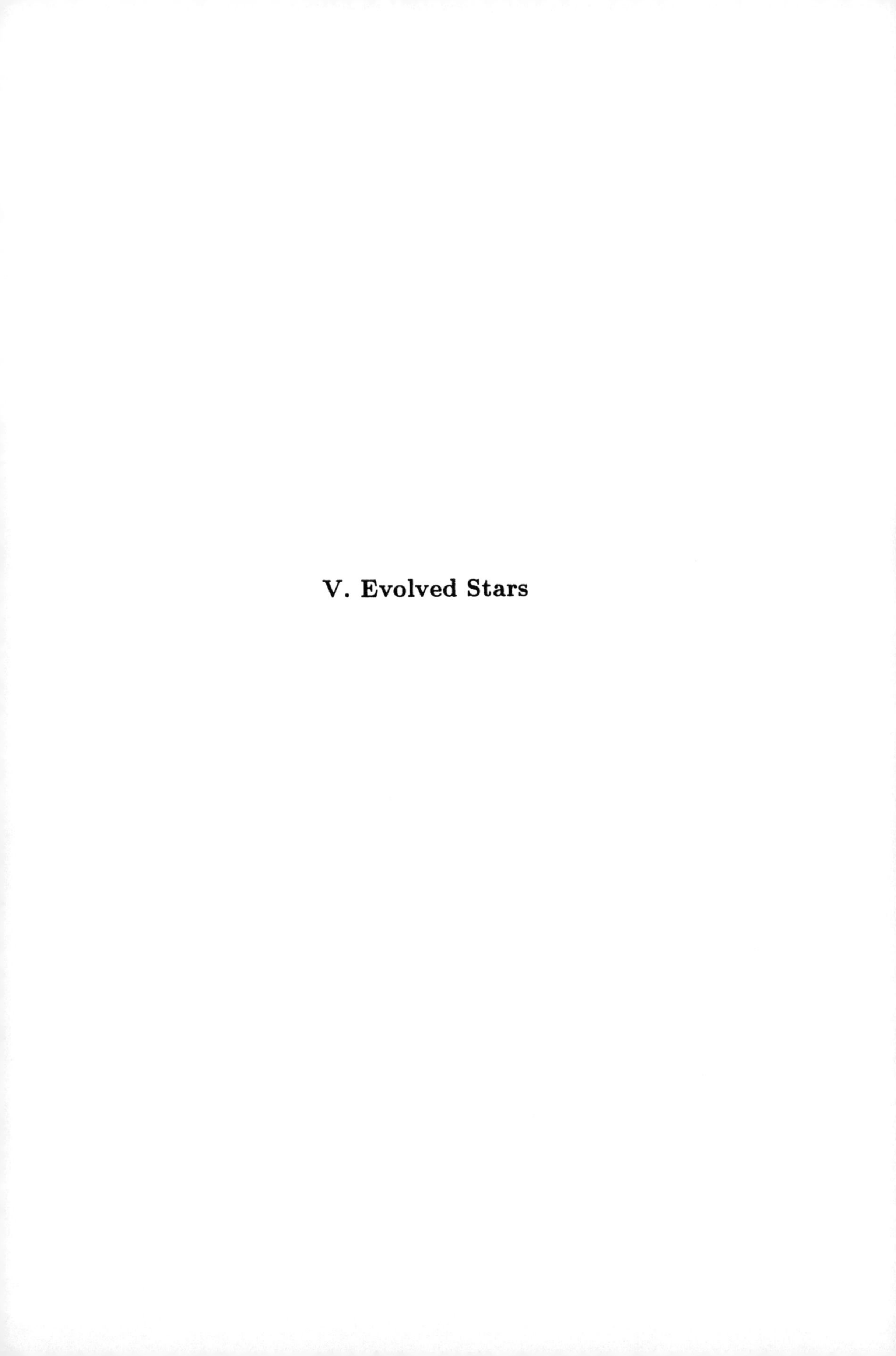

V. Evolved Stars

Hot Subluminous Stars

Ulrich Heber
Institut für Theoretische Physik und Sternwarte der Universität Kiel
Olshausenstr. 40, W-2300 Kiel, Germany

Abstract: Spectroscopic analyses of blue horizontal branch stars, subluminous B- and O-stars are reviewed. These classes of stars trace stellar evolution from the horizontal branch towards the white dwarf cooling sequence. The resulting atmospheric parameters and abundances are used to obtain constraints on the evolutionary status of the different classes of stars. The sdB stars form a homogeneous group and can be identified with models of the extended horizontal branch. Abundance anomalies (deficiency of helium and some metals, enrichment of ^{3}He) observed in Horizontal Branch stars and sdB stars are caused by atmospheric diffusion. The class of subluminous O stars is much less homogenous and two subclasses can be defined: the "compact" sdO stars probably evolved from the extended horizontal branch and are hence successors of the sdBs, whereas some sdOs of relatively low gravity are in a post-AGB stage of evolution. Hot subdwarfs in binary systems can be formed by case B or case C mass transfer.

1 Introduction

The hot subluminous stars are of great importance for our understanding of the late stages of stellar evolution since they belong the immediate progenitors of the white dwarfs. Recent extensive sky surveys (especially the Palomar-Green survey, PG, Green, Schmidt & Liebert, 1986) have shown that the hot subdwarfs are not "rare freaks" but even outweigh the white dwarfs, at least at apparent magnitudes brighter than V = $16\overset{m}{.}2$. While prior to 1986 only about 200 hot subdwarfs were known, the recent compilation of spectroscopically identified hot subdwarfs lists 1225 stars (Kilkenny, Heber & Drilling, 1988). Because they are quite numerous in the halo of the Galaxy, the hot subluminous stars have to be considered as sources of radiation ionizing interstellar gas at high galactic latitudes (de Boer, 1985) and have been proposed as good candidates to explain the ultraviolet excess found in elliptical galaxies (Greggio & Renzini, 1991).

Subluminous stars form two spectroscopic sequences, a helium poor sequence consisting of the horizontal branch stars of spectral type B (HBB), the subdwarf B stars (sdB) and the sdOB stars - a sequence of increasing effective temperature - and a helium-rich sequence (sdO stars). This bimodality is somewhat reminiscant of the white dwarf atmospheres which show either a pure hydrogen spectrum (spectral type DA) or a pure helium spectrum (DB) with only few exceptional cases (DAB- and DBA-stars). The segregation into two disjunct spectroscopic sequences, though, is not as strict for the subdwarfs as it appears to be for the white dwarfs, since most hydrogen dominated subdwarfs display weak helium lines in their spectra, while, vice versa, in many helium dominated subdwarfs hydrogen can be identified.

This paper reviews the results of spectroscopic analyses of HBB, sdB, sdOB and sdO stars*. The implications of these analyses for understanding the subdwarf's evolutionary status and origin will be emphasized.

Various aspects of the hot subdwarf research have been reviewed previously, e.g. by Vauclair & Liebert (1987), Heber (1987a, b, 1991) and Kudritzki (1987). The evolutionary considerations are reviewed by Groth, Kudritzki & Heber (1985, GKH) and Tutukov & Yungelson (1990).

2 Spectral classification

A uniform classification scheme for hot subluminous stars that allows spectral features to be related to atmospheric parameters (e.g. effective temperature, absolute brightness) has not yet been established. Because of the variety of spectral pecularities in these stars it is very difficult to set up such a system, especially for the O-type subdwarfs. Up to now atmospheric parameters can only be inferred from detailed spectroscopic analyses.

Originally the classes "sdB" and "sdO" were recognized by broad Balmer line absorption and the early confluence of the Balmer series; B-types were classified by weak or non-existent helium lines and O-types by the presence of He II, 4686 Å or other He II lines (see e.g. Sargent & Searle, 1968). More recently, Baschek & Norris (1975) introduced the "sdOB" subtype which is also helium-weak but has colours typical of O stars. He II, 4686Å is often weakly present in sdOB spectra. It has been recognized for some time that there is a great range of surface helium abundance in the sdO stars (e.g. Greenstein, 1966) and the suffix (or prefix) "He" has been used by several authors to indicate unusually strong helium lines or the complete absence of any hydrogen lines. Vauclair & Liebert (1987) accepted the sdB, sdOB, sdO sequence as one of increasing temperature and helium abundance but propose a new "hot sd" category with $T_{\rm eff}>$ 60000 K which overlaps with central stars of planetary nebulae, PG 1159 stars and hot white dwarfs. Spectra of intermediate resolution (2 to 3 Å/mm) reveal that the spectroscopic inhomogenity of the sdO class is not restricted to the strengths of helium lines but also to that of carbon and nitrogen lines. In particular, several hot sdO stars display C IV lines of considerable strength (Dreizler *et al.*, 1990) whereas HZ 44 is an example of a nitrogen-rich sdO (Greenstein, 1966).

The horizontal-branch B stars (HBB) have spectra similar to those of the sdBs with Balmer lines visible to n = 10 to 12. In addition they show the Mg II, 4481Å with a strength often comparable to that of He I, 4471Å, thereby indicating somewhat lower temperatures and gravities than the sdB stars.

In the Palomar-Green survey, Green, Schmidt & Liebert (1986) used a scheme that differs from the one described above, especially for the sdO stars, for which they introduced the subtypes "sdOA" (strong Balmer absorption with He I, 4471Å and often 4026Å),

*Although some central stars of Planetary Nebulae are well known to be of spectral type sdO, the discussion of these stars (except for a star discussed in section 5) is beyond the scope of this review and the reader is referred to Mendez (1991).

"sdOB" (dominated by He I and He II lines; often Balmer absorption present), "sdOC" (dominated by He II absorption; possibly weak He I 4471Å and Balmer series blended with He II Brackett lines) and "sdOD" (pure He I absorption; weak or absent H and He II). As already mentioned by Green, Schmidt & Liebert (1986) their coding is not part of any published classification scheme and might lead to some confusion since the PG classification "sdOB" is a helium-rich sdO star quite different from the Baschek & Norris helium-weak "sdOB".

While the PG subclasses sdOA to sdOC describe spectra known before, the PG subclass sdOD appears to describe a new small subclass (see Heber *et al.*, 1988) which has received little attention up to now. Since their spectra are indicative for temperatures typical for sdB stars but unlike the latter are helium dominated, Moehler *et al.* (1990) proposed the designation "He-sdB" for this new subclass.

Template spectra for the various spectral subtypes are given by Moehler *et al.* (1990).

3 The hydrogen-rich subluminous stars (HBB, sdB and sdOB)

Since the pioneering work of Newell (1973) and Greenstein & Sargent (1974) great improvements have been achieved both in observational techniques and in the modelling of subdwarf atmospheres. The IUE-satellite observatory has opened a new spectral range - the ultraviolet - for analysis. UV spectrophotometry allowed effective temperatures to be determined with unprecedented precision for these B-type stars. The He I/He II ionisation equilibrium is likewise an additional precise $T_{\rm eff}$-indicator for the OB-types. Unlike the sdO stars, the HBB/sdB/sdOB atmospheres can be modelled realistically within the classical LTE approach if metal line blanketing is fully accounted for, because relatively low temperatures ($T_{\rm eff} < 40,000$K) are combined with quite large gravities ($5.0 < \log g < 6.0$). This enormously simplifies the analysis compared to the Non-LTE techniques that have to be employed for the sdOs (see section 4). Therefore, a large sample of B-type subdwarfs has already been analysed from medium resolution spectra for their basic atmospheric parameters ($T_{\rm eff}$, $\log g$ and helium abundance). Four major projects have been conducted in recent years:

(i.) The Kiel group (Hunger *et al.*, 1981, Heber *et al.*, 1984a, Heber & Langhans, 1986 and Heber, 1986) analysed 26 stars selected from south galactic pole sky surveys. $T_{\rm eff}$ was determined from IUE plus optical spectrophotometry, while gravity and helium abundance resulted from H_γ and He I, 4471Å line profile and equivalent width fitting of medium resolution (2 to 3 Å) optical spectra. Line blanketed LTE model atmosheres (Heber *et al.*, 1984a) and unblanketed NLTE models (Hunger *et al.*, 1981) were used.

(ii.) Moehler, Heber & de Boer (1990), Theissen *et al.* and Schmidt *et al.* (both in these proceedings) extended this work to analyse 48 sdBs and 14 HBBs from the PG catalogue using the same model atmospheres and optical spectra of similar resolution as the previous analyses (i). Since, however, UV spectrophotometry was unavailable for most of their sdBs, effective temperatures were determined from Stroemgren photometry.

(iii) The Tucson group (Saffer *et al.*, 1991) studied an even larger sample (72 stars) of PG subdwarf candidates but used an analysis technique different from that used in (i) and (ii). They determined the basic atmospheric parameters from high S/N optical spectra of somewhat lower resolution (6Å). All three parameters were derived simultaneously by fitting all the Balmer lines from H_β to H_8 and He I lines with Koester's hydrogen line blanketed model atmospheres without having recourse to photometry. This project is based on a very homogeneous set of observations.

(iv) The fourth project (Bixler, Bowyer & Laget, 1991) concerns a sample of subdwarfs selected from a balloon bourne UV sky survey. Effective temperatures and gravities of 28 apparently single sdB stars were derived from combined (UV-, b-, y-photometry ($T_{\rm eff}$) and low resolution (6 to 10 Å) optical spectroscopy of moderate S/N using Wesemael *et al.* (1980) pure hydrogen model atmospheres. Helium abundances were derived from He I, 4471Å, 4922Å and He II, 4686Å lines using published tables of theoretical equivalent widths.

Although some stars are common to studies (ii) and (iii) it is difficult to intercompare the results for individual objects since Saffer *et al.* (1991) did not yet publish their results in detail. Despite of considerable differences of the four projects with respect to spectral resolution and quality of the data, analysis techniques and model atmospheres, the main results agree and the following basic conclusions can be drawn:

(i) A break exists at an effective temperature of about 40,000K, i.e. none of the hydrogen-rich subdwarfs analysed has $T_{\rm eff}$ exceeding 40,000K. Note, however, that a few hydrogen-rich sdO stars were found at extremely high effective temperatures ($T_{\rm eff}> 60,000$K, Schönberner & Drilling, 1984).

(ii) a gap exits close to $T_{\rm eff}=$ 22000K which separates the (hotter) sdBs from the (cooler) HBBs. This confirms the early findings of Newell (1973) and Newell & Graham (1976). Hence the term "Newell gap" has been coined.

(iii) The sdB and sdOB stars form a single homogeneous group which, in the ($\log g$, $T_{\rm eff}$)-diagram, lies close the socalled "extended horizontal branch" (EHB). This term was already coined by Greenstein & Sargent (1974) to describe the fact that the hot subdwarfs seem to extend horizontal branches of globular clusters to the blue. It is worthwhile to note that the width of the distribution in the ($\log g$, $T_{\rm eff}$)-diagram is smaller for the Tucson sample than for the others. This is probably due to the superior homogeneity of the data used by Saffer *et al.* (1991).

These spectroscopic results have been used to constrain the evolutionary status and origin of the sdB/sdOB stars. Following the ideas outlined by Heber *et al.* (1984a) and Heber (1986) the sdB/sdOB stars can be identified with models for EHB stars, which differ markedly from those for normal HB stars. While the luminosity of the HB star's hydrogen burning shell equals or even exceeds that of the helium burning core, an EHB star gains its luminosity almost solely from helium core burning. The hydrogen envelope is very thin ($M < 0.02\ M_\odot$) and the luminosity of the hydrogen burning shell therefore negligibly small. Hence the internal structure of an EHB star bears great resemblance to a helium main sequence star of half a solar mass† and its further evolution should proceed similar to that of the latter. Recent calculations by Caloi (1989) show that this is indeed the case. The EHB models evolve bluewards directly to the white dwarf cooling sequence thereby omitting the second giant branch. For this reason it is justified to distinguish EHB models from HB models. The positions of most analysed sdB/sdOB stars nicely agree with the theoretical model predictions if evolutionary effects and widening of the observed distribution due to observational errors are considered (see Figure 1 of Heber, 1991 and Figure 4 of Saffer *et al.*, 1991). It is important to note that no sdBs are found to lie below the zero-age extended horizontal branch (ZAEHB) in the ($\log g$, $T_{\rm eff}$)-diagram (to within observational limits). On the other hand, some sdB/OBs and also some HBBs are found at somewhat lower gravity than predicted by theory - a phenomenon that is also observed in several globular cluster HBB stars (Crocker *et al.*, 1989).

The origin of the sdB/OB stars is still strongly under debate. Their space distribution and kinematical properties indicate that they belong to the intermediate to old disk population. The question then arises, how the stellar mass can be reduced to half a solar mass during the pre-EHB evolution. Two scenarios for single sdBs have been put forward:

(i) Enhanced mass loss on the RGB or during the core helium flash may remove almost the entire hydrogen-rich envelope. The physical reason for such strong mass loss on the RGB, of course, is not understood. Birthrate estimates (Heber, 1986) indicate that only 2% of the RGB stars need to experience such enhanced RGB mass loss. Evidence that it is indeed possible comes from the existence of RR Lyrae stars of population I which must also have lost half of their mass during their evolution. Of course, also in this case the physical reason is not yet understood.

(ii) An alternate scenario was proposed by Iben (1990), who pointed out that single hot subluminous stars can be formed in the evolution of close binary systems leading to mergers of He-He cores. After merging, core helium burning stars with masses from 0.3 $M_\odot$ (the minimum mass for a helium main sequence star) to about 0.9 $M_\odot$ result. Hence, the predicted mass distribution is much wider than that of the "enhanced mass loss scenario". In particular the formation of low mass sdBs ($0.3M_\odot < M < 0.5M_\odot$)) is favoured in this scenario, which would lie somewhat below the theoretical zero-age extended horizontal branch.

The existence of the Newell gap can easily be explained by the "merger scenario" (Bailyn & Iben, 1989), but is more difficult to explain within the "enhanced mass loss scenario". Heber *et al.* (1984a) speculated that it may be an evolutionary phenomenon, because the morphology of HB evolutionary tracks changes near the gap temperature from evolution at constant $T_{\rm eff}$ to lower gravity for cooler stars to blueward evolution at roughly constant gravity for the hotter ones.

An empirical determination of the sdB's mass distribution would allow these scenarios to be tested. First attempts were undertaken by Heber *et al.* (1986), Crocker *et al.* (1989) and Moehler *et al.* (these proceedings) using globular cluster subdwarfs (see section 5).

†This fact led Heber *et al.* (1984a) to term the sdB and sdOB stars "generalized helium main sequence stars". This term, however, is misleading since the masses of the sdB/OB stars have to lie in a very narrow range according to this scenario and no mass sequence is involved.

Although the cluster results are inconclusive up to now, the existing analyses of field subdwarfs, though biased observationally against faint low mass stars, display a clear cutoff of the stellar distribution - no star lies below the ZAEHB in the ($\log g$,$T_{\rm eff}$)-diagram - as already pointed out which appears to be at variance with the "merger scenario".

3.1 Abundance anomalies and atmospheric diffusion

Helium is deficient in the vast majority of the analysed HBB, sdB and sdOB stars by factors ranging from 2 to more than 300. A few exceptions have been reported by Heber *et al.* (1988), de Boer, Heber & Richtler (1988), Viton *et al.* (1991), Saffer *et al.* (1991) and Schulz, Heber & Wegner (1991). Trends for the helium abundance with $T_{\rm eff}$ or $\log g$ cannot be established (see e.g. Schulz, Heber & Wegner, 1991). It has long been realized that the helium deficiency in their atmospheres is caused by atmospheric diffusion (Greenstein, Truran & Cameron, 1967). However, parameterfree diffusion models cannot reproduce the observed abundance anomalies, at least for the sdOB subclass (Michaud *et al.*, 1989).

Besides the helium deficiency other anomalies with respect to isotopic abundance ratios and metal abundances are to be expected because of differences in their respective atomic weights. These can be studied only from high resolution spectra, for which the number of accessible subluminous stars is limited due to their brightness. One such isotopic ratio which is easily accessible to high resolution spectroscopy is the ^{3}He/^{4}He ratio. Feige 86 was the first HBB star to show evidence for the presence of ^{3}He (Hartoog, 1979). Heber (1991) systematically searched for ^{3}He in 14 HBB and sdB stars and found that in three objects (SB 290, PHL 25 and PHL 382) ^{4}He is almost entirely replaced by ^{3}He [‡].

Let us now turn to the metal abundance pattern. The first metal abundance analyses were based on high resolution UV spectra obtained with the IUE satellite, which allowed the strong resonance absorption lines of e.g. C IV, N V and Si IV to be measured. The analyses mainly aimed at the nucleogenetically important elements carbon and nitrogen and at silicon (Baschek *et al.*, 1982a,b; Lamontagne *et al.*, 1985, 1987; Heber *et al.*, 1984b, Lynas-Gray *et al.*, 1984. The results for are summarized by Lamontagne *et al.* (1985, 1987). The helium deficiency is accompanied by large deficiencies of carbon and silicon whereas nitrogen is almost normal (see Fig. 6 of Lamontagne *et al.*, 1985). There appears to be a trend for the C and Si abundance to decrease with inceasing $T_{\rm eff}$. These findings are corroborated by recent analyses of optical CASPEC spectra of eight HBB and sdB stars by Kügler & Heber (in prep.). The observed C, N and Si abundance can be reproduced by diffusion models if mass loss at a small rate (of the order of the solar one, i.e. 10^{-14}M$_\odot$/yr) is considered (Michaud *et al.*, 1985).

However, the diffusion process does not necessarily lead to a depletion of all the metals. An overabundance of phosphorus in the HBB star Feige 86 for example has already been noted by Baschek & Sargent (1976). Recently, high resolution blue optical spectra obtained with ESO-CASPEC revealed the existence of another abundance anomaly. The HBB stars PHL 382, PHL 25 and PHL 1434 showed an enrichment of chlorine by factors

[‡]These stars display isotopic line shifts of He I, 6678Å that are larger than found previously in any other ^{3}He star (Hartoog & Cowley, 1979).

ranging from 3 to 170. In addition Ar was found to be slightly enriched in PHL 382 and PHL 25. Cl and Ar lines are not detected in the CASPEC spectra of any of the other subdwarfs. The chlorine anomaly is illustrated in Fig. 3 of Heber (1991). The Cl, Ar and P enrichment is probably caused by the radiative acceleration being large for these elements ("radiative levitation").

The observed abundance pattern of HBB and sdB stars are still a challenge to theoreticians in the field of diffusion processes.

4 Subluminous O-stars

The sdO stars apparently extend the sdB/sdOB sequence to higher temperatures ($T_{\rm eff}>$ 35000 K) and have, in general, helium rich atmospheres (exceptions are reported by Schönberner & Drilling, 1984 and Bauer, 1991) in contrast to the sdB and HBB stars discussed in section 3. Early LTE analyses (e.g. Richter, 1971) revealed that, because of the large effective temperatures in question, NLTE effects become very important. Two characteristic effects become apparent (see e.g. Fig. 2 of Werner, Heber & Hunger , 1991): the NLTE line profile is considerably stronger than the LTE one and occassionally displays a central emission reversal.

Following up on the pioneering work of Kudritzki & Simon (1978), only very few NLTE analyses of sdO stars have been performed (Giddings, 1980; Kudritzki *et al.*, 1980; Hunger *et al.*, 1981, Simon, 1982, Gruschinske *et al.*, 1983, Kilkenny, Heber & Hunger, 1986) up to recently. The reason for this is that problems with the classical NLTE technique, the complete linearisation method of Auer & Mihalas (1969), arose. Hardly any helium-rich model could be converged in a parameter range relevant to the sdO stars (see Husfeld *et al.*, 1989). Moreover, the complete linearisation technique allowed only a very limited number of atomic levels to be treated in NLTE. Both drawbacks have now been overcome by a new method to construct NLTE model atmospheres, the socalled Accelerated Lambda Iteration developed by Werner & Husfeld (1985) and Werner (1986, 1988). For a review see Hubeny (these proceedings). This technique has been applied by Dreizler *et al.* (1990), Rauch *et al.* (1991), Dreizler and Rauch (both in these proceedings) to analyse sdO stars. Up to now, some thirty subluminous O stars have been analysed for their atmospheric parameters. The class of sdO stars is far less homogenous than the sdB class with respect to their atmospheric parameters since they are spread out all over the ($T_{\rm eff}$, log g)-plane (see Fig. 4 of Heber, 1991). Many of the sdO stars lie close to the post-EHB tracks and therefore might have evolved from the sdB stars. However, several sdOs (of relatively low gravities) cannot be post-EHB stars but are identified as descendants of the AGB stars because they lie close to the post-AGB tracks. These stars are found in the same region of the ($T_{\rm eff}$, log g)-plane as the central stars of Planetary Nebulae (CSPN). Hence the sdO stars form two distinct groups with respect to their evolutionary history: The "compact" sdOs evolving from the EHB and the "low gravity" sdOs being post-AGB objects similar to the CSPNs.

4.1 Abundances

First attempts to derive metal abundances (N, Si) for three sdO stars from photographic optical and UV spectra were carried out by Simon *et al.* (1980) while Gruschinske *et al.* (1980) determined upper limits for the carbon abundances in theses three stars from UV resonance line profiles. Recently, metal abundance analyses based on high resolution optical spectra (ESO-CASPEC) have been carried out for 13 helium-rich sdOs.

Before we discuss the results, the problems inherent in these analyses are worth mentioning as exemplified in the case of KS 292 (Rauch *et al.*, 1991). The most important spectral lines of carbon and nitrogen are due to C IV and N V and arise mainly from highly excited levels (n = 5 or higher). This fact renders their analysis difficult for two reasons: i) Very elaborate model atoms are required in NLTE line formation calculations in order to obtain reliable occupation numbers for these high lying levels. ii) The highly excited levels are close to degeneracy rendering the treatment of the line broadening difficult. The broadening mechanism ranges between the linear and the quadratic Stark effect. A gradual change from quadratic to linear Stark effect occurs with increasing principal quantum number. Unfortunately, an exact theory for this intermediate case is not available. An approximate treatment has been proposed by Werner, Heber & Hunger (1991).

Most of the stars analysed sofar are helium stars with no H detectable even at high spectral resolution. Abundances of C, N, Si[§] have been derived for six members of the "low gravity" sdO subclass (HD 49798, Simon *et al.*, 1980, Gruschinske *et al.*, 1980; LSE 159, LSE 259, LSE 263, Husfeld *et al.*, 1989; KS 292, Rauch *et al.*, 1991, BD+37°442, Bauer, 1991) and eight members of the "compact" subclass (BD+75°325, Simon *et al.*, 1980, Gruschinske *et al.*, 1980, LS IV+10°9, LSS 1274, UV 0832-01, UV 0904-02, Dreizler, these proceedings; CPD-31°1701, HD 127493, Feige 46 and SB 884, Bauer, 1991). The emerging picture is very confusing, to say the least. Both subgroups are very inhomogenous with respect to their photospheric abundance pattern. Feige 46 and SB 884 which show very peculiar abundances shall be discussed in section 5. In all other stars N is overabundant (by factors between 8 and 30) while carbon is also enriched except for LSE 263, CPD-31°1701, HD 49798, HD 127493 and BD+75°325 which are C deficient by a factor of 10 or more. It is particularly striking that the "compact sdOs" LS IV+10°9, LSS 1274, UV 0832-01, UV 0904-02 and CPD-31°1701 have almost identical atmospheric parameters and the first four are C rich while the latter is strongly C deficient. The same situation is encountered amongst the "low gravity" sdOs LSE 159, LSE 259 (C rich) and LSE 263 (C deficient).

How can we interprete these abundance anomalies? Diffusion is unlikely to be important for the helium-rich sdO atmospheres since they are convectively unstable in a wide parameter range (Winget & Cabot, 1980; Wesemael *et al.*, 1982; GKH) and some "low gravity" sdOs loose mass at rates as high as 10^{-10} to $10^{-9} M_\odot$/yr (Hamann *et al.*, 1981). Convection and/or strong mass loss can suppress diffusion efficiently.

It appears to be more likely that we see material at the stellar surface that has been processed by the CN cycle (N enrichment, C deficiency) as well as the products of the

[§] O, Ne, Mg, and Al abundances are also available for some of these stars

3 α-process (C enrichment) in some cases. Of course in those stars where H has been detected there is also a large fraction of unprocessed material.

How can these peculiar abundance pattern as well as the stars' position in the HRD be explained in the context of stellar evolution? Various evolutionary scenarios are conceivable (see Dreizler *et al.*, 1990, Husfeld *et al.*, 1989 and Rauch *et al.*, 1991 for a detailed discussion). Since the analysed stars are spread out over the ($\log g$, $T_{\rm eff}$)-diagram, it is unlikely that a single evolutionary scenario can explain both subgroups of sdO stars. For the "low gravity" subgroup, Heber & Hunger (1987) suggested that the "born-again post-AGB-star" scenario of Iben *et al.* (1983) is the most promising one. According to this scenario a post-AGB star suffers a late helium shell flash bringing it back onto the AGB for a second time. Most of its hydrogen envelope reduced to 10^{-4}M$_\odot$ during the first descent from the AGB can now be removed if there is a second superwind phase. During the second descent a modest mass loss (10^{-8}M$_\odot$/yr is sufficient to strip off the remaining hydrogen layer. The star is now powered by helium burning for which evolutionary time scales are so long that a planetary nebula, which has possibly been ejected during the first departure from the AGB, has long been dispersed. This explains why no nebulosities could be detected. The "compact" sdOs are more likely to have evolved from the EHB (sdB/OB stars, see Dreizler *et al.*, 1990). In this scenario, however, it is difficult to explain how a He-deficient sdB star can evolve into a He-rich sdO. It was shown by GKH that photospheric convection zones, as originally suggested by Winget & Cabot, 1980 and Wesemael *et al.* (1982), cannot cause this transition. Alternatively, mixing triggered by He-shell flashes has been suggested by Dreizler *et al.* (1990).

In summary, we can identify two evolutionary channels leading to the sdO stage: one starting from the extended horizontal branch and explaining the "compact" sdO and another one starting from the asymptotic giant branch and leading to the "low gravity" sdOs possibly via a second loop in the HRD caused by a final helium shell flash.

4.2 Microturbulent velocity

Recent UV observations with the Hubble Space Telescope and the GHRS (Hubeny, Heap & Altner, 1991) of the sdO star BD+75°325 suggested a high microturbulent velocity of $v_t = 20$ km/sec, close to the sound velocity. Taken at face value such a high microturbulent velocity leads to a significant turbulence pressure in the atmosphere. Since this additional pressure term was neglected in previous NLTE analyses described above, Hubeny, Heap & Altner (1991) showed that the gravities determined would be systematically too low by 0.1 to 0.15 dex, if all sdOs have such high microturbulent velocities. Therefore it is important to know whether other sdOs also show indications for a high v_t or whether BD+75°325 is an unusual case. Since the optical spectrum contains only few lines it is difficult to derive the microturbulent velocities from it. Bauer (1991) and Dreizler (these proceedings) succeeded to derive upper limits to v_t for five sdOs: in no case the microturbulent velocity exceeded 10 km/s. Hence, BD+75°325 seems to be atypical and an independent estimate of v_t from a high resolution optical spectrum is highly desirable as a consistency check.

5 Hot subdwarfs of Population II

Most of the hot subdwarfs belong to the old disk population according to their kinematics and space distribution (scale heights). However a few subdwarfs are known to belong to population II. Of course the members of galactic globular clusters have to be named. The bluest horizontal branch stars in NGC 6752 have been shown to be analogs to the field sdB stars (Heber *et al.*, 1986). Amongst the sdO stars only very few are known to be members of globular clusters (see e.g. Glaspey *et al.*, 1985, Drukier, Fahlman & Richer, 1989, Cohen & Gillet, 1989), the best studied ones are ROB 162 in NGC 6397 and K 648 in M 15 the latter being a CSPN. The NLTE analysis of ROB 162 revealed that it has solar helium abundance and belongs to the "low gravity" subgroup and is in a post-AGB phase of evolution (Heber & Kudritzki, 1986). No metal lines can be detected in its high resolution optical spectrum. A medium resolution, high S/N optical spectrum of K 648 has recently been obtained by the author with the 3.5m telescope at the Calar Alto observatory. It allowed the atmospheric Balmer absorption line wings to be resolved from the nebular emissions and the atmospheric He II- and C III, C IV-lines to be measured with high precision. Metal lines other than C could not be detected. The presence of C lines is no surprise since the nebula is known to be C-rich (Adams *et al.*, 1984). A NLTE abundance analysis is under way.

Amongst the field sdOs, two stars have to be considered as bona fide population II stars due to their high radial velocities: LS IV-12°1 (-178km/s, Drilling and Heber, 1987) and BD+39°3226 (-280km/s, Giddings, 1980). This classification is corroborated by high resolution optical spectra which do not show any traces of metal lines. A NLTE analysis of LS IV-12°1 (Heber & Hunger, 1987) showed this star to be a spectroscopic twin of ROB 162. Recently, Dreizler & Heber (in prep.) analysed the high resolution IUE spectrum and found C, N, O and Si to be underabundant by 2.0 ± 1 dex consistent with the upper limit for C and N (1/10 solar) from the CASPEC spectrum. This removes any doubt about the classification of LS IV-12°1. BD+39°3226 is a helium star and belongs to the "compact" subclass according to the NLTE analysis of Giddings (1980).

Two other field sdOs (Feige 46 and SB 884) are suspected to be population II stars as well, because of their peculiar metal abundances: Bauer (1991) found deficiencies of Mg, Al and Si (by up to a factor of 50: Si in Feige 46) whereas N is nearly solar in Feige 46 and overabundant by a factor 7 in SB 884. Similarly, C is nearly normal in the former and underabundant by more than a factor 5 in the latter. One might speculate that the low abundances of Mg, Al and Si observed in these high galactic latitude stars might be primordial, whereas C and N are enriched by nuclear burning and have been mixed to the stellar surface.

Hence there is a large spread in the abundance pattern from star to star: from apparently unmixed atmospheres (ROB 162 and LS IV-12°1) to atmospheres that must have undergone strong mixing (K 648). This is indicative of variable mixing efficiency during the progenitors' evolution.

6 Subdwarfs in binary systems

Hot subluminous stars in binary systems could provide an important tool for checking the evolutionary scenarios discussed above since they possibly allow stellar masses to be determined. Visual binaries and eclipsing spectroscopic binaries are of utmost importance in this respect.

Several dozens of objects with composite spectra consisting of a hot subdwarf and a dwarf or subgiant G-K star have been discovered (e.g. Ferguson, Green & Liebert, 1984; Unglaub & Bues, 1990; Viton *et al.*, 1991; Theissen *et al.*, these proceedings).

In order to test the single star scenarios the separation of the members have to be large enough to exclude that mass transfer could have occured during their pre-subdwarf evolution.

Table I lists the two visual binaries and (to the author's knowledge) all spectroscopic binaries with known periods.

Table 1: Specroscopic binaries with known periods and visual binaries containing a hot subdwarf

star	sp. types	period [d]	type	reference
BD-7°3477 (HW Vir)	sdB +?	0.116719651	eclipsing	Menzies & Marang, 1986
LB 3459 (AA Dor)	sdOB +?	0.261539726	eclipsing	Kilkenny *et al.*, 1978 Kilkenny, 1986
HZ 22 (UX CVn)	sdB +wd	0.573703		Young *et al.*, 1972 Young & Wentworth, 1982
HD 49798	sdO +?	1.5477		Thackeray, 1970
BD-3°5357 (FF Aqr)	G8III +sdO	9.207755		Dworetsky *et al.*, 1977
HD 185510 (V1379 Aql)	K0III-IV +sdB	20.658	eclipsing	Fekel & Simon, 1985 Balona *et al.*, 1987
HD 128220	GIII-IV +sdO	871.78		Wallerstein & Wolff, 1966; Howarth, 1987
HD 113001 (ADS 8734)	F2V +sdO		vis. binary	Wallerstein & Spinrad, 1960
HD 17576 (DAW 35)	G0IV-V +sd		vis. binary	Olsen, 1980

However, even for HD 128220, the system with the longest period (870 days) known, interaction cannot be ruled out (Howarth & Heber, 1991). Hence, the visual binaries HD 113001 and HD 17576 would be the best candidates. However, their orbits have not yet been determined.

Hence all spectroscopic binaries listed in Table I are likely to be the product of close binary evolution. Most of these sdBs can be explained by case B mass transfer (Mengel, Norris & Gross, 1976, Iben & Tutukov, 1986, Tout & Eggleton, 1988a), while HD 128220 is likely to be the product of case C mass transfer (Howarth & Heber, 1991). The hot subdwarfs formed by case B mass transfer are predicted to have masses well below 0.5 $M_{\odot}$, the canonical value for the single subdwarfs (Iben & Tutukov, 1986, Tout & Eggleton, 1988a). Indeed, such low masses have been found for LB 3459 (0.3$M_{\odot}$, Kudritzki *et al.*, 1982), HZ 22 (0.39 $M_{\odot}$, Schönberner, 1978) and HD 185510B (0.3$M_{\odot}$, Jeffery, Simon & Lloyd Evans, 1992).

Systems like FF Aqr and UX CVn have evolved from Algol-like binaries which have begun their Roche lobe overflow when the mass of the primary was already smaller than that of the secondary due to mass loss from the primary. Enhanced mass-loss rates on the RGB by a factor of about 100 are required, which might be triggered by tidal forces (Tout & Eggleton, 1988b). For HD 128220, also substantial mass loss is required before the onset of case C mass transfer (Howarth & Heber, 1991). The binaries with periods below one day will probably evolve into cataclysmic binaries while those with longer periods might be pre-symbiotics (Tout & Eggleton, 1988b).

Spectroscopic abundance analyses are difficult due to the composite nature of many binary systems containing a hot subdwarf. They can be done most efficiently from ultraviolet spectra, since in this spectral range the cool component does not contribute. Lynas-Gray *et al.* (1984) determined C and Si abundances of LB 3459 from UV resonance line profiles. A NLTE analysis of HD128220B is in progress based on a superpostion of 30 IUE high resolution spectra (see Rauch, these proceedings).

REFERENCES

Adams, S., Seaton, M.J., Howarth, I.D., Auriere, M., Walsh, J.R. 1984: *MNRAS* **207,** 471.

Auer, L.H., Mihalas, D. 1969: *ApJ* **158,** 641.

Balona, L.A., Lloyd Evans, T., Simon, T., Sonneborn, G. 1987: *IBVS*, 3061.

Baschek, B., Norris, J. 1975: *ApJ* **199,** 694.

Baschek, B., Sargent, A.I. 1976: *A&A* **53,** 47.

Baschek, B., Kudritzki, R.P., Scholz, M., Simon, K.P. 1982a: *A&A* **108,** 387.

Baschek, B., Höflich, P., Scholz, M. 1982b: *A&A* **112,** 76.

Bailyn, C.D., Iben, I., Jr. 1989: *ApJ* **347,** L21.

Bauer, F. 1991: *PhD thesis, LMU München*

Bixler, Bowyer, S., Laget, M. 1991: *A&A* **250,** 370.

Cohen, J.G., Gillet, F.C. 1989: *ApJ* **346,** 803.

Caloi, V. 1989: *A&A* **221,** 27.

Crocker, D.A., Rood, R.T., O'Connell, R.W. 1989: *ApJ* **332,** 236.

de Boer, K.S. 1985: *A&A* **142,** 321.

de Boer, K.S., Heber, U., Richtler, T. 1988: *A&A* **202,** 113.

Dreizler, S., Heber, U., Werner, K., Moehler, S., de Boer, K.S. 1990: *A&A* **235,** 234.

Drilling, J.S, Heber, U. 1987: *IAU Coll. 95: The Second Conference on Faint Blue Stars,* eds. A.G.D. Philip, D.S. Hayes and J. Liebert, Schenectady: Davis Press, p. 603.

Drukier, G.A., Fahlman, G.G., Richer, H.B. 1989: *ApJ* **342,** L27.
Dworetsky, M.M., Lanning, H.H., Etzel, P.B., Patenaude, D.J. 1977: *MNRAS* **181,** 13P.
Fekel, F.C., Simon, T. 1985: *AJ* **90,** 812.
Ferguson, D.H., Green, R.F., Liebert, J. 1984: *ApJ* **287,** 320.
Giddings, J.R. 1980: *Ph.D. thesis, UCL, London*
Glaspey, J.W., Demers, S., Moffat, A.F.J., Shara, M. 1985: *ApJ* **289,** 326.
Green, R.F., Schmidt, M., Liebert, J. 1986: *ApJS* **61,** 305.
Greenstein, G.S., Truran, J.W., Cameron, A.G.W. 1967: *Nature* **213,** 871.
Greenstein, J.L. 1966: *ApJ* **144,** 496.
Greenstein, J.L., Sargent, A.I. 1974: *ApJS* **28,** 157.
Greggio, L., Renzini, A. 1990: *ApJ* **364,** 35.
Groth, H.G., Kudritzki, R.P., Heber, U. 1985: *A&A* **152,** 107 (GKH).
Gruschinske, J., Hunger, K., Kudritzki, R.P., Simon, K.P. 1980: ESA SP-**157,** 311.
Gruschinske, J., Hamann, W.-R., Kudritzki, R.-P., Simon, K.P., Kaufmann, J.P. 1983: *A&A* **121,** 85.
Hamann, W.-R., Gruschinske, J., Kudritzki, R.P., Simon, K.P. 1981: *A&A* **104,** 249.
Hartoog, M.R. 1979: *ApJ* **231,** 161.
Hartoog, M.R., Cowley, A.P. 1979: *ApJ* **228,** 229.
Heber, U. 1986: *A&A* **155,** 33.
Heber, U. 1987a: *Mitt. Astron. Ges.* **70,** 71.
Heber, U. 1987b: *IAU Coll. 95: The Second Conference on Faint Blue Stars,* eds. A.G.D. Philip, D.S. Hayes and J. Liebert, Schenectady: Davis Press, p. 79.
Heber, U. 1991: *Proc. IAU symp. No. 145,* Kluwer, p. 363.
Heber, U., Kudritzki, R.P. 1986: *A&A* **169,** 244.
Heber, U., Langhans, G. 1986: in: *New Insights in Astrophysics,* ESA-SP **263,** 279.
Heber, U., Hunger, K. 1987: *IAU Coll. 95: The Second Conference on Faint Blue Stars,* eds. A.G.D. Philip, D.S. Hayes and J. Liebert, Schenectady: Davis Press, p. 599.
Heber, U., Hunger, K., Jonas, G., Kudritzki, R.P. 1984a: *A&A* **130,** 119.
Heber, U., Hamann, W.-R., Hunger, K., Kudritzki, R.P., Simon, K.P., Mendez, R.H. 1984b: *A&A* **136,** 331.
Heber, U., Kudritzki, R.P., Caloi, V., Castellani, V., Danziger, J., Gilmozzi, R. 1986: *A&A* **162,** 171.
Heber, U., Dreizler, S., de Boer, K.S., Moehler, S., Richtler, T. 1988: *Astron. Ges. Abs. Ser.* **1,** 16.
Howarth, I.D. 1987: *MNRAS* **226,** 249.
Howarth, I.D., Heber, U. 1991: *PASP* **102,** 912.
Hubeny, I., Heap, S.R., Altner, B. 1991: *ApJ* **377,** L33.
Hunger, K., Gruschinske, J., Kudritzki, R.P., Simon, K.P. 1981: *A&A* **95,** 244.
Husfeld, D., Butler, K., Heber, U., Drilling, J.S. 1989: *A&A* **222,** 150.
Iben, I., Jr. 1990: *ApJ* **353,** 215.
Iben, I., Jr., Kaler, J.B., Truran, J.W., Renzini, A. 1983: *ApJ* **264,** 605.
Iben, I., Jr., Tutukov, A.V. 1986: *ApJ* **311,** 742.
Jeffery, C.S., Simon, T., Lloyd Evans, T. 1992: *MNRAS* , submitted.
Kilkenny, D. 1986: *Observatory* **106,** 160.
Kilkenny, D., Hilditch, R.W., Penfold, J.E. 1987: *MNRAS* **183,** 523.
Kilkenny, D., Heber, U., Hunger, K. 1986: *A&A* **155,** 175.
Kilkenny, D., Heber, U., Drilling, J.S. 1988: *SAAO Circulars* **12,** 1.
Kudritzki, R.P. 1987: *IAU Coll. 95: The Second Conference on Faint Blue Stars,* eds. A.G.D. Philip, D.S. Hayes and J. Liebert, Schenectady: Davis Press, p. 177.

Kudritzki, R.P., Simon, K.P. 1978: *A&A* **70,** 653.
Kudritzki, R.P., Hunger, K., Gruschinske, J., Simon, K.P., Kaufmann, J.P. 1980: ESA SP-**157,** 307.
Kudritzki, R.P., Simon, K.P., Lynas-Gray, A.E., Kilkenny, D., Hill, P.W. 1982: *A&A* **106,** 254.
Lamontagne, R., Wesemael, F., Fontaine, G., Sion, E.M. 1985: *ApJ* **299,** 496.
Lamontagne, R., Wesemael, F., Fontaine, G. 1987: *ApJ* **318,** 844.
Lynas-Gray, A.E., Heber, U., Kudritzki, R.P., Simon, K.P. 1984: ESA SP- **218,** 285.
Mendez, R.H. 1991: in: *Stellar Atmospheres: Beyond Classical Models* , eds L. Crivellari, I. Hubeny, D.G. Hummer, Kluwer, p. 331.
Mengel, J.G., Norris, J., Gross, P.G. 1976: *ApJ* **204,** 488.
Menzies, J.W., Marang, F. 1986: in: *Instrumentation and Research Programmes for small telescopes,* Kluwer, p.305.
Michaud, G., Bergeron, P., Wesemael, F., Fontaine,G. 1985: *ApJ* **299,** 741.
Michaud, G., Bergeron, P., Heber, U., Wesemael, F. 1989: *ApJ* **338,** 417.
Moehler, S., Heber, U., de Boer, K.S. 1990: *A&A* **239,** 265.
Moehler, S., Richtler, T., de Boer, K.S., Dettmar, R.J., Heber, U. 1990: *A&AS* **86,** 53.
Newell, E.B. 1973: *ApJS* **26,** 37.
Newell, B., Graham, J.A. 1976: *ApJ* **204,** 804.
Olsen, E.H. 1980: *IBVS* , No. 1770.
Rauch, T., Heber, U., Hunger, K., Werner, K., Neckel, Th. 1990: *A&A* **241,** 457.
Richter, D. 1971: *A&A* **14,** 415.
Saffer, R.A., Bergeron, P., Liebert, J., Koester, D. 1991: in: *White Dwarfs,* eds. G. Vauclair and E. Sion, Kluwer, p. 53.
Sargent, W.L.W, Searle, L. 1968: *ApJ* **152,** 443.
Schönberner, D. 1978: *A&A* **70,** 451.
Schönberner, D., Drilling, J.S. 1984: *ApJ* **278,** 702.
Schulz, H., Heber, U., Wegner, G. 1991: *PASP* **103,** 435.
Simon, K.P. 1982: *A&A* **107,** 313.
Simon, K.P., Gruschinske, J., Hunger, K., Kudritzki, R.P. 1980: ESA SP-**157,** 305.
Thackeray, A.D. 1970: *MNRAS* **150,** 215.
Tout, C.A., Eggleton, P.P. 1988a: *ApJ* **334,** 357.
Tout, C.A., Eggleton, P.P. 1988b: *MNRAS* **231,** 823.
Tutukov, A.V., Yungelson, L.R. 1990: *Sov. Astron.* **34,** 57.
Unglaub, K., Bues, I. 1990: *A&A* **233,** 159.
Vauclair, G., Liebert, J. 1987: in: *Exploring the Universe with the IUE Satellite* ed. Y. Kondo, Reidel, Dordrecht, 355.
Viton, M., Deleuil, M., Tobin, W., Prevot, L., Bouchet, P. 1991: *A&A* **242,** 175.
Wallerstein, G., Spinrad, H. 1960: *PASP* **72,** 486.
Wallerstein, G., Wolff, S.C. 1966: *PASP* **78,** 390.
Werner, K. 1986: *A&A* **161,** 177.
Werner, K. 1988: *A&A* **204,** 159.
Werner, K., Husfeld, D. 1985: *A&A* **148,** 417.
Werner, K., Heber, U., Hunger, K. 1991: *A&A* **244,** 437.
Wesemael, F., Auer, L.H., Van Horn, H.M., Savedoff, M.P. 1980: *ApJS* **43,** 159.
Wesemael, F., Winget, D.E., Cabot, W., van Horn, H.M., Fontaine, G. 1982: *ApJ* **254,** 221.
Winget, D.E., Cabot, W. 1980: *ApJ* **242,** 1166.
Young, A., Nelson, B., Mielbrecht, R. 1972: *ApJ* **174,** 27.
Young, A., Wentworth, S.T. 1982: *PASP* **94,** 815.

THE SPECTRA OF BLUE HORIZONTAL BRANCH STARS

KLOCHKOVA V.G., PANCHUK V.E., GALAZUTDINOV G.A.

Special Astrophysical Observatory AS USSR
Nizhnij Arkhyz, Stavropol Territory, 357147, USSR

Last time the attempts are undertaken to ascertain the nature of the EHB-stars, i.e. the hottest stars of globular clusters. The helium abundance determination for globular clusters is a crucial experiment for verification of notions about primary helium. However, for reliable conclusions in this problem the spectroscopy of the large sample of stars of different location on the HB for some clusters is needed.

The spectra of 23 stars of horizontal branch in M13, M15, M71 were obtained with echelle-spectrometer ZEBRA with the photon counting system (Gajur et al., 1990, Klochkova, Panchuk, 1991). Spectral resolution $\delta\lambda$=1.5A is sufficiently for reliable measurements of strong (W>0.5A) lines (Fig. 1).

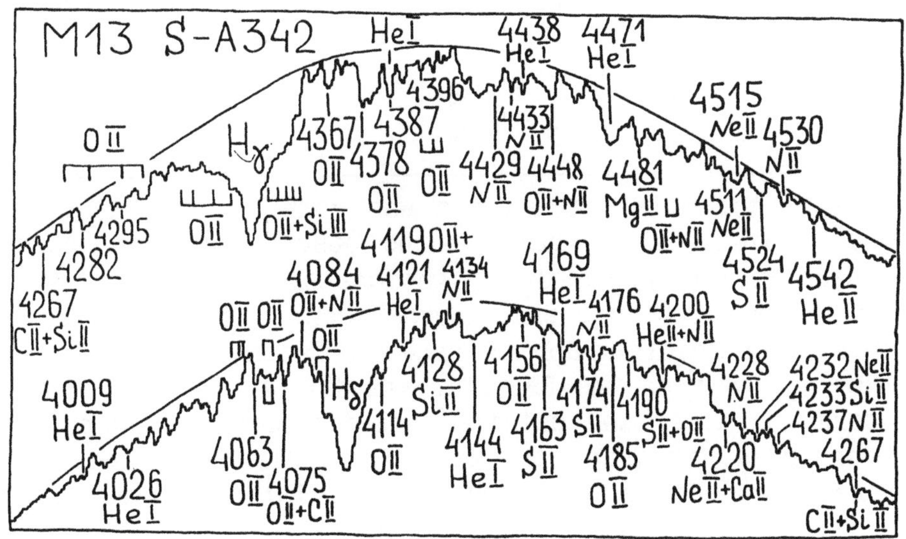

Fig.1. An example of line identification in the two from 13 echelle orders of EHB-star M13-SA-342 spectrum.

Some results of the spectra processing are presented in Tabl.1.

Table 1. The equivalent widths W (A)

Star	HI			HeI				CII
	H_δ	H_γ	H_β	4026	4144	4388	4471	4267
MI3-I-7I	6.IO	6.00	7.50	-	0.56	0.46	0.55	0.50
M13-IV-52	9.35	8.05	IO.IO	0.68	0.65	0.50	0.64	0.53
M13-IV-55	6.65	6.55	6.65	0.59	0.48	0.65	0.54	0.70
M13-S-A19	3.55	3.85	2.70	1.07	0.70	0.46	0.72	0.46
M13-S-A29	5.87	5.41	5.17	1.13	0.59	0.71	0.86	0.56
M13-S-A342	7.20	7.17	7.20	0.55	0.60	0.60	1.13	0.44
M13-S-B708	4.42	3.97	5.90	0.72	0.46	-	0.62	0.47
M15-II-36	8.70	8.40	8.15	0.60	-	0.50	0.44	0.64
M15-IV-40	7.75	7.65	10.45	0.55	0.78	0.42	0.48	0.60
M15-IV-19K	9.50	9.65	10.00	0.51	0.36	0.57	0.53	0.50
M15-IV-25K	11.20	8.27	10.45	0.80	0.62	0.89	-	0.62
M15-83	2.15	1.60	2.15	0.48	0.73	-	0.48	0.40
M15-84	6.57	6.47	4.65	0.88	0.64	0.79	0.72	0.52
M15-143	5.40	5.07	5.12	0.51	0.86	0.52	0.50	0.40
M15-154	5.97	5.25	5.75	-	0.48	0.62	0.58	0.42
M15-203	5.15	5.15	4.80	0.56	0.47	0.42	0.51	0.44
M15-339	4.32	4.87	3.30	0.64	-	0.74	0.59	0.64
M15-374	6.72	6.62	5.55	0.98	0.64	0.50	0.50	0.42
M15-476	4.50	3.60	3.20	0.72	0.57	0.65	0.55	0.47
M71-31	1.30	1.30						
M71-L	0.90	1.20	2.60					
M71-Z	4.00	3.70	4.50					

Moreover, the atlas of spectra was compiled and recorded on the magnetic disk. To standardize the observations, we have carried out also spectroscopic investigations of 20 bright ($8\text{-}9^m$) B-stars in the open clusters, which have been studied earlier with the Main stellar spectrograph (Klochkova, Panchuk, 1987).

Having compared the BHB-star spectra with those of normal O-, B-stars and after the detailed identification of echelle spectra we concluded, that for the spectrum analysis by the model atmosphere method it is necessary to calculate the atmosphere models for stars with chemical composition, noticeably differing from the solar one by both average metallicity and relative element abundances. The study of the atlas allows to trace the origin and disappearance of the selected spectral details during the transfer to hotter BHB-stars.

The absolute luminosities M_V and half-sums of equivalent widths of H_γ and H_δ lines - W(H) are compared in Fig.2. It follows that W(H) and luminosities of BHB-stars are not consistent with the relation, that was established for normal B, A-stars with hydrogen burning in the core. The accuracy of hydrogen line profiles in spectra of hot BHB-stars drops due to strong blending of the profile by light element lines (C, N, O). However, this influence cannot secure the spread of W(H), which is seen in Fig.2, because the spread value does not decrease during the transfer to colder BHB-stars.

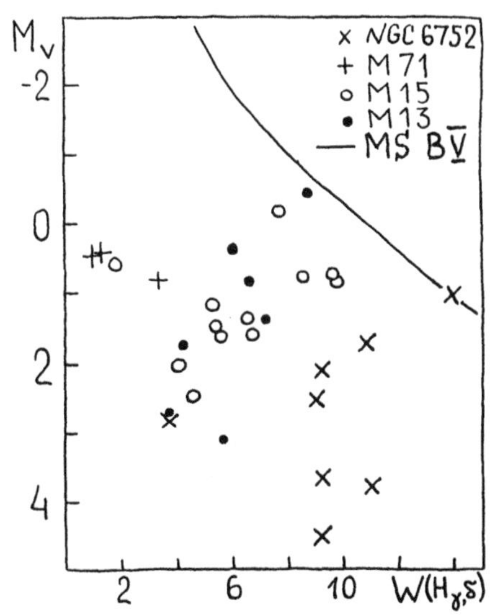

Fig.2. Data for NGC6752 from Heber et al. (1986). The line is luminosity calibration of the main sequence stars, obtained by spectra of B-stars in the open clusters (Klochkova, Panchuk, 1990).

In Fig.3 the dependence of W(4471) on $(U-B)_0$ of BHB-stars is compared with that for disk stars (Klochkova, Panchuk, 1989a). We concluded that He-weak stars among the studied BHB-stars were not detected. Earlier Klochkova and Panchuk (1989a,b) have obtained the spectral criterion for finding of He-weak stars: the ratio of equivalent widths of singlet and triplet series in the spectra of He-weak and normal B-stars is confidently different. In Tabl.2 is given a comparison of ratio W(4388)/W(4471) for different types of hot stars with low luminosity.

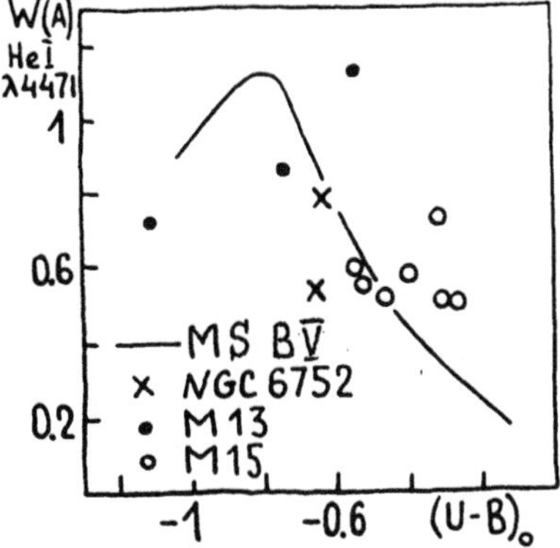

Fig.3. A dependence of W(4471) on $(U-B)_0$ for EHB-stars. The line is an averaged relation from Klochkova and Panchuk (1989a).

Table 2. The ratio W(4388)/W(4471) for B-stars

Type of objects	W(4388)/W(447I)	σ	Number of stars
EHB-stars	0.947	0.0737	10
B_e-stars	0.763	0.0257	25
BV (MS)	0.675	0.0069	306
He-weak	0.467	0.0286	55

The anomalous values of W_s/W_t ratio in spectra of CP-stars are caused by different sensitivity of singlet and triplet lines to departure of atmosphere conditions from LTE. The main difference between triplet and singlet levels is that the former are overpopulated, but the latter are underpopulated. It is quite logical to assume, that for the He-weak stars, which have the deficit of energy in UV-region, ionization processes are carried out with lesser departures from LTE. As a result, overpopulations of singlet levels and insufficient density of population of triplet levels will be decreased and on the whole the ratio W_s/W_t will be decreased in comparison with the normal stars.

By means of developing this assumption for the EHB-stars, from Tabl.2 we may conclude that the degree of departure from LTE in the atmospheres of EHB-stars is higher, than in those of normal B-stars of the main sequence. As to the atmospheres of Be-stars it is intermediate case.

REFERENCES

Gazhur E.B., Klochkova V.G., Panchuk V.E.: I990. Pis'ma Astron. Zh., *16*, p.473.

Heber U., Kudritzki R.P., Caloi V., et al.: I986. Astron. & Astrophys., *162*, p.171.

Klochkova V.G., Panchuk V.E.: I987. Pis'ma Astron Zh., *13*, p.56.

Klochkova V.G., Panchuk V.E.: I989a. Astrophys. issled. (Izv. SAO), *27*, p.25.

Klochkova V.G., Panchuk V.E.: I989b. Preprint SAO, No. 3I.

Klochkova V.G., Panchuk V.E.: I990. Pis'ma Astron. Zh., *16*, p.435.

Klochkova V.G., Panchuk V.E.: I99I. Preprint SAO, No. 70.

Weight Watching in M 15

S. Moehler [1], K.S. de Boer [1], U. Heber [2]

[1]Sternwarte der Universität Bonn, Auf dem Hügel 71, D-5300 Bonn, FRG
[2]Institut für theoretische Physik und Sternwarte der Universität, Olshausenstrasse 40, D-2300 Kiel, FRG

Lonely Stars ...

The stars on the Blue Horizontal Branch (BHB) are thought to be successors to red giants. They consist of a helium-burning core of 0.50 $M_{\odot}$ and a hydrogen shell. The position on the BHB depends mainly on the final mass of the star and thus on the mass loss of the red giant. The smaller the mass of the hydrogen-rich envelope the hotter is the star. The sdB stars forming the blueward extension of the HB (Extended Horizontal Branch, EHB, Heber, 1987) in this scenario have an inert hydrogen-rich envelope of less than 0.02 $M_{\odot}$. In this case the history of a sdB star would be essentially the same as that of a BHB star.

... or Social Stars ?

According to the theory of Iben (1990) a hot subdwarf is formed by very close binary evolution. It is thought that the binary system finally consists of two helium-rich white dwarfs. Due to gravitational radiation they may lose so much of their angular momentum that a merging of the two components can take place. The mass distribution of the resulting sdBs would extend from 0.3 $M_{\odot}$ up to about 0.9 $M_{\odot}$. In this case sdB and BHB stars would be the results of completely different evolutionary tracks. This fact offers an explanation for the gap found in several globular clusters between BHB and EHB (e.g. M 15 and NGC 6752).

Weight–Watching

The two hypotheses predict stars with almost identical physical parameters (T_{eff}, log g), but very different mass distributions. A determination of the masses of sdB stars thus offers a possibility to distinguish between the two models. We try to determine masses of EHB stars in the globular cluster M 15, using photometric and spectroscopic data to derive T_{eff} and log g. Together with the distances of the stars we then get the masses. The spectroscopic data were obtained at the Calar Alto Observatory in 1989 and 1990, using the 3.5m telescope with the Cassegrain Twin Spectrograph. Photometry was taken from the literature (Buonnano et al., 1983).

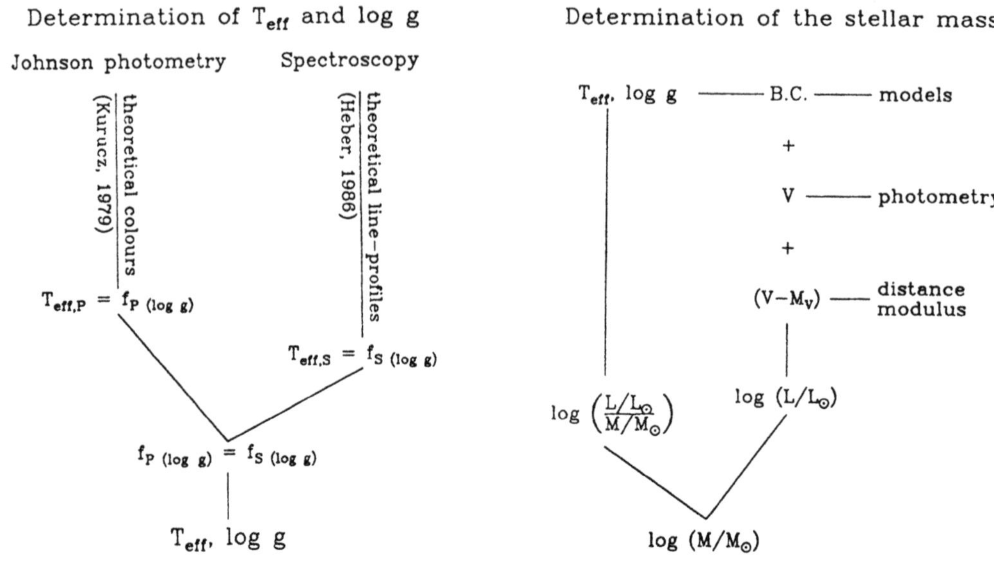

Fig. 1. The determination of effective temperatures, surface gravities and masses of the stars.

Results

Nr*	T_{eff}	log g	B.C.^{K}	V*	M[$M_\odot$]	Type
B 18	9300	3.45	-0.24	16.10	0.92	BHB
B 27	12300	3.45	-0.51	16.66	0.31	EHB
B 258	10800	3.20	-0.54	16.49	0.21	EHB
B 325	16900	4.00	-1.64	16.78	0.58	EHB
B 348	11800	4.00	-0.74	16.69	1.15	EHB
B 440	8400	2.45	-0.10	15.79	0.16	BHB
B 484	8400	2.70	-0.10	16.04	0.23	BHB

*: Buonnano et al. (1983) ^{K}: Kurucz (1979)

Cluster	< M_{BHB} >	< M_{EHB} >
M 15	$(0.43 \pm 0.24)\ M_\odot$	$(0.56 \pm 0.21)\ M_\odot$
NGC 6752*	$(0.39 \pm 0.09)\ M_\odot$	$(0.82 \pm 0.29)\ M_\odot$

*: Heber et al. (1986)

The errors given above are only the errors of the mean. Each single mass value has a minimum error of about 50%, due mainly to the error of 0.2 in log g.

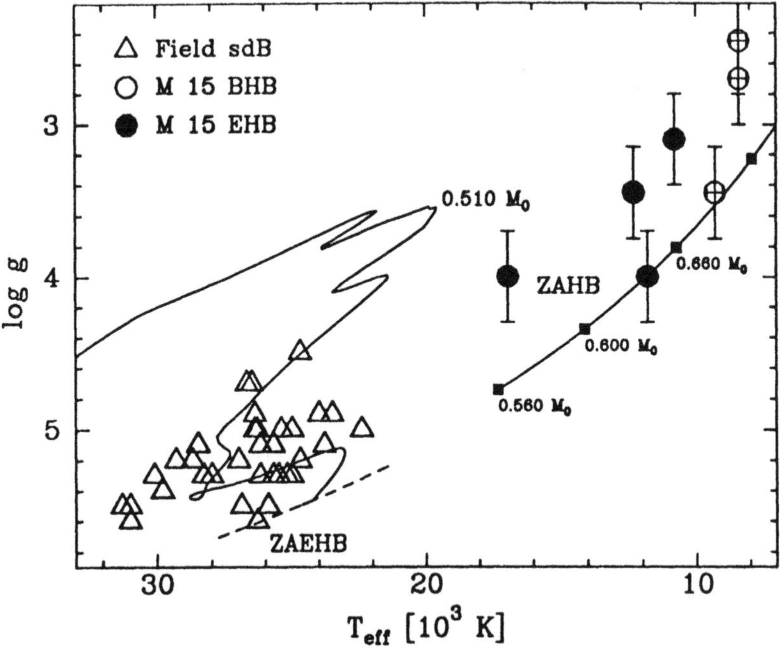

Fig. 2. The distribution of the BHB and EHB stars of M 15 in the (T_{eff}, log g)–diagram.

Conclusions

– Within the errors the mean mass of the BHB and EHB stars in M 15 is the same.
– According to their physical parameters the observed EHB stars are not sdB stars (as supposed before) **but belong to the BHB.**
– As all observed stars belong to the BHB there is no simple explanation for the gap found in the CMD of M 15. One way out would be a bimodal mass distribution (Dorman et al., 1991) for which we find no indication.
– The gap found in the CMD of M 15 is not reproduced in the (T_{eff}, log g)–diagram (in contrast to NGC 6752, where it separates BHB from sdB stars). It seems to be a quite common phenomenon, that the gaps found within the BHB of globular clusters do not show up in the physical parameters of the stars (cf. Crocker et al., 1988).

References

Buonnano R., Buscema G., Corsi C.E., Iannicola G., Fusi Pecci F., 1983, A&AS 51, 83
Crocker D.A., Rood R.T., O'Connell R.W., 1988, ApJ 332, 236
Dorman B., Lee Y.-W., VandenBerg D.A., 1991, ApJ 366, 115
Heber U., Kudritzki R.P., Caloi V., Castellani V., Danziger J.,
Gilmozzi R., 1986, A&A 162, 171
Heber U., 1987, Mitt. Astron. Ges. 70, 79
Iben I., 1990, ApJ 353, 215

Looking for gaps

J. H. Schmidt, S. Moehler, A. Theissen, K. S. deBoer

Sternwarte der Universität Bonn, D-5300 Bonn, Fed. Rep. of Germany

The distribution of stars on the horizontal branch (HB) in CMD's of globular clusters shows in several cases gaps. The gaps can be found between the blue horizontal branch (BHB) and the extended horizontal branch (EHB). But in other cases the gaps are found within the BHB. These gaps can also be seen in the distribution on the HB of stars of the field of the Milky Way.

We investigated a sample of field horizontal branch (FHB) stars, observed at optical and ultraviolet wavelengths. Using Strömgren photometry and IUE spectra, if available, we determined the effective temeratures. The surface gravities were then obtained by fitting theoretical line profiles to the Balmer lines in our optical spectra.

Star	$c1$	$b-y$	T_{eff}	$\log g$
PG 1258-030	+0.562	−0.056	13100	3.73
PG 1343+578	+0.203	+0.054	19400	4.00
PG 1432+004	+0.095	−0.084	21950	4.85
PG 1451+492	+0.229	−0.079	18400	4.45
PG 1704+222	+0.332	−0.015	17000	2.55
PG 1705+537	+0.302	−0.044	17000	4.00
PG 2111+023	+0.423	−0.028	15200	4.00
PG 2229+099	+0.306	−0.012	16700	3.50
PG 2301+259	+0.289	−0.023	17000	3.50
PG 2318+239	+0.309	−0.030	15700	3.40
PG 2345+247	+0.252	−0.037	17100	3.50
PG 2351+198	+0.380	+0.009	16100	3.80
+36 2242	+0.758	−0.031	11433	4.27
F86	+0.311	−0.062	15183	4.00

Using also a number of field sdB stars (Moehler et al., 1990; Theissen, 1991) and of cooler BHB stars (Huenemoerder et al., 1984), we were able to collect a sample of stars spanning a large range of parameters. Fig.1. and Fig.2. are showing the gaps in the $[c1, (b-y)]$-diagram and in the $[\log g, T_{eff}]$–diagram.

It was long thought that the EHB and BHB are products of the same evolutionary path, where the position of the stars on the HB is determined by the final mass after the giant branch phase. The stars on the HB have helium cores of $M_{core} \approx 0.5\ M_{\odot}$, so, the bigger the mass loss on the giant branch, the thinner the hydrogen shell, leading to hotter effective temperatures. However, this scenario cannot explain the clumpy structure on the HB. It is therefor likely, that the stars follow different evolutionary paths (Iben, 1990) after the giant branch, e.g., a different mass loss history might be an explanation for this distribution on the HB (Heber et al., 1985). Iben desrcibes a scenario in which the sdB stars are the final product of merging helium-rich white dwarfs. Stars from classical evolution are then the cooler HB stars.This could explain the gap at $T_{eff} = 22000$ K.

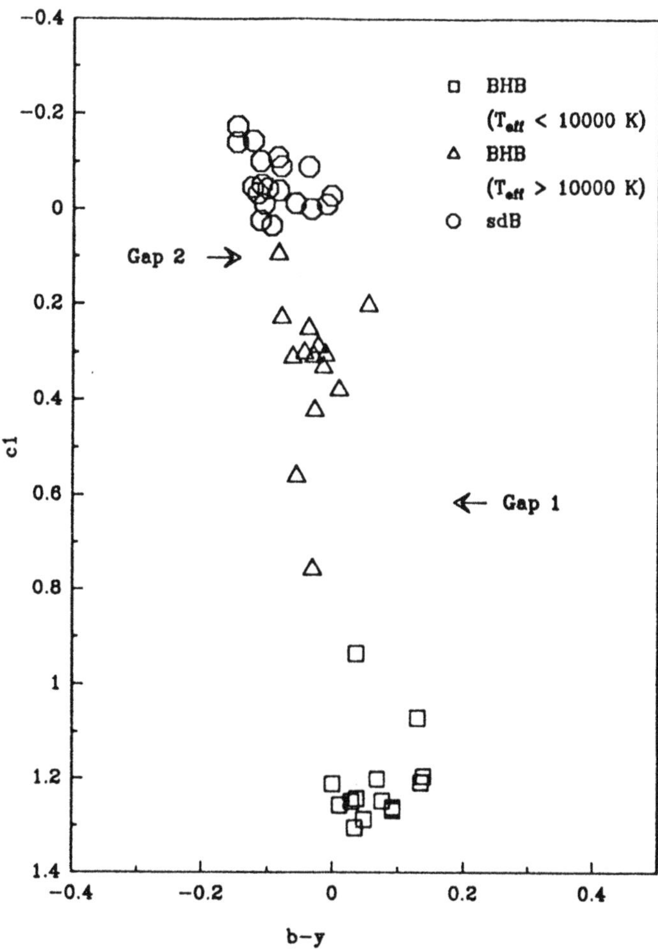

Fig. 1. The $[c1, (b-y)]$-diagram of our whole sample of field stars shows a noncontinous distribution in the photometry. There are three regions of higher density, separated by regions of a lower one. In these regions of lower density the gaps "Gap 1" and "Gap 2" are located, which have been determined by Newell and Graham (1976) from their $[c1, (b-y)]$-diagram of blue field horizontal branch star. "Gap 1" is located at $c1 \approx 0.6$, which corresponds to $T_{eff} = 13000$ K, and "Gap 2" at $c1 \approx 0.1$ corresponding to $T_{eff} = 22000$ K.

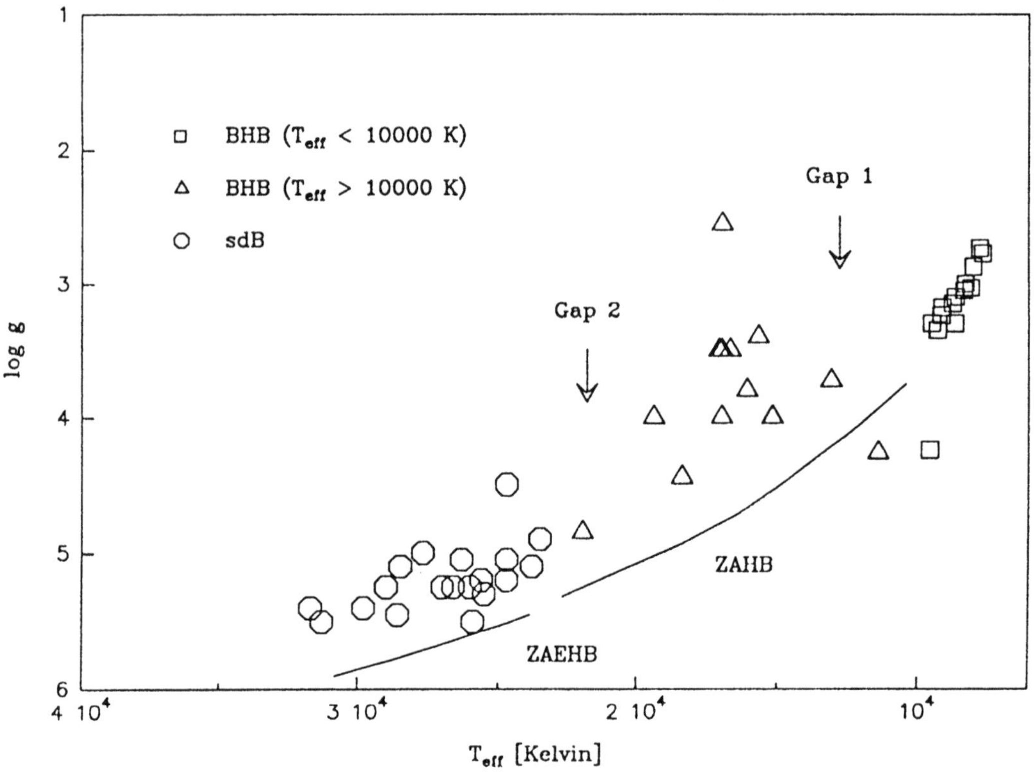

Fig. 2. The $[\log g, T_{\rm eff}]$-diagram shows the subdivision of the HB more clearly. "Gap 1" separates the cooler BHB stars from the hotter ones and "Gap 2" separates the BHB stars from the sdB stars.

References

Heber U. ,Kudritzki R.P., Castellani V., Danziger J., Gilmozzi R., (1986), AA**162**, 171

Huenemoerder D.P., de Boer K.S., Code A.D., 1984, AJ **89**, 851

Iben I., 1990, ApJ **253**, 215

Moehler S., Heber U., de Boer K.S., 1990, AA, **239**, 265

Newell B., Graham J.A., 1976, ApJ **204**, 804

Theissen A., 1991, Rhein. Friedr. Wilh. Universität Bonn, 45

More on the Progenitors of White Dwarfs

J.S. Drilling

Department of Physics and Astronomy, Louisiana State University
Baton Rouge

Abstract: A study of a complete sample of 22 very hot subdwarf O stars indicates that the main evolutionary channel to the white dwarf state is represented by the central stars of planetary nebulae detected in surveys for planetary nebulae, and that all other channels combined provide only about 0.2 times as many white dwarfs as the main channel.

In 1984, when I was on a sabbatical here in Kiel, Prof. Weidemann suggested that Detlef Schönberner and I undertake a study of the birthrates of the immediate progenitors of white dwarfs. The study was based on the HR diagram given by Schönberner and Drilling (1984) for a complete sample of 12 subdwarf O stars with effective temperatures greater than 50,000 K which had been obtained by follow-up spectroscopy by Drilling (1983) of more than 600 stars classified as OB^+ in the Case-Hamburg-LSU objective prism surveys (see also Stephenson and Sanduleak 1971). The OB^+ stars are those which show nearly continuous spectra on the survey plates, and these surveys cover the entire Milky Way plus galactic latitudes ±30° for longitudes ±60° down to photographic magnitude 12. Effective temperatures were determined by fitting model atmospheres to IUE and visible continuum observations, and absolute magnitudes were determined using distances derived by comparing values of E_{B-V} estimated from the strength of the 2200 Å interstellar feature with published maps of E_{B-V} versus distance.

Comparison of the positions of these stars in the HR diagram with published evolutionary tracks revealed three groups of stars: central stars of planetary nebulae (Group 1), other solar composition or helium-poor post-AGB stars which were not quite massive enough to evolve fast enough to become hot enough to illuminate their planetary nebulae until after the nebulae had dispersed (Group 2), and stars which have evolved from the extended horizontal branch (Group 3). The birthrates which we then derived for these three groups of stars indicated that the Group 3 birthrate was negligible compared to that of Group 1, and that the Group 2 birthrate was intermediate to those of Groups 1 and 3 (Drilling and Schönberner 1985).

A number of new developments have occured since then which I would like to report on today. First, the size of our sample has increased from 12 to 22 stars. In a few cases, we had overlooked previously known subdwarf O stars which had been classified as OB^+ in the Case-Hamburg surveys, but most of these are new discoveries (Drilling 1986). Second, NLTE fine analyses of high-resolution spectra of 11 of the stars have been carried out at Kiel and Munich (Hunger, Gruschinske, Kudritzki, and Simon 1981; Heber and Hunger 1987; Mendez, Kuritzki, Herrero, Husfeld, and Groth 1988; Heber, Werner, and Drilling 1988; Husfeld, Butler, Heber, and Drilling 1989; Rauch, Heber, Hunger, Werner, and Neckel 1991; and Dreizler 1991), providing independent estimates of the effective temperatures, absolute magnitudes, and distances. Third, there have been a number of new theoretical developments which effect the calculations of the birthrates, which will be discussed below. Fourth, I believe now that it was incorrect to identify the three stars in our original sample which show surrounding nebulosity with other known central stars of planetary nebulae. The reason is that the spectroscopic peculiarities of the latter were discovered after it was realized that these stars were the central stars of planetary nebulae, whereas it was the spectroscopic peculiarities of the former which led to the discovery of the surrounding planetary nebulae, *which were too faint to be detected in existing surveys for planetary nebulae.* On the other hand, the central stars of planetary nebulae discovered in surveys for planetary nebulae will normally not be included in the OB star surveys because of the confusion caused by the overlapping of the bright nebular and stellar spectra on the objective-prism plates.

I have therefore redone the study of Drilling and Schönberner (1985) using the larger sample of stars given by Drilling (1986). The effective temperatures determined by the NLTE fine analysis of high-resolution line spectra have been found to be consistent with those determined by Schönberner and Drilling (1984) from stellar continuum measurements. Schönberner and Drilling (1984), on the other hand, have found that the effective temperatures determined from stellar continuum measurements are strongly correlated with the quantity

$$R = \int_{1240}^{1945} F_\lambda^\circ d\lambda \Big/ \int_{1945}^{3120} F_\lambda^\circ d\lambda, \tag{1}$$

the ratio of the integrated, de-reddened fluxes in the two IUE cameras. The following calibration of $T_{\rm eff}$ versus R *for He-rich stars* was determined using the 11 stars in our sample plus the four additional objects listed in Table 2 of Schönberner and Drilling (1984) for which effective temperatures had been determined by NTLE fine analyses of high-resolution line spectra:

$$\log T_{\rm eff} = 4.06 + 0.28R \pm 0.04, \tag{2}$$

where I have used the result of Schönberner and Drilling (1984) that the effective temperatures of He-rich stars are 25% higher than those of normal or He-poor stars for a given value of R. This calibration was then used to determine the effective temperatures of all but one of the other stars in the sample, LSV+22°38, which has not yet been observed with IUE.

The color excesses in B-V were determined from the strength of the 2200 Å interstellar feature using the standard reddening law of Seaton (1979), or from the observed B-V color, assuming the intrinsic color to be -0.35. The results were consistent in those cases where both methods were used, and the color excesses were compared to published maps of the interstellar reddening versus distance (FitzGerald 1968; Lucke 1978; Nandy, Thompson, Carnochan, and Wilson 1978; Perry and Johnston 1982; and Perry and Christodoulou 1991) in order to estimate the distances of the stars. In those cases where NLTE analyses of high resolution spectra were available, the distances were also determined by means of the following equation:

$$M_V = B.C. + M_{bol\odot} - 2.5\log(4\pi GM\sigma T_{eff}{}^4/gL_\odot), \quad (3)$$

where M is the mass of the star corresponding to the evolutionary track which passes through it in the $\log g$ - $\log T_{eff}$ diagram. A plot of the distances determined from the color excesses against those obtained from the fine analyses shows that the distances obtained from the fine analyses are sytematically larger than those obtained from the color excesses. We interpret this effect as being due to the fact that many of the stars in our sample lie outside the dust layer of the Galaxy, so that the distances estimated from the color excesses are actually lower limits. Indeed, a detailed study by C.L. Perry and myself of reddening versus distance for A and F stars immediately surrounding a number of our subdwarf O stars in the sky shows that in all cases, the color excesses of the subdwarfs are consistent with the distances determined using the results of the fine analyses. I have therefore determined M_V using the above equation for all of the stars in the sample (except LSV+22°38). In those cases where NLTE analyses have not been carried out, the effective temperatures were determined as described above, the surface gravities estimated from the appearence of the low-resolution optical line spectra, and the bolometric corrections determined from the relation defined by Tables 3 and 4 of Schönberner and Drilling (1984).

When the resulting absolute magnitudes and effective temperatures are plotted on the HR diagram of Schönberner and Drilling (1984), one can identify four groups of stars which lie on evolutionary tracks leading to the white dwarf state: central stars of planetary nebulae *which were discovered because of the spectroscopic peculiarities of their central stars* (Group 1); other solar composition or helium-poor post-AGB stars (Group 2); helium-rich post-AGB stars (Group 3); and post-EHB stars (Group 4). The resulting space densities, evolutionary times, and birth rates for these four groups of stars are given in the following table. For Groups 1, 2, and 3, the times required to cross the observed loci of these stars in the HR diagram were estimated from the post-AGB evolutionary tracks of Schönberner (1981) and Wood and Faulkner (1986). Group 3 is new, and consists of helium-rich stars which lie within that region of the HR diagram occupied by the central stars of planetary nebulae, but show no trace surrounding nebulosity. For this reason Heber and Hunger (1987) have suggested that these stars are 'born-again' post-AGB stars, i.e. central stars of planetary nebulae which have experienced a final thermal pulse just before entering the white dwarf state which returned them to the tip of the AGB minus the planetary nebula (which has by this time dispersed) and the hydrogen-rich outer layers (which have mixed with the helium-burning shell). For group 4, I have used the Paczynski (1971) evolutionary track for a pure helium star of 0.5 solar masses. The evolutionary time given by this track does not differ greatly from those given by the more recent tracks of Caloi (1989) for post-EHB stars. The resulting birth rates are to be compared to that estimated for the central stars of planetary nebulae detected in surveys for planetary nebulae, 2.3 x 10^{-12} pc^{-3} yr^{-1} according to Phillips (1989). I conclude that the latter must be the main evolutionary channel to the white dwarf state, and that all of the other channels taken together provide only about 0.2 times as many white dwarfs as the main channel.

Table 1. Birth rates for white dwarf progenitors not found in surveys for planetary nebulae

	No. Stars	Space Density pc^{-3}	Evolution Time (yr)	Birth Rate $pc^{-3}\ yr^{-1}$
Group 1	3	1×10^{-9}	5×10^{3}	3×10^{-13}
Group 2	3	3×10^{-9}	5×10^{4}	6×10^{-14}
Group 3	5	5×10^{-10}	5×10^{3}	1×10^{-13}
Group 4	10	1×10^{-7}	1×10^{7}	1×10^{-14}

Acknowledgements: This work was supported in part by grants from NASA (NAG 5-71) and the National Science Foundation (AST-8514574).

References

Caloi, V. (1989): Astron. Astrophys. **221**, 27
Dreizler, S. (1991): these proceedings.
Drilling, J.S. (1983): Astrophys.J. **270**, L13
Drilling, J.S. (1986): IAU Coll. 95, Second Conference on Faint Blue Stars, eds. A.G.D. Philip, D.S. Hayes, and J.W. Liebert (L.Davis Press: Schenectady), p.489
Drilling, J.S., and Schönberner, D. (1985): Astron. Astrophys. **146**, L23
FitzGerald, M.P. (1968): Astron.J. **78**, 983
Heber, U., and Hunger, K. (1987): The Messenger **47**, 36
Heber, U, Werner, K. and Drilling J.S. (1988): Astron. Astrophys. **194**, 223
Hunger, K., Gruschinske, J., Kudritzki, R.P., and Simon, K.P. (1981): Astron. Astrophys. **95**, 244
Husfeld, D., Butler, K., Heber, U., and Drilling, J.S. (1989): Astron. Astrophys. **222**, 150
Lucke, P.B. (1978): Astron. Astrophys. **64**, 367
Mendez, R.H., Kudritzki, R.P., Herrero, A., Husfeld, D., and Groth, H.G. (1988): Astron. Astrophys. **190**, 113
Nandy, K., Thompson, G.I., Carnochan, D.J., and Wilson, R. (1978): M.N.R.A.S. **184**, 733
Paczynski, B. (1971): Acta. Astr. **21**, 1
Perry, C.L., and Christodoulou, D.M. (1991): submitted to Publ. Astron. Soc. Pac.
Perry, C.L., and Johnston, L. (1982): Astrophys.J. Suppl. **50**, 451
Phillips, P. (1989): IAU Symp.131, Planetary Nebulae, ed. S. Torres-Peimbert (Kluwar), p.425
Rauch, T., Heber, U., Hunger, K., Werner, K., and Neckel, T. (1991): Astron. Astrophys. **241**, 457
Schönberner, D. (1981): Astron. Astrophys. **103**, 119
Schönberner, D., and Drilling, J.S. (1984): Astrophys.J. **278**, 702
Seaton, M.J. (1979): M.N.R.A.S. **187**, 73P
Stephenson, C.B., and Sanduleak, N. (1971): Publications of the Warner and Swasey Observatory **1**, 1
Wood, P.R., and Faulkner, D.J. (1986): Astrophys.J. **307**, 659

Non-LTE analysis of the Palomar Green subdwarf O stars

Peter Thejll [1], Franziska Bauer [2], Rex Saffer [3], Dietmar Kunze [2],

Harry Shipman [4] and James Liebert [3].

[1]NORDITA, Blegdamsvej 17, DK-2100 Copenhagen , Denmark
[2]Universität der München, Scheinerstrasse 1, D-8000, BRD.
[3]Steward Observatory, University of Arizona, Tucson, AZ 85721, USA.
[4]Department of Physics, University of Delaware, Newark, DE 19716, USA.

Abstract: Using NLTE model spectra, the sdO subdwarfs found in the Palomar-Green (PG) survey, are analyzed in order to determine their relative contribution to the white dwarf cooling sequence. We can confirm that only a small fraction of the WD's, and thus also of the DB WD's, can be related to the sdO channel.

1 Goal and Method

Which role do the helium-rich sdO stars play in late stages of stellar evolution? In particular, do they form an important link to the helium-rich white dwarfs.

By fitting optical and UV-spectra to H and He-line blanketed NLTE models we find Teff, log(g) and composition. Then we calculate distances and find population descriptors such as scale height, density in plane and rate of evolution, and finally compare to WD's. Our model grid covers from 35,000K to 65,000K, log(g)=4.0 to 6.5 and from 50 to 99% helium. The observed spectra are assumed to form a random selection of objects from the statistically complete PG survey.

2 Results

There are about 230 hot subdwarfs, showing at least traces of helium, in the PG survey of blue high-galactic latitude objects. They are labeled as sdO, sdOA, sdOB, sdOC or sdOD in the survey and will in this work be referred to as "sdO's". We have about 130 low to high quality medium dispersion sdO spectra. We attempted to fit 70 of these and were able to fit 35. Most spectra that we failed to fit were more rich in hydrogen than the maximum of 50% that limits the present grid. We calculated the distances above the galactic plane - the results are shown in figure 1. Best fitting (using the Vmax test) scale height is 460 to 1150 parsec, density in the galactic plane is 3.3 to 6.5 x 10^{-7} stars per cubic parsec. The temperatures and gravities found in our sample of stars is very similar to those found by Dreizler, et al., (1990), in their smaller study of sdO stars.

The positions in the Teff/log(g)-plane tentatively identify the stars with post-HB stars that are burning He in a shell (Caloi, 1989). Lifetime is then 2 to 3 x 10^{7} years and rate of evolution is thus from 1.1 to 3.3 x 10^{-14} stars per cubic parsec per year. The total rate of evolution represents from 8 to 25% of the non-DA birthrate measured by Fleming, Liebert, and Green (1986). 24% of the 70 stars have an abundance of Helium above 99%. Since H and He will separate very quickly as g increases when/if the sdO contracts to a DB it may only be possible to link the helium pure fraction of sdO's to the DB stage - or just 24%. That fraction of the total birthrate Then gives the rate of evolution of helium-rich sdO's: 0.8 - 1.6 x 10^{-14} stars per cubic parsec per year, or about 2 to 8% of the non-DA birthrate. Downes (1986) found densities in the plane of about 7 x 10^{-7} stars per cubic parsec in his study of the galactic plane sdO's, which is very similar to what we find, but the corresponding scale height is much lower than our range of values: 270 parsec.

Gilmore, and Reid (1983) identified a "thick disc" population in the galaxy, with a scale height near 1500 parsec. A thick disc should contain a population of white dwarfs, and their precursors, that shared this large scale height. Habing (1988) describes a disc thickness of from 1200 to 2800 parsec, based on IRAS data. Others have measured scale heights for white dwarfs: 500 parsec (Chiu, 1980), and 250 parsec (Fleming, Liebert, and Green, 1986).

3 Future plans

Since our data set is not homogeneous in quality we will improve the spectra, by additional observations, until they form a complete and uniform set. The better spectra should lower the uncertainty on the population parameters - this we know from our work on the high-quality part of our data. We will also improve on the sometimes inaccurate magnitudes of some of the subdwarfs. We will extend the NLTE grid to cover higher temperatures and larger abundances of H. This will enable us to distinguish between hot and H-rich objects which all show strong lines at the positions of the Balmer series. Sophisticated calculations of the evolution from the HB and the EHB are needed in order to accurately identify what stage of evolution the subdwarfs are in so that appropriate lifetimes can be assigned to the various evolutionary stages. Attention should be paid to the effects of the H-flash that appears at the final approach to the WD cooling sequence, to understand if this might be a way of ridding an otherwise He dominated photosphere of hydrogen.

References

Caloi, V. (1989). A&A, 221, 27-35.
Chiu, L.-T. G. (1980). Astron. J. 85, 812.
Dreizler, S., Heber, U., Werner, K., Moehler, S., and de Boer, K.S. A&A, 235, 234-241.
Downes, R. A. (1986). Ap. J. Supplement Series, 61, 569-584.
Fleming, T. A., Liebert, J., and Green, R. F. (1986). Ap. J. 308, 176-189.
Gilmore, G., and Reid, N. (1983). M.N.R.A.S., 202, 1025-1047.
Habing, H. J. (1988). A&A, 200, 40-50.
Koester, D., and Schönberner, D. (1986). A&A, 154, 125- 134.
Werner, K., Heber, U., and Hunger, K. (1991). A&A., 244, 437-461.

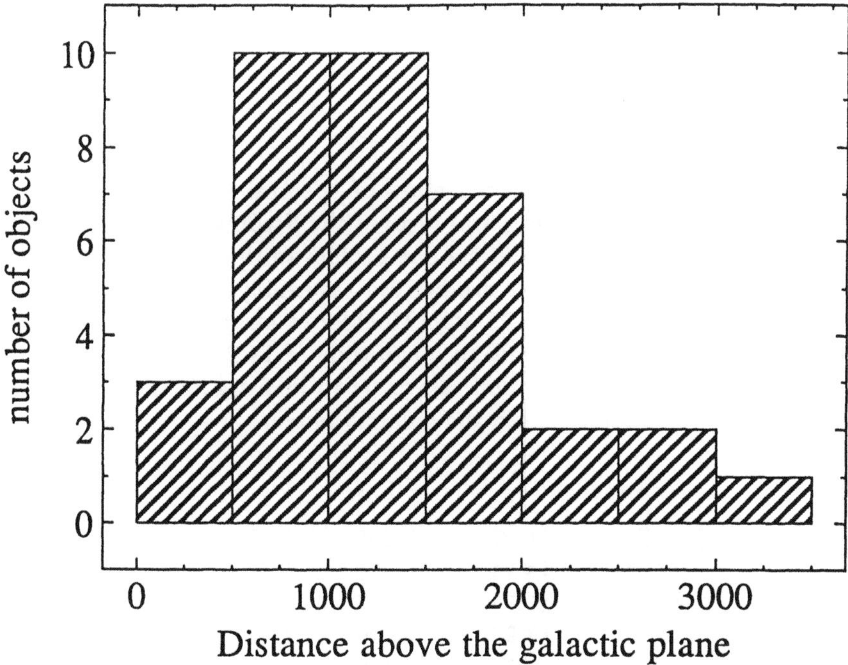

Fig. 1. A histogram of the measured distances above the galactic plane. Using the Vmax test (a statistical test that defines the complete part of a sample) we 'fit' this histogram to an exponentially decaying density function with scale height between 460 and 1150 parsec, and a density in the galactic plane between 3.3 and 6.5 $\times 10^{-7}$ stars per cubic parsec. An exponentially decaying density distribution, sampled by a cone, as the PG survey did, is expected to peak at $3z_0$.

A table of the fitting results is available upon request. Interested readers should write to P. Thejll, or use e-mail **thejll@nordita.dk**.

A Statistical Complete Sample of Hot Subdwarfs

A. Theissen [1], S. Moehler [1], J. H. Schmidt [1], U. Heber [2]

[1]Sternwarte der Universität Bonn, Auf dem Hügel 71, D-5300 Bonn, Fed. Rep. of Germany
[2]Institut für theoretische Physik und Sternwarte der Universität, Olshausenstrasse 40, D-2300 Kiel, Fed. Rep. of Germany

Distribution in Space ...

We analysed a sample of hot subdwarf stars of spectral type sdB and sdOB. Using the reddening free colour indices [c1] and [u-b] of the Strömgren system we determined effective temperatures. Fitting theoretical line profiles to the observed Balmer lines gave us the surface gravity. The distances are derived assuming the masses of the subdwarfs to lie in the small range of 0.50 ± 0.02 $M_\odot$ (Moehler et al., this volume). Our sample is statistically complete down to a limiting magnitude of 14.2 mag in an area defined by $35° \leq l \leq 95°$ and $28° \leq b \leq 40°$. This allows us to investigate the distribution of the hot subdwarfs with the distance from the galactic plane.

Object	SpType	y	T_{eff}	log g	r[pc]	z[pc]
PG 1647+253	sdOB	14.07	36200	5.65	1200	720
PG 1656+318	sdOB	14.14	29900	5.25	1390	840
PG 1701+359	sdB	13.15	27700	5.00	1050	630
PG 1710+490	sdB	12.90	28600	5.45	600	350
PG 1716+426	sdB	13.97	25600	5.20	1240	700
PG 1718+519	sdB	13.69	28200	4.90	1020	590
PG 1722+286	sdOB	13.24	31700	5.40	870	440
PG 1725+252	sdB	12.89	26000	5.25	660	320
PG 1738+505	sdB	13.15	24700	5.05	970	510
PG 1739+489	sdB	13.05	24700	5.15	810	420
PG 1743+477	sdB	13.79	27400	5.45	860	440

The number of stars decreases exponentially in z with a scale height of $z_0 = 200$ pc as can be seen in Fig. 1 and Fig. 2 where we test a distribution like $N_i \sim z^2 \exp(-z/z_0)$. Therefore the sdB– and sdOB–stars in the field belong to the old disk population.

Because of the small statistics of only 11 stars we cannot be sure about the scale height derived above. Using Monte Carlo simulations we test our result by investigating how the distribution looks like if the *a priori* scale height of the exponential distribution would be another one (Fig. 3).

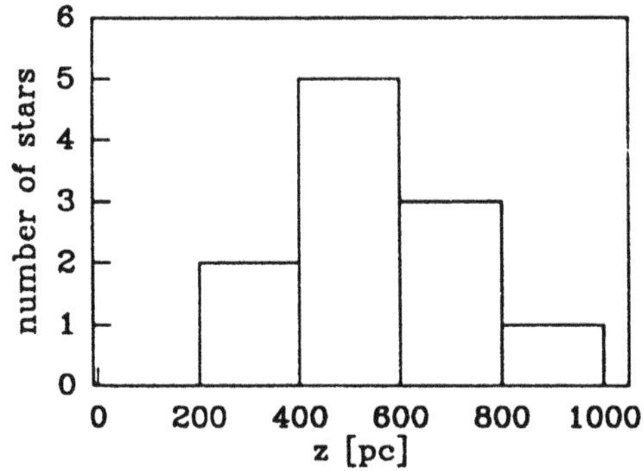

Fig. 1. Distribution of the sdB- and sdOB-stars in the z-direction, shown in a histogram.

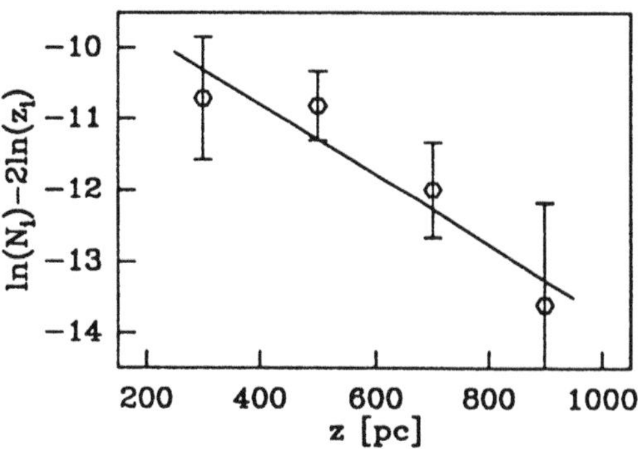

Fig. 2. $\ln N_i - 2\ln z_i$ plotted against z_i. The error bars are due to statistical errors of the histogram. The inverse slope of the straight line is the scale height of the exponential distribution which turns out to be 200 pc.

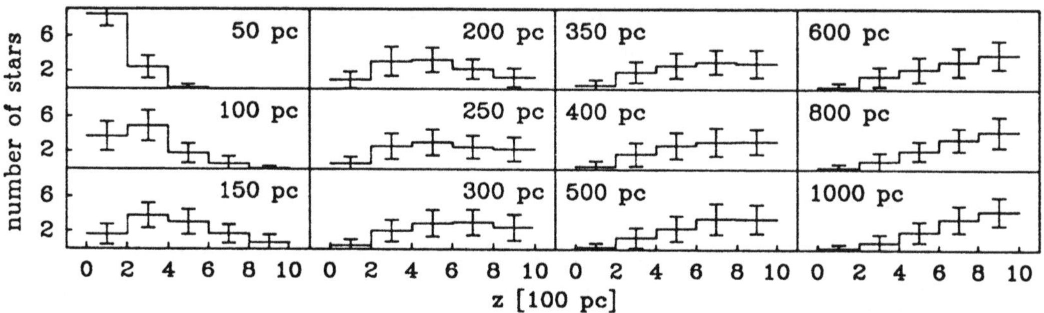

Fig. 3. Average distributions of 100 Monte Carlo simulations for different a priori scale heights. The simulations are done in such a way that the number of stars up to $z = 1000$ pc equals 11 corresponding to our observations. Only the simulations with scale heights between about 150 and 300 pc are similar to them. The decrease of the 50 pc–simulation is much too strong, and for scale heights greater than 400 pc one only sees the quadratic increase with z.

In the end we cannot fix the scale height to a sharp value, but we can determine a range of $150 \leq z_0 \leq 300$ pc which supports our interpretation of the field–subdwarfs as members of the old disc population.

... Hot Subdwarfs in Binary Systems ...

The spectra of PG 1718+519 and PG 2110+127 (shown in Fig. 4 in comparison with a spectrum of a typcial sdB–star) exhibit binary nature. This is interesting in respect to the subdwarfs' evolutionary history. Maybe all subdwarfs have got cool companions, some not detected like M dwarfs, which are quite faint. The hot subdwarfs' progenitors could then be interpreted as red giants which suffered a heavy mass loss due to Roche lobe overflow in a close binary system.

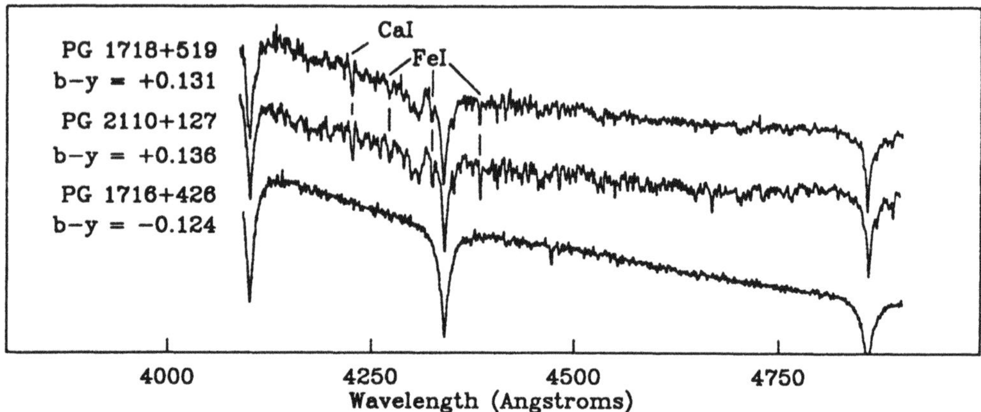

Fig. 4. Comparison of the spectra of 2 binaries containing a hot subdwarf with a single subdwarf. The binary nature becomes apparent in some spectral features and a flattened continuum which are due to a cooler companion

... and a Runaway Star

On our search for hot subdwarfs we happened to meet a main sequence star of spectral type B (Fig. 5), which is located about 9 kpc above the galactic plane. Due to some mechanism like cluster ejection or a supernova explosion in a binary system it is possible that such a young star gets a high velocity. In this way the star is able to travel far away from it's place of birth, i.e. the galactic disk, and is called a "runaway star".

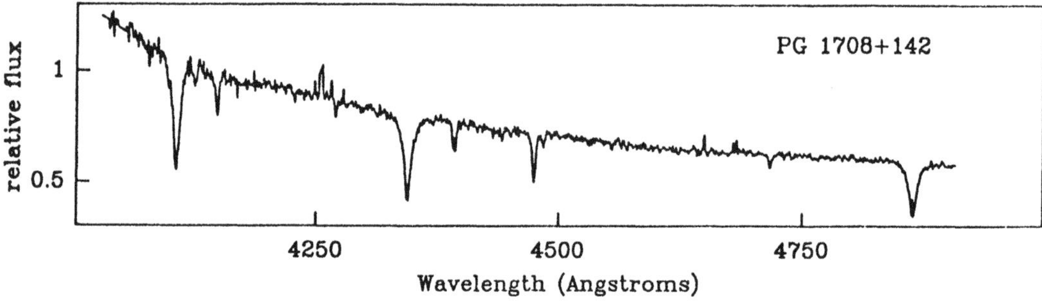

Fig. 5. Spectrum of PG 1708+142, a main sequence B–star 9 kpc above the galactic plane. The z–component of the star's radial velocity is 100 km s^{-1}

NLTE ANALYSIS OF A SDO BINARY: HD128220

T. RAUCH
Institut für Theoretische Physik und Sternwarte der Universität zu Kiel
Olshausenstraße 40-60, W-2300 Kiel, Germany

ABSTRACT. We present first results of a NLTE analysis of the sdO component of the binary system HD128220. Based on new IUE spectra we carried out a detailed NLTE analysis by means of the Accelerated Lambda Iteration method.

1. Introduction

In 1986 Werner applied the Accelerated Lambda Iteration method to the calculation of NLTE-model atmospheres. Since then many analyses have shown that this technique has become an accurate and fast mean to derive stellar parameters like $T_{\rm eff}$, $\log g$ and abundances for several elements. In the analysis of the sdO star KS 292, Rauch et al. (1991) have shown the possibility to derive these parameters with high accuracy by using several lines of different elements and high resolution ESO-CASPEC spectra. In an ongoing investigation the sdO component of the binary system HD128220 shall be analysed.

HD128220B was expected to be a probable supernova candidate when Wallerstein & Wolff (1966) estimated $M \cdot \sin^3 i = 4.3 M_\odot$. This was recently revised by Howarth & Heber (1990), who derived $M \cdot \sin^3 i = (0.5 \pm 0.2) M_\odot$. In a first NLTE analysis by means of "classical" model atmosphere calculations (see Rauch & Werner, 1991), Gruschinske et al. (1982) restricted the stellar parameters by two models to the range 37.5kK $< T_{\rm eff} <$ 47.5kK and $4 < \log g < 5$ with $n_{\rm He}/n_{\rm H} = 0.7$ (by mass) for the lower limits and $n_{\rm He}/n_{\rm H} = 0.45$ for the higher limits, respectively.

The spectral analysis is very difficult because the spectrum is composite and the visual range is dominated by the cool component. In the case of HD128220 the flux components are shown in figure 1. A small error in the radii will drastically change the flux fractions preventing any analysis of the visual spectrum.

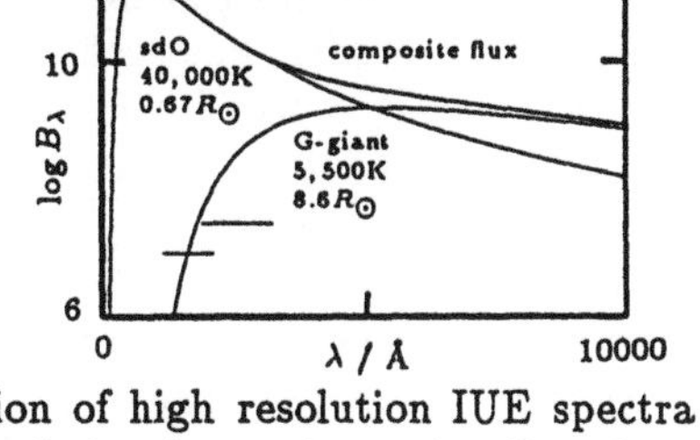

Figure 1: Flux components of a binary system. In the case of HD128220 Planck functions are used to demonstrate the fractions. The radii are adopted from Howarth & Heber (1990). Note that for $\lambda <$ 3000Å the composite flux is almost the sdO flux. The IUE SWP and LWR ranges are marked by the two bars.

New observational data is available now. A superposition of high resolution IUE spectra by Howarth (priv. comm.) yields a very good resolution and signal-to-noise ratio. Based on this spectra a new analysis is carried out in order to derive the stellar parameters and to give information about the evolutionary state of the system.

2. The "classical" approach

In order to determine the basic parameters $T_{\rm eff}$ and $\log g$ and the helium abundance we calculated firstly a grid of model atmospheres with 36kK $\leq T_{\rm eff} \leq$ 50kK, $4.3 \leq \log g \leq 5.0$ and $0.001 \leq n_{\rm He}/n_{\rm H} \leq 0.99$ by number. From our strategic He II lines $\lambda\lambda$ 1640, 2252, 2306, 2385, 2511, 2733Å we could derive only a lower limit for the helium abundance ($n_{\rm He}/n_{\rm H} \geq 0.1$). For higher abundances all examined lines are almost saturated and show only little changes even if one compares the features of $n_{He}/n_H = 0.1$ and $n_{He}/n_H = 0.99$ (figure 2). Unfortunately the He II lines are not very sensitive to changes in T_{eff} and $\log g$ (figure 3), too. An increase of T_{eff} results in a slight decrease of the equivalent widths of the lines. $\log g = 4.5^{+0.1}_{-0.2}$ is estimated.

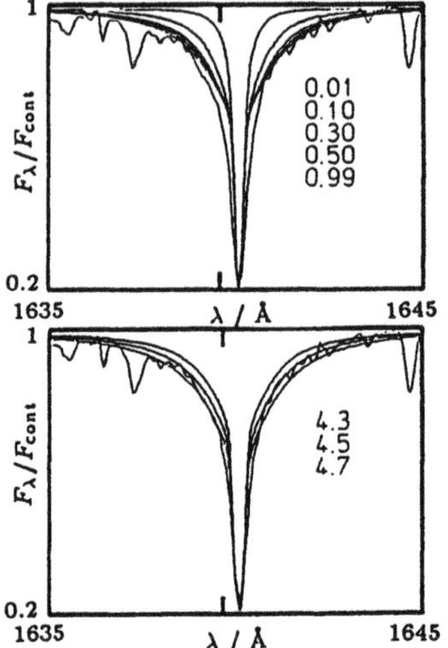

Figure 2: Theoretical He II λ 1640Å line profiles, calculated from model atmospheres with $T_{eff} = 45$kK, $\log g = 4.5$ and $n_{He}/n_H = 0.01, 0.10, 0.30, 0.50$ and 0.99 by number. The theoretical profiles are compared to the IUE spectrum. The line profile is virtually saturated for $n_{He}/n_H \geq 0.3$.

Figure 3: Theoretical He II λ 1640Å line profiles, calculated from model atmospheres with $T_{eff} = 45$kK, $\log g = 4.3, 4.5$ and 4.7 and $n_{He}/n_H = 0.50$. Note that the line cores appear broader than in figure 2 because the fine structure has been considered in detail, see also section 3.

It is worthwhile to note that a fit of the theoretical helium line profiles to the IUE spectra is not unambiguous — T_{eff}, $\log g$ and n_{He}/n_H have to be adjusted simultaneously. The IUE spectrum displays several lines of C III as well as of C IV. This allows to determine T_{eff} from the NLTE ionisation equilibrium with much higher precision than by any other method. A small grid of models has been calculated with carbon in addition (solar abundance relative to hydrogen has been assumed). The results of this analysis are presented in section 4. In view of the small changes of the He II lines over the examined range of stellar parameters we made some test calculations in order to check for the influence of line broadening. The results are reported in the following section.

3. Line broadening

For the line broadening of He II lines tables from Schöning & Butler (1989) are available. We calculated theoretical line profiles of He II λ 1640Å using these tables and for comparison using a formula given by Unsöld: $\kappa(\Delta\lambda) = \frac{\pi e^2}{mc^2}\lambda^2 f \frac{1}{s_n^* F_0} U(\frac{\Delta\lambda}{s_n^* F_0})$. (For details see Rauch et al. 1991.) If there are many ions considered in the atmosphere simultaneously the correct microfield F_0 is given by $F_0 = 2.61 \cdot e \cdot [\sum_{ions} z_i^{3/2} n_i]^{2/3}$. Theoretical line profiles of He II λ 1640Å were calculated using a real microfield distribution, $z = 1$ to simulate the case of a pure hydrogen atmosphere with protons as perturbers and $z = 2$ for a pure helium atmosphere, respectively. In figure 4 the profiles are compared. The line profile calculated with the real microfield distribution appears deeper than the one with $z = 1$. The same effect can be expected if other perturber ions besides protons are included in broadening theories.

Another lack in broadening theory is that no fine structure is considered. In figure 2 we see that the line core of the He II λ 1640Å is not broad enough. For this line the fine structure splitting amounts to 0.2Å . If the fine structure is considered, the line core is expected to be broader resulting in a much better fit. In figure 3 the same line profile is shown as in figure 2 but with fine structure splitting. Obviously, fine structure splitting should be included in future calculations of line broadening tables.

The line broadening of C III is well described by the quadratic Stark effekt. For some C IV lines new broadening parameters are available (Dimitrijević et al. 1991, and priv. comm.). For the other C IV lines we assume the linear Stark effect to be dominant and use the Unsöld formula (see above).

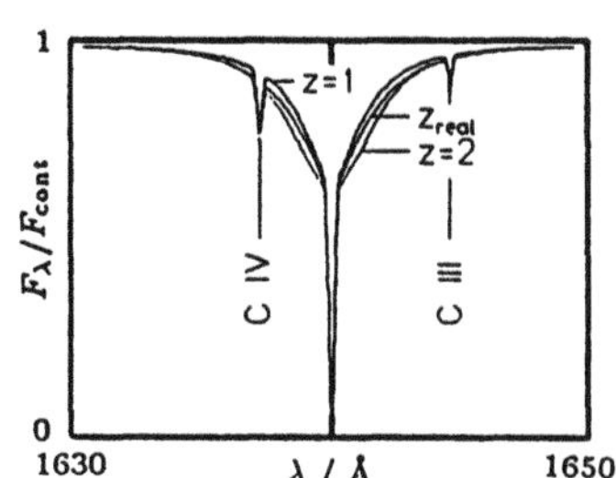

Figure 4: Theoretical line profiles of He II λ 1640Å using the Unsöld formula (see above) with different microfield distributions. The model parameters are the same as in figure 2 with $n_{\rm He}/n_{\rm H} = 0.5$ and $n_{\rm C}/n_{\rm H} = 0.005$. Note that in case of our model parameters the effect of the real microfield distribution instead of $z = 1$ increases the equivalent width by a factor of 1.15.

4. Carbon

The results of the first small grid are of great promise. Using several C III and C IV lines the ionisation equilibrium gives $T_{\rm eff} = 40,600$ K. For the fact that the C IV λ 1550 Å equivalent width is increasing with decreasing $T_{\rm eff}$ in our temperature range (in contrary to all other C IV lines), we can determine the excitation temperature from C IV lines alone: 40,700 K. In addition a detailed line profile fit with numerous C III and C IV lines results in $T_{\rm eff} = 40,000$ K. Adopting this $T_{\rm eff}$, the equivalent widths of both C III and C IV lines are too small by a factor of 1.7. In addition the theoretical He I $2\,^3S - n\,^3P^o$ line profiles (λλ2696, 2722, 2762, 2827, 2945Å) are too deep. This indicates that $n_{\rm He}/n_{\rm H}$ is smaller than 0.5 — with a lower helium abundance the carbon lines will appear deeper because the continuous opacity will be smaller. In praxi this can be used to determine the helium abundance. In figures 5 and 6 the C III UV 4 triplet and the C IV UV 1 doublet are shown for our derived effective temperature $T_{\rm eff} = 40,000$K.

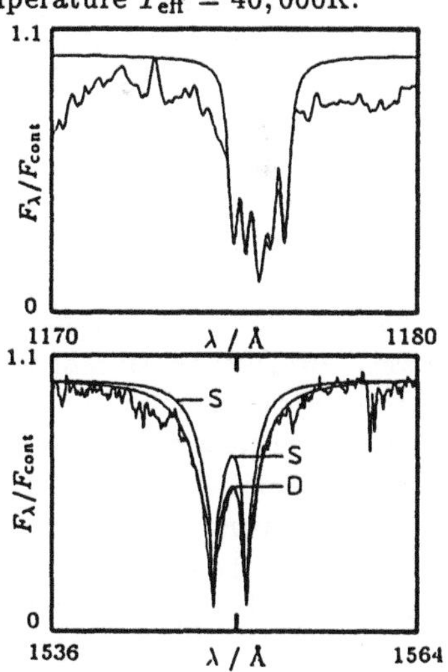

Figure 5: Theoretical line profiles of C III λ 1175Å . Note that the structure of this blend (6 lines) is reproduced in detail. The model parameters are: $T_{\rm eff} = 40,000$K, $\log g = 4.5$, $n_{\rm He}/n_{\rm H} = 0.5$ and $n_{\rm C}/n_{\rm H} = 0.005$.

Figure 6: Theoretical line profile of C IV λ 1550Å . S denotes the theoretical line profile which is calculated with broadening data by Sahal-Bréchot & Segre (1971), for D new data of Dimitrijević et al. (1991) is used. (model parameters see figure 6)

REFERENCES

Dimitrijević, M.S., Sahal-Bréchot, S., Bommier, U. 1991, ApJS, 89, 581

Gruschinske, J., Hamann, W.-R., Kudritzki, R.-P., Simon, K. P., Kaufmann, J. P. 1983, A&A, 121, 85

Howarth, I. D., Heber, U. 1990, PASP, 102, 912

Rauch, T., Werner, K. 1991, in Stellar Atmospheres: Beyond Classical Models, NATO ASI Series C, Vol. 341, eds. L.Crivellari, I.Hubeny, D.G.Hummer, Kluwer, Dordrecht, p. 165

Rauch, T., Heber, U., Hunger, K., Werner, K., Neckel, T. 1991, A&A, 241, 457

Sahal-Bréchot, S., Segre, E. R. A. 1971, A&A, 13, 161

Schöning, T., Butler, K. 1989 ApJS, 78, 51

Unsöld, A. 1968, Physik der Sternatmosphären, 2nd edition, Springer, Berlin, p. 324

Wallerstein, G., Wolff, S. C. 1966, PASP, 75, 61

Werner, K. 1986, A&A, 161, 177

NLTE ANALYSIS OF HELIUM RICH SUBDWARF O STARS

S. DREIZLER

Institut für Theoretische Physik und Sternwarte der Universität Kiel
Olshausenstr. 40, D-2300 Kiel 1, Germany

The analysis of four extremely He rich subdwarf O-stars is presented. Two stars (LSS 1274 and LS IV +10°9) are taken from Drilling's survey (Drilling 1983, 1987) (blue stars of the galactic disc), the other two (UV 0832-01 and UV 0904-02) are located just outside the boundary of that survey (Carnochan and Wilson, 1983). These stars are among the few subdwarfs which are bright enough to be studied with high spectral resolution in the optical and UV spectral ranges.

The analysis is based on ESO-CASPEC and IUE high resolution spectra covering the wavelength ranges from 3950 Å to 4950 Å and from 1150 Å to 2100 Å with a resolution of 0.25 Å and 0.1 Å, respectively. All four stars have nearly identical spectra. Therefore the analysis is performed for LSS 1274 representatively. The following ions can be identified in the optical: H I, He I-II, (C II), III- IV, N III-IV, O III, Ne II, (Al III), Mg II, (Si III), IV; and in the UV: He II, C III- IV, N III-V, O IV-V, Si IV, Fe IV-VI. The lines of the ions in brackets are not suitable for analysis. In order to determine the effective temperature and gravity a grid of NLTE model atmospheres including H, He, C, and N was calculated using the ALI-code of Werner (1986). Element abundances have been determined from subsequent line formation calculations using very detailed atomic models for C, N, O, Ne, Mg, and Si.

Mainly new atomic data from the Opacity Project are used. The number of NLTE levels and line transitions included in the calculations are listed in Table 1. Dielectronic recombination of highly excited levels (lying above the ground state of the next ionisation stage) is taken into account following Hummer & Mihalas (1973). All ions are treated in LS-coupling. However, Ne II states with high n or l are better described in jl-coupling. For the spectrum synthesis of the Ne II 3d – 4f transition the correct coupling is used. The level energies are taken from Persson (1971), the oscillator strengths were calculated by K. Butler (priv. comm.).

Theoretical line profiles are calculated with the program LINE1 by T. Rauch. Line profiles of H and ionized He are calculated using VCS Stark tables (Schöning & Butler 1989). Broadening tables for neutral He are taken from Barnard, Cooper and Smith (1974) etc. For lines of other H like ions arising from highly excited levels (C IV, Si IV, $n \geq 5$) a semi-empirical formula is used to mimic the transition from quadratic to linear Stark broadening (for details see Werner et al. 1991). Line profiles with pure quadratic Stark broadening are calculated according to Griem (1968).

Table 1: Number of NLTE levels (NLTE) and line transitions (RBB) included in model atmosphere and line formation calculations.

		model atmosphere		line formation				line formation	
Element	Ions	NLTE	RBB	NLTE	RBB	Element	Ions	NLTE	RBB
H	I, II	10	36	10	36	O	III-VI	78	185
He	I-III	37	99	44	169	Ne	I-IV	76	232
C	II-V	24	60	118	394	Mg	I-IV	28	42
N	II-VI	29	/	126	396	Si	III-V	40	65

In general the theoretical line profiles reproduce the observed ones quite well (Figs. 1, 2). Even

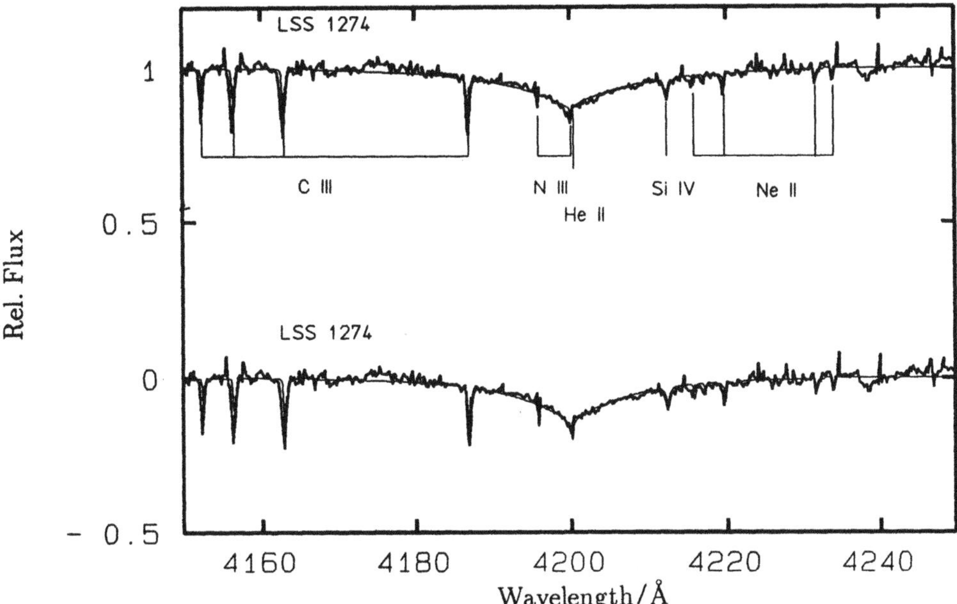

Figure 1: Spectrum of LSS 1274 around 4200 Å compared with two theoretical spectra: top: $T_{\rm eff} = 45,000$ K, $\log g = 5.7/({\rm cm\,s^{-2}})$; bottom: $T_{\rm eff} = 44,000$ K, $\log g = 5.4/({\rm cm\,s^{-2}})$; abundances according to Table 2.

for rather complex ions like C III, N III the agreement is good. Problems, however, occur for He II 4686 Å: The fit of the He II 4686 Å line has also been difficult in previous analyses of He rich subdwarfs (Dreizler et al. 1990). The theoretical profile is too shallow in the transition between core and wing (Fig. 2). There could be different reasons for this discrepancy: **1.)** The VCS tables do not take the ion dynamic into account. **2.)** Iron line blanketing might change the level populations in He II or the formation depth of the line. (This is discussed by Dreizler & Werner 1992). **3.)** The fine structure splitting of the He II lines are also not taken into account. This results in a widening of the core of He II 4686 Å by 0.5 Å. For Ne II the scatter in abundances derived from different lines is larger than for other ions. The discrepancy in the Ne II spectrum might be reduced with a more detailed model atom (in the present calculations up to $n = 5$ in NLTE) or with the consideration of the jl-coupling also for the line formation calculations.

Table 2: Element abundances of LSS 1274 given as number ratios with respect to He and as mass fractions (β). The solar mass fraction is also given. Uncertain results are marked with a colon.

X	n_X/n_{He}	$\log\beta$	$\log\beta^{\odot}$	$\Delta = \log\beta - \log\beta^{\odot}$
He	1	-0.001	-0.544	+0.54
C	$5\cdot10^{-3}\pm5\cdot10^{-4}$	-1.83	-2.400	+0.63
N	$1\cdot10^{-3}\pm3\cdot10^{-4}$	-2.46	-3.010	+0.55
O	$1\cdot10^{-3}\pm3\cdot10^{-4}$	-2.40	-2.022	−0.38
Ne	$7\cdot10^{-4}\pm2\cdot10^{-4}$	-2.52:	-3.111	+0.59 :
Mg	$8\cdot10^{-5}\pm2\cdot10^{-5}$	-3.31	-3.231	−0.08
Si	$9\cdot10^{-5}\pm2\cdot10^{-5}$	-3.20	-3.197	0.0

An effective temperature and a gravity between 45,000 K, $10^{5.7}$ cm/s^2 and 44,000 K, $10^{5.4}$ cm/s^2

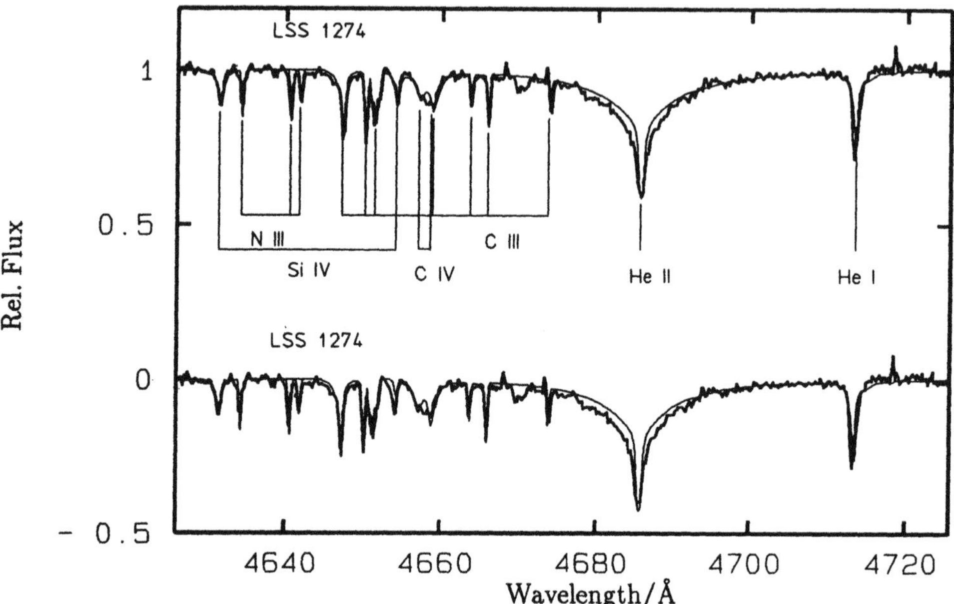

Figure 2: Spectrum of LSS 1274 around 4650 Å compared with two theoretical spectra: top: $T_{\rm eff} = 45,000$ K, $\log g = 5.7/({\rm cm\,s^{-2}})$; bottom: $T_{\rm eff} = 44,000$ K, $\log g = 5.4/({\rm cm\,s^{-2}})$; abundances according to Table 2.

have been determined. An upper limit for the hydrogen abundance of 10% results. Besides the extreme enrichment of He in the atmosphere, C, N, and Ne are overabundant while O is slightly underabundant. Mg and Si have abundances near the solar value (Table 2). The microturbulence velocity is smaller than 10 km/s. Due to spectral similarity these results are also valid for LS IV +10°9, UV 0832-01, and UV 0904-02.

The atmospheric abundance pattern of the four sdOs under investigation can be interpreted as a mixture of CNO and triple α processed material. In this picture He and N enrichment results from the CNO cycle which simultaneously destroys C and O. C is produced in the triple α process. By subsequent α captures O, Ne, Mg, and Si may be produced depending on the temperature in the burning region. Due to the mixture of the CNO and triple α processed material the depletion of O and C is nearly compensated or even changed to enrichment. While Ne is slightly enriched, Mg and Si abundances are solar.

Acknowledgement. This work is supported by the DFG (grant Hu 39/29-2).

References

Barnard, A.J., Cooper, J., Smith, E.W. 1974, JQSRT, 14, 1025
Dreizler, S., Heber, U., Werner, K., Moehler, S., de Boer, K.S. 1990, A&A, 235, 234
Carnochan, D.J., Wilson, R. 1983, MNRAS, 202, 317
Dreizler, S., Werner, K. 1992, these proceedings
Drilling, J.S. 1983, ApJ Letters, 270, L13
Drilling, J.S. 1987, IAU Coll. No. 95, p. 489
Griem, H.R. 1968, Phys. Rev., 165, 258
Hummer. D.G., Mihalas, D. 1973, ApJ, 179, 827
Persson, W. 1971, Physica Scripta, 3, 133
Schöning, T., Butler, K. 1989, A&A Suppl., 78, 51
Werner, K. 1986, A&A, 161, 177
Werner, K., Heber, U., Hunger, K. 1991, A&A, 244, 437

ANALYSIS OF PG 1159 STARS

K. WERNER
Institut für Theoretische Physik und Sternwarte der Universität Kiel
Olshausenstr. 40, W-2300 Kiel, Germany

ABSTRACT. The analysis of spectroscopic and pulsational properties of the PG 1159 stars is reviewed. Detailed NLTE model atmosphere analyses have shown that the PG 1159 stars are extremely hot, hydrogen deficient pre-white dwarfs. Generally, the atmospheres are dominated by carbon and helium with a significant amount of oxygen being present, too. These abundances indicate that the stars are exhibiting deep intershell layers which have been layed bare by strong [post-] AGB mass loss events. The abundance pattern is in concordance with the idea that the non-radial g-mode pulsations which are exhibited by about half of the objects are driven by cyclic ionisation of carbon and oxygen.

1. Introduction

In 1979 McGraw et al. announced the detection of a new, hot, non-DA pulsating white dwarf: PG 1159-035. The first optical spectrum revealed a broad absorption trough due to He II 4686 and neighbouring C IV lines, accompanied by central emission reversals. The absence of any H and He I absorptions indicated a hot hydrogen deficient photosphere. The observed light curve revealed multi-periodic variations which were readily interpreted as non-radial g-mode pulsations. It was suspected that highly ionised carbon could be driving the instabilities.

PG 1159-035 has become the prototype of a new spectroscopic class of H-deficient pre-white dwarfs, being characterised by the absorption trough mentioned above. Today the total number of PG 1159 stars amounts to 18. Out of these, 7 were shown to be low-amplitude non-radial g-mode pulsators whereas 8 of them are definitely stable against pulsations. Eight PG 1159 stars are central stars of a planetary nebula (CSPN). A search for remnant PNs around several of the other stars (including the prototype) was unsuccessful (Kwitter et al. 1989, Méndez et al. 1988). Except for the high luminosity central stars (K1-16, NGC 246), there is no sign of ongoing mass loss (in particular from PG 1159-035, Fritz et al. 1990). Table 1 lists all known PG 1159 stars.

Since the detection of PG 1159-035, these peculiar stars have received a lot of attention as they represent the immediate progenitors of the hottest non-DA white dwarfs (spectral type DO). It is hoped that their analysis provides important clues to the question, to what extent the existence of two separate white dwarf sequences (H- and He-rich) is determined by the previous post-AGB evolution. This analysis is based on two complementary approaches. First, the spectroscopic analysis reveals the atmospheric parameters, effective temperature, surface gravity, and the chemical surface composition. Second, the pulsational analysis reveals stellar parameters (ideally, under consideration of the photospheric properties as boundary conditions), e.g., the total mass and the interior structure. Both approaches require detailed modeling of the stellar atmosphere and the envelope, respectively, and our knowledge about the PG 1159 stars is therefore closely related to the progress achieved in the construction of such models.

In the following we describe the pulsational properties of PG 1159 stars and the analysis of their power spectra (Sect. 2). Then we turn to the spectroscopic analysis based on model atmospheres (Sect. 3). The results are discussed in Sect. 4 and an outlook is given in the last section.

Table 1: The known PG 1159 stars, their spectroscopic subclassification (see Sect. 3.1; p=peculiar, :=uncertain) and their photometric periods. Column 4 denotes if the star has a planetary nebula.

star	spectroscopic subclass	photometric period	PN ?	reference
PG 1159-035	E	125 periods (385 s to 1000 s)	no	1,a
PG 0122+200	A	7 or 8 bands (330 s to 600 s)	no	9,d
PG 1707+427	A	2 periods (448 s and 335 s)	no	9,c
PG 2131+066	A	6 to 8 bands (340 s to 450 s)	no	2,f
K 1-16	lgE	2 main bands (1500 s and 1700 s)	yes	10,b
VV 47	E:	irregular variable	yes	4
Longmore 4		9 or 11 bands (830 s to 2300 s)	yes	3,e
Longmore 3		no variability detected	yes	3,e
NGC 246	lgE	no variability detected	yes	3,b
MCT 0130-1937	A	no variability detected	no	5
PG 1144+005	Ep	no variability detected	no	2,b
PG 1151-029		no variability detected	no	9,b
PG 1424+535	A	no variability detected	no	9,b
H 1504+65	Ep	no variability detected	no	6,b
PG 1520+525	E	no variability detected	no	9,b
IW 1	A:	no data available	yes	7
Jn 1	A:	no data available	yes	7
Abell 21	A:	no data available	yes	8

References for spectral type: [1] McGraw et al. 1979 [2] Green et al. 1986 [3] Méndez et al. 1986 [4] Liebert et al. 1988 [5] Demers et al. 1990a,b [6] Nousek et al. 1986 [7] Schönberner & Napiwotzki 1990 [8] Napiwotzki 1992 [9] Wesemael et al. 1985 [10] Grauer & Bond 1984. References for pulsation data: [a] Winget et al. 1991 [b] Grauer et al. 1987 [c] Grauer et al. 1992 [d] Hill et al. 1987 [e] Bond & Meakes 1990 [f] Bond et al. 1984

2. Analysis of non-radial pulsations

The "Whole Earth Telescope" (WET) project (Nather et al. 1990) has been proven to be a most powerful tool to observe stellar variabilities. It constitutes a world wide network of observation sites which allows continuous monitoring over several days. The first PG 1159 type target was the prototype itself (Winget et al. 1991). The light curve was obtained from 9 sites during 12 days and the ensueing power spectrum is of outstanding quality. 125 frequencies were resolved, 101 of them could be identified with specific pulsation modes. (The strongest periods were also detected in the X-ray region, Barstow et al. 1986). These excellent data were used to determine stellar parameters for PG 1159-035. Let us shortly discuss the models on which these results are based upon.

Soon after the detection of the pulsating prototype PG 1159-035 first pulsational models were constructed yielding the right periods. Starrfield et al. (1980, 1983, 1984) showed that cyclic ionisation (κ–γ effect) of C and O can drive non-radial g-mode pulsations in hot degenerate stars. The existence of a new instability strip in the HR diagram was postulated, which was supported by the subsequent detection of other variable stars of the PG 1159 type (Grauer & Bond 1984, Bond et al. 1984). Enclosing the so-called DO variables (DOV), this new instability strip represents the third one known for post-AGB stars. The first two are established by the DA and the DB variable white dwarfs, in which non-radial pulsations are driven in the H and He partial ionisation zones, respectively.

The C/O driving mechanism works at temperatures of about 10^6 K, therefore the pulsations are driven in layers very close to the stellar surface ($\approx 10^{-12}$ M_*). A low helium abundance (<25-30%, Cox 1986, Starrfield 1987) and the complete absence of hydrogen are prerequisites for the operation of the C/O mechanism, otherwise the pulsations are "poisoned". Furthermore a high oxygen abundance is required to sustain the pulsations at very high effective temperatures ($T_{\rm eff} > 100\,000$ K). The detection of O VI lines in the spectra of several PG 1159 stars (Sion et al. 1985) and, finally, the high C and O abundances found in the photospheres (see below) strongly imply that it is in fact the κ–γ mechanism of C and O that is responsible for the instabilities.

An alternative model for the driving of PG 1159 pulsations is obviously in contradiction with the photospheric abundances. Assuming that helium dominates the composition in the envelope and that diffusion is responsible for the levitation of heavy elements, Vauclair (1990) suggests that the κ–γ mechanism of metal clouds in diffusion equilibrium can drive g-mode pulsations. However, far too small C and O abundances are predicted by diffusion theory to explain the observed composition.

A different driving mechanism, which is due to instabilities in the energy release of a nuclear burning shell (ϵ-mechanism), may presently be ruled out because the pulsation periods predicted by the models are too short by a factor of 2-5 (Kawaler 1987).

As already noted by McGraw et al. (1979), the fast evolution of PG 1159-035 to the white dwarf configuration should result in a period change observable within a few years. In fact, Winget et al. (1985, 1991) detected a *decrease* of the 516 s periodicity ($\dot{P} = -2.5 \cdot 10^{-11}\,{\rm s\,s^{-1}}$), which was in agreement with first theoretical expectations (Winget et al. 1983). Two competing processes have to be considered, on one hand cooling tends to increase the pulsation periods while on the other hand contraction yields a decrease. However, repeating the calculations with more realistic post-AGB stellar models (Kawaler et al. 1985) showed that the period of PG 1159-035 should *increase*, in total disagreement with observation. Very recently the secular period change of another DOV star, PG 1707+427, was determined (Grauer et al. 1992). The two strongest modes near 448 s decrease in time as $\dot{P} = -5.9 \cdot 10^{-11}{\rm s\,s^{-1}}$, again being in qualitative contradiction with current models. One possible solution of this problem, rotational spin-up (Kawaler et al. 1985), can be ruled out as it requires a much faster rotation than observed in the case of PG 1159-035 ($P_{\rm rot}$ = 1.38 days, Winget et al. 1991). Another solution might be offered by mode trapping (see below).

The power of pulsational analysis lies with the possibility to derive stellar parameters and to probe the stellar interior ("asteroseismology"). The mean period spacing essentially depends on the stellar mass alone (Kawaler 1987). Mass determinations based on pulsational models were performed for three stars and yielded 0.585 $M_\odot$ (PG 1159-035), 0.7 $M_\odot$ (PG 1707+427), and 0.73 $M_\odot$ (PG 0122+200) (Winget et al. 1991, Fontaine et al. 1991, Kawaler 1987). While the mass for PG 1159-035 is in perfect agreement with the spectroscopic mass determination (0.57 $M_\odot$), relatively large differences occur for PG 1707+427 (spectroscopic mass 0.53 $M_\odot$). We note that only the value for PG 1159-035 is based on high precision WET data. In principle, the mass derived from pulsational analyses is of unprecedented accuracy, however, systematic errors due to unrealistic assumptions in the stellar models still need to be investigated (Winget et al. 1991). For example, inadequately stratified models with "thick" H and He envelopes ($10^{-4} M_\odot$ and $10^{-2} M_\odot$, respectively) were employed. These are not adequate because i) due to the high temperature beneath these envelopes the C/O driving mechanism cannot work and ii) they are at variance with the high C and O abundances found in the photosphere.

However, some kind of "thick" envelope on top of the C/O core or, more precisely, a steep composition gradient well beneath the surface ($10^{-3} M_\odot$), is required in order to ensure *mode trapping.* At least the strongest of the observed modes are probably "trapped" modes, i.e., modes with reduced amplitudes in the stellar core, because:

- Mode trapping has to occur to select the observed periods from the full g-mode spectrum. Trapped modes are more easily excited due to their lower kinetic energy.

- Only trapped modes show the observed departures from uniform period spacing. It is this phenomenon that is indicative for a stratified composition of the star.
- The observed secular decrease of the 516 s period in PG 1159-035 can only be explained if the corresponding mode is trapped (Kawaler 1990). For untrapped modes cooling dominates the period change. Trapped modes, however, with their amplitudes being more confined to the contracting envelope, more strongly react on the envelope contraction and therefore yield the observed period decrease.

A possible model that is compatible with both pulsational and spectroscopic observations has been suggested by Stanghellini & Cox (1991): A "thick" envelope above the C/O stellar core (in the order of a few % of M_*, being capable to trap modes) that is composed of He, C and O (in concordance with the photospheric abundances) and which comprises the C/O driving region.

Despite all remaining problems in the analysis of non-radial pulsations, it appears safe to state that the requirements for the C/O driving mechanism to operate imply that the PG 1159 stars must have lost the entire H-rich and a part of the He-rich envelope. Hence, for further progress in pulsation analysis it is important to determine the surface abundances by spectroscopic analysis.

3. Spectroscopic analysis

The first model atmosphere approach to analyse optical and UV data of PG 1159 stars was attempted by Wesemael et al. (1985, henceforth "WGL 85"). Using line blanketed LTE model atmospheres composed of H and He they roughly concluded for 7 examined objects that $T_{\rm eff}$ exceeds 80 000 K and $\log g \approx 7$. From the weakness of the Balmer lines they estimated that the atmospheres are helium rich (He/H>1, number ratio). In strong contradiction to that, an analysis of EXOSAT data from four PG 1159 stars performed with unblanketed (H-He-CNO) LTE models arrived at essentially solar abundances (Barstow & Tweedy 1990, Barstow & Holberg 1990). A NLTE analysis of the central star of NGC 246 (Husfeld 1987), which we now classify as PG 1159, also gave a strong hydrogen deficiency (He/H>10) and an extraordinarily high carbon abundance (50% by mass). Pure H-He unblanketed NLTE models were used in this study, but the neglect of carbon in the model atmospheres casts some doubt on the results.

The need for line blanketed NLTE model atmospheres including metal opacities in a self-consistent way became highly desireable, however, the "classical" numerical method (Complete Linearization, Auer & Mihalas 1969) is not capable to compute models with the required sophistication. A new numerical approach to the NLTE model atmosphere problem was developed (mainly by the Kiel group), the so-called *Accelerated Lambda Iteration* method (Werner & Husfeld 1985, Werner 1986-1989, Dreizler & Werner 1991; for the latest review see also Hubeny, 1992), allowing the construction of models appropriate for the exotic chemical composition of PG 1159 atmospheres. In parallel, high-S/N optical spectra of most PG 1159 stars were obtained in order to perform quantitative analyses.

Preliminary results were presented by Werner, Heber & Hunger (1989, 1990). Final analyses of altogether six PG 1159 stars were published by Werner, Heber & Hunger (1991, henceforth "WHH 91"), Werner & Heber (1991a, henceforth "WH 91"), and Werner (1991, henceforth "W 91"). Most of the remaining part of this review is based on these papers.

3.1 Observational data and spectral classification

High-S/N medium resolution spectra are needed for an analysis because the spectral features are broad and shallow. We have observed most of the PG 1159 stars accessible with the 3.5m telescope on Calar Alto (Spain). A collection of spectra covering the characteristic absorption trough region is presented in Fig. 1. All known PG 1159 stars are shown, except for the two southern central stars Lo 3 and Lo 4, (spectra of these stars may be found in Méndez et al.

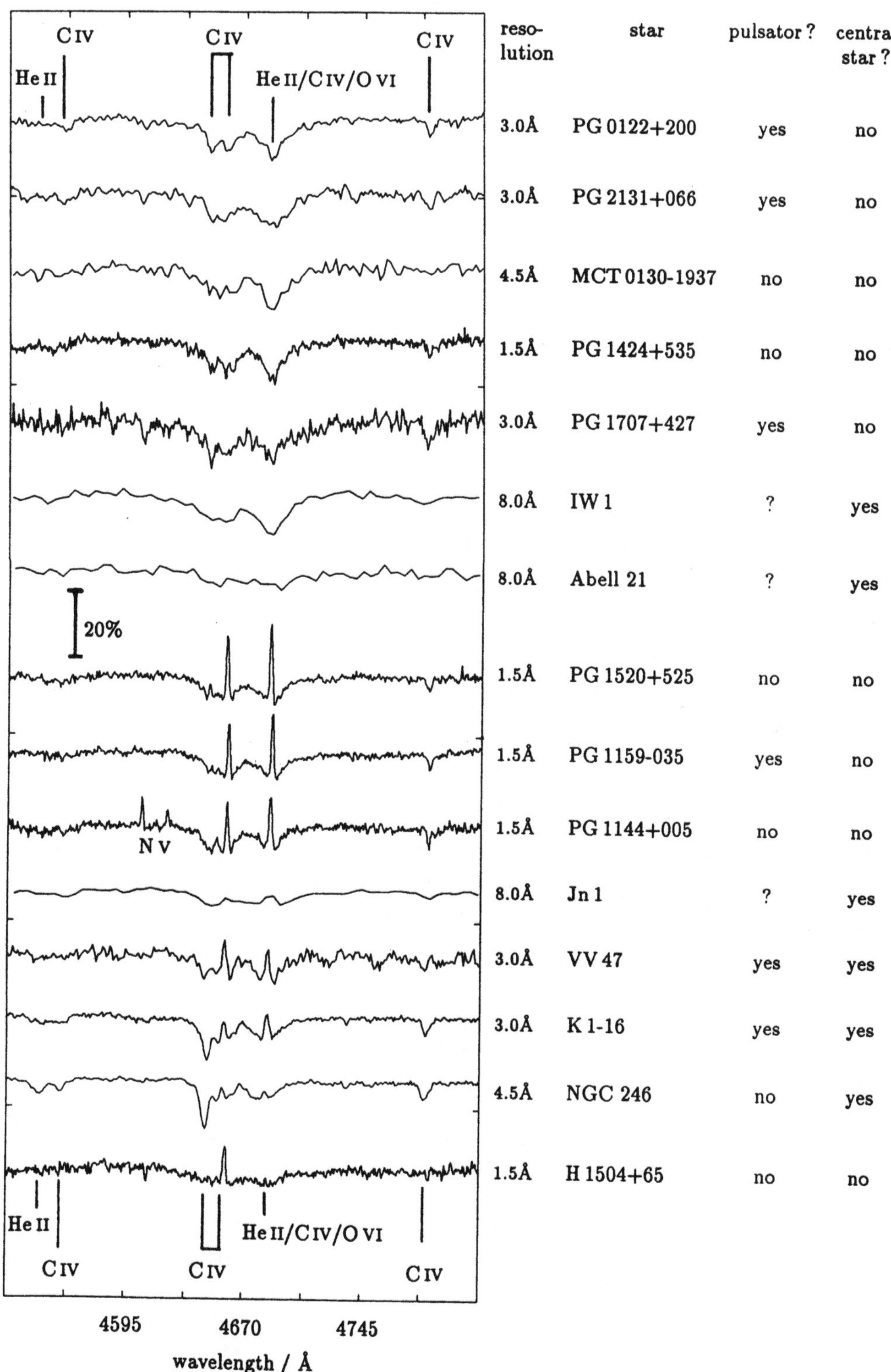

Figure 1: Normalized spectra of PG 1159 stars around the characteristic absorption trough.

1985) and PG 1151-029 (to our knowledge no spectra have been published). The resolution is not uniform but varies from star to star (from 1.5Å to 8Å), which needs to be considered when narrow emission features are compared. The broad trough is a complex blend of more than a dozen of lines with He II 4686Å and high-l C IV (5→6) transitions being the strongest contributors. Further opacity sources are several C IV (6→8) and (7→11) lines. In the particular case of H 1504+65 even highly excited lines of oxygen (9→12 in O VI) can be identified.

According to Fig. 1, a PG 1159 star can be assigned to one of the following three groups.

- GROUP 1 = **A**(bsorption), shows pure absorptions in the trough (first six spectra).
- GROUP 2 = **E**(mission), (PG 1520+525, PG 1159-035, PG 1144+005) exhibits emission cores in He II 4686Å, C IV 4659, and, rather weak, in C IV 4647Å.
- GROUP 3 = **lgE** (low gravity Emission), represented by K 1-16 and NGC 246, comprises (relatively) low gravity central stars (no such object without a PN has been found). The absorption wings are narrower than in the other objects. Like the stars in group E they exhibit two strong central emissions but C IV 4647Å is in absorption.

The spectrum of VV 47 is of poor quality (E:) and those of A 21, IW 1 Jn 1 are of too low resolution, preventing a clear grouping. The lack of the O VI 3811Å/3834Å (Liebert et al. 1988) and C IV 5801Å/5812Å emissions as well as of the O VI 5291Å feature (unpublished spectra) indicates that VV 47 might be a transition object between groups A and E. We note that the relative weakness of the K 1-16 and NGC 246 emissions compared to, e.g., PG 1159-035 arises from the lower spectral resolution (3.5Å instead of 1.5Å). A K1-16 spectrum of higher resolution (Fig. 4) shows emission strengths similar to PG 1159-035. We realize that all groups comprise pulsators and non-pulsators and that groups A and E comprise CSPN as well as non-CSPN.

Three objects deserve short comments here. H 1504+65, displayed at the bottom of Fig. 1, is a single extraordinary case. The peculiar appearance of its spectrum, the presence of the C IV 4659Å emission but the lack of the He II 4686 emission and of any other helium feature, gave rise to a controversy if the star is a PG 1159 object at all. Abundance estimates ranged from essentially solar values (Barstow & Tweedy 1991) up to complete absence of H and He (Nousek et al. 1986). A recent NLTE analysis verified the latter suspicion (W 91) so H 1504+65 may be called an extreme PG 1159 star. Noteworthy is the occurrence of optical nitrogen features in PG 1144+005 (N V 4604Å/4620Å), which is a unique phenomenon among the PG 1159 stars and which indicates a relatively high nitrogen abundance (WH 91). PG 2131+066 has a composite spectrum (WGL 85). The companion (a K7 dwarf) dominates the red spectral region and contributes about 10% to the continuum flux in the trough region.

For one star out of the three groups we show typical spectral features in three other wavelength regions (Fig. 2). Together with Fig. 1, PG 1159 stars that will be detected in future may be assigned to one of the groups.

The O VI 3s→3p doublet (3811Å/3834Å) cannot be detected PG 1159 stars of group A (bottom of Fig. 2a) but it appears in absorption in some stars of group E (e.g. in PG 1159-035, center of Fig. 2a, but not in PG 1144+005). It is seen as a prominent emission line in the low gravity objects (K 1-16, top of Fig. 2a).

Another oxygen feature, near 5291Å, arises from highly excited O VI lines (high-l 7→8 transitions), see Fig. 2b. PG 1424+535 (being from group A) does not show this feature. In contrast, all stars from group E present a strong central emission line which is sometimes encompassed by broad, shallow absorption wings (e.g. in PG 1159-035 but not in PG 1144+005). The low gravity objects (K 1-16) also show an emission line, however, it is less obvious.

Very conspicuous in many PG 1159 stars is the C IV 3s→3p doublet (5801Å/5812Å, Fig. 2c). It is in emission in the stars from group E (e.g. PG 1159-035) and is even more strongly present

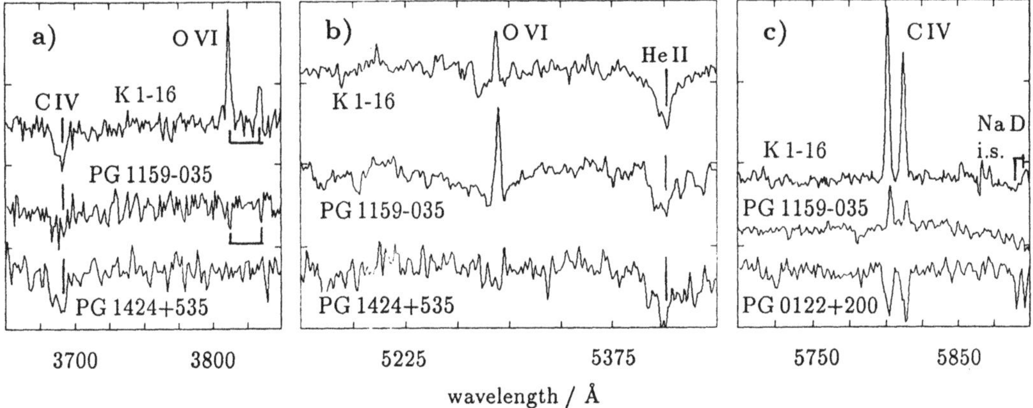

Figure 2: The spectra of representative PG 1159 stars in three different wavelength regions comprising: **a)** the O VI doublet 3811Å/3834Å, **b)** the O VI 5291Å complex, **c)** the C IV doublet 5801Å/5812Å. Each panel shows the spectra of three objects which represent groups 1-3 (bottom to top) as introduced in the text. Tick marks at ordinates are spaced by 25%, 12.5%, 25%, respectively. The resolution is 2Å-3Å.

in the low gravity objects. Within the group A it is either invisible (PG 1424+535) or it is in absorption (PG 0122+200, bottom of Fig. 2c).

It remains to emphasize that no lines of ionisation stages as low as He I, C III, N IV, O V were detected in any of our optical spectra. Ultraviolet low resolution (5Å) IUE spectra of several PG 1159 stars exist and C IV and O VI lines were identified besides He II (see compilation in WHH 91). O V 1371Å is seen in the cool stars only, but also (much weaker) in a (noisy) high resolution spectrum of the prototype (Liebert et al. 1989). Much information – unexploited up to now – contains a comparably good IUE high resolution spectrum of NGC 246 which is the brightest PG 1159 star.

In summary, the PG 1159 stars can be divided into three subclasses, characterized by the appearance of the absorption trough and a few other spectral features. There are i) stars not showing any emissions, ii) stars exhibiting several emission features, in particular O VI 5291Å and C IV 5801Å/5812Å, and iii) low gravity stars with narrower absorption lines, several emission features, in particular strong O VI 3811Å/3834Å and C IV 5801Å/5812Å emission doublets.

3.2 Model atmospheres

Quantitative analyses of PG 1159 stars could not be performed successfully until the advent of a new generation of line blanketed NLTE model atmospheres. In retrospect it was shown that departures from LTE are large in PG 1159 stars and strongly affect optical and UV line profiles as well as the EUV flux distribution (WHH 91). NLTE effects prevail because of the extremely high temperatures encountered, although the gravities are high. Proper modeling of PG 1159 atmospheres is a challenging task, especially because of the "exotic" chemical composition. Complex model atoms of the most abundant chemical species (He, C, O, Ne) comprising all relevant ionisation stages must be included in the calculations. The statistical equations for the population of more than 100 atomic levels have to be solved simultaneously under the constraints of hydrostatic and radiative equilibrium, considering hundreds of radiative line and bound-free transitions and thousands of electron collisional rates. Together with more than 1000 radiative transfer equations one ends up with a highly non-linear system of equations coupled in depth, frequency, and angle. The solution is far beyond the scope of the classical complete linearization approach (Auer & Mihalas 1969).

From a more general point of view, the solution of the NLTE metal line blanketing problem, the computation of NLTE models for O stars comparable in sophistication to Kurucz's (1991) LTE models, seemed to be an unattainable aim for a long time. The situation has changed fundamentally within the last few years. A model atmosphere code based on the *Accelerated Lambda Iteration* method was developed in Kiel, being capable to compute the desired PG 1159 models. Even more, blanketing by millions of lines from iron group elements can be included now (Dreizler & Werner 1992). In parallel, a completely different NLTE method was invented by Anderson (1989), culminating in the construction of metal line blanketed solar models.

What particular difficulties are encountered when models for PG 1159 stars are constructed? It was demonstrated (Werner & Heber 1991b; Barstow, Tweedy & Werner 1991) that the model atoms must be detailed enough otherwise unrealistic fluxes result. On the other hand too large model atoms may waste scarce computer time (a complete model calculation requires ≈ 3 Cray X-MP hours). Test computations are needed to work them out. As a consequence all models of our basic grid include H, He, C, O represented in total by 100 NLTE levels and 240 line transitions. For a proper treatment of all opacities more than 1500 frequency points are necessary. A few exploratory models (tailored to H 1504+65) also consider neon, bringing the number of NLTE levels to 127. This high degree of sophistication is still not sufficient to reproduce the observed line profiles, because many of the lines arise from highly excited levels. Therefore the calculations are continued with even larger model atoms for C and O, however, keeping fixed the atmospheric structure (temperature, pressure). Such line formation iterations greatly reduce the computational effort and the neglect of back-reactions onto the model structure is justified. Trace elements (like N) may be completely treated in this manner.

Line broadening is the next problem. Accurate broadening theories exist for H and He II lines, however, published tabulations were calculated for a pure hydrogen plasma, whereas the PG 1159 stars are H-deficient! Even more serious is the lack of broadening data for lines of C IV, N V, and O VI arising from highly excited levels. These are close to degeneracy and line broadening ranges between the linear and quadratic Stark regime. Recent work by Dimitrijević et al. (1991) and Dimitrijević & Sahal-Bréchot (1991a,b) on lines of C IV, N V, and O VI is helpful in the cases where the quadratic Stark effect is dominant (but again, CNO perturbers are not considered). In order to deal with these difficulties we assume the linear Stark effect to be prevailing. This overestimates the broadening and we correct for this by comparison with experimental data. That procedure allows us to account for the "correct" electric microfield, i.e., the highly charged perturbing ions (He III, C V, O VII). Details of this approximate treatment and an error discussion for the analysis are given in WHH 91. It must be emphasized that the lack of a good broadening theory is the greatest source of uncertainty in the spectral analysis of PG 1159 stars. Fortunately many spectral features are independent from these uncertainties and help to constrain the photospheric parameters.

Line blending is another problem. Since every H line is blended by a He II line, only an upper limit on the hydrogen abundance can be given. In addition every He II line coincides with a C IV line, which must be kept in mind when considering background opacities for line profile calculations. To make things even worse, some of the He II and C IV lines coincide with an O VI line. The most complex blend is the characteristic absorption trough around He II 4686Å. The reason for this bother is that all these ions have one-electron-spectra.

3.3 Determination of atmospheric parameters

The atmospheric parameters of PG 1159 stars are determined from detailed line profile fits. The fit procedure is complicated because six model parameters have to be adjusted: $T_{\rm eff}$, $\log g$ and the abundance ratios of H, He, C, N, O. An extended grid of model atmospheres was computed ranging from 80 000 K to 300 000 K in $T_{\rm eff}$ and from 5 to 9 in $\log g$. A large number of different element mixtures was considered.

Each analysis is started with that grid model which most closely reproduces the observed line profiles. Then the "fine-tuning" of parameters begins with the computation of models with slightly different parameters. Let us outline the principle steps only. Most sensitive temperature indicators are i) the He II and C IV line cores in the trough region, ii) the C IV 5801Å/5812Å doublet, and iii) the O VI 5291Å complex. In view of the described problems with line broadening, the gravity determination not only relies on the line absorption wings: The line cores of the above listed regions may drastically change with $\log g$. Concerning the abundances, the first step must be the determination of the C/He ratio as these elements are most abundant (except for H 1504+65). Wings and cores of the He II/C IV trough are used, together with numerous other isolated C IV lines. Then the O abundance is determined from the O VI 5291Å line or, due to the absence of O VI lines in the cool objects, it must be estimated from the O V 1371Å line. Variation of one of the four parameters determined up to this point ($T_{\rm eff}$, $\log g$, C/He, O/He) always affects the determination of the others and requires some iterative approach to a consistent solution. This procedure is guided by experience because a rigorously systematic way requires prohibitively much computer time for models. Once these parameters have been fixed it is straightforward to determine the H and N abundances (or upper limits). The back-reaction on the other parameters is small either because of the low abundance or because of the low opacities (in the case of H, which is almost completely ionised). It is this very last point that does not allow to put a tight limit on the H abundance (H/He < 1 for five analysed objects, H/He < 0.1 for NGC 246). Only indirect arguments (high C and O abundances) involving evolutionary considerations can be given for a complete absence of hydrogen.

Table 2: Atmospheric parameters and derived quantities for PG 1159 stars. Abundances are given in % mass fractions, distances in kiloparsecs.

star	$T_{\rm eff}$/K	$\log g$	He	C	O	$M/M_\odot$	$\log L/L_\odot$	distance	pulsator?
PG 1424+535	100 000	7.0	33	50	17	$0.53^{+.04}_{-.04}$	$2.2\pm.7$	$1.1^{+.7}_{-.5}$	no
PG 1707+427	100 000	7.0	33	50	17	$0.53^{+.04}_{-.04}$	$2.2\pm.7$	$1.3^{+1.}_{-.6}$	yes
PG 1159-035	140 000	7.0	33	50	17	$0.57^{+.10}_{-.07}$	$2.8\pm.7$	$0.8^{+.6}_{-.4}$	yes
PG 1520+525	140 000	7.0	33	50	17	$0.57^{+.10}_{-.07}$	$2.8\pm.7$	$1.1^{+.8}_{-.5}$	no
PG 1144+005	150 000	6.5	39	58	1.5	$0.57^{+.11}_{-.03}$	$3.4\pm.7$	$2.8^{+2.2}_{-1.2}$	no
H 1504+65	170 000	8.0	<1	50	50	$0.87^{+.15}_{-.15}$	$2.1\pm.7$	$0.6^{+1.5}_{-.3}$	no

Up to now six PG 1159 stars have been studied with appropriate model atmospheres and the results are summarized in Tab. 1. Maximum errors are 10% for $T_{\rm eff}$ and 0.5 dex for $\log g$ and abundances. All objects are extremely hot ($T_{\rm eff} \geq 100\,000$ K). The first four stars in Tab. 1 have the same high surface gravities ($\log g = 7$) and equal abundances. Carbon and helium are the dominant species and oxygen is also present in a significant amount. The next star, PG 1144+005, is slightly hotter than the prototype, has a lower surface gravity, and the O abundance is remarkably lower. Along with the high nitrogen abundance (1.5%), the N/O ratio is more than a 100 times larger than in the other PG 1159 stars! (An upper limit has been derived for the other objects from the lack of N V lines: N < 0.4%). H 1504+65 is the hottest star ever analysed by model atmosphere techniques. It has a very high surface gravity and, being most exciting, it is completely devoid of H *and* He. The atmosphere is composed of C and O to equal parts. Masses are determined by comparison with evolutionary tracks (see Sect. 4.2). The luminosities within our sample range from $100\,L_\odot$ to a few $1000\,L_\odot$. The average distance of the analysed objects (1.3 kpc, calculated from model fluxes and visual magnitudes) indicates that the PG 1159 stars are not nearby objects as most of the known white dwarfs. Hence, the space density is much lower. According to our analyses, the WGL85 estimate ($0.6 \cdot 10^{-7}\,\mathrm{pc}^{-3}$) is still too high (WHH91).

3.4 Interpretation of spectral classification

In light of the spectral analyses we may interpret the spectral classes introduced in Sect. 3.1. Stars from group **A** are relatively "cool" objects, i.e. $T_{\rm eff} \lesssim 100\,000$ K, with a surface gravity of $\log g \approx 7$. They do not show any optical O VI lines because O V is the dominant ionisation stage. The C IV 3s→3p doublet (5801Å/5812Å) is very gravity and temperature sensitive. Within the group of cool stars it is either invisible (e.g. in PG 1424+535, $T_{\rm eff}$=100 000 K) or, at rather low temperatures (say $T_{\rm eff} \lesssim 90\,000$ K), it is seen in absorption (PG 0122+200, bottom of Fig. 2c).

Stars from group **E** have emission cores in the absorption trough lines He II 4686Å, C IV 4659, and, rather weak, in C IV 4647Å. The C IV 5801Å/5812Å doublet is in pure emission. Despite of the high gravity ($\log g \approx 7$) the emissions are caused by NLTE effects at the extremely high effective temperatures ($T_{\rm eff} \gtrsim 140\,000$ K). O VI lines are visible (because O VI is stronger populated than in the cooler stars), provided the oxygen abundance is high enough like in PG 1159-035 ($\gtrsim$10%) and provided that $T_{\rm eff}$ does not exceed $\approx 150\,000$ K, otherwise the doublet disappears (e.g. PG 1144+005 has too low an O abundance, 1.5%, and lacks this line).

The low gravity objects (**lgE** $\log g \lesssim 6$) do, like stars in group E, exhibit two strong central emissions within the absorption trough but C IV 4647Å is in absorption. Because of the lower densities in the photospheres, downward cascades within the C IV ion preferably occur between levels with the highest angular quantum number, explaining the (6h→5g) 4659Å emission and the simultaneous (6g→5f) 4647Å absorption. The O VI 3811Å/3834Å and C IV 5801Å/5812Å doublets are strong gravity indicators and are seen in very strong emission.

4. Discussion

4.1 Non-radial pulsations and atmospheric parameters

The photospheric abundances strongly imply that the pulsations are in fact driven by the κ–γ mechanism of carbon and oxygen. Remember that a high O abundance is necessary to sustain the pulsations in the hot PG 1159 stars ($T_{\rm eff} \geq 100\,000$ K) and that the He abundance should not exceed about 25%-30%. Both is in agreement with the photospheric abundances. What conclusions can be drawn concerning the limits of the DOV instability strip? The first four stars in Tab. 1 form two spectroscopic twins (with $T_{\rm eff}$=100 000 K and 140 000 K); one star out of each pair pulsates and the other does not. The reason for this behaviour is unclear, however, one might speculate that very small undetected differences in the atmospheric parameters could be responsible for that. In this case the instability strip is roughly confined between $100\,\mathrm{kK} < T_{\rm eff} < 140\,\mathrm{kK}$. PG 1144+005 does not pulsate but this is not necessarily because of the higher $T_{\rm eff}$, but probably because of the O abundance which might be too low to drive pulsations. H 1504+65 is not variable, however, the completely different surface composition causes a different location of the instability strip in the HRD (Stanghellini et al. 1990). Stability analyses for pure C-O compositions with models accounting for the high mass of H 1504+65 have not yet been performed. More analyses of PG 1159 stars and pulsation calculations are necessary for a conclusive picture.

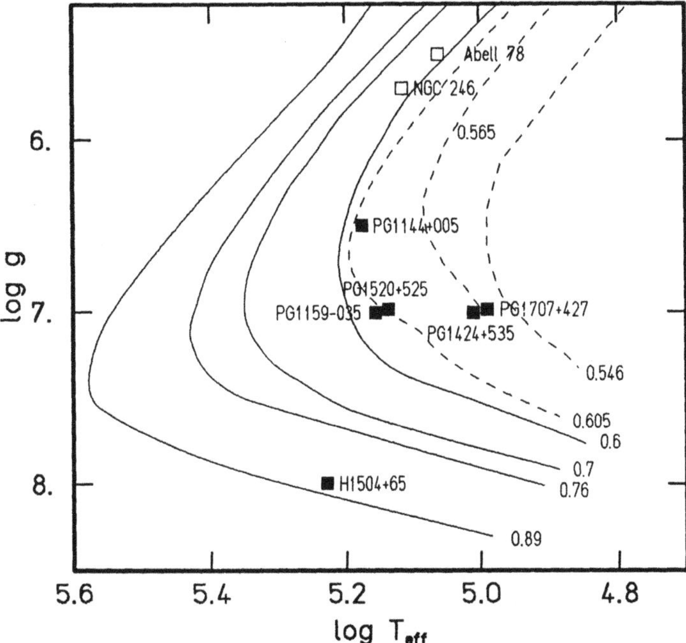

Figure 3: Position of six analysed PG 1159 stars in the $\log g$–$\log T_{\rm eff}$ diagram. Also given are (preliminary) positions of the central stars Abell 78, and NGC 246. Evolutionary tracks are labelled with the respective stellar mass (in $M_\odot$). Dashed: H burners, solid: He burners.

4.2 The evolutionary status of the PG 1159 stars

Figure 3 shows the positions of the analysed stars in the $\log g$–$\log T_{\rm eff}$ diagram together with evolutionary tracks for helium burning post-AGB stars from Wood & Faulkner (1986) and for H burners from Schönberner (1983) and Blöcker & Schönberner (1990). The stellar masses given in Tab. 2 are inter- and extrapolated from the He burning tracks. This is particularly problematic for the objects with the lowest masses, because no track for $M < 0.6\,M_\odot$ has been published. Note that the numbers in Tab. 2 slightly differ from those given in WHH 91, WH 91, W 91 who throughout employed H burning tracks (in part contrary to their claim). The masses are just below $0.6\,M_\odot$, close to the mean value for white dwarfs, except for H 1504+65 which is significantly more massive. The quoted errors for M only regard the uncertainties in $T_{\rm eff}$ and $\log g$, but small systematic errors in the mass determination must be accounted for ($\approx 0.03\,M_\odot$), judging from small deviations of tracks computed by different authors.

The most likely explanation for the observed abundance pattern is that nuclear processed matter is seen at the surface, as is evident from the products of the 3α process, the elements C and O. Together with the results from pulsational analysis we must conclude that the stars have lost entirely their H-rich envelope and a part of their He-rich envelope, too. Obviously H 1504+65 represents the most extreme case and it is interpreted as a bare C-O stellar core. These findings are in contradiction with standard stellar evolution theory which predicts that post-AGB stars retain "thick" H- and He-rich envelopes (10^{-4} and $10^{-2}\,M_\odot$, respectively) on top of the C-O core (see e.g. Schönberner 1979). Stripping off mass from the stars at rates higher than anticipated in canonical theory only speeds up the evolution but does not change the surface abundances.

We have interpreted the PG 1159 stars as late He flash objects according to a scenario proposed by Iben et al. (1983). In this scenario a post-AGB star suffers a late He shell flash during its descent from the AGB to the white dwarf cooling track. The star re-expands to giant dimensions ("born-again" AGB star) and experiences a second superwind phase. Together with modest mass loss during the second descent from the AGB the H-rich layer may be removed completely. If the mass loss is high enough then deep intershell layers become visible, which are enriched in C and O by pulse triggered convection. Most extreme is the case of H 1504+65 which has lost its entire He envelope. The predicted surface abundances (Iben 1984) are in qualitative agreement with observations, however, a quantitative comparison appears premature because several details (e.g. efficiency of the pulse triggered convection) are not yet well understood. The late He flash scenario can also explain why many of the PG 1159 stars lack a PN: Helium burners evolve more slowly than hydrogen burners and the PN may have dispersed.

The detection of small amounts of photospheric nitrogen (N<0.4% in PG 1159-035, but N is definitely present as N V 1240Å is visible) or even large amounts (N=1.5% in PG 1144+005) seems to be at variance with this scenario, because N is immediately destroyed during 3α burning and should not be present in the helium buffer layer. Traces of N could be explained from diffusion processes in the interior of the precursor star (see discussion in WHH 91), but the high N abundance in PG 1144+005 provides an important clue to details of the final He flash. Iben et al. (1983) have pointed out a possibility how N might be preserved or even enriched within this scenario. In short, H is burned via $^{13}C(p,\gamma)^{14}N$ during the pulse peak in a convective shell that is detached from the He burning convective shell by a radiative layer. The base of the H burning shell lies within the former intershell region explaining the C-He dominance and the O content (see details in WH 91). In summary, all analysed PG 1159 stars can be interpreted as late He flash objects with subsequent thermonuclear destruction/wind ejection of H and N.

Which stars can be identified as progenitors and successors of the PG 1159 stars? According to Méndez (1991) the H-deficient, C-rich central stars may be grouped into three different classes:

1. WC; the Wolf-Rayet central stars ranging from subtypes WC 12 (V348 Sgr) to WC 2 (NGC 5189). Note that only one object (NGC 6751, WC 4) is classified as "N strong", or, "WCN" (Méndez et al. 1986).

2. Of(C), Of-WR(C); a transition group (comprising two stars only: A 30 and A 78) which shows a mixed spectrum of absorption as well as (strong) emission lines.

3. O(C); showing predominantly absorption lines of He II and C IV. These objects are considered as PG 1159 stars by WHH 91.

An evolutionary link between groups 2 and 3 (including the PG 1159 stars that lack a PN) which ends with the formation of a non-DA white dwarf, was suggested by Sion et al. (1985). This picture is now confirmed by model atmosphere analyses and, moreover, the evolutionary link can be convincingly traced back until the WR central stars (Méndez 1991, WHH 91, previously suggested by Méndez et al. 1986) because:
i) a few (coarse) abundance analyses exist for WR CSPN (see WHH 91). They are H-deficient and the C/He number ratio ranges between 0.3 and 0.5 (PG 1159-035: 0.5). In the case of Sanduleak 3 (WC 3, Barlow & Hummer 1982) O/He=0.05 was determined (PG 1159-035: 0.13).
ii) A NLTE analysis of the transition object A 78 is currently performed. Preliminary results (Werner & Koesterke 1992) indicate that the abundances, in particular the high N content, are very similar to PG 1144+005 (the N-rich PG 1159 star).
The existence of such young post-AGB objects (especially the late type WCs) with already high C abundances implies that the removal of the H-rich layer already takes place while the star is on the AGB (Méndez 1991, WH 91). In this context we quote Méndez who emphasizes that

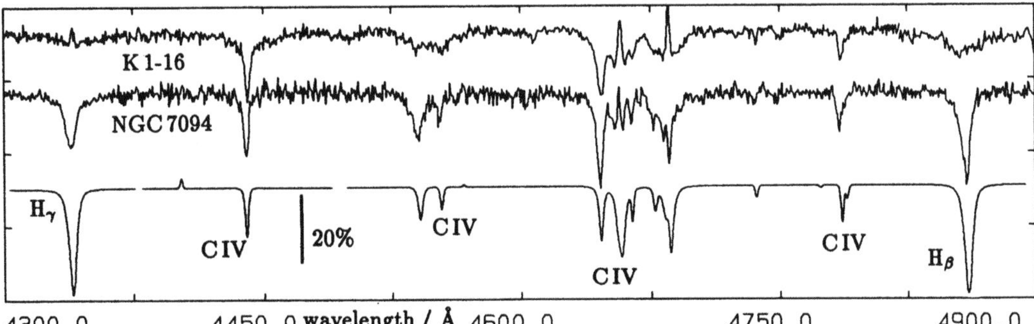

Figure 4: Normalized spectra of two central stars, K 1-16 and NGC 7094, representing the PG 1159 and H-C-rich types, respectively. The C IV/He II 4686Å regions look similar but H_β and H_γ are very strong in NGC 7094. Also shown is a synthetic spectrum, not thought as a fit but rather as a first approach, from a model with T_{eff}=100 000 K, $\log g = 5$ and He/H=0.5, C/H=0.17 (number ratios). Accordingly, NGC 7094 is H-rich and has a remarkably high amount of carbon. The resolution is 1.5Å.

the late He flash scenario is not the only possible interpretation. One should not yet rule out the possibility of producing a H-deficient, C-rich post-AGB star already at the time of the first departure from the AGB. It seems obvious that the PG 1159 stars evolve into DO (i.e. hot He-rich) white dwarfs as soon as gravitational settling removes C and O from the photospheres, however, one may not forget that only upper limits for the H abundance could be derived. The presence of H even by small amounts would turn them into DA white dwarfs. In analogy this holds also for H 1504+65 which might have retained undetectable small amounts of He in the envelope, perhaps being high enough to result in a thin, pure He layer in the future.

In summary, the PG1159 stars represent an episode in the life of a H-deficient post-AGB star evolving as WR → PG 1159 → DO. It is not clear how the few N-rich objects fit into this scenario. One might speculate (WH 91) about another sequence WCN → A 78 → PG 1144+005 → DO.

5. Future work

In order to clarify the evolutionary scenario outlined above we need precise spectral analyses of more PG 1159 stars and from related objects. Abundance determinations of early and late type WR central stars from detailed NLTE models are desireable; in particular, it is unknown if the coolest subtypes contain hydrogen or not. (An evolutionary link from late to early WR subtypes is appealing, however, the lack of any WC 5 to WC 7 stars is still embarrassing, Méndez 1991). Furthermore we need analyses of the hottest DO white dwarfs in order to determine when gravitational settling turns PG 1159 stars into DOs. Analyses of objects from all these groups are under way in Kiel.

In light of the results summarized in this review it appears that the separation of the white dwarfs into two spectral sequences (DA, DB) can be traced back close to the AGB. However, apart from well known problems with the simple picture of two strictly separated channels (DB gap, lack of DAs hotter than ≈ 80 000 K), another problem came up very recently.

As noted by Méndez (1991) a wide variety of C abundances among the hydrogen rich central stars exists. The first hint for a very strongly C enriched object was given by Kaler & Feibelman (1985) who mention an (unpublished) optical spectrum of the central star NGC 7094, which is similar to that of K 1-16 (PG 1159 type), but shows strong Balmer lines. We have taken a spectrum of NGC 7094 at the Calar Alto 3.5m telescope and we compare it with K 1-16 (Fig. 4). In fact strong C IV *and* H lines are seen, and from a first model calculation we may *roughly* infer He/H ≈ 1 and C/H $\gtrsim 0.1$ (by number). Two other C-rich stars were discovered, the central stars S 68 and A 43 (Napiwotzki & Schönberner 1991, Napiwotzki 1992) and another candidate is NGC 4361 (Méndez 1991). These objects imply that 3α processed matter was mixed into unprocessed surface material. The same conclusion was drawn in order to explain the abundance pattern in KS 292, which is a post-AGB star lacking a PN and which is also interpreted as a late He flash object (Rauch et al. 1991). The question arises if there is a continuous transition from H-rich over H-C-rich to H-deficient CSPN.

ACKNOWLEDGEMENTS. I thank my colleagues U. Heber and K. Hunger for their collaboration on the PG 1159 project, which is supported by the BMFT (grant 50 OR 9007 3). Thanks go to the Calar Alto staff for the support during observations.

REFERENCES

Anderson, L.S. 1989, ApJ, 339, 558

Auer, L.H., Mihalas, D. 1969, ApJ, 158, 641

Barlow, M.J., Hummer, D.G. 1982, IAU Symp. 99: Wolf-Rayet Stars: Observations, Physics, Evolution, eds. C.W.H. de Loore and A.J.Willis, Reidel, Dordrecht, p. 387

Barstow, M.A., Tweedy, R.W. 1990, MNRAS, 242, 484

Barstow, M.A., Holberg, J.B. 1990, MNRAS, 245, 370

Barstow, M.A., Holberg, J.B., Grauer, A.D., Winget, D.E. 1986, ApJ, 306, L25

Barstow, M.A., Tweedy, R.W., Werner, K. 1991, in White Dwarfs, NATO ASI Series C, Vol. 336, eds. G.Vauclair and E.Sion, Kluwer, Dordrecht, p. 17

Blöcker, T., Schönberner, D. 1990, A&A, 240, L11

Bond, H.E., Meakes, M.G. 1991, AJ, 100, 788

Bond, H.E., Grauer, A.D., Green, R.F., Liebert, J.W. 1984, ApJ, 279, 751

Cox, A.N. 1986, in Highlights of Astronomy, ed. J.-P.Swings, 7, 229

Demers, S., Wesemael, F., Irwin, M.J., Fontaine, G., Lamontagne, R., Kepler, S.O., Holberg, J.B. 1990a, ApJ, 351, 271

Demers, S., Wesemael, F., Irwin, M.J., Fontaine, G., Lamontagne, R., Kepler, S.O., Holberg, J.B. 1990b, ApJ, 357, 291

Dimitrijević, M.S., Sahal-Bréchot, S. 1991a,b ApJS submitted

Dimitrijević, M.S., Sahal-Bréchot, S., Bommier, U. 1991, ApJS, 89, 581

Dreizler, S., Werner, K. 1991, in Stellar Atmospheres: Beyond Classical Models, NATO ASI Series C, Vol. 341, eds. L.Crivellari, I.Hubeny, D.G.Hummer, Kluwer, Dordrecht, p. 155

Dreizler, S., Werner, K. 1992, these proceedings

Fontaine, G., Bergeron, P., Vauclair, G., Brassard, P., Wesemael, F., Kawaler, S.D., Grauer, A.D., Winget, D.E. 1991, ApJ, 378, L49

Fritz, M.L., Leckenby, H.J., Sion, E.M. 1990, AJ, 99, 908

Green, R.F., Schmidt, M., Liebert, J. 1986, ApJS, 61, 305

Grauer, A.D., Bond, H.E. 1984, ApJ, 277, 211

Grauer, A.D., Bond, H.E., Green, R.F., Liebert, J. 1987, IAU Coll. 95: The Second Conference on Faint Blue Stars, eds. A.G.D. Philip, D.S. Hayes and J. Liebert, Schenectady: Davis Press, p. 231

Grauer, A.D., Green, R.F., Liebert, J. 1992, ApJ, in press

Hill, J.A., Winget, D.E., Nather, R.E. 1987, IAU Coll. 95: The Second Conference on Faint Blue Stars, eds. A.G.D. Philip, D.S. Hayes and J. Liebert, Schenectady: Davis Press, p. 627

Hubeny, I. 1992, these proceedings

Husfeld, D. 1987, IAU Coll. 95: The Second Conference on Faint Blue Stars, eds. A.G.D. Philip, D.S. Hayes and J. Liebert, Schenectady: Davis Press, p. 237

Iben, I., Jr. 1984, ApJ, 277, 333

Iben, I., Jr., Kaler, J.B., Truran, J.W., Renzini, A. 1983, ApJ, 264, 605

Kaler, J.B., Feibelman, W.A. 1985, ApJ, 297, 724

Kawaler, S.D. 1987, IAU Coll. 95: The Second Conference on Faint Blue Stars, eds. A.G.D. Philip, D.S. Hayes and J. Liebert, Schenectady: Davis Press, p. 297

Kawaler, S.D. 1990, in Confrontation Between Stellar Pulsation and Evolution, The ASP Conference Series, eds. C.Cacciari and G.Clementini, p. 494

Kawaler, S.D., Hansen, C.J., Winget, D.E. 1985, ApJ, 295, 547
Kurucz, R.L. 1991, in Stellar Atmospheres: Beyond Classical Models, NATO ASI Series C, Vol. 341, eds. L.Crivellari, I.Hubeny, D.G.Hummer, Kluwer, Dordrecht, p. 441
Kwitter, K.B., Massey, P., Congdon, C.W., Pasachoff, J.M. 1989, AJ, 97, 1423
Liebert, J., Fleming, T.A., Green, R.F., Grauer, A.D. 1988, PASP, 100, 187
Liebert, J., Wesemael, F., Husfeld, D., Wehrse, R., Starrfield, S.G., Sion, E.M. 1989, AJ, 97, 1440
McGraw, J.T., Starrfield, S.G., Liebert, J., Green, R. 1979, IAU Coll. 53: White Dwarfs and Variable Degenerate Stars, eds. H.M. van Horn and V.Weidemann, Rochester, p. 377
Méndez, R.H. 1991, IAU Symp. 145: Evolution of stars: The atmospheric abundance connection, eds. G.Michaud and A.Tutukov, Kluwer, Dordrecht, p. 375
Méndez, R.H., Kudritzki, R.P., Simon, K.P. 1985, A&A, 142, 289
Méndez, R.H., Miguel, C.H., Heber, U., Kudritzki, R.P. 1986, IAU Coll. 87: Hydrogen Deficient Stars and Related Objects, eds. K.Hunger, D. Schönberner, N.Kameswara Rao, Reidel, Dordrecht, p. 323
Méndez, R.H., Gathier, R., Simon, K.P., Kwitter, K.B. 1988, A&A, 198, 287
Napiwotzki, R., Schönberner, D. 1991, A&A, 249, L16
Napiwotzki, R. 1992, these proceedings
Nather, R.E., Winget, D.E., Clemens, J.C., Hansen, C.J., Hine, B.P. 1990, ApJ, 361, 309
Nousek, J.A., Shipman, H.L., Holberg, J.B., Liebert, J., Pravdo, S.H., White, N.E., Giommi, P. 1986, ApJ, 309, 230
Rauch, T., Heber, U., Hunger, K., Werner, K., Neckel, T. 1991, A&A, 241, 457
Schönberner, D. 1979, A&A, 79, 108
Schönberner, D. 1983, ApJ, 272, 708
Schönberner, D., Napiwotzki, R. 1990, A&A, 231, L33
Sion, E.M., Liebert, J., Starrfield, S.G. 1985, ApJ, 292, 471
Stanghellini, L., Cox, A.N. 1991, in White Dwarfs, NATO ASI Series C, Vol. 336, eds. G.Vauclair and E.Sion, Kluwer, Dordrecht, p. 205
Stanghellini, L., Cox, A.N., Starrfield, S.G. 1990, in Confrontation Between Stellar Pulsation and Evolution, The ASP Conference Series, eds. C.Cacciari and G.Clementini, p. 524
Starrfield, S., Cox, A.N., Hodson, S.W. 1980, Space Science Reviews, 27, 621
Starrfield, S. 1987, IAU Coll. 95: The Second Conference on Faint Blue Stars, eds. A.G.D. Philip, D.S. Hayes and J. Liebert, Schenectady: Davis Press, p. 309
Starrfield, S.G., Cox, A.N., Hodson, S.W., Pesnell, W.D. 1983, ApJ, 268, L27
Starrfield, S., Cox, A.N., Kidman, R.B., Pesnell, W.D. 1984, ApJ, 281, 800
Vauclair, G. 1990, in Progress of Seismology of the Sun and Stars, Proc. OJI International Seminar, eds. Y.Osaki and H.Shibahashi, Springer, Berlin, p. 437
Werner, K. 1986, A&A, 161, 177
Werner, K. 1987, in Numerical Radiative Transfer, ed. W.Kalkofen, Cambridge University Press, p. 67
Werner, K. 1988, A&A, 204, 159
Werner, K. 1989, A&A, 226, 265
Werner, K. 1991, A&A, 251, 147 (W91)
Werner, K., Husfeld, D. 1985, A&A, 148, 417
Werner, K., Heber, U. 1991a, A&A, 247, 476 (WH91)
Werner, K., Koesterke, L. 1992, these proceedings
Werner, K., Heber, U. 1991b, in Stellar Atmospheres: Beyond Classical Models, NATO ASI Series C, Vol. 341, eds. L.Crivellari, I.Hubeny, D.G.Hummer, Kluwer, Dordrecht, p. 341
Werner, K., Heber, U., Hunger, K. 1989, IAU Coll. 114: White Dwarfs, ed. G.Wegner, Springer, Berlin, p. 194
Werner, K., Heber, U., Hunger, K. 1990, in Properties of Hot Luminous Stars, ed. C.D.Garmany, The ASP Conference Series, 7, 86
Werner, K., Heber, U., Hunger, K. 1991, A&A, 244, 437 (WHH91)
Wesemael, F., Green, R.F., Liebert, J. 1985, ApJS, 58, 379 (WGL85)
Winget, D.E., Hansen, C.J., Van Horn, H.M. 1983, Nature, 303, 781
Winget, D.E., Kepler, S.O., Robinson, E.L., Nather, R.E., O'Donoghue, D. 1985, ApJ, 292, 606
Winget, D.E. et al. 1991, ApJ, 378, 326
Wood, P.R., Faulkner, D.J. 1986, ApJ, 307, 659

NLTE ANALYSIS OF THE HYDROGEN-DEFICIENT CENTRAL STAR OF THE PLANETARY NEBULA ABELL 78

K. WERNER* and L. KOESTERKE
Institut für Theor. Physik u. Sternwarte d. Univ. Kiel, Olshausenstr. 40, 2300 Kiel, Germany

1. Properties of Abell 78

The planetary nebulae (PNe) Abell 30 and Abell 78 are the only known objects which display strongly hydrogen-depleted inner regions embedded into large old (H:He normal) nebulosities. The hydrogen-deficient central stars show unique optical spectra. These stars were suggested as prime examples for objects that have suffered a late helium shell flash during their evolution from the Asymptotic Giant Branch (AGB) to the white dwarf configuration. As a consequence, the star re-expands to giant dimensions and suffers for the second time a superwind phase as well as post-AGB mass loss. This scenario ("born-again AGB star", Iben et al. 1983) can explain the unique properties of the PNe and their nuclei. In particular,

i) Abell 78 is a multiple shell PN with an inner shell expanding with at least two different velocities and a more slowly expanding H-rich external layer (Manchado et al. 1988a). The innermost part of the PN ($\leq 10''$) is hydrogen free (Jacoby & Ford 1983). While an enhancement of the He abundance is a common phenomenon in PNe that can be interpreted as expelled matter with a normal stellar composition enriched by dredged-up processed material, the H-depletion in Abell 78 means that we see material that was completely processed in a H-burning shell. Even more interesting, very high N, O, Ne abundances relative to He were found (5%, 5%, 1%, respectively [number ratios]; Manchado et al. 1988b), implying that deep intershell matter has been expelled from the star.

ii) UV spectra of the central star of Abell 78 exhibit powerful P Cygni profiles of C IV, N V, and O V, indicating still ongoing mass loss (Heap 1979). In the optical range the spectrum of the central star is dominated by photospheric *and* wind lines of He II, C IV, O VI (Cohen et al. 1977), and it shows several similarities to the photospheres of the PG 1159 stars (see Werner & Heber 1991b, for a comparison).

We started a model atmosphere analysis in order to determine the photospheric parameters, the chemical composition as well as the mass loss rate of the nucleus of Abell 78. For this purpose new observations were carried out.

2. Observations

Medium-resolution (1.5Å–2.5Å) high-S/N optical spectra were obtained at the 3.5m telescope on Calar Alto (Spain) covering the range from 3350Å to 6700Å. The characteristic features are:

- Strong wind emission doublets of O VI 3811Å/3834Å and C IV 5801Å/5812Å
- A complex C IV/He II blend near 4686Å with emission wings due to He II
- Photospheric absorption lines from He II and C IV
- Emission lines of N V (the 4604Å/4620Å doublet and 4945Å)
- Weak emission features of O V (e.g. 4124Å, 4499Å, 4930Å)
- Stronger emission features of O VI (3434Å, 5291Å)

A high resolution IUE SWP spectrum was retrieved from the IUE data archive.

3. Model atmospheres and results

The observations are analysed by detailed line profile fits. Synthetic spectra are calculated using NLTE model codes for spherically expanding atmospheres (Hamann et al. 1991) and for

*Visiting astronomer, German-Spanish Astronomical Center, Calar Alto, operated by the Max-Planck-Institut für Astronomie Heidelberg jointly with the Spanish National Commission for Astronomy

plane-parallel static atmospheres (Werner & Heber 1991a). Both codes employ detailed model atoms of He I-III, C III-V, N IV-VI, and O IV-VII. The plane-parallel version computes fully Stark-broadened line profiles for the photospheric features, whereas Doppler broadening suffices for the computation of the wind emissions. For the calculations presented here, the element abundances were kept fixed at values that closely resemble those of the PG 1159 stars (except for the high N content which is observed in only one of these objects, PG 1144+005):

He = 33% C = 50% N = 2% O = 15% (mass fractions).

The effective temperature is quite strictly constrained by the oxygen ionisation balance which strongly affects the O V 1371Å and O VI 3810Å/3834Å lines, and by the fact that N V appears in emission. Both (completely independent) codes yield the best line fits at slightly different values for $T_{\rm eff}$ (110 000 K and 120 000 K from the wind and from the photospheric lines, respectively) and we regard $T_{\rm eff}$=115 000 K as a compromise value. The gravity is determined from the photospheric absorption lines. Fig. 1 shows line profile fits (other model parameters are $\log g$ =5.5, v_∞=3700 km/s, and log $\dot{M}$ =-7.6 ($M_\odot$/yr). Although the grid of model atmospheres is not complete yet, the following preliminary results can be stated by now:

▷ $T_{\rm eff} = 115\,000\,{\rm K} \pm 15\,000\,{\rm K}$ $\log g = 5.5 \pm 0.5$ (cgs)

▷ $v_\infty = 3700\,{\rm km/s} \pm 200\,{\rm km/s}$ $\log \dot{M} = -7.6 \pm 0.3$ ($M_\odot$/yr)

We may already conclude that:
- $T_{\rm eff}$ is markedly higher than the most recent estimates (80 000 K, Kaler & Feibelman 1984).
- The terminal velocity is in agreement with earlier determinations, however, the mass loss rate is considerably higher than hitherto assumed.
- Both, v_∞ and $\dot{M}$, agree reasonably well with the predictions from *radiation driven wind* theory, which yields (after slight extrapolation in the Pauldrach et al. (1988) diagrams; applying the graphs for 0.644 $M_\odot$) v_∞ = 4000 km/s and log $\dot{M}$=-7.95 $M_\odot$/yr.
- As previously suspected (Werner & Heber 1991b), Abell 78 can be identified as an immediate progenitor of the relatively N-rich PG 1159 star PG 1144+005. The surface and inner PN abundances can be explained by a late He flash event followed by thermonuclear destruction/wind ejection of H and N. The question to what degree the He-rich envelope has been eroded can only be adressed after a precise abundance determination. (For a more detailed discussion on the evolutionary link to the WC central stars and the PG 1159 stars see Werner, these proceedings.)

Comparison with evolutionary tracks implies a stellar mass of 0.62 $M_\odot$, however, from the uncertainty of the gravity determination a mass as high as 0.8 $M_\odot$ cannot be excluded.

ACKNOWLEDGEMENTS. L.K. and K.W. are supported by the DFG (grant Ha 1455/4-1) and by the BMFT (grant 50 OR 9007 3), respectively.

REFERENCES

Cohen, M., Hudson, H.S., O'Dell, S.L., Stein, W.A. 1977, MNRAS, 181, 233
Hamann, W.-R., Koesterke, L., Wessolowski, U. 1991, in Stellar Atmospheres: Beyond Classical Models, NATO ASI Series C, Vol. 341, eds. L.Crivellari, I.Hubeny, D.G.Hummer, Kluwer, Dordrecht, p. 69
Heap, S.R. 1979, IAU Symp. 83, p. 99
Iben, I., Jr., Kaler, J.B., Truran, J.W., Renzini, A. 1983, ApJ, 264, 605
Jacoby, G.H., Ford, H.C. 1983, ApJ, 266, 298
Manchado, A., Pottasch, S.R., Mampaso, A. 1988a, IAU Symp. 131, p. 184
Manchado, A., Pottasch, S.R., Mampaso, A. 1988b, A&A, 191, 128
Pauldrach, A., Puls, J., Kudritzki, R.P., Méndez, R.H., Heap, S.R 1988, A&A, 207, 123
Werner, K., Heber, U. 1991a, in Stellar Atmospheres: Beyond Classical Models, NATO ASI Series C, Vol. 341, eds. L.Crivellari, I.Hubeny, D.G.Hummer, Kluwer, Dordrecht, p. 341
Werner, K., Heber, U. 1991b, A&A, 247, 476

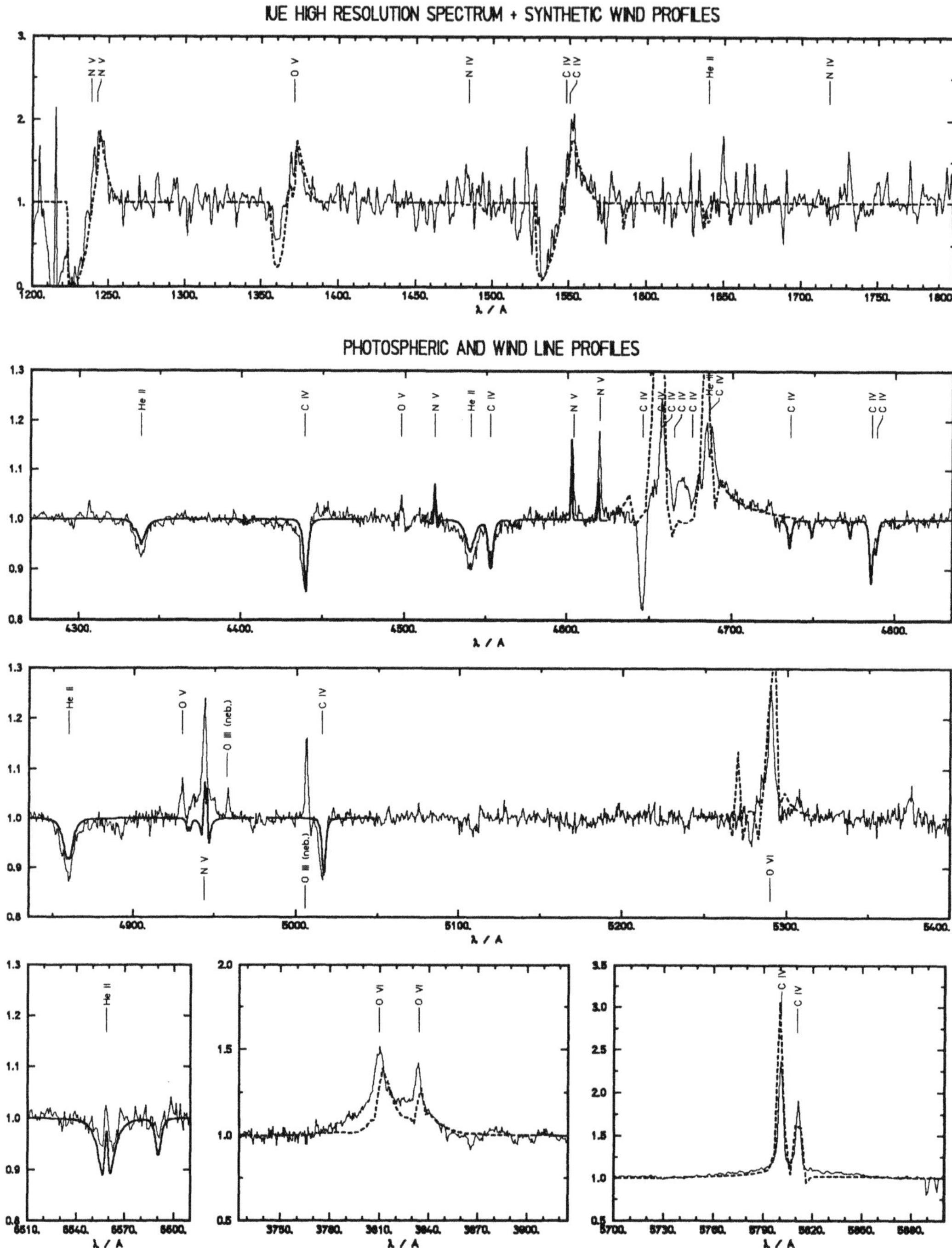

Figure 1: Line profile fits for the central star of Abell 78. Dashed and solid theoretical profiles were computed by the wind code and by the static photosphere code, respectively.

OPTICAL OBSERVATIONS OF THE ULTRAHIGH-EXCITATION PRE-WHITE DWARF KPD 0005+5106

K. WERNER* and U. HEBER

Institut f. Theoretische Physik u. Sternwarte d. Universität Kiel, Olshausenstr. 40, 2300 Kiel, Germany

KPD 0005+5106 was discovered by Downes et al. (1985) as a very hot helium-rich pre-white dwarf, which exhibits a unique optical spectrum dominated by He II absorption lines. In addition, emission lines were detected in He II 4686Å and at 4658Å. An emission line at 4340Å remained mysterious and was tentatively attributed to H_γ by Downes et al. (1985, 1987) although no other Balmer emission could be detected. We present medium-resolution high-S/N optical spectra obtained at the Calar Alto 3.5m telescope, covering 3350Å-6700Å (Fig. 1). We note:
i) A new emission feature not detected before (at 6068Å) as well as the "mysterious" emission at 4340Å are identified as due to high-n transitions of O VIII (n=10 → 9 and n=9 → 8, respectively; cf. line list by Garcia & Mack 1965).
ii) The 4658Å emission line can be identified with O VIII (n=12 → 10). Identification with a coinciding C IV line is rather improbable because of the lack of any other C IV line.
iii) Another emission feature at 4945Å is possibly due to C V. At the same wavelength a N V line is located, however, the presence of this line seems unprobable as it is usually accompanied by a strong N V 4604Å/4620Å emission doublet (e.g. in the hot pre-white dwarf PG 1144+005, see Werner & Heber 1991). The latter cannot be identified without doubt.
iv) It is important that the O VI 3811Å/3834Å doublet is weakly present while the O VI 5291Å emission feature (which is usually the strongest photospheric O VI line) is not visible.
v) We have also searched for other ultrahigh-excitation features; ions which are isoelectronic to O VIII, i.e. C VI and N VIII are *not* detected, neither are any O VII lines present.

With the detection of O VIII emission lines, KPD 0005+5106 displays the *highest degree of excitation* found in any pre-white dwarf star. It is of even higher excitation than the ultrahigh-ionisation (Pop. II) Wolf-Rayet star Sanduleak 3 (Barlow et al. 1980). The origin of the high-n O VIII emissions is not yet clear. Since they are formed by O^{+8} recombination, the extremely high ionization energy of O^{+7} (871 eV) might point at a coronal origin of the X-rays that photoionize the emitting gas. Any modeling of the spectrum has to account for these facts: i) No photometric variability, neither on long terms (Downes et al. 1985), nor on short terms (Grauer et al. 1987). ii) No spectroscopic variability on the time scale of days (own observations). iii) No current mass outflow (Grauer et al. 1987). iv) No evidence for binarity; neither a red excess (Downes et al. 1987) nor radial velocity variations (Saffer & Liebert 1989) were found.

Another hint to an explanation of the ultrahigh-excitation features comes from a conclusion of Fritz et al. (1990): The detected UV resonance lines of C IV, N V, and Si IV are attributed to an expanding H II region around KPD 0005+5106. Finally we note that the effective temperature of KPD 0005+5106 is probably higher than previously assumed (80 000 K, Downes et al. 1987). A non-LTE analysis is under way.

REFERENCES

Barlow, M.J., Blades, J.C., Hummer, D.G. 1980, ApJ, 241, L27
Downes, R.A., Liebert, J., Margon, B. 1985, ApJ, 290, 321
Downes, R.A., Sion, E.M., Liebert, J., Holberg, J.B. 1987, ApJ, 321, 943
Fritz, M.L., Leckenby, H.J., Sion, E.M. 1990, AJ, 99, 908
Garcia, J.D., Mack, J.E. 1965, J.Opt.Soc.Am., 55, 654
Grauer, A.D., Bond, H.E., Liebert, J., Fleming, T.A., Green, R.F. 1987, ApJ, 323, 271
Saffer, R.A., Liebert, J. 1989, IAU Coll. 114: White Dwarfs, ed. G.Wegner, Springer, Berlin, p. 408
Werner, K., Heber, U. 1991, A&A, 247, 476

*Visiting astronomer, German-Spanish Astronomical Center, Calar Alto, operated by the Max-Planck-Institut für Astronomie Heidelberg jointly with the Spanish National Commission for Astronomy

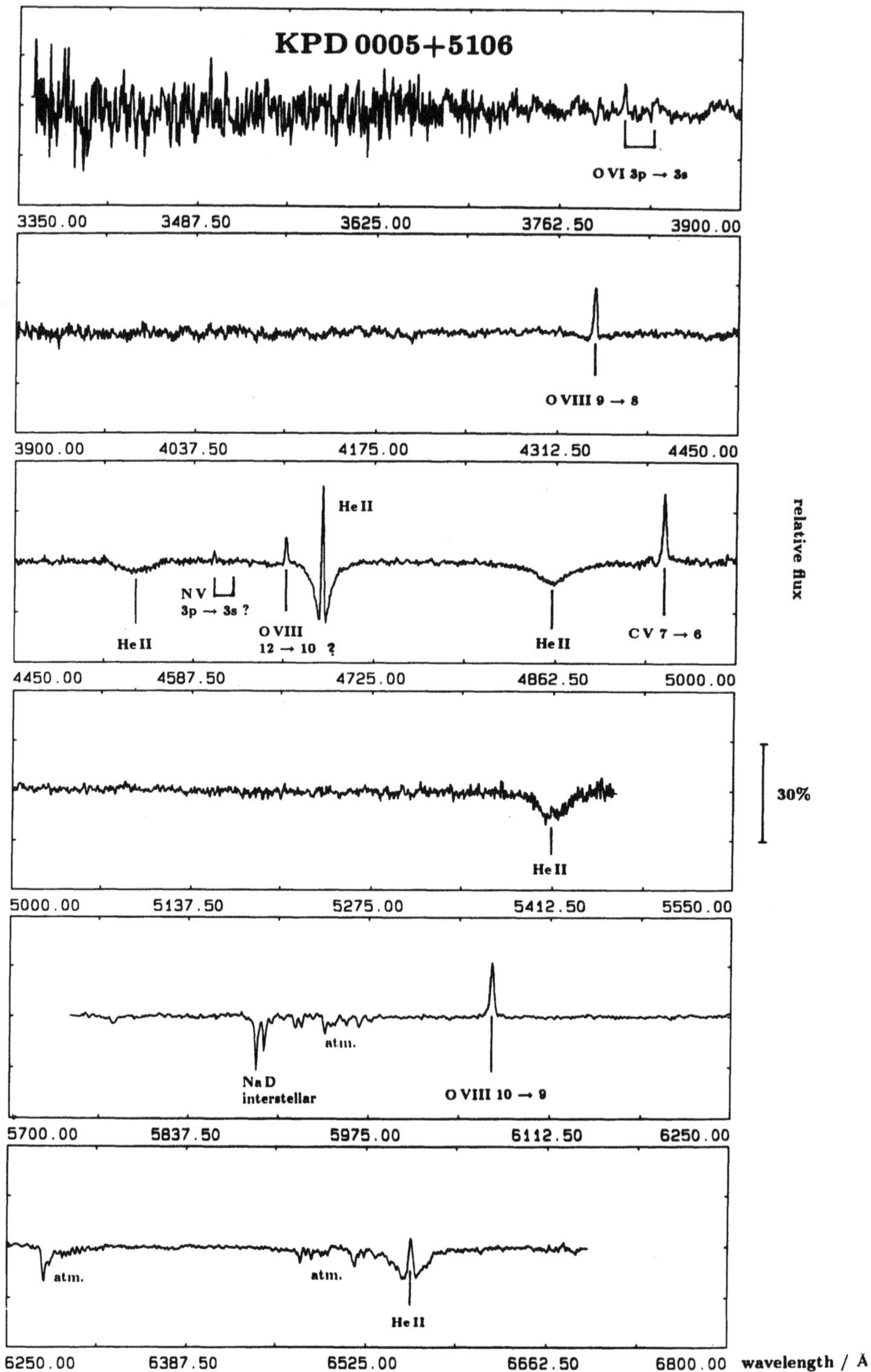

Figure 1: Optical spectrum of KPD 0005+5106. Note particularly the ultrahigh-excitation features of O VIII at 4340Å and 6068Å.

The Atmospheres of Extreme Helium Stars

C. Simon Jeffery

Dept of Physics & Astronomy, University of St Andrews, St Andrews,
Fife KY16 9SS, Scotland

Abstract: Extreme helium (EHe) stars are low-mass early-type supergiants with hydrogen-deficient and carbon-rich atmospheres. Progress and recent results in the quantitative analysis of their atmospheres are reviewed, including the first results to include the effects of line-blanketing in the model atmospheres. The hot RCrB star DY Cen has been found to be relatively hydrogen-rich, but extremely metal-poor. A comparison of EHe star abundances with other H-deficient stars is made.

1 Introduction

In his review of spectroscopic analyses of hot extreme helium stars (EHe's), Heber (1986a) observed that these objects represent the most obvious cases for large atmospheric depletion of hydrogen and enrichment of helium amongst all classes of hydrogen-deficient stars. However, analyses of only four objects showed considerable abundance variations from star to star, whilst a variety of other properties (emission line spectra, pulsations) were being discovered (eg Heber 1986b, Jeffery 1986, Jeffery & Malaney 1985, Jeffery et al. 1985). The inhomogenity of the class dictated that a more systematic study of abundances and properties was necessary before the origin of the hydrogen deficiency and the evolutionary status of the objects could be diagnosed.

Such studies of EHe's and related objects have been in progress since 1985 (eg Heber et al. 1986, Jeffery et al. 1986, 1988, Jeffery & Heber 1992a,b). This paper reports on recent results.

2 Model atmospheres and spectroscopic analyses

A complete list of EHe's was compiled by Drilling & Hill (1987, Table A1), together with a number of related objects (sdO's, binaries and V652 Her). Nearly all have now been observed with the Cassegrain Echelle Spectrograph (CASPEC) on the ESO 3.6m telescope at Chile during observing runs in 1985 and 1987, providing a valuable database of high S/N spectra at a resolution of 0.25Å in the wavelength range 4000 – 4900 Å.

The most significant problem encountered in the analysis of EHe star atmospheres is caused by the absence of hydrogen. Electron scattering provides most of the opacity in the outer atmosphere, whilst carbon and nitrogen are the dominant opacity sources in the deeper layers. The latter problem was overcome (Schönberner & Wolf 1974) by including continuous opacities from C and N calculated by Peach (1970) into their static plane-parallel LTE model atmospheres code. Heber & Schönberner (1981) introduced partial metal-line blocking to improve these models, but the most significant step forward has been the recent introduction of fully line-blanketed model atmospheres (Möller 1990).

The method of analysis requires that the three parameters $T_{\rm eff}$, log g and $n_{\rm C}/n_{\rm He}$ have to be determined simultaneously (cf. Heber 1986a). Given a value for $n_{\rm C}/n_{\rm He}$, $T_{\rm eff}$ is determined from the ionisation equilibria of various elements, log g by matching the line profiles of the Stark

Table 1: Atmospheric parameters and abundances for extreme helium stars. Abundances have been normalized to $\log \Sigma_i \mu_i n_i = 12.15$

	HD168476	MV Sgr	HD124448	BD+10°2179	DY Cen	BD−9°4395	γ Peg
$T_{\rm eff}$	13 700	15 400	15 500	16 800	20 000	22 700	
$\log g$	1.35	2.5	2.5	2.55	2.2	2.55	
H	< 7.8		< 7.5	8.5	10.55	8.74	12.00
He	11.54	11.6	11.53	11.53	11.50	11.54	10.96
C	9.4	7.8	9.46	9.54	9.41	9.17	8.45
N	8.9	8.0	8.83	8.11	8.08	7.97	7.82
O	8.4		8.5	8.1	8.85	7.90	8.66
Ne	9.3:					8.76	8.54
Mg	7.7		8.2	8.02	6.9	7.25	7.45
Al	7.2:		6.2	6.25	5.9	5.55	6.45
Si	7.7	6.9	7.51	7.32	8.0	7.83	7.38
P	6.3		5.6	5.5	5.6	6.21	5.20
S	7.0	6.5	7.2	7.12	7.2	7.83	7.21
Ar			6.6	6.4	< 6.1	7.24	6.70
Fe	7.5		7.4	6.49	5.5	6.57	7.44

References:

HD168476	Walker & Schönberner 1981	γ Peg	Peters 1976
BD+10°2179	Heber 1983	HD124448	Schönberner & Wolf 1974
MV Sgr	Jeffery et al. 1988	BD−9°4395	Jeffery & Heber 1992a
		DY Cen	This paper

broadened HeI lines. $n_{\rm C}/n_{\rm He}$ has then to be determined by iteration for consistency between the assumed and the derived value. Once a final model has been determined, other elemental abundances may be derived.

3 Atmospheric parameters and abundances

Before 1985, only four EHe's had been analysed using model atmospheres: HD124448 (Schönberner & Wolf 1974), BD−9°4395 (Kaufmann & Schönberner 1977), HD168476 (Walker & Schönberner 1981) and BD+10°2179 (Heber 1983). Between 1985 and 1988 a preliminary analysis of the pulsating helium star V652 Her (Jeffery et al. 1986) and an analysis of the hot RCrB star MV Sgr (Jeffery et al. 1988) were carried out.

BD−9°4395

The present series of studies started by reanalysing BD−9°4395 in order to ensure that the analysis procedures were operating correctly, to investigate the effect of fully line-blanketed model atmospheres, and to incorporate the most modern atomic data. Also, this star had been the subject of variability studies (Jeffery & Heber 1991), and it was desirable that the atmospheric parameters should be checked before interpreting the variations.

The analysis is described in detail by Jeffery & Heber (1992a), and summarised here in Table 1. The main consequence of using fully line-blanketed model atmospheres was to reduce $T_{\rm eff}$ by $\sim 2\,000$K, and to remove the discrepancy between $T_{\rm eff}$ obtained from ionisation equilibrium and from the flux distribution (Drilling et al. 1984). Minor abundance differences

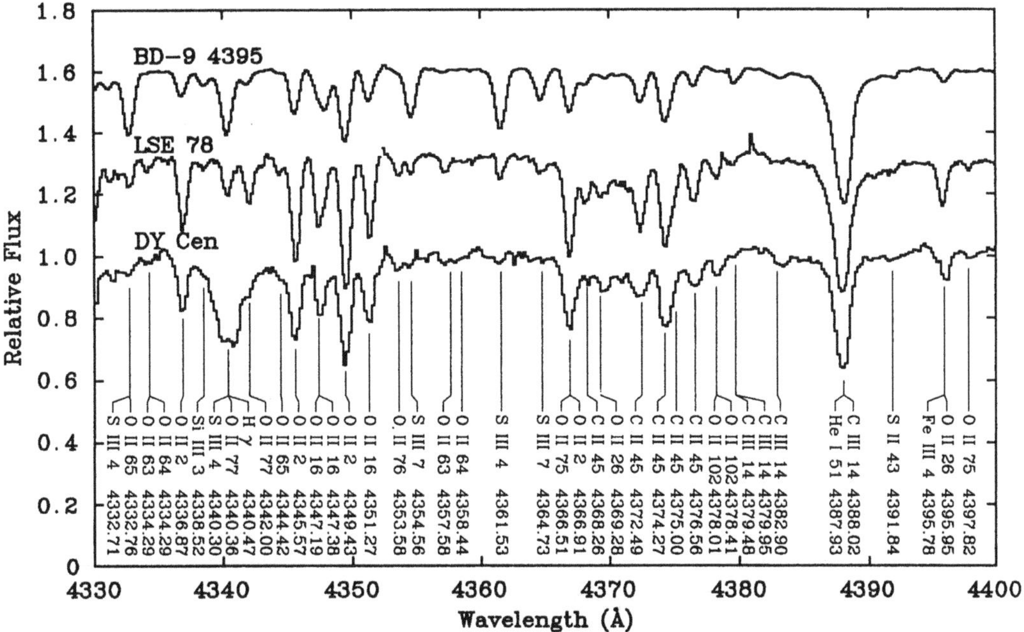

Figure 1: Spectra of three EHe stars, illustrating the features described in the text.

cf. the previous analysis (Kaufmann & Schönberner 1977) are also noted, due in main to improvements in the quality of the observational and atomic data.

DY Cen

After BD−9°4395, the object most urgently requiring analysis was DY Cen. This star has been classified as a hot RCrB star by Hoffleit (1930), although it has not shown recent minima (Bateson 1975, 1978). The star has featured extensively in recent literature, but estimates of $T_{\rm eff} \sim 14\,000$K, $n_{\rm H}/n_{\rm He} \lesssim 0.9$ (Pollacco & Hill 1989) and $n_{\rm C}/n_{\rm He} \sim 1$ (Rao et al. 1991) contrast starkly with the CASPEC spectrum. The $T_{\rm eff}$ estimate was obtained by comparison with LSE 78, whose $T_{\rm eff}$ was based (Drilling et al. 1984) on an incorrect measurement of the interstellar reddening (Jeffery & Heber 1992b).

A comparison of the CASPEC spectrum of DY Cen with those of other EHe's confirms that DY Cen is virtually the spectroscopic twin of LSE 78, except that the Balmer lines in DY Cen are substantially stronger (Fig. 1). A central emission is detected in Hβ and Hγ. Both spectra are similar to BD−9°4395, with the difference that more highly ionized species (eg CIII, NIII, SIII, SiIV) show weaker lines, indicating a lower temperature.

The method of fine analysis was used to determine $T_{\rm eff}$ and $\log g$ from the ionisation equilibrium of CII/ CIII and SiII/ SiIV and the profiles of HeI 4471, 4026, 4921, and 4388 Å. The atmospheric parameters were determined to be $T_{\rm eff} = 20\,000 \pm 500$K, $\log g = 2.20 \pm 0.1$. This measurement had an additional complication compared with BD−9°4395, as indicated by the stronger Balmer lines. Since the hydrogen abundance is non-negligible, model atmospheres had to be calculated with appropriate values for $n_{\rm H}$ **and** $n_{\rm C}$, both of which had to be determined by iteration. To overcome the presence of emission, the equivalent width of Hδ and profile fits to the wings of Hβ and Hγ were used to determine the hydrogen abundance.

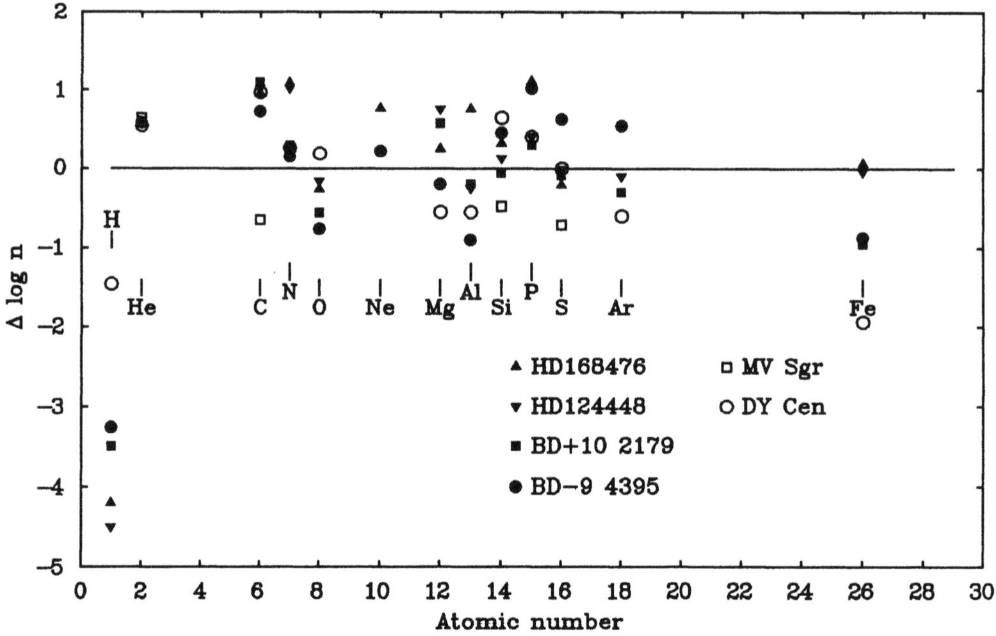

Figure 2: Abundances of EHe stars relative to the B star γ Peg (Peters 1976).

The final abundances are given in Table 1. DY Cen clearly has properties ($T_{\rm eff}$, log g, CNO abundances) very similar to those of true EHe stars. The high H abundance (2-3 dex) is remarkable and violates the usual EHe criterion ($n_{\rm H}/n_{\rm He} \lesssim 10^{-3}$). However, all other properties indicate that DY Cen should rightly be considered as a H-rich extreme helium star. The Fe abundance (also Mg and Ar) indicates severe metal depletion ($\sim$ 2 dex).

4 Abundance patterns

Heber (1986a) suggested that the EHe's may form two abundance groups. HD168476 and HD124448 have low $n_{\rm H}$, high $n_{\rm Fe}$, and small $n_{\rm C}/n_{\rm N}$. BD$-9°4395$ and BD$+10°2179$ have high $n_{\rm H}$, low $n_{\rm Fe}$, and large $n_{\rm C}/n_{\rm N}$ (Fig. 2). Abundances of C, Si and S in MV Sgr are probably underestimated due to emission-filling of absorption lines. Whilst the $n_{\rm C}/n_{\rm N}$ ratio in DY Cen is typical of the second group, $n_{\rm H}$ and $n_{\rm Fe}$ reinforce the suggestion of an anti-correlation between residual hydrogen and metallicity, but do not support the division into subgroups.

It is interesting to compare the abundance properties of EHe's with those of other hydrogen-deficient low-mass stars (eg R CrB's, He-sdO's and DO's). In particular, the EHe's share the CNO abundance characteristics of normal RCrB stars (Lambert 1986) and helium-rich subdwarf O stars (Heber 1992). In all of these stars, a mixing process has produced an atmosphere in which residual hydrogen, CNO-cycled and 3α processed material coexist.

The DO (or PG1159 stars, Werner et al. 1991) show a radically different form of H-deficiency, with extreme C and O enhancements indicating 3α and $C^{12}(\alpha,\gamma)O^{16}$ processed material from the CO core, suggesting that any H-rich envelope has been completely removed. There is currently no evidence for a connection between extremely H-deficient DO's and the cooler EHe's.

Although more abundance studies are required, it is already clear that the production mechanism for EHe's (cf. Schönberner 1986) will have to allow for a wide range of residual hydrogen in the remnant atmospheres, as well as to account for the presence of 3α and CNO-cycle products and a range of metallicities.

Acknowledgments: The author expresses particular appreciation to Uli Heber for his contribution to these investigations. CCP7 has provided substantial support.

References

Bateson F.M., 1975. Publ. New Zealand Astr. Soc. 3, 1
Bateson F.M., 1978. Publ. New Zealand Astr. Soc. 6, 39
Drilling J.S. & Hill P.W., 1986. Hydrogen-deficient stars and related objects, IAU Coll. 87, p. 499, Hunger K., Schönberner D. & Rao N.K., Reidel, Dordrecht, Holland
Drilling J.S., Schönberner D., Heber U. & Lynas-Gray A.E., 1984. ApJ 278, 224
Heber U., 1983. A&A 118, 39
Heber U., 1986a. IAU Coll. 87, 33
Heber U., 1986b. IAU Coll. 87, 73
Heber U., 1992. Atmospheres of early-type stars, Heber U., & Jeffery C.S., Springer-Verlag; these proceedings
Heber U., Jonas G. & Drilling J.S., 1986. IAU Coll. 87, 67
Heber U. & Schönberner D., 1981. A&A 102, 73
Hoffleit D., 1930. Harvard Bulletin 874, 1
Jeffery C.S., 1986. IAU Coll. 87, 81
Jeffery C.S. & Heber U., 1991. Rapid Variability of OB stars: Nature and Diagnostic Value, p. 73, Baade,D., ESO, Garching bei München, Germany
Jeffery C.S. & Heber U., 1992a. A&A submitted
Jeffery C.S. & Heber U., 1992b. in preparation
Jeffery C.S., Heber U., & Hill P.W., 1986. IAU Coll. 87, 101
Jeffery C.S., Heber U., Hill P.W. & Pollacco D., 1988. MNRAS 231, 175
Jeffery C.S. & Malaney R.A., 1985. MNRAS 213, 61P
Jeffery C.S., Skillen I., Hill P.W., Kilkenny D., Malaney R.A. & Morrison K., 1985. MNRAS 217, 701
Kaufmann J.P. & Schönberner D., 1977. A&A 37, 87
Lambert D.L., 1986. IAU Coll. 87, 127
Möller R.U., 1990. Diplom. Thesis, Universität Kiel
Peach G., 1970. Mem. R. astr. Soc. 73, 1
Peters G.J., 1976. ApJS 30, 551.
Pollacco D. & Hill P.W., 1991. MNRAS 248, 572
Rao N.K, Giridhar S. & Lambert D.L., 1991. preprint
Schönberner D. & Wolf R.E.A., 1974. A&A 37, 87
Schönberner D., 1986. IAU Coll. 87, 471
Walker H.J. & Schönberner D., 1981. A&A 97, 291
Werner K., Heber U. & Hunger K., 1991. White Dwarfs, p. 219, Vauclair G. & Sion E., Kluwer, Dordrecht, Holland

Mass Loss from υ Sgr and Other Helium Stars

Richard E. Dudley & C. Simon Jeffery

Dept of Physics & Astronomy, University of St Andrews, St Andrews,
Fife KY16 9SS, Scotland.

Abstract: Mass-loss rates have been determined for 6 hydrogen-deficient stars by modelling their wind line profiles in the UV from high resolution spectra. Line profiles were calculated with using the SEI method (Sobolev approximation with Exact Integration of the transfer equation: Lamers, Cerruti-Sola and Perinotto 1987). Optimum fits to the observations were obtained using a least squares procedure. Results for single helium stars are in good agreement with previous studies. The stellar wind from the binary star υ Sgr has been modelled succesfully for the first time by adopting an extended two component wind.

1 The SEI Method and Mass-Loss Rates

The Sobolev (1947) approximation allows the solution of the radiative transfer and scattering equations to be computed simply by assuming that large velocity gradients in the wind keep the photon-interaction region small. Lamers *et al.* (1987) extended the method by integrating the transfer equation exactly. The SEI (Sobolev with Exact Integration) method gives good agreement with the co-moving frame method.

The SEI code was automated using a least squares procedure to determine best fit parameters for many lines of many stars while retaining the objectivity of the data. The velocity and optical depth laws were in the form specified by Lamers *et al.* and the same notoation is used here. Photospheric lines were included, where neccesary, as a lower boundary condition for the solution of the equation of transfer. Limb-darkening was not considered as the effects are small (Castor and Lamers 1979; Groenewegen and Lamers 1991).

Mass-loss rates were determined from [1], where q_i is the ionisation fraction of ionisation stage i of element E. Y_E is the mass fraction of that element. Mass-loss rates in solar masses per year, velocity in $\mathrm{km\,s^{-1}}$ and wavelengths in Å. Mass-loss rates were determined at the line center ($w = 0.5$) and the adopted mass-loss rates taken from the most reliable line profiles, C IV (when not saturated) and Si IV.

$$\dot{M}q_i = 8.70 \times 10^{-19} \left(\frac{m_E R_\star}{\lambda_0 f Y_E}\right) v_\infty^2 \left(\frac{dw}{dx}\right) x^2 w \tau(w) \qquad (1)$$

2 Extreme Helium Stars

Several classes of extremely hydrogen-deficient stars ($n_H/n_{He} \leq 10^{-3}$) show evidence for a stellar wind in their UV resonance line profiles. These include the extreme helium stars (low mass supergiants, EHe's) and hydrogen-deficient binaries (*e.g.* υ Sgr). Two EHe stars are sometimes classified as sdO stars.

The SEI method was used on these stars; initially the value of β (gradient of the velocity law) allowed to be free, a large range of final values were found. However uncertainties are large and good fits were also found with $\beta = 1$ fixed. The exception to this was BD $+10^\circ$ 2179 where a very low value of β (a constant velocity wind) was required to give a good fit.

The two hottest stars (sdO stars BD +37° 1977, BD +37° 442) have large terminal velocities and give good profile fits. Derived mass-loss rates are typical for the sdO stars ($-\log \dot{M} q_i = 9 - 10$). Profile fits for the cooler stars become more difficult, especially where the Sobolev approximation breaks down (eg BD +10° 2179). Adopted mass-loss rates, based on maximum values of $\dot{M} q_i$, and sample profiles are given in Figure 1. These mass-loss rates are in good agreement with those obtained by Hamann, Schönberner and Heber (1982: HSH) using the co-moving frame method.

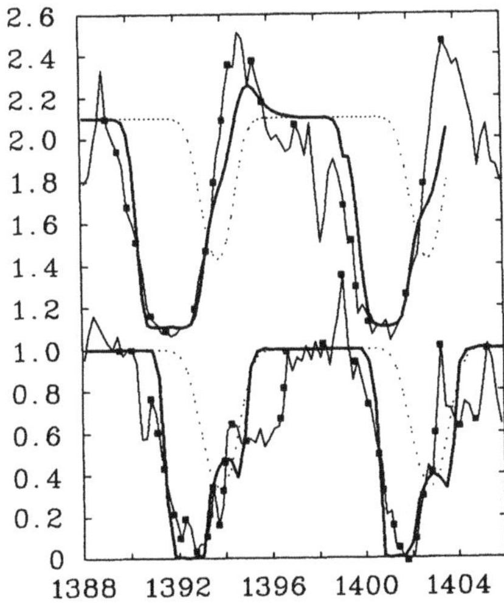

Star	MLR	HSH
BD -9° 4395	-10.5	-9.9
BD +10° 2179	-14.4	-11.4
HD160641	-10.8	-9.4
BD +37° 442	-9.4	
BD +37° 1977	-9.8	

Figure 1: Left: Sample final profiles of Si IV for BD-9 4395 (lower) and HD 160641 (upper). The IUE spectra (light line), including the points adopted in the fit procedure (filled squares), are shown together with the theoretical profiles (heavy line) and the adopted photospheric profiles (dotted line). Right: Adopted mass-loss rates for the EHe stars in log $M_\odot\,yr^{-1}$. Published values (HSH) are given for comparison.

3 υ Sgr

This hydrogen-deficient binary shows prominent wind line profiles for a number of ions. The system consists of a $2.5\,M_\odot$ A-type primary filling its Roche lobe and a hotter secondary of $4.0\,M_\odot$ (Dudley and Jeffery 1990). Although very little variation of wind line profile with phase is noted a single star approximation does not give a good fit to observed profiles.

We set out to test the hypothesis that Roche lobe overflow from the primary feeds material into a wind which is driven by the harder radiation field of the secondary. A simple binary model was constructed consisting of two concentric spherically symmetric regions, the boundary being defined by the orbit of the primary around the secondary ($r = 240\,R_\odot$). The optical depth law of Lamers *et al.* was again used but with the exponents α_1 and α_2 for the inner region and two different parameters α_3 and α_4 replacing them for the outer region. Interaction of radiation between the two regions was accounted for.

Initial solutions with β free showed that a low value of β was indicated (uniform velocity field), however when $\beta = 0.1$ was fixed convergent solutions were not found for most lines. Fixing $\beta = 1$ gave good profile fits. The adopted mass-loss rate for the star was -10.6, sample profiles are given in Figure 2.

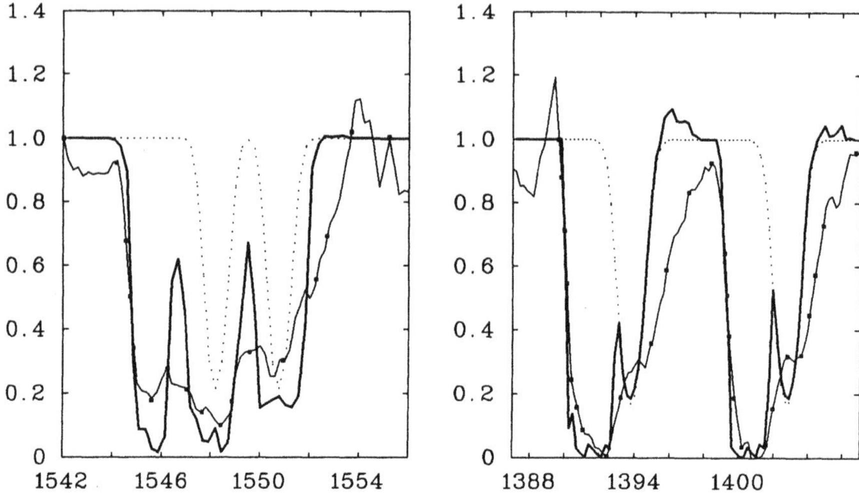

Figure 2: Sample final profiles for v Sgr, C IV (left) and Si IV (right). The lines and symbols are as in Fig. 1. Note that the observed spectrum of v Sgr consists of 16 coadded IUE spectra.

This mass-loss rate is a lower limit while the upper limit of -5 is constrained by the non-detection of orbital parameter changes (Dudley and Jeffery 1990). A value for β of 1 was adopted as the SEI method does not appear to be a good discriminator of different values of β in these cases. v Sgr appears to be undergoing substantial mass loss at this time, but the mass-loss rate does not appear to be high enough to allow the system to avoid a type Ib supernova explosion.

We would like to thank Dr. Ian Howarth (University of London) for some very helpful comments on the original poster paper.

4 References

Castor, J.I. & Lamers, H.J.G.L.M., 1979. *Astr. Astrophys.*, **144**, 87.
Dudley, R.E. & Jeffery, C.S., 1990. *Mon. Not. R. astr. Soc.*, **247**, 400.
Groenewegen, M.A.T. & Lamers, H.J.G.L.M., 1991. *Astr. Astrophys. Suppl.*, **88**, 625.
Hamann, W.-R., Schönberner, D. & Heber, U., 1982. *Astr. Astrophys.*, **116**, 273.
Lamers, H.J.G.L.M., Cerruti-Sola, M. & Perinotto, M., 1987. *Astrophys. J.*, **314**, 726.
Sobolev, V.V., 1947. *Moving Envelopes of Stars*, Harvard University Press, trans. 1960.

Blue Post-Asymptotic Giant Branch Stars at High Galactic latitude

R.J.H. McCausland, E.S. Conlon, P.L. Dufton, F.P. Keenan

Department of Pure and Applied Physics, Queens University of Belfast, Belfast BT7 1NN, United Kingdom

Abstract: Model atmosphere analyses are presented for high resolution spectra of six stars at high galactic latitude. Although their derived atmospheric parameters are consistent with their previous classification as early B-type stars, their metal abundances are significantly different from those expected for Population I objects. However both their chemical composition and atmospheric parameters appear consistent with a Post Asymptotic Giant Branch evolutionary status. Additional evidence for this hypothesis is present in the spectra of one star (LS IV-12°111), where its higher effective temperature is sufficient to excite emission lines from the surrounding nebula.

1 Introduction

As part of a long term project we have been investigating the nature of B-type stars at high galactic latitude; in particular we have been attempting to identify young hydrogen burning early B-type objects (see Conlon et al., 1988 and references therein). From low dispersion spectra, such objects can be differentiated from those with higher gravities, such as subdwarfs and white dwarfs. Unfortunately there exist several groups of stars for which this is not the case, including those at a Post Asymptotic Giant Branch (PAGB) evolutionary stage. Such objects have surface gravities similar to those of (near) main sequence objects for effective temperatures appropriate to early B-type spectral types (Groth et al, 1985). Hence it only becomes possible to differentiate between evolved PAGB and hydrogen burning stars using high dispersion spectra where an accurate abundance analyses should reveal abundance anomalies in the former associated with the processing which occurs in the later stages of stellar evolution (see for example Iben and Renzini 1983).

Here we present the results of such an analysis for six stars previously identified from either photometric or spectroscopic data by Kilkenny and co-workers (see Kilkenny and Pauls, 1990 and references therein) to have atmospheric parameters consistent with them being early B-type stars. We deduce similar atmospheric parameters but their peculiar chemical compositions suggest that they are PAGB stars.

2 Observation

All the stars were observed with the 3.9 m Anglo Australian Telescope; the RGO spectrograph was used for PHL 1580, the UCLES echelle spectrograph for the other five stars. In all cases, an Image Photon Counting System (IPCS) was used as the detector. The observations were reduced and analysed using the STARLINK supported computer packages FIGARO (Shortridge 1986) and DIPSO (Howarth and Murray, 1988). For all the program stars spectra were obtained from approximately 3900 to 4650 Å. Further details can be found in Conlon et al (1991) and McCausland et al (1991).

3 Method of Analysis

For all six stars the observed spectral features have been compared with the results of Local Thermodynamic Equilibrium (LTE) radiative transfer calculations using model atmospheres generated by the ATLAS 6 code of Kurucz (1979). Strömgren photometric colours were used to determine effective temperatures, while surface gravities were deduced from the observed hydrogen line profiles of Hϵ, Hδ and Hγ. The atmospheric parameters and results of the abundance analysis are summarized in table 1.

Table 1. Atmospheric parameters and mean logarithmic abundances of suspected Post-Asymtopic-Branch-Stars together with normal B-star composition.

	PHL 1580	LSIV-12111	LB 3219	PHL 174	LB 3193	LSIV-4 01
T_{eff}	24000	23750	21250	18000	13000	11000
log g	3.6	2.7	2.8	2.7	2.2	2.0
V_T	8	18.0	16	10	15	15
He	11.00±0.2	11.10±0.2	11.00±0.3	10.80±0.2	10.90±0.3	10.40±0.6
C	6.40±0.2	6.70±0.6	6.70±0.3	≤6.10±0.1	≤6.50±0.1	≤7.20±0.1
N	7.50±0.2	7.80±0.3	7.60±0.1	6.80±0.2	-	-
O	8.20±0.3	8.80±0.2	7.60±0.2	7.90±0.5	-	-
Mg	6.70±0.2	7.30±0.3	7.00±0.2	6.20±0.2	5.50±0.2	5.60±0.2
Al	6.00±0.2	≤ 5.40±0.4	≤5.20±0.10	≤5.70±0.1	-	-
Si	7.00±0.6	7.60±0.4	6.60±0.2	6.50±0.4	5.40±0.4	5.50±0.1
S	6.70±0.2	6.60±0.1	5.90±0.2	≤6.50±0.1	-	-
Ca	-	-	-	-	3.70±0.4	3.30±0.5
Ti	-	-	-	-	≤4.20±0.4	≤3.40±0.4
Fe	≤6.80±0.2	≤6.70±0.2	≤6.70±0.10	≤7.20±0.1	≤6.50±0.2	≤5.90±0.2

The chemical compositions are also shown in figure 1, normalised to that found in normal B-type stars.

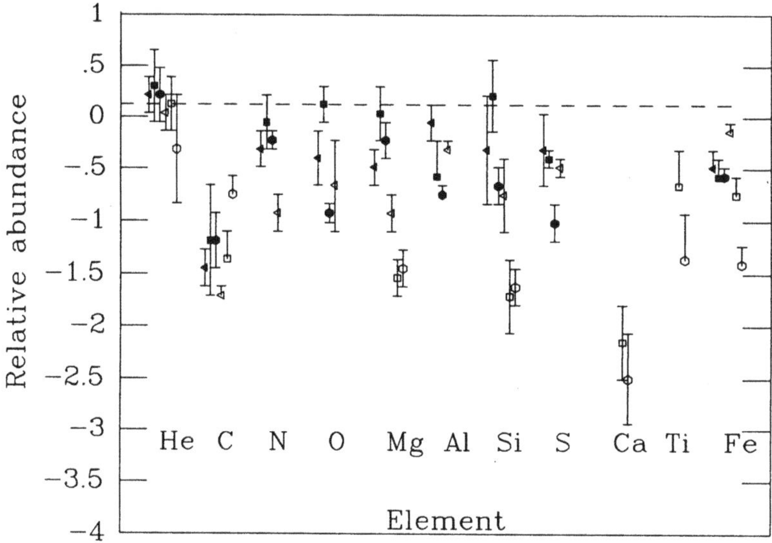

Fig. 1. Element to hydrogen abundances for the six suspected Post-Asymptopic-Giant-Branch Stars on a logarithmic scale. A value of zero (shown by the dotted line) represents a normal B-type star abundance. When no downwards error bar is shown the abundance is an upper limit. The stars are designated as follows: ◂ — PHL 1580; ■ — LSIV -12 111; • — LB 3219; ◃ — PHL 174; ⊔ — LB 3193; ∘ — LSIV -4 01.

4 Discussion

The atmospheric parameters for all six targets imply that they could be evolving from the hydrogen burning main sequence; however their chemical compositions appear to be inconsistent with this evolutionary phase. Both carbon and iron are depleted, the former by more than 1.0 dex in all cases. There is some evidence for two distinct groups, with the hotter stars showing no consistent abundance pattern while the cooler show increasing degrees of depletions from magnesium through to calcium.

Figure 2 is adapted from Groth et al (1985) and shows the position of our stars in an effective temperature-surface gravity diagram, together with the position of the hydrogen burning zero age main sequence, horizontal branch and OB-type subdwarfs. The gravities of our stars are incompatible with them being either horizontal branch or subdwarf stars; however they lie on the Post Asymptotic Giant Branch tracks of Schönberner (1983). Hence we believe that they are intermediate between the cooler A- and F-type PAGB stars (eg HR 4049) discussed by Trams et al (1991) and the hotter PAGB objects such as the planetary nebula DDDM-1 (Clegg et al, 1987) and RWT 152 (Ebbets and Savage, 1982). Although these PAGB stars show a wide variety of chemical compositions, we note that the abundances found here are at least qualitatively in agreement with those derived for the cooler objects HR 4912 and HD 46703 and the hotter object DDDM-1.

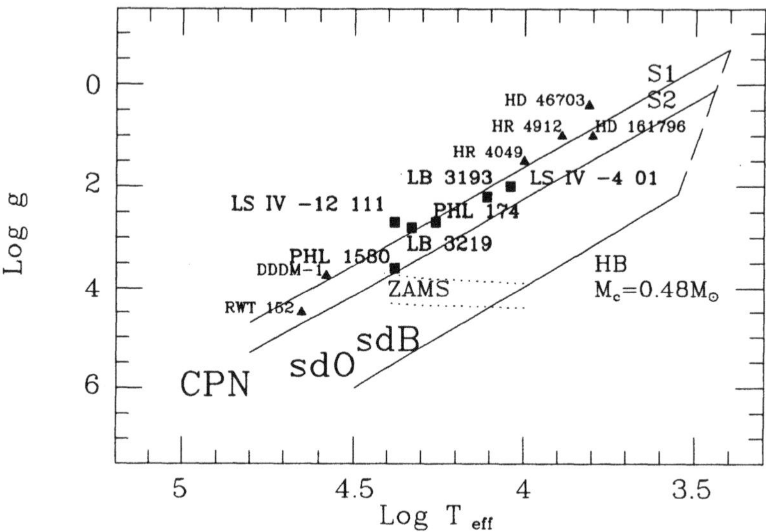

Fig. 2. Positions of the six suspected Post-Asymptopic-Giant-Branch Stars in an effective temperature-gravity diagram. Also shown are cooler and hotter stars identified as being PAGB objects. The areas of the diagram populated by B and O-type subdwarfs and the central stars of planetary nebulae are also indicated. The extent of the zero age hydrogen burning main sequence is shown by dotted lines and the horizontal branch by a solid line. Two PAGB evolutionary tracks of Schönberner (1983) are plotted for masses of $0.565 M_\odot$ (S1) and $0.546 M_\odot$ (S2).

Additional evidence for a PAGB evolutionary stage is the presence of emission lines of neutral hydrogen, helium, [N II], [O II] and [S II]in the spectrum of LS IV -12 111. These low excitation lines indicate that a planetary nebula is beginning to develop around this object.

References

Clegg, R.E.S., Peimbert, M., Torres-Peimbert, S. 1987, Mon. Not. R. astr. Soc., **224**, 761.

Conlon, E.S., Brown, P.J.F., Dufton, P.L., Keenan F.P. 1988, Astr. Astrophys., **200**, 168.

Conlon, E.S., Dufton, P.L., Keenan, F.P., McCausland R.J.H. 1991, Mon. Not. R. astr. Soc., **248**, 820.

Ebbets, D.C., Savage, B.D. 1982, Astrophys. J., 262, 234.

Groth, H.G., Kudritzki, R.P., Heber, U. 1985, Astr. Astrophys., **152**, 107.

Howarth, I.D., Murray, J.M. 1988, SERC Starlink User Note, No. 50.

Iben, I., Renzini, A. 1983, Ann. Rev. Astr. Astrophys., **21**, 271.

Kurucz, R.L. 1979, Astrophys. J. Suppl., **40**, 1.

Kilkenny, D., Pauls, L. 1990, Mon. Not. R. astr. Soc., **724**, 133.

McCausland R.J.H., Conlon, E.S., Dufton, P.L., Keenan, F.P. 1991, Astrophys. J., submitted.

Schönberner, D. 1983, Astrophys. J., **272**, 708.

Shortridge, K. 1986, SERC Starlink User Note, No. 86.

Trams, N.R., van Hoof, P.A.M., van de Steene, G.C.M. 1991, Astr. Astrophys., in press.

OLD PLANETARY NUCLEI AND THEIR EVOLUTIONARY CONNECTIONS

D. SCHÖNBERNER AND T. BLÖCKER
Institut für Theoretische Physik und Sternwarte der Universität Kiel
Olshausenstr. 40, W-2300 Kiel, Germany

Abstract

According to recent observations a significant fraction of all so far known "PG 1159" objects reside within planetaries. Thus, their evolution may also give hints to the evolutionary status of those "PG 1159" objects without a nebular shell. We present some new evolutionary calculations with extensive mass loss to explain the very peculiar compositions of PG 1159 stars in general, and of the even more exotic object H1504 in particular. Our conclusion is that a simple mass loss scenario does not work and that one has very likely to invoke burning of the residual hydrogen envelope and its mixing with 3α processed matter.

1 Introduction

It appears now established that, according to spectral appearances, two major evolutionary sequences do exist for central stars of planetary nebulae (CSPN): One sequence consists of objects with solar helium-to-hydrogen abundance ratio, whereas the other seems to be completely devoid of hydrogen, but enriched in carbon and, sometimes, also in oxygen. There seems to exist a third evolutionary channel, consisting of objects with a surface made of a mixture of hydrogen, helium and carbon. In an ongoing spectroscopic survey of central stars residing within very old planetaries (PN) performed by Napiwotzki and Schönberner, the majority of them, namely 21 (= 78%), belongs to the hydrogen-rich sequence, 3 (= 11%) to the hydrogen-deficient sequence, and also 3 (= 11%) are of the "mixed" type as descibed above (cf. Napiwotzki and Schönberner 1991 a,b; Napiwotzki, these proceedings). Using a list mainly containing brighter objects Méndez (1991) gets similiar results: 71 (= 61%) hydrogen-rich and 38 (=33%) hydrogen-poor objects, with a small number of objects (6 or 5%, resp.) not belonging to either group.

Even if one admits that both compilations are based on severely biased observational material, they contain the most extensive spectroscopic information about central stars presently available and confirm the finding of Schönberner (1986) which was based on the sparse spectroscopic information for a local, spatially limited sample of CSPN: only 4 out of 27 (= 15%) have a hydrogen-deficient surface, i.e. concerning hydrogen and helium the majority of the central star population has a normal (solar) surface composition*. The four hydrogen-deficient CSPN found by Napiwotzki and Schönberner belong to the (spectroscopically defined) class of "PG 1159" objects, giving now rise to a total number of 5 high-gravity central stars that belong to that group. Thus, a substantial fraction of this stellar population resides within well-known planetaries, and it appears reasonable to assume that *all* "PG 1159" stars share a common origin, i.e. that they are all remnants of the Asymptotic Giant Branch (AGB) which managed to exposed their interior layers by the action of mass loss and/or mixing.

*Some of the high-gravity CSPN on the white dwarf "cooling" track show signs of gravitational separation between hydrogen and helium, as it is known to occur at the surface of white dwarfs. Nevertheless, they are assigned to the hydogen-rich group.

2 Observations vs. evolutionary calculations

From the very fact that *all* PN, regardless of the spectral appearances of their central stars, have about solar helium-to-hydrogen ratios in their main nebular shells, we conclude that the branching into the different spectroscopic classes is ruled by the internal properties of the stars while they are leaving the AGB. From detailed evolutionary calculations it is known that the post-AGB evolutionary behaviour is ruled by the relative strength of the hydrogen and helium burning shells (Schönberner 1979, Iben 1984). The most important phase is that in which the star is powered by hydrogen burning (in a shell), and most central stars with normal composition can be explained by this evolutionary phase (Schönberner 1981, 1986; Méndez 1991).

Calculations show that mass loss alone cannot remove the hydrogen-rich layers while the remnant is still burning hydrogen. Instead, the horizontal evolution in the HR-diagram is accelerated by mass loss until the mass limit is reached ($\approx 10^{-4}M_{\odot}$) below which the envelope cannot sustain hydrogen burning anymore. Then, the luminosity (and mass loss) will decrease rapidly. Indeed, the observations indicate that the hydrogen-rich central stars evolve directly towards white dwarfs of spectral type DAO/DA (hydrogen-dominated atmospheres, cf. Napiwotzki, these proceedings). The very existence of so many high-gravity central stars with traces of helium indicate that the gravitational settling time of the latter is of the order of the ages of the oldest PN, i.e. close to 10^5 years!

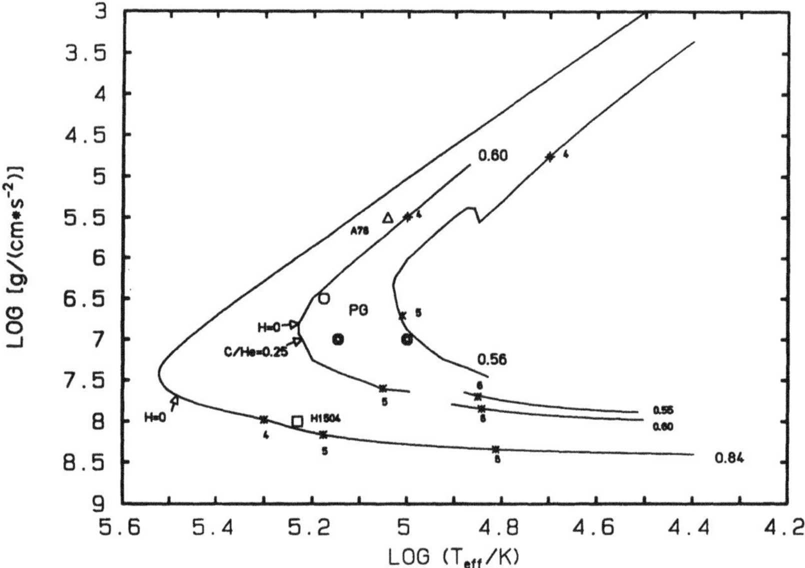

Figure 1: Evolutionary tracks (labeled with the corresponding remnant masses) of shell helium-burning,mass-loosing post-AGB models from different sources: $0.56M_{\odot}$ (Schönberner, unpublished), $0.6M_{\odot}$ (Iben 1984) and $0.84M_{\odot}$ (Blöcker, unpublished) supplemented by white dwarf evolutionary calculations of Koester and Schönberner (1986) for 0.55 und $0.6M_{\odot}$. The apparent discontinuity between tracks of similiar remnant masses is due to the inclusion of non-ideal plasma effects in the latter calculations. Note that the phase before the last thermal pulse has been omitted in this figure for the $0.60M_{\odot}$ and $0.84M_{\odot}$ track. The asterisks refer to ages of $\log\frac{t}{\mathrm{yr}} = 4, 5$ and 6, counted from the tip of the AGB. The loci of five analyzed PG 1159 stars are from Werner et al. (1991), that of H1504 from Werner (1991). The preliminary position of the central star of A78 is also given. (see these proceedings).

The situation is different if the AGB remnant is powered by the helium-burning shell during the aftermath of a thermal pulse. This can happen either at the tip of the AGB, followed by the rapid contraction of the stellar envelope, or when the star is already on its way towards a compact configuration. Iben (1984) showed that a sufficiently strong stellar wind is able to remove the hydrogen-rich layers and even to expose carbon-rich matter of the former flash-driven convective shell without changing the evolutionary path and speed too much.

Fig. 1 gives a collection of helium burning post-AGB models from different sources. The 0.56 $M_{\odot}$ track is from Schönberner (unpublished) and incorporates mass loss according to Reimers (1975), which, however, is too small to remove hydrogen. The hook corresponds to the rekindling of hydrogen. The 0.6 $M_{\odot}$ track has been computed by Iben (1984) with a constant mass-loss rate of $10^{-8} M_{\odot} \mathrm{yr}^{-1}$. Hydrogen is totally removed when the model approaches the turn-around point, and later also carbon appears at the surface (C/He $\approx$ 0.25 by mass). The 0.84 $M_{\odot}$ track is provided by Blöcker (still unpublished) who used mass loss rates according to the theory of radiatively driven stellar winds (Pauldrach et al. 1988). This model looses its hydrogen envelope shortly after the turn-around point, but then the strength of the stellar wind decreases rapidly and the deeper, carbon-rich layers remain untouched. The loci of so far analyzed hydrogen-deficient AGB remnants suggest the following: The five PG 1159 objects have masses slightly below the typical white-dwarf mass of 0.6 $M_{\odot}$ whereas A78 ($\approx 0.7 M_{\odot}$) and H1504 ($\approx 0.8 M_{\odot}$) are apparently more massive objects. Their post-AGB ages range from $\approx 10^4$ up to about $\approx 10^5$ years. Note that the kinematical age of A78, estimated from the size of the nebular shell and its expansion velocity, is consistent with that age, viz. about $2 \cdot 10^4$ years.

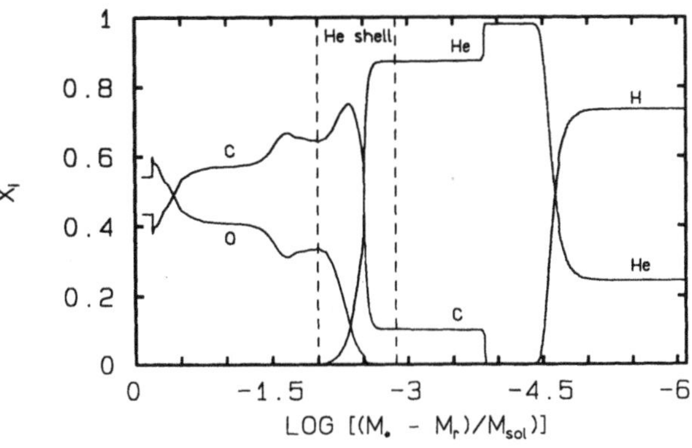

Figure 2: Internal composition of the $0.84 M_{\odot}$ shell helium-burning AGB remnant after a late thermal pulse when the hydrogen-burning shell is extinct. The position of the helium-burning shell is indicated. To expose carbon-rich intershell matter with C/He$\approx$0.1 (by mass) at least about $10^{-4} M_{\odot}$ of the envelope are to be removed.

However, a severe defect is apparent: the analyzed objects have very high carbon abundances, C/He>1(by mass), and H1504 is essentially helium free but oxygen-rich (C$\approx$O). These abundances are typically for the layers within, or even beneath (H1504), the helium burning shell (see Fig. 2). Obviously the mass-loss rates used in the calculations were too small. Fig. 2 suggests that about $3 \cdot 10^{-3} M_{\odot}$ are to be removed, equivalent to a mean mass-loss rate of $10^{-8} M_{\odot} \mathrm{yr}^{-1}$ acting over $3 \cdot 10^5$ years. Therefore, larger mass-loss rates of, say, $10^{-7} M_{\odot} \mathrm{yr}^{-1}$ are necessary. We thus recomputed the evolution of a shell helium-burning AGB remnant of $0.84 M_{\odot}$, fixing the mass-loss rate at $10^{-7} M_{\odot} \mathrm{yr}^{-1}$ as soon as the rate according to Pauldrach et al. (1988) drops beneath that value, until the C/O core became exposed (Fig. 3). The track is essentially the same as that in Fig. 1, only the evolutionary time scale is somewhat different. A high carbon-to-helium ratio, as observed in the PG 1159 objects, is not reached before the helium-burning shell becomes extinct and the model enters the "cooling" phase of white dwarf evolution. A virtually helium-free surface is reached after $9.3 \cdot 10^4$ years, and the position of this model nicely coincides with that of H1504.

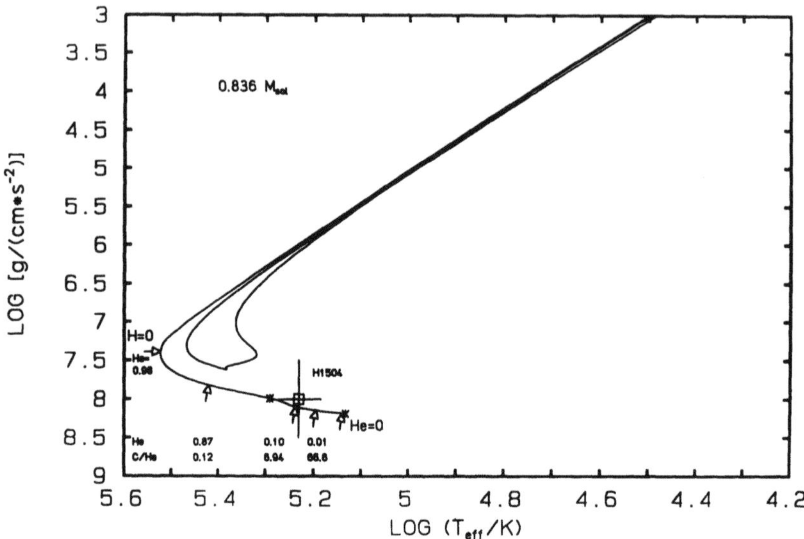

Figure 3: Evolutionary track of a shell helium-burning $0.84 M_{\odot}$ AGB remnant with constant mass loss rate of $10^{-7} M_{\odot} \mathrm{yr}^{-1}$. Asterisks indicate post-AGB ages of $\log \frac{t}{\mathrm{yr}} = 4$ and 5, the arrows where hydrogen and helium, resp., vanish from the surface. The square refers to the position of H1504 (with error bars) according to Werner (1991). The He content and the C/He ratio are also given for some selected points.

Despite of this success this relatively simple mass-loss scenario has severe drawbacks:

i) one has to invoke rather high mass-loss rates for long times, especially at rather low stellar luminosities ($\approx 10^2 L_{\odot}$) since the evolution is slowest there. This poses the question how these powerful winds are driven, and why we do not see any wind signatures in the spectra of PG 1159 objects.

ii) a carbon-to-helium ratio greater than unity is not reached before entering the white dwarf "cooling phase". Some PG 1159 objects appear to be much more luminous (cf. Fig. 1), and particularly the central star of A78, even more luminous, cannot be explained since the corresponding model has still a hydrogen-rich surface. Also, numerous cooler, carbon-rich Wolf-Rayet central stars do exist.

3 Mixing and mass loss

In the last section we already presented arguments that do not support the simple mass-loss scenario. The most convincing one is, however, the fact that some PG 1159 objects show detectable nitrogen lines in their spectra (Werner and Heber 1991) which is unexpected because at helium-burning temperatures nitrogen is quickly destroyed. The solution might be mixing *and* burning. It is well known since the computations by Schönberner (1979) that, if a helium shell flash occurs very late at already declining stellar luminosities, the flash driven convective shell cuts into hydrogen-rich layers and mixes protons downwards towards very high temperatures. Pilot computations (Sweigart 1974, Caloi 1990) indicated that then powerful hydrogen burning sets in just on top of the helium-burning layers. The further downward mixing of hydrogen (and nitrogen!) is prevented by a small radiative layer between both shells. No detailed computations exist yet but we may expect that, due to the large nuclear energy release, the model expands to giant dimensions very quickly, and that finally the surface contains a mixture of He, C *and* N, and, if hydrogen-burning was incomplete, also traces of hydrogen. After this short excursion to the red ("born-again central star"), the model resumes its contraction to the blue on typical time scales of shell helium-burning AGB remnants (Iben et al. 1983). This scenario is able to explain central stars with hydrogen-deficient surfaces already at rather low effective temperatures. The predicted C/He ratio is, however, $\approx$ 0.1...0.2 and corresponds to that of the intershell region since mixing of $\approx 10^{-4} M_{\odot}$ hydrogen into $2 \cdot 10^{-3} M_{\odot}$ of helium-carbon matter with immediately following burning does not change the abundances.

It must be noticed in this connection that the intershell carbon content depends on how convection is being treated. Standard evolution theory , as used here, assumes Schwarzschild's criterion for the determination of the boundaries of convectively unstable regions. During a helium shell flash the lower boundary of the flash-driven convective shell cuts into carbon-rich layers at the lower half of the helium burning region and dredges up freshly processed carbon. The carbon abundance profile is very steep (cf. Fig. 2), and if we would allow for some "undershooting" of the convective elements, the amount of dredged-up carbon may increase substantially, and hence also the C/He ratio that later becomes visible. However, it must also be said that the molecular-weight gradient may act in the opposite sense.

Concluding one can say that mixing and burning following a very late helium shell flash provides a promising scenario to explain hydrogen-deficient central stars, and especially also peculiar objects like the PG 1159 objects. Quantitative evolutionary calculations are badly needed in order to proof this scenario.

Acknowledgements

One of us (T.B.) thanks the DFG for financial support (grant Scho/1-1).

References

Caloi, V.: 1990, A&A 232, 67
Iben, I., Jr.: 1984, ApJ 277, 333
Iben, I., Jr., Kaler, J.B., Truran, J.W., Renzini, A.: 1983, ApJ 264, 605
Koester, D., Schönberner, D.: 1986, A&A 154, 125
Méndez 1991, IAU Symp. 145: Evolution of stars: The atmospheric abundance connection, eds. G.Michaud and A.Tutukov, Kluwer, p. 375
Napiwotzki, R., Schönberner, D.: 1991a, Proc. 7th European workshop on white dwarfs, eds. G.Vauclair and E.Sion, Kluwer, p. 39
Napiwotzki, R., Schönberner, D.: 1991b, A&A 249, L16
Napiwotzki, R. 1992, these proceedings
Pauldrach, A., Puls, J., Kudritzki, R. P., Méndez, Heap, S. R.: 1988, A&A 207, 123
Schönberner, D.: 1979, A&A 79, 108
Schönberner, D.: 1981, A&A 103, 119
Schönberner, D.: 1986, A&A 169, 189
Sweigart, A. V.: 1974, ApJ 189, 289
Werner, K.: 1991, A&A 251, 147
Werner, K., Heber, U.: 1991, A&A 247, 476
Werner, K., Heber, U., Hunger, K.: 1991, A&A 244, 437

Analysis of central stars of old planetary nebulae: Problems with the Balmer lines

R. Napiwotzki*
Institut für Theoretische Physik und Sternwarte der Universität Kiel
Olshausenstr. 40, D-2300 Kiel, Germany

Abstract: New observations of the central stars of old planetary nebulae are presented and discussed. The attempts to fit the observed Balmer lines of the hydrogen rich central stars by theoretical profiles and their failure is described. Possible reasons for this failure are discussed.

The spectra and their interpretation

The observations were performed in three runs in October 1989, November 1990, and July 1991 at the Calar Alto observatory, Spain. We used the 3.5 m telescope together with the Cassegrain Twin spectrograph equipped with CCDs as detectors. The dispersion used was 144 Å/mm in the blue and 160 Å/mm in the red channel. Depending on the seeing conditions the resulting spectral resolution ranges from 3 Å to 9 Å (FWHM).

Table 1: Spectral types and Shklovskii distances vs. distances derived from Na D lines of the newly observed central stars together with the stars presented in Napiwotzki & Schönberner (1991a+b). The absolute magnitude, M_V, includes correction for interstellar absorption, and the kinematical ages are preferentially based upon measured expansion velocities (Weinberger, 1989, Hippelein & Weinberger, 1990).

				Shkl. dist.		"Sodium" distances			
Object	PK	Type	V	d/pc	M_V	d/pc	M_V	R/pc	$\frac{t_{kin}}{1000\,yrs}$
S 68	30+06	hybrid	16.59	300	8.0	560	6.6	0.54	70
HW 13	34−13	sdO	16.6	4460	3.4	> 1000	< 6.6	> 0.11	> 5
A 43	36+17	hybrid	14.71	2100	2.3				
NGC 7293	36−57	DAO	13.43	150	7.3				
A 39	47+42	sdO/DAO	15.8	1180	5.4				
A 46	55+16	close binary	15.07	2220	3.3				
NGC 6853	60−03	sdO/DAO	13.82	240	6.9	550	5.1	0.55	17
NGC 6720	63+13	sdO/DAO	15.00	810	4.9	900	4.6	0.16	6
NGC 7094	66−28	hybrid	13.61	1560	2.6				
A 74	72−17	DAO	17.11	230	9.7	750	7.5	1.51	64
Jn 1	104−29	PG 1159	16.13	700	6.9	400	8.1	0.32	21
IW 2	107+07	DAO	17.71	260	10.0	1490	6.5	3.24	230
DHW 5	111+11	DA	15.6	400	7.5	440	7.4	0.56	110
S 176	120−05	DA(O?)	18.6	270	11.4	> 2000	<7.1	>3.48	>180
EGB 1	124+10	DA	16.35			480	7.4	0.31	15
S 188	128−04	DAO	17.44	310	9.1	600	7.7	0.79	39
HFG 1	136+05	close binary	15.0						
EGB 4	144+24	close binary	12						
IW 1	149−03	PG 1159	16.56	230	8.4	480	7.8	0.91	74
HW 4	149−09	DAO	17.05	400	8.5	1130	6.4	1.31	64
HW 6	156+12	DA	16.7	2000	5.2				
PW 1	158+17	DAO	15.4	240	8.7	470	7.0	1.37	56
S 216	158+00	DAO	12.3	40	9.3	110	7.1	1.60	390
WDHS 1	197−06	DA(O?)	17.4	320	9.6	880	7.7	1.98	110
K 2-2	204+04	sdO	15.0	520	6.4	1170	4.7	1.17	110
A 21/YM 29	205+14	PG 1159	15.99	290	8.7	600	7.1	0.93	45
A 7	215−30	DAO	15.43						
A 31	219+31	DAO	15.51	240	8.6	240	8.6	0.56	24

*Visiting astronomer, German-Spanish Astronomical Center, Calar Alto, operated by the Max-Planck-Institut für Astronomie Heidelberg jointly with the Spanish National Commission for Astronomy

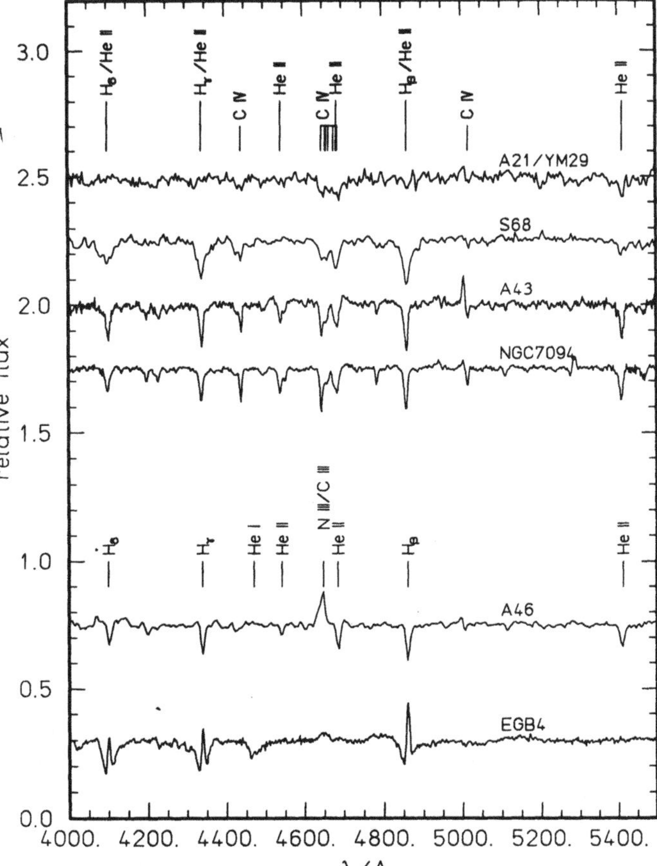

Fig. 1: Blue spectrograms of the newly observed PG 1159, hybrid, and binary central stars.

The new observations (Table 1) confirm the results already presented in Napiwotzki & Schönberner (1991a): Most of the central stars of old planetary nebulae are hydrogen rich and resemble either high-gravity sdO stars or white dwarfs. The helium content ranges from several percents to below the observational limit (He/H$< 10^{-3}$). This indicates that gravitational separation of elements is still going on and may even continue beyond the planetary nebula phase. Note the confirmation of S 216 as a planetary nebula, likely to be the nearest and oldest planetary known.

Additional to the two PG 1159 central stars presented in Schönberner & Napiwotzki (1990) one new PG 1159 object was detected (cf. Fig. 1). Now about half of the known PG 1159 stars are central stars of planetary nebulae (cf. the lists in Werner, 1992). Notably is the lack of central stars with nearly pure helium atmospheres like K 1-27 or DO white dwarfs in our sample. It seems that a mixture of helium and carbon is more common.

Additional two new "hybrid" objects were detected. These objects contain hydrogen, helium, *and* strongly enriched carbon. The first object of this class (S 68) was presented in Napiwotzki & Schönberner (1991b). But according to their spectral appearance both new hybrid central stars seem to have a lower gravity. Some wind features are visible in their spectrograms.

The sample also contains three central stars already known as members of close binary systems.

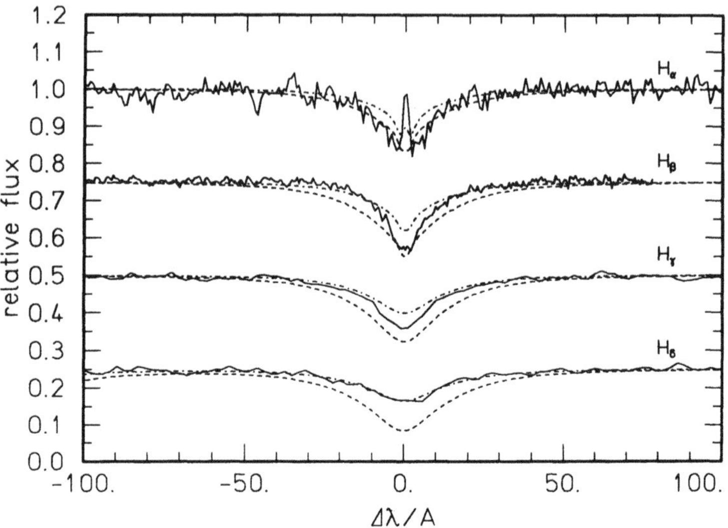

Fig. 2: Balmer lines of S 216 compared to theoretical profiles: T_{eff} = 50000 K (dashed line) and T_{eff} = 90000 K (dashed-dotted line). For all profiles $\log g$ = 7.0 and He/H = 0.01 was assumed. The spectra of H_α and H_β were taken by K. Werner with 1.5 Å and 1.8 Å resolution respectively

Distances, absolute magnitudes, and ages:

For an approximate determination of the present evolutionary status of the central stars, we estimated the distances from the interstellar Na D lines as described in Napiwotzki & Schönberner (1991a). The results are listed in Table 1. Our new distances are in general larger than the distances derived by the Shklovskii method ($M_i = 0.2\,M_\odot$) and the large M_V often quoted for these central stars are removed. Their position in the age-M_V diagram is now consistent with a mean stellar mass of about 0.6 $M_\odot$ (Schönberner, 1981, see also Weidemann, 1990, for a review).

Model analysis and the "mystery" of white dwarf central stars

For the model analysis of the hydrogen rich stars a grid of NLTE model atmospheres containing H and He calculated with the ALI code of Werner (1986) was used. For H/HeI/HeII we took 10/29/14 levels with 32/91/91 lines into account. For the two first series of hydrogen (Lyman and Balmer) and the four first of HeII we included the effect of stark broadening in the calculation of the atmospheric structure. The resulting change in the line profiles relative to a model computed without stark broadening is small but detectable even at our low spectral resolution. A comparison with line profiles calculated with the LTE code of D. Koester for DA white dwarfs in the region 30000 K... 60000 K showed good agreement except for H_α, where deviations caused by NLTE effects are important. Thus we claim that the model atmospheres used by us are the most sophisticated to analyse the spectrograms of hot, hydrogen-rich degenerates.

But this is not enough as it is obvious from Fig. 2: the spectra of the DAO central star of S 216 are compared to model calculations. One can see that for every Balmer line a different temperature is necessary for a fit: ranging from 50000 K to 90000 K. The Balmer lines of the central star of NGC 7293, already analysed by Mendez et al. (1988), showed the same trend. We conclude: **It is not possible to fit all Balmer lines of DAO central stars consistently with today state-of-the-art model atmospheres.**

It is obvious that the "mystery" of white dwarf central stars, i.e. their tendency to have too large "cooling" ages as compared to the kinematical ages of the nebular shells (Napiwotzki & Schönberner, 1991a) is, at least partly, caused by the problems with the Balmer lines described here. **Perhaps the whole distance scale of hot DA white dwarfs which is based on Balmer line fits may be affected.**

Now, what is wrong?

Line blanketing: The model atmospheres take only hydrogen and helium into account, but as has been demonstrated by Schönberner & Drilling (1985) in the UV spectra of hot subdwarfs often lines of iron are visible. First test calculations taking lines of iron into account in the region of hot stars are presented by Dreizler & Werner (1992).

Line broadening: For the calculations of our line profiles the VCS tables of Schöning and Butler (1989 and priv. comm.) were used. It is known that the VCS approximation is not strictly valid even down to 13000 K (cf. Thomsen & Helbig, 1991). Obviously these defects are not significant for B-type stars. Maybe they become important for the very hot stars discussed here. Better theories are available (e.g. the MMM theorie) but extensive tabulations, as necessary for line formation calculations, have not yet been published until now.

Wind effects: At the high gravities encountered here no signs of a stellar wind are visible and we expect these effects to be completely unimportant.

Magnetic fields: Not too strong magnetic fields broaden (and deepen) the cores of the Balmer lines. Though it is unlikely that this may be a common phenomen in DAO central stars, this cannot completely be ruled out at the moment.

What to do now? We try to use the data available from the nebula itself to fix the temperature of the central stars by employing photoionization models. From comparison of fluxes from the different model atmospheres and black body fluxes (often used for photoionization models) the importance of realistic atmospheric models for the nebular analysis became evident. For instance, even at high gravities ($10^7\,cm/s^2$), non-LTE effects are important. This is also relevant for the interpretation of ROSAT and EXOSAT measurements of hot degenerates, however, recent analysis are based on LTE model atmospheres (e.g. Jordan et al., 1987).

We plan to incorporate the effect of iron line blanketing in DAO model atmospheres in order to clarify the whole situation.

Acknowledgements: Thanks to K. Werner, T. Rauch, and S. Dreizler for their support of my NLTE calculations, D. Koester for allowing to use his LTE code, and R. Weinberger for his help to prepare some of the finding charts. This work was supported by the BMFT and by DFG travel grants. Last not least thanks go to D. Schönberner who initiated this investigation.

References:

Dreizler, S., Werner, K.: 1992, These Proceedings
Hippelein, H., Weinberger, R.: 1990, A&A 232, 129
Jordan, S., Koester, D., Wulf-Mathies, C., Brunner, H.: 1987, A&A 185, 253
Mendez, R.H., Kudritzki, R.P., Herrero, A., Husfeld, D., Groth, H.G.: 1988, A&A 190, 113
Napiwotzki, R., Schönberner, D.: 1991a, in: White Dwarfs, eds. G. Vauclair and E. Sion, NATO ASI Series, Mathematical and Physical Sciences No. 336, Kluwer, Dordrecht, p. 39
Napiwotzki, R., Schönberner, D.: 1991b, A&A 249, L16
Schönberner, D.: 1981, A&A 103, 119
Schönberner, D., Drilling, J.S.: 1985, ApJ 290, L49
Schönberner, D., Napiwotzki, R.: 1990, A&A 231, L33
Schöning, T., Butler, K.: 1989, A&AS 78, 51
Thomsen, C., Helbig, H.: 1991, Spectrochim. Acta 46B, 1215
Weidemann, V.: 1990, ARA&A 28, 103
Weinberger, R.: 1989, A&AS 78, 301
Werner, K.: 1986, A&A 161, 171
Werner, K.: 1992, These Proceedings

Hot White Dwarfs

Detlev Koester [1] and David Finley [2]

[1]Department of Physics and Astronomy, Louisiana State University
Baton Rouge
[2]Center for EUV Astrophysics, University of California, Berkeley

1 Introduction

For the purpose of this review, we define "hot white dwarfs" as covering the range of effective temperatures from about 12000 to 200000 K. In the following we give a brief description of the different types of white dwarfs that fall into this temperature range.

2 Atmospheric Parameters for Helium-rich Objects: T_{eff}, log g, and abundances

2.1 PG1159 Stars

In the "official" classification scheme (Sion et al. 1983), which we follow closely for the other types, these objects are considered as a subtype of the DO. Spectroscopically, the distinguishing signature is a characteristic absorption trough between 4650 and 4690 Å, formed by HeII and many CIV lines. The class has about 20 members, if one includes K1-16, IW1, VV-47, and Jn1 (central stars of planetary nebulae) and the peculiar object H1504+65.

Although the prototype of this class was discovered a long time ago (McGraw et al. 1979), the composition of the atmosphere remained a mystery until very recently. The reasons are that the extremely high effective temperatures ($\geq$ 100000 K) absolutely demand the use of line blanketed NLTE atmospheres, which in this case requires the construction of elaborate model atoms for CIV, CV, NV, OVI, etc. In addition, the enormous numerical task could only be tackled after the development of new methods (approximate lambda operator techniques).

Results from the recent NLTE analyses by Werner, Heber, and Hunger (1991, WHH), and Werner and Heber (1991, WH) are presented elsewhere in this volume; we will therefore give only a very brief summary:

Four objects analyzed in WHH have identical abundances (He 0.61, C 0.31, N 0.0012, O 0.08 by number). PG1159-035 and PG1520+525 have an effective temperature of 140000 K, PG1424+535 and PG1707+427 of 100000 K.

PG1144+055 — analyzed in WH — is slightly hotter (150000 K), but the major difference is a higher nitrogen abundance of 0.0066 and the same very low oxygen abundance of 0.0066.

2.2 The DO Stars

This group is characterized by strong HeII lines and the presence of H or HeI. Depending on the relative abundances of H and He, it has been further subdivided into cool DO, hot DO, and DAO groups. Contrary to the amount of recent work on the exciting PG1159 objects, not much seems to have been done on DO recently. An early attempt to determine atmospheric parameters from observational data of relatively low quality compared to today's standard was Koester, Liebert, and Hege (1979); the classical paper now is Wesemael, Green, and Liebert (1985, WGL), which is based on optical and UV (IUE) data. These authors introduced a classification into the subtypes "cool DO" (HeI + HeII/H blends, prototype HZ21), "hot DO" (HeII, prototype PG1034+001), and "DAO" (hydrogen-rich, HZ34). They include as a fourth class the PG1159 objects, which we prefer to treat separately here. Results from the analyses by WGL, Sion, Guinan, and Wesemael (1982), Liebert et al. (1981), and Mendez et al. (1981) are summarized below:

The hot DO have (not well determined) effective temperatures around 80000 K and H/He number ratios ≥ 100.

The cool DO are found in a relatively narrow range between 47500 and 55000 K and are also dominated by helium, with abundance ratios ranging from 3 to ≥ 100.

The DAO have temperatures between 50000 and 65000 K, and He/H ratios ≤ 0.02, that is, the atmospheres are dominated by hydrogen.

A very interesting recent development is the work by Napiwotzki and Schönberner (1991) on 11 central stars of old, low surface brightness planetary nebulae. They find that nine of these are degenerate stars, the majority being hydrogen-dominated but with traces of He, and thus DAO. The helium abundances seem to cover the range from 10^{-2} to 0 continuously, very suggestive of ongoing gravitational settling in these stars (see below). An unsolved puzzle is, however, the age discrepancy: while the preliminary analysis gives temperatures around 60000 K for all DAO and DA, and thus a cooling age of 10^6 years, the kinematic ages of the nebulae are at least a factor of ten shorter!

2.3 The DB Stars

The majority of these stars show only spectral lines of HeI, although in recent years an increasing number with weak hydrogen lines has been detected and classified DBA. With the exception of this subclass, which may be as large as 20% of the DB (Shipman, Liebert, and Green 1987), and a few DBZ with traces of metals, not much more than some upper limits is known about abundances other than He in the DB. Attention has therefore focused on the temperature scale for these objects, which has posed some problems in the past, but is of considerable general interest for at least two reasons:

1. At the low temperature end for this class — about 11000 K — the limit is determined by the fading of HeI lines into invisibility due to decreasing excitation of the high levels from which the optical lines arise. There is no doubt that this class of

He-dominated atmospheres shows a continuous transition to the DC at lower temperatures, which are not a subject of this study. At the high end, however, there is a discontinuity first pointed out by Liebert et al. (1986): no He-rich objects are known between about 28000 K and 47000 K where the DOs begin. The precise location of the lower limit has been a matter of debate throughout the eighties.

2. A small subclass of hot DB shows non-radial oscillations similar to the ZZ Ceti (hydrogen-rich DA) at lower temperature. The location of the instability strip could provide important clues, e.g., on the efficiency of convection in these stars (Winget et al. 1983). Unfortunately, the results for the prototype GD358 differed by as much as 4000 K (Koester et al. 1985; Liebert et al. 1986).

This discrepancy has recently been resolved by Thejll, Vennes, and Shipman (1991), who trace the higher temperatures obtained by Liebert et al. (1986) back to some incorrect atmospheric models used in their analysis. They confirm the temperature of 24000 K for the variable GD358, and find an instability strip ranging from 21500 to 24000 K, with uncertainties of 1000 K on both sides. Comparing this with theoretical calculations for the instability strip, they deduce that convection is best described by the so-called ML2 parametrization, that is, it is less efficient than found by Fontaine, Tassoul, and Wesemael (1984). They also determine the upper limit of the DB range to be 28000 K, somewhat lower than assumed previously.

2.3.1 The DBA

The only members of this class until 1979 were LDS785A (Wickramasinghe and Whelan 1977) and G200-39 (Liebert et al. 1979). In a study of the DB stars in the Palomar-Green Survey Shipman, Liebert, and Green (1987) found 5 new objects with weak hydrogen lines in the range of T_{eff}= 14000 to 19000 K. They estimate that the total fraction that could be found with better observations among all DB could be as high as 20%, a number confirmed by Wegner and Nelan (1987). In both studies the derived hydrogen abundances range from 0.3 to 3 10^{-4}. Because the lower numbers are at the detection limit, it is possible that there is a continuous range of hydrogen abundances from zero to a few 10^{-4}.

2.3.2 Metals in DB

Until recently only two DB were known to show lines of metals: GD40 (Wickramasinghe et al. 1975) and CBS78 (Sion et al. 1986). This number has now increased considerably: Kenyon et al. (1988) found Ca in the DBA G200-39, and Sion, Aannestad, and Kenyon (1988) report the detection of weak Ca lines in two DBA and two DB stars, and possible detection in 3 DB. All these features are extremely weak and in some cases at the detection limit.

3 Atmospheric Parameters for the Hydrogen-rich Hot DA

Considering the whole range of temperatures where white dwarfs are observed, this is by far the dominant class. Contrary to all types listed above — which are all deficient in hydrogen relative to solar abundances — these objects seem to have almost pure hydrogen atmospheres. Helium abundances are very difficult to determine from optical spectra, but EINSTEIN and EXOSAT observations of the soft-X ray region seem to demand the presence of heavier trace elements in the hotter members. The most likely candidate is of course helium, but problems with this interpretation remain in some cases. A very small subclass shows stronger He lines (DAB) with the prototype being GD323, and a few more members reported recently.

The DA spectral type is not only the largest but also a very homogeneous class and therefore in most cases used to derive statistical properties for all white dwarfs as, for example, mean masses, mass distributions, and luminosity functions. We first concentrate on the temperature scale, because we have done some not yet published work in this area, and also because it is of more general interest for two reasons:

1. The absolute calibration of the IUE satellite for the "final archive" will be based on three hot DA (Finley and Koester 1991a). The calibration standards are theoretical models, predicting the UV flux from the effective temperature and gravity determined from optical spectra. It is obvious that these values should therefore be as accurate as possible.
2. In contrast to the sequence of helium-rich stars, which extends to well over 100000 K, the DA sequence terminates somewhere between 60000 and 70000 K. This has been puzzling for a long time and is still not understood fully. It is important to confirm that this deficiency of very hot objects is indeed real and not an artifact caused by errors in the temperature scale.

3.1 New Results on Temperatures and Surface Gravities for Hot DA from Balmer Line Profiles (Finley and Koester 1991b)

High quality spectra have been taken of some three dozen hot DA between about 20,000 K and 60,000 K using the CCD spectrometer at the Cassegrain focus of the 3m reflector at Lick Observatory. The spectra covered the range of 3500-5100 Å, with 2 Å bins and a resolution of 8 Å FWHM. Exposure times were such that a S/N of 50-100 or greater was achieved.

The analysis was performed by fitting the spectra with models calculated using the white dwarf model atmosphere and radiative transfer codes of Koester. The major improvements in these codes compared to descriptions in, e.g., Koester, Schulz, and Weidemann (1979, KSW) are the use of the Hummer-Mihalas occupation number formalism (Hummer and Mihalas 1988; Daeppen et al. 1988) and of new Stark broadening tables calculated with the the unified theory of Vidal, Cooper, and Smith (1973) by Schoening and Butler (1990). These tables include the first 8 lines of the Balmer series and are extended to much higher temperatures than the published tables.

We have compared our temperature determinations with previously published results of KSW, McMahan (1989), Holberg, Wesemael, and Basile (1986, HWB), and Finley,

Basri, and Bowyer (1990, FBB). In all cases, our temperatures are slightly lower (by about 5%) than previous values, a result we attribute partly to calibration effects and partly to the new physics being used in our models.

The new Finley and Koester IUE absolute flux calibration, which is based on Balmer line temperatures for three hot DA, results in IUE fluxes which are consistent with the Balmer line temperatures for the stars that were studied by FBB.

3.2 Masses and Mass Distribution of White Dwarfs

The historical papers on which much of our knowledge about white dwarf masses was based until recently were KSW, Weidemann and Koester (1983), and Shipman (1979). The picture that emerged from these studies was that the mean mass is 0.58 $M_{\odot}$— with fairly large uncertainties due mostly to calibration problems of photometry and spectrophotometry — and, with higher certainty, a very narrow distribution. About two thirds of all objects were found in an interval of ± 0.10 $M_{\odot}$ around the mean. These values were essentially confirmed independently by the modern generation of gravitational redshift measurements (Koester 1987a; Wegner 1989; Wegner and Reid 1991; Wegner, Reid, and McMahan 1991), although the Hyades white dwarfs seem to have a slightly larger average (Wegner, Reid, and McMahan 1989).

McMahan (1989) derives an (unweighted) mean of 0.523 $M_{\odot}$ from surface gravities for 53 stars, which agrees well with the corresponding KSW result (KSW preferred the parallax stars for their final mean value quoted).

A significant step forward is the recent work of the Tucson group (Bergeron, Saffer, and Liebert 1991, BSL). Using excellent observations of a large sample and improved model atmosphere techniques, they obtain extremely accurate (at least in a relative sense) surface gravities and masses for DA in the temperature range 15000 to 40000 K. Their masses were determined from the temperatures and surface gravities using the C-O evolutionary sequences calculated by Wood (1990). One of the important new results is the appearance of a secondary peak in the mass distribution below 0.35 $M_{\odot}$. Apart from this peak and a few objects with high masses, they confirm the extremely narrow distribution.

Liebert (1991a) has kindly provided us with their latest results in advance of publication, which are now based on 129 objects. The mean mass they obtain is 0.559 $M_{\odot}$. The existence of the secondary peak is confirmed, as is that of a few high mass objects: for GD50 they obtain 1.20 $M_{\odot}$ (Bergeron et al. 1991). We confirm this mass for GD50. Liebert also reports excellent agreement with gravitational redshift measurements: their average value for redshift objects is only 0.01 $M_{\odot}$ lower than the gravity result.

We have calculated the mass distribution for our sample of 35 hot stars using the same evolutionary sequences as BSL. Our mean mass including GD50 is 0.54 $M_{\odot}$; without, it is 0.52 $M_{\odot}$. The mean mass obtained by BSL without GD50 would be 0.551 $M_{\odot}$. The mean masses reflect differences in the mass distributions. Both samples have comparable distributions for masses less than 0.55 $M_{\odot}$, and both have approximately 10% with masses greater than 0.65 $M_{\odot}$. However, fully a third of the stars in the BSL sample have masses which lie between 0.55 and 0.65 $M_{\odot}$, while our sample includes only 2 of 35 stars in that range. This deficit may be due to observational selection effects.

The narrow distribution reported in the earlier works for cooler DA is also seen in this hotter sample; we do not find, however, any stars in our sample with masses less than 0.40 $M_{\odot}$.

Taken at face value, these recent results clearly argue for a somewhat lower mean mass than the value of around 0.58 $M_{\odot}$ that has been indicated by the earlier studies, and the Tucson results indicate that this can now be reconciled with masses derived from parallaxes and redshifts.

3.2.1 Masses of Helium-rich White Dwarfs

Because of the much smaller number of objects and the difficulties of obtaining accurate log g values from He spectra, the mean masses and the distribution are not well established. There is, however, no indication that they are different from those of the DA stars (Shipman 1979; Wickramasinghe 1983; Oke, Weidemann, and Koester 1984). This is confirmed by the recent determination of highly accurate masses from the pulsation periods for PG1159+035 (0.586±0.003, Winget et al. 1991) and PG1707+427 (0.69 ± 0.01, Fontaine et al. 1991). These results may also remind us that the generally smaller surface gravities found spectroscopically in very hot white dwarfs do not imply smaller masses, but a configuration that is not yet completely degenerate.

3.3 Helium in DA: Homogeneous Versus Stratified Atmospheres

With the exception of the DAO, which show generous and easily detectable amounts of helium, not much is known directly about helium abundances. Holberg et al. (1989) have recently studied optical and UV spectra of six hot DA. In 3 objects they find weak He lines, leading to abundance estimates for He/H of $\approx 3\ 10^{-3}$, which places them in the DAO class. The other 3 show no detectable helium with upper limits around 10^{-3}.

The evidence for traces of helium in many hot DA is rather indirect. It rests on the soft X-ray observations of the EINSTEIN and EXOSAT satellite, which definitely demand an absorber below the HeII threshold at 228 Å (Kahn et al. 1984; Petre, Shipman, and Canizares 1986; Jordan et al. 1987; Petre and Shipman 1987; Paerels and Heise 1989). The most likely candidate is helium; the abundances derived range from 10^{-5} to $5\ 10^{-3}$.

The interpretation of these results raises severe problems. In some cases the He abundances are in direct conflict with optical and UV data (e.g. G191-B2B, Holberg et al. 1989, Koester 1991a). In view of our current ideas about gravitational settling and radiative forces, the presence of helium in the atmosphere is also not understandable theoretically (Vennes et al. 1988). In a few cases (e.g. Feige 24) another alternative explanation for the EUV observations might be the assumption that the absorber is not helium but a mixture of heavier metals (Vennes et al. 1989), although this possibility can apparently be ruled out in other cases due to the absence of metal lines in high-resolution IUE spectra (Vennes, Thejll, and Shipman 1991, VTS).

Several authors have therefore advocated an alternative explanation (Koester 1989, Vennes, Fontaine, and Wesemael 1989b). In this picture the atmosphere is not homogeneous, but consists of a very thin hydrogen layer of the order of 10^{-15} to 10^{-13} $M_{\odot}$, with a smooth transition into the underlying helium. In the soft X-ray regime the opacity of hydrogen becomes very low, and the helium layers become "visible". This scenario can explain the EXOSAT observations as well as the homogeneous atmospheres; it is also supported by some of the ideas about the origin of white dwarf spectral types and the relatively thin hydrogen layers demanded by ZZ Ceti pulsation theories. On the other hand, it is in conflict with results from stellar evolution calculations (Iben and Tutukov 1984; Koester and Schönberner 1986), which predict a remaining hydrogen layer of the order of 10^{-4} $M_{\odot}$. The question remains unsettled for the moment; in the near future the puzzle might be resolved by the "Rosetta stone" G191-B2B, where homogeneous H/He atmospheres are ruled out, because the HeII 4686 line would be much stronger than observed.

3.3.1 Multiwavelength Analysis of the Hot DA White Dwarf G191-B2B

A comprehensive analysis of the available data for G191-B2B was recently carried out, motivated by our intent to use this star as an absolute flux standard in the ultraviolet. Details of this analysis will be reported elsewhere (Finley and Koester 1991c); here we will briefly describe the results derived from optical spectra and EXOSAT photometry.

Despite all the attention G191-B2B has received, its properties remain somewhat elusive. It has long been known as one of the hotter white dwarfs ($T_{eff} \approx$ 60000 K) and has been classified as a DA on the basis of the absence of any helium features in its spectrum.

Trace metal abundances in G191-B2B (see below) have been a subject of controversy, although the detection of an emission component in Hα with the same velocity as the metal lines makes a photospheric origin likely (Reid and Wegner 1988).

New clues to the composition arose when G191-B2B was observed with EXOSAT. G191-B2B was found to have a significant flux deficit below 300 Å compared to that expected if the atmosphere were composed of pure hydrogen. The EXOSAT data have been analyzed by a number of authors using either uniform H/He or stratified models and the results have been discussed extensively (Jordan et al. 1987; Vennes et al. 1988; Paerels and Heise 1989; Koester 1989; Finley, Basri, and Bowyer 1990; Green, Jelinsky, and Bowyer 1990; MacDonald and Vennes 1991). The EXOSAT observations were seen to be consistent with either uniform or stratified models.

Our reanalysis was primarily motivated by the new observation of G191-B2B with the Hopkins Ultraviolet Telescope, which measured the EUV flux between 420 and 650 Å (Kimble et al. 1991). These authors calculated the neutral interstellar columns by measuring the He I 504 Å edge and comparing the observed flux longward of 504 Å with model predictions. We therefore analyzed the optical and EUV data with the new constraint of measured interstellar column values. We also allowed a more conservative error for the Al/P bandpass measurements to account for the problems due to the very soft spectrum of G191-B2B (Paerels and Heise 1989).

We found that the data can be fit by stratified H/He models, with a temperature of 53-54,000 K and a hydrogen layer mass of between 6 10^{-15} and 8 10^{-15} $M_{\odot}$; no other trace elements are required. However, this is not a firm conclusion and may change with the inclusion of data which already exist.

The Hopkins Ultraviolet Telescope spectrum of G191-B2B includes the ultraviolet region longward of 912 Å, and has very good S/N in the Lyman line region. Visual inspection of the spectrum indicates that the HeII 1085 Å line is completely absent. Our initial estimate is that a fit to the Lyman region including the He lines may result in a lower limit of about 2 10^{-14} $M_\odot$, which would then be incompatible with the EXOSAT result.

Additional EUV photometric data have been taken with the ROSAT Wide Field Camera (WFC). When these new data are analyzed in conjunction with the other data discussed here, it may require rejection of the stratified hypothesis. In that case, it would have to be assumed that the EUV opacity is provided by trace metals, as is evidently the case with Feige 24 (Vennes et al 1989a; Vennes, Thejll, and Shipman 1991). The presence of photospheric C, N, and Si lines in G191-B2B with strengths equivalent to those observed in Feige 24 certainly makes the metal option an attractive hypothesis.

3.4 Heavy Metals in Hot DA

Weak lines of C, N, Si, and Fe have been detected in the IUE spectra of a number of hot DA (Bruhweiler and Kondo 1981, 1983; Dupree and Raymond 1982; Holberg et al. 1985; Sion, Liebert, and Starrfield 1985; Sion 1991; VTS; Bruhweiler and Feibelman 1991). The interpretation of these lines is very complex, because it is not clear whether they really arise in the photosphere, in a circumstellar region, or in the interstellar matter along the line of sight. A decision is possible in principle using the redshift of the lines and comparing them with Balmer line redshifts. The results of such studies are mixed (see also VTS): in G191-B2B the lines seem to be photospheric (Reid and Wegner 1988; Bruhweiler and Feibelman 1991), in Feige 24 some lines are and others not (Vennes, Thejll, and Shipman 1991, Vennes et al. 1991). Under the *assumption* that the lines are photospheric, VTS derive upper limits for C/H (10^{-7} to 10^{-9}) and Si/H (10^{-6} to 10^{-9}) in 9 DA.

4 How can we understand the observed abundance patterns?

If we assume that some white dwarfs reach their final stages without any hydrogen left in the surface layers, the most basic pattern — the division into almost pure hydrogen and almost pure helium atmospheres – can be understood as a result of gravitational settling (Schatzman 1949, 1958). All heavy elements leave the atmosphere at the bottom, and only the lightest remains floating on the top. Accepting this as the basic process at work, the following problems need an explanation:

1. Why do some objects reach the final stages apparently without hydrogen? Or is that not true, do all have some H left?
2. The absence of hydrogen-rich atmospheres at high temperatures
3. The existence of mixed H/He atmospheres around 60000 K
4. The DB gap between 28000 and 47000 K
5. The DAB, stars with strong H and clearly visible He lines, around 28000 K
6. The DBA, DB stars with small traces of hydrogen
7. Metals in PG1159, DO, DA, and DB stars

The list of problems is long enough; it would get much longer if we included cooler white dwarfs, but fortunately that is outside the scope of this review. Tentative explanations and speculative scenarios have been discussed over the last years by many authors. A certainly non-exhaustive list includes Sion (1986), Vauclair and Liebert (1987), Shipman (1987), Koester (1987b), Fontaine and Wesemael (1987), Shipman (1989), Vauclair (1989), MacDonald and Vennes (1991), Sion (1991), Liebert (1991b), Fontaine and Wesemael (1991).

It would be impossible to discuss all these papers in detail. In the following we will rather try to present the main ideas; almost all come from the papers mentioned above, although they will not be attributed in detail. The mixture and emphasis is, however, biased by our own prejudices; none of the authors of the previous reviews should be blamed for this particular presentation.

4.1 H-rich Versus He-rich and the H-layer Thickness

Stellar evolution theory of AGB and post-AGB evolution gives some clues to the possible origin of the dichotomy observed in white dwarfs (Iben 1984; Renzini 1989). It is expected that about 80% of all AGB stars go through the planetary nebula stage with at least 10^{-4} $M_{\odot}$ hydrogen left at the surface; they derive their luminosity from a hydrogen-burning shell. The remaining 20% leave the AGB with very little or no hydrogen and are He-burning nuclei. Hydrogen-rich and hydrogen-poor objects are indeed observed among PNN; this suggests the idea to associate the two classes directly with the DA and non-DA white dwarfs. However, it is not quite that easy.

The question is not yet settled for the extremely hot PG1159 stars, but some DO certainly show hydrogen, although helium is dominant. One could assume that the hydrogen left is not exactly zero, that, for example, a small amount might be hidden in the outer envelope. The range from 150000 K to 50000 K would then be the era of gravitational settling, at the end of which all stars show a pure H atmosphere (remember the DB gap: no helium-dominated atmosphere in any white dwarf between 47000 and 28000 K). The DAO would then represent the hot end of the DA sequence, with He still just observable.

The theory of gravitational settling, however, predicts diffusion time scales of the order of a few years, whereas white dwarfs at 60000 K have cooling ages of 10^6 years. Why does diffusion take so long? The only plausible explanation is that there must be an effect counteracting gravitational settling, which could at least theoretically be mass loss in a stellar wind.

On the other hand, if there is a hydrogen-rich sequence, where are the H-rich white dwarfs between 70000 and 120000 K? Even if diffusion is not effective at these temperatures, we should at least find objects with solar H/He ratios! There may be a few H-rich PNN around 100000 K (e.g. NGC7293), but their numbers appear to be smaller than those of the DO and PG1159 objects.

At about 28000 K the He-rich sequence suddenly reappears in the form of the DB. The only plausible explanation that has been suggested is that the He convection zone, which starts to expand at this temperature, reaches into the photosphere and dilutes the hydrogen layer (Liebert et al. 1986; MacDonald 1989). This idea is strongly supported by the existence of a few DBA at just this temperature (Koester 1991b, Holberg, Kidder, and Wesemael 1990), which might be objects in the transition from a DA to a DB. However, this mixing process will only work if the total hydrogen mass present on top of the helium-rich envelope is less than about 10^{-15} $M_\odot$! If we accept the possibility that about 20% of all DA have such a thin hydrogen layer, we get a completely new perspective. At higher temperatures a layer with $M_H \leq 10^{-15}$ $M_\odot$ would leave HeII lines visible and could be an alternative explanation at least for the DAO and possibly for the DO.

A thin hydrogen layer could also explain the EXOSAT soft X-ray observations of many DA, and would avoid problems with the visibility of HeII lines in the optical spectra (see the discussion for G191-B2B above).

We can borrow more arguments from the cooler white dwarfs: studies of the pulsation properties of ZZ Ceti stars give an upper limit to the total hydrogen mass of 10^{-8} $M_\odot$ for *all* DA, although this result is not universally accepted. The number ratios of H-rich versus He-rich white dwarfs below 10000 K strongly suggest an upper limit of the order of 10^{-10} $M_\odot$. There is growing evidence that most if not all DA white dwarfs have very thin hydrogen layers.

Stellar evolution theory, however, insists that at the end of hydrogen-shell burning, which occurs in the late PN phase, the amount of hydrogen left must be of the order of 10^{-4}. To remove this remaining hydrogen within about 10^6 years requires a very strong mass loss of 10^{-10} $M_\odot$ per year.

Mass loss rates of this magnitude are observed in the earlier, much hotter and more luminous phases of PN evolution. The long time apparently needed to establish the He/H diffusion equilibrium in the white dwarfs could be explained by a mass loss $\geq$ 10^{-13} $M_\odot$/yr, but the direct evidence for stellar winds is very weak. The shortward shifted lines of highly ionized metals in a handful of hot DA points to the presence of static or slowly expanding halos, which may indicate a stellar wind (present or past). However, even if this interpretation is correct, the mass loss rates would very likely be smaller than 10^{-13} $M_\odot$/year.

For the helium-rich stars (including PG1159+035 itself) searches for mass loss have had only negative results (Fritz, Leckenby, and Sion 1990; Sion 1991).

4.1.1 The DBA and DAB

These are objects in the lower temperature range below 30000 K, which show traces of H in a He-dominated atmosphere or vice versa. As mentioned above, the DB gap and the location of the DAB around 28000 K strongly suggest a connection to the mixing process that transforms a DA with a thin H layer into DB. It is conceivable that the DBA are one possible result of such a mixing for objects with a slightly higher hydrogen mass. A DB star around T_{eff}= 20000 K has about 10^{-10} $M_\odot$ in its convective envelope. If this is completely mixed with the hydrogen, the resulting H abundance would be 10^{-5}, in agreement with observed values. In this scenario the surviving majority of DA would have "thicker" H masses in the range 10^{-15} to 10^{-10} $M_\odot$. However, the mass in the He convection zone increases steeply with decreasing temperature, reaching 10^{-6} at 15000 K. This should result in a correlation of H abundance with temperature, which is not observed.

One would also expect to find a range of H abundances extending below the canonical 10^{-5}, if the observations are good enough to detect it. In a recent study of 4 DB, Shipman et al. (1991) found only upper limits, whereas Weidemann and Koester (1991) found indeed a very low abundance in GD408. Shipman et al. (1991) use their result to reject the mixing process and to favor the alternative: accretion. A much stronger argument for accretion is the detection of CaII — which definitely cannot be the result of mixing — in several DBA and DB (Kenyon et al. 1988; Sion, Aannestad, and Kenyon 1988).

4.1.2 Metals in hot white dwarfs

The most likely process to compete with gravitational settling for some elements has been known for a long time: selective radiative acceleration or radiative levitation (Vauclair, Vauclair, and Greenstein 1979; Vauclair 1989). These calculations have been refined and extended to include more elements during the last ten years (Morvan, Vauclair, and Vauclair 1986; Chayer, Fontaine, and Wesemael 1989, 1991). Qualitatively, the predictions of these calculations are confirmed by the observations in some cases; in others, discrepancies remain. The comparison is hampered by the lack of consistent determinations of observational abundances, which take into account the inhomogeneous distribution of the heavy elements ("metal clouds") predicted by theory.

A clear disagreement between diffusion theory and observation is noted by WHH for the PG1159 stars, where the abundances predicted by Vauclair (1989) do not agree with the result of their analysis. These authors interpret the abundances therefore as the outcome of nuclear burning processes in previous stages, which are now exposed at the surface. WH present an extensive discussion on the connection of the PG1159 with subtypes of PN nuclei and the cooler DO. For the C and O rich PG1159 objects, they believe that an evolutionary connection between central stars of types WC - O(C) - PG1159 - DO is possible. The basic assumption is that the present atmospheres expose the results of the H-burning in the former He buffer layer of the red giant. The high N content in PG1144+005 would be most likely the result of a late He shell flash, which caused complete burning of the remaining hydrogen in a convection zone, without destroying the nitrogen (Renzini 1982; Iben et al. 1983).

4.2 Final Comment on Abundances and Spectral Types

We have seen that the summary of present ideas attempted above invokes gravitational settling, radiative levitation, convective mixing, accretion and stellar winds. There are not many more possibilities that one could imagine, and still we are far from a complete explanation. Yet, as complicated as it may appear, there is in our opinion a unifying thread: the idea that the hydrogen layers even in the DA are probably very thin. Many pieces of the puzzle seem to fall into their place, although many details remain to be worked out. The major open problem, the missing link, is: how do the white dwarfs manage to get rid of their 10^{-4} $M_\odot$ of hydrogen without showing clear signs of stellar winds?

In concluding this chapter, we want to emphasize, however, that not all of our colleagues agree with this picture, and that especially the thin vs. thick layer discussion is currently a very hot topic.

5 Prospects for the Future: EUV Observations of White Dwarfs

A new era in hot white dwarf studies is currently opening up with the EUV all-sky surveys being performed by the ROSAT Wide Field Camera (WFC) and the EUV Explorer (EUVE), and the EUV spectroscopy which will be performed with EUVE.

The WFC survey was completed early this year, and analysis of the data is currently under way. Preliminary results are reported elsewhere at this workshop (Barstow et al. 1991); the last numbers we heard indicate that about 120 white dwarfs have been detected so far, about half of them known previously, the rest new discoveries. The projected total number of hot white dwarfs detected is about 250.

The EUVE sky survey is expected to take place in the first half of 1992, and should also detect a few hundred white dwarfs. The two surveys are complementary in that the WFC survey was done in two bandpasses: Lexan and Lexan/beryllium, which do not extend longward of about 200 Å. The EUVE survey will include four bandpasses: Lexan, aluminum/carbon, which covers 170-300 Å, as well as two other bandpasses between 400 and 800 Å. The latter will provide unambiguous EUV detections of the several EUV sources which are expected to have extremely low hydrogen columns. Another enhancement provided by EUVE is a deep survey in Lexan and aluminum/carbon, which will cover a 2° wide swath over half the ecliptic with a sensitivity an order of magnitude greater than that of the EUVE or WFC all-sky surveys.

The EUVE spectroscopic observations, with a planned minimum duration of three years, will begin at the completion of the sky survey and continue until the hardware fails or funding is discontinued. Expectations are that about 200 spectroscopic observations will be performed per year. The spectrometer simultaneously covers the range of 70 to 760 Å in three independent channels, with a resolution of 300.

Some major benefits are likely to be realized from the new data which will be forthcoming for hot white dwarfs. The number of hot (> 30,000 K) white dwarfs will be expanded by about a factor of 4, giving us many new stars to study, including WD in binaries. The photometric data will provide an indication of the distribution of trace elements in hot WD. The spectroscopic observations will provide an unprecedented opportunity for measuring accurate temperatures, gravities, and trace element compositions for white dwarfs.

We should also not forget at the end to mention that the Space Telescope, in spite of all its problems, is working on a white dwarf project that has almost everybody in the world with an interest in white dwarfs among the PI's. Very exciting times for the white dwarf community are ahead!

Acknowledgements: This work was partly supported by grants from NASA (D.K. and D.F.) and from the National Science Foundation (D.K).

References

Barstow, M.A., Fleming, T.A., Diamond, C.J., and Finley, D.S. (1991): these Proceedings

Bergeron, P., Saffer, R.A., Liebert, J. (1991): NATO ASI Series C, Vol. 336, White Dwarfs, eds. G.Vauclair and E.Sion (Kluwer: Dordrecht), p. 75 (BSL)

Bergeron, P., Kidder, K.M., Holberg, J.B., Liebert, J., Wesemael, F., Saffer, R.A. (1991): Astrophys.J. **373**, 267

Bruhweiler, F.C., Feibelman, W.A. (1991): preprint

Bruhweiler, F.C., Kondo,Y. (1981): Astrophys.J. **248**, L123

Bruhweiler, F.C., Kondo,Y. (1983): Astrophys.J. **269**, 657

Chayer, P., Fontaine, G., Wesemael, F. (1989): IAU Coll. 114, White Dwarfs, ed. G.Wegner (Springer: Berlin), p. 253

Chayer, P., Fontaine, G., Wesemael, F. (1991): NATO ASI Series C, Vol. 336, White Dwarfs, eds. G.Vauclair and E.Sion (Kluwer: Dordrecht), p. 249

Daeppen, W., Mihalas, D., Hummer, D.G., Mihalas, W. (1988): Astrophys.J. **332**, 261

Dupree, A.K., Raymond, J.C. (1982): Astrophys.J. **263**, L63

Finley, D.S., Basri, G., Bowyer, S. (1990): Astrophys.J. **359**, 483

Finley, D.S., Koester, D. (1991a): in preparation

Finley, D.S., Koester, D. (1991b): in preparation

Finley, D.S., Koester, D. (1991c): in preparation

Finley, D.S. (1991): these proceedings

Fontaine, G., Tassoul, M., Wesemael, F. (1984): Proc. 25th Liege Coll, Theoretical Problems in Stellar Stability and Oscillations (Universite de Liege: Liege), p. 328

Fontaine, G., Wesemael, F. (1987): IAU Coll. 95, The Second Conference on Faint Blue Stars, eds. A.G.D.Philip, D.S.Hayes, and J.Liebert (L.Davis Press: Schenectady), p.319

Fontaine, G., Wesemael, F. (1991): IAU Symp. 145, Evolution of Stars: The Photospheric Abundance Connection, eds. G.Michaud and A.Tutukov (Kluwer: Dordrecht), p.421

Fontaine, G., Bergeron, P., Vauclair, G.,Brassard, P., Wesemael, F., Kawaler, S.D., Grauer, A.D., Winget, D.E. (1991): Astrophys.J. **378**, L49

Fritz, M.L., Leckenby, H.J., Sion, E.M. (1990): Astron.J. **99**, 908

Green, J., Jelinsky, P., Bowyer, S. (1990): Astrophys.J. **359**, 499

Holberg, J.B., Kidder, K., Liebert, J., Wesemael, F. (1989): IAU Coll. 114, White Dwarfs, ed. G.Wegner (Springer: Berlin), p. 188

Holberg, J.B., Kidder, K.M., Wesemael, F. (1990): Astrophys.J. **365**, L77

Holberg J.B., Wesemael, F., Basile, J. (1986): Astrophys.J. **306**, 629

Holberg, J.B., Wesemael, F., Wegner, G., Bruhweiler, F.C. (1985): Astrophys.J. **293**, 294

Hummer, D.G., Mihalas, D. (1988): Astrophys.J. **331**, 794

Iben, I.Jr. (1984): Astrophys.J. **277**, 233

Iben, I.Jr., Kaler, J.B., Truran,J .W.,Renzini, A. (1983): Astrophys.J. **264**, 605

Iben, I.Jr., Tutukov, A.V. (1984): Astrophys.J. **282**, 615

Jordan, S., Koester, D., Wulf-Mathies, C., Brunner, H. (1987): Astron. Astrophys. **185**, 253

Kahn, S., Wesemael, F., Liebert, J., Raymond, J., Steiner, J., Shipman, H.L. (1984): Astrophys.J. **278**, 255

Kenyon, S.J., Shipman, H.L., Sion, E.M., Aannestadt, P.A. (1988): Astrophys.J. **328**, L65

Kimble, R.A. et al. (1991): BAAS **23**, 883

Koester, D. (1987a): Astrophys.J. **322**, 852

Koester, D. (1987b): IAU Coll. 95, The Second Conference on Faint Blue Stars, eds. A.G.D. Philip, D.S.Hayes and J.Liebert (L.Davis Press: Schenectady), p.329

Koester, D. (1989): Astrophys.J. **342**, 999

Koester, D. (1991a): IAU Symp. 145, Evolution of Stars: The Photospheric Abundance Connection, eds. G.Michaud and A.Tutukov (Kluwer: Dordrecht), p.435
Koester, D. (1991b): NATO ASI Series C, Vol. 336, White Dwarfs, eds. G.Vauclair and E.Sion (Kluwer: Dordrecht), p. 343
Koester, D., Liebert, J., Hege, E.K. (1979): Astron. Astrophys. **71**, 163
Koester,D., Schönberner, D. (1986): Astron. Astrophys. **154**, 125
Koester, D., Schulz, H., Weidemann, V. (1979): Astron. Astrophys. **76**, 262 (KSW)
Koester, D., Vauclair, G., Dolez, N., Oke, J.B., Greenstein, J.L., Weidemann, V. (1985): Astron. Astrophys. **149**, 423
Liebert, J. (1991a): private communication
Liebert, J. (1991b): IAU Symp. 145, Evolution of Stars: The Photospheric Abundance Connection, eds. G.Michaud and A.Tutukov (Kluwer: Dordrecht), p.411
Liebert, J., Green, R.F., Wesemael, F., Margon, B. (1981): Astron.J. **84**, 1612 (LGWM)
Liebert, J., Gresham, M., Hege, E.K., Strittmatter, P.A. (1979): Astron.J. **84**, 1612
Liebert, J., Wesemael, F., Hansen, C.J., Fontaine, G., Shipman, H.L., Sion, E.M., Winget, D.E., Green, R.F. (1986): Astrophys.J. **309**, 241
MacDonald, J. (1989): IAU Coll. 114, White Dwarfs, ed. G.Wegner (Springer: Berlin), p. 172
MacDonald, J., Vennes, S. (1991): Astrophys.J. **371**, 719
McGraw, J.T., Starrfield, S., Liebert, J., Green, R.F. (1979): Proc. IAU Coll. 53: White Dwarfs and Variable Degenerate Stars, eds. H.M. Van Horn and V. Weidemann (University of Rochester: Rochester), p.337
McMahan, R.K. (1989): Astrophys.J. **336**, 409
Mendez, R.H., Kudritzki, R.P., Gruschinske, J., Simon, K.P. (1981): Astron. Astrophys. **101**, 323
Morvan, E., Vauclair, G., Vauclair, S. (1986): Astron. Astrophys. **163**, 145
Napiwotzki, R., Schönberner, D. (1991): NATO ASI Series C, Vol. 336, White Dwarfs, eds. G.Vauclair and E.Sion (Kluwer: Dordrecht), p. 39
Oke, J.B., Weidemann, V., Koester, D. (1984): Astrophys.J. **281**, 276
Paerels, F.B.S., Heise, J. (1989): Astrophys.J. **339**, 1000
Petre, R., Shipman, H.L. (1987): BAAS **19**, 1041
Petre, R., Shipman, H.L., Canizares, C.R. (1986): Astrophys.J. **304**, 356
Reid, I.N., Wegner, G. (1988): Astrophys.J. **335**, 953
Renzini, A. (1982): IAU Symp. 99, Wolf-Rayet Stars: Observations, Physics, Evolution, eds. C.W.H. de Loore and A.J.Willis (Reidel: Dordrecht), p. 413
Renzini, A. (1989): IAU Symp. 131, Planetary Nebulae, ed. S.Torres-Peimbert (Reidel: Dordrecht), p. 391
Schatzman, E. (1949): Publ.Koebenhavns Obs. No. 149
Schatzman, E. (1958): White Dwarfs, (North-Holland Publishing Company: Amsterdam)
Schoening, T., Butler, K. (1990): private communication to K.Werner
Shipman, H.L. (1979): Astrophys.J. **228**, 240
Shipman, H.L. (1987): IAU Coll. 95, The Second Conference on Faint Blue Stars, eds. A.G.D.Philip, D.S.Hayes and J.Liebert (L.Davis Press: Schenectady), p. 273
Shipman, H.L. (1989): IAU Coll. 114, White Dwarfs, ed. G.Wegner (Springer: Berlin), p. 220
Shipman, H.L., Liebert, J., Green, R.F. (1987): Astrophys.J. **315**, 239
Shipman, H.L., Thejll, P., Bhatia, S., Liebert, J. (1991): NATO ASI Series C, Vol. 336, White Dwarfs, eds. G.Vauclair and E.Sion (Kluwer: Dordrecht), p. 229
Sion, E.M. (1986): Publ. Astron. Soc. Pac. **98**, 821
Sion, E.M. (1991): NATO Advanced Research Workshop, Mass Loss and Angular Momentum Loss in Hot Stars, eds. R.Stalio and L.A.Willson (Kluwer: Dordrecht) in press
Sion, E.M., Aanestad, P.A., Kenyon, S.J. (1988): Astrophys.J. **330**, L55

Sion, E.M., Greenstein, J.L., Landstreet, J.D., Liebert, J., Shipman, H.L., Wegner, G. (1983): Astrophys.J. **269**, 253
Sion, E.M., Guinan, E.F., Wesemael, F. (1982): Astrophys.J. **255**, 232 (SGW)
Sion, E.M., Liebert, J., Starrfield, S.G. (1985): Astrophys.J. **292**, 471
Sion, E.M., Shipman, H.L., Wagner, R.M., Liebert, J., Starrfield, S.G. (1986): Astrophys.J. **308**, L67
Thejll, P., Vennes, S., Shipman, H.L. (1991): NATO ASI Series C, Vol. 336, White Dwarfs, eds. G.Vauclair and E.Sion (Kluwer: Dordrecht), p. 257
Vauclair, G. (1989): IAU Coll. 114, White Dwarfs, ed. G.Wegner (Springer: Berlin), p. 176
Vauclair, G., Liebert, J. (1987): in Exploring the Universe with the IUE Satellite, ed. Y.Kondo (Reidel : Dordrecht), p. 355
Vauclair, G., Vauclair, S., Greenstein, J.L. (1979): Astron. Astrophys. **80**, 79
Vennes, S., Chayer, P., Fontaine, G., Wesemael, F. (1989): Astrophys.J. **336**, L25
Vennes, S., Fontaine, G., Wesemael, F. (1989a): IAU Coll. 114, White Dwarfs, ed. G.Wegner (Springer: Berlin), p. 368
Vennes, S., Fontaine, G., Wesemael, F. (1989b): IAU Coll. 114, White Dwarfs, ed. G.Wegner (Springer: Berlin), p. 363
Vennes, S., Pelletier, C., Fontaine, G., Wesemael, F. (1988): Astrophys.J. **331**, 876
Vennes, S., Thejll, P., Shipman, H.L. (1991): NATO ASI Series C, Vol. 336, White Dwarfs, eds. G.Vauclair and E.Sion (Kluwer: Dordrecht), p. 235 (VTS)
Vennes, S., Thorstensen, J.R., Thejll, P., Shipman, H.L. (1991): Astrophys.J. **372**, L37
Vidal, C.R., Cooper,J., Smith, E.W. (1973): Astrophys.J. Suppl. **25**, 37
Wegner, G. (1989): IAU Coll. 114, White Dwarfs, ed. G.Wegner (Springer: Berlin), p. 401
Wegner, G., Nelan, E.P. (1987): Astrophys.J. **319**, 916
Wegner, G., Reid, I.N. (1991): Astrophys.J. **375**, 674
Wegner, G., Reid, I.N., McMahan, R.K. (1991): Astrophys.J. **376**, 186
Weidemann, V., Koester, D. (1983): Astron. Astrophys. **121**, 77
Weidemann, V., Koester, D. (1991): Astron. Astrophys. **249**, 389
Werner, K., Heber, U., Hunger, K. (1991): Astron. Astrophys. **244**, 437 (WHH)
Werner, K., Heber, U. (1991): Astron. Astrophys. **247**, 476 (WH)
Wesemael, F., Green, R.F., Liebert, J. (1985): Astrophys.J. Suppl. **58**, 379 (WGL)
Wickramasinghe, D.T., Hintzen, P., Strittmatter, P.A, Burbidge, E.M. (1975): Astrophys.J. **202**, 191
Wickramasinghe, D.T., Whelan, J.A.J. (1977): M.N.R.A.S. **178**, 119
Winget, D.E. et al. (1991): Astrophys.J. **378**, 326
Winget, D.E., Van Horn, H.M., Tassoul, M., Hansen, C.J., Fontaine, G. (1983): ApJ **268**, L33
Wood, M. (1990): PhD Thesis, Univ. of Texas, Austin

ROSAT Observations of White Dwarfs

M.A.Barstow [1], T.A.Fleming [2], C.J.Diamond [3] and D.S.Finley [4]

[1]Department of Physics and Astronomy, University of Leicester, UK.
[2]Max Planck Institüt Für Extraterrestriche Physik, Garching, FRG.
[3]School of Physics and Space Research, Birmingham University, UK.
[4]Centre For EUV Astronomy, University of California, Berkeley, USA.

Abstract: We present here the first results from the ROSAT all-sky EUV and X-ray survey of white dwarfs. Detailed study of a sample of 21 DA stars shows that below ≈40,000K their atmospheres are well described by homogeneous models comprising H + trace He. However, above ≈40,000K additional sources of opacity are required to explain the observed EUV and X-ray fluxes. Hence, we can clearly see compositional/structural evolution in these stars as they cool.

1 Introduction

It has been demonstrated, from observations made with the EXOSAT and Einstein satellites, that the emergent EUV and X-ray fluxes from hot DA white dwarfs are very sensitive to the presence of trace elements (notably He and possibly heavier elements) in their atmospheres (Paerels and Heise, 1989). If the temperature of a particular star is known, the EUV/soft X-ray data can be used to estimate the abundance of He in the photosphere, making the assumption that He is homogeneously mixed in the envelope. From a sample of around 20 DAs there is weak evidence for a trend of increasing He abundance with increasing T_{eff}. An alternative interpretation of the data can be made using a stratified model for the photosphere, with a thin layer of H overlying a predominantly He atmosphere. In such an analysis, the free parameter is the mass of the H layer. No single valued relationship between M_H and T_{eff} is found but there is a general tendency for thinner layers in the hotter stars (Koester, 1989). Both of these authors' work make the implicit assumption that He is the only significant opacity source in the EUV/X-ray. However, this is not necessarily true. An EUV spectrum of the DA white dwarf Feige 24 cannot be explained either by homogenous H+He or stratified models (Paerels et al, 1986; Vennes et al, 1989). However, Vennes et al (1989) show that a photospheric model with trace metals can fit what is observed.

So far, there is insufficient evidence to decide between which physical model, or combination of models, is the correct representation of DA white dwarf photospheres. The ROSAT all-sky EUV and X-ray surveys presents us with an opportunity of resolving this issue. Potentially, there is a much larger sample of objects to study improving the statistical basis of any composition/T_{eff} correlation. Secondly, the photometric bands available yield improved spectral resolution over earlier instruments. We present here a summary of the first results of the ROSAT survey of DA white dwarfs, discussing the outcome of an analysis of the data using homogeneous H+He model atmospheres.

2 The ROSAT All-sky Survey

The instruments flown aboard the ROSAT observatory have been well described elsewhere but those details that are relevant to this discussion are briefly reiterated here. ROSAT has two coaligned telescopes. The X-ray telescope plus position sensitive proportional counter (PSPC; Pfefferman et al, 1986) covers the energy range $\approx 0.1-2.4$keV (5.2-100Å). Although the PSPC has modest energy resolution, which we will exploit in the future, for this preliminary study we use just the integrated count rate in the band between $\approx 0.1-0.4$kev (25-100Å). The second telescope, the Wide Field Camera (WFC; Wells et al, 1990), spans the extreme ultraviolet (EUV) range. For survey observations two broad band filters were employed - S1 covering the range 90-200eV (60-140Å) and S2 the band 60-110eV (112-200Å). Two additional filters are available for the post survey part of the mission, extending the coverage to lower energies - P1 spans 56-83eV (150-220Å) and P2 17-24eV (500-730Å).

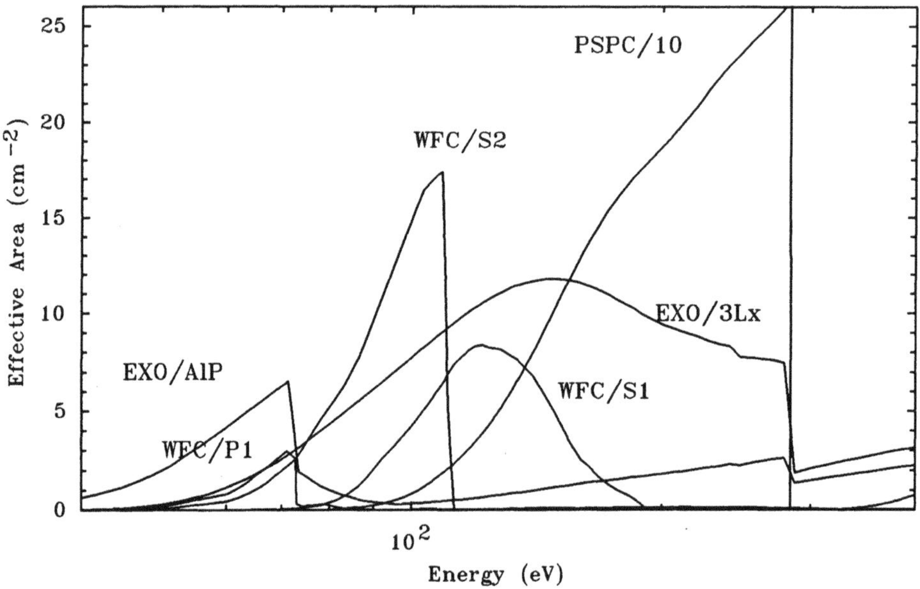

Fig. 1. Effective area curves for the EXOSAT thin lexan (3Lx) and aluminium/parylene (AlP) filters, the ROSAT PSPC and the WFC S1, S2 and P1 filters.

The ROSAT all-sky survey was conducted from July 1990 to January 1991. The majority of the data presented here derive from this phase of the mission but a few observations were also obtained during the pre-survey calibration phase. In effect we have obtained three photometric EUV/X-ray 'colours', analogous to UBV optical bands, for each of the white dwarfs detected by the instruments. Those stars also observed for calibration purposes have an additional 'colour' measurement with the P1 filter. Figure 1 shows the effective areas of each of these spectral bands. The responses of the prime filter bands of the earlier EXOSAT telescopes are shown for comparison. It can be seen that the ROSAT intruments yield $\approx$ a factor three better spectral resolution than available with EXOSAT.

3 Numbers of white dwarfs detected in the survey

An estimate of the total number of known white dwarfs exceeding 25,000K (ROSAT is not sensitive to stars below this temperature) suggests that between 100 and 150 known stars should have been detected. In addition, it has been estimated (Barstow, 1988) that approximately twice as many DA white dwarfs should be detected by the PSPC compared to the WFC. The WFC detected 55 known DA white dwarfs during the sky survey. By comparison 46 DAs were detected by the PSPC. The fact that we only see less than 1/2 the objects that we might expect is not necessarily a surprise. The EUV and soft X-ray fluxes incident at the Earth are strongly influenced by interstellar absorption. Since interstellar material is rather unevenly distributed there may be many lines of sight where this absorption prevents us being able to detect white dwarfs. Indeed, study of the distribution of white dwarf detections may well yield new information about the distribution of interstellar material.

More significant is the ratio of PSPC to WFC detections of DA stars, running at $<\approx 1:1$ rather than the predicted 2:1. This result cannot be explained by the effects of interstellar absorption, since this would tend to *decrease* the number of WFC detections with respect to those in the PSPC. Predictions of the ratio of white dwarfs detected make simplistic assumptions regarding the expected composition of their atmospheres. Our calculations (Barstow, 1988) assumed a canonical He:H content of 10^{-4} in a homogeneous mixture. The apparent shortfall of PSPC detections is strong circumstantial evidence for the presence of additional opacity in DA white dwarf atmospheres.

4 The photospheric composition of DA white dwarfs

Studies of DA white dwarf atmospheres with EUV/X-ray photometric observations require three main independent parameters to be determined - T_{eff}, N_H and He abundance (or H layer mass for a stratified model). The emergent fluxes are also dependent on log g, but the effects are small over the ranges covered by DA white dwarfs and it can be conveniently fixed at a canonical value of 8.0 in the absence of any other information. For this a minimum of four independent observations are required. However, often only two or three are available. The value of the EUV/X-ray data can be maximised if one parameter is already known. For example, Paerels and Heise (1989) fix the value of T_{eff} in their modelling, from optical or UV determinations, which allows the EUV/X-ray data to constrain the He abundance.

In this work we have selected 21 DAs for which we have accurate determinations of T_{eff} and log g from the fitting of synthetic Balmer line profiles to optical spectra. In addition to the high quality of the data these results, based on models supplied by D. Koester, have the advantage of uniformity compared to miscellaneous published sources. We have folded homogeneous NLTE H+He models through the instrument responses and compared the predicted count rates with the observed values. For a given model composition it is possible to determine values of T_{eff} and N_H that correspond to the observed count rates in each filter/instrument combination (eg. see Barstow and Holberg, 1990). If this value of T_{eff} agrees with that determined optically then the model is a reasonable interpretation of the data. To obtain agreement the model He abundance can be adjusted appropriately. However, if we try to do this with the ROSAT data we encounter a problem. A number of stars require the He abundance to be increased to bring the EUV/X-ray measurement into line with the optical value but when we need to consider models with He:H$> 3 \times 10^{-5}$ there are no values of T_{eff} or N_H that are consistent with **all** the observed ROSAT count rates.

Studying in detail all of the stars in our chosen sample, we find that in general the count rates from those DA white dwarfs with T_{eff} less than 40,000K can be explained simply by H + trace He atmospheres with He:H in the range $0 - 3 \times 10^{-5}$. Formally, we can only place an upper limit on the He abundance of 3×10^{-5}, given the experimental errors, but we note that a pure H model is also adequate for these objects. Above 40,000K, it is not possible to find H+He models that yield temperatures in agreement with the optical values, with the exception of HZ43 and GD 2 To illustrate this, figure 2 shows the temperature ranges allowed by the EUV/X-ray data, for the acceptable H+He models, and the corresponding optical values for each star. Those stars that lie on the equal temperature line can be adequately interpreted by H+He models. In contrast those objects that lie below, ie. where the EUV/X-ray temperature is less than the optical value, must have an additional source of opacity in their photospheres that is not accounted for in our simple models. We note that for five stars - G191-B2B, Feige 24, WD1123+189, WD1109+244 and EG70 - we do not find any H+He models that can explain the ROSAT observations. Since, except in the case of EG70, these have $T_{eff} \approx 40,000 - 60,000$K it would appear that they are extreme cases of the effect we observe.

5 Conclusion

From our sample of 21 well-known DA white dwarfs we find that the atmospheres of these stars do not appear to be as simple as previously believed. Most of the stars below 40,000K show abundances of He=3×10^{-5} or less. This is in keeping with the earlier EXOSAT results. In contrast to the EXOSAT study, we find that for temperatures above 40,000K we require an additional source of opacity in the stellar photospheres to explain the ROSAT observations. We can explain this difference in the required interpretation as resulting from the improved spectral resolution achieved by ROSAT.

We must now investigate alternative models for the atmospheres of the hotter objects. The additional opacity may arise from the effects of compositional stratification. Alternatively, it could be caused by traces of large numbers of elements heavier than

He, as seems to be required to explain the EUV spectrum of Feige 24 (which is not detected by the PSPC!), observed by the EXOSAT TGS (Vennes, 1989). Whichever explanation is eventually found to be the most reasonable, it is clear that the ROSAT data unequivocally reveal compositional or structural evolution in DA white dwarfs for the first time.

The authors acknowledge the support of the ROSAT project through BMFT (FRG), SERC (UK) and NASA (USA). MAB is a SERC Advanced Fellow. We would like to thank Dr D.Koester for providing the model atmosphere data used to reduce the optical spectra.

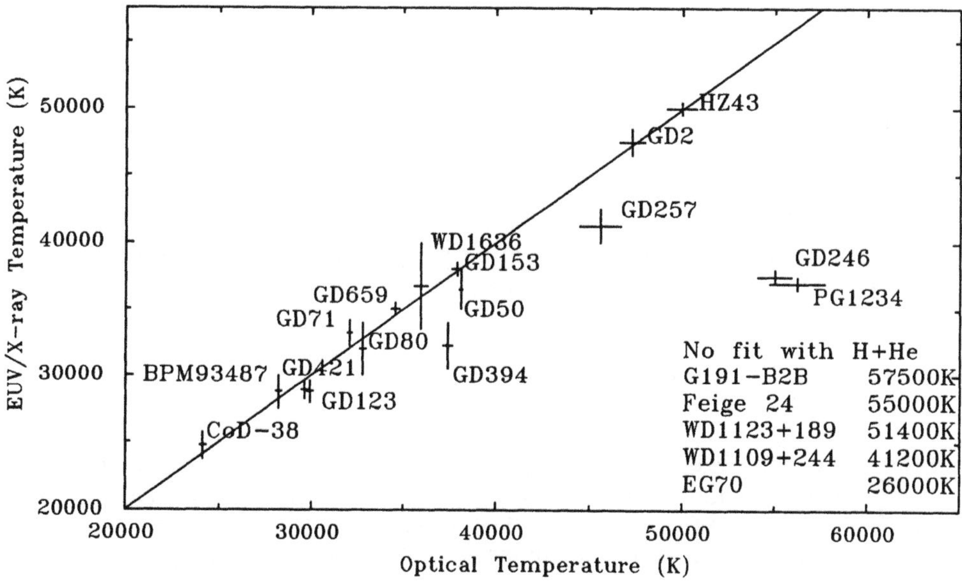

Fig. 2. The temperature range allowed by the ROSAT data and H+He models compared with the optically determined value. The solid line corresponds to equal ROSAT and optical temperatures.

References

1. F.B.S.Paerels and J.Heise: Ap.J., **339**, 1000 (1989).
2. D.Koester: Ap.J., **342**, 999 (1989).
3. F.B.S.Paerels, J.A.M.Bleeker, A.C.Brinkman and J.Heise: Ap.J., **309**, L33 (1986).
4. S.Vennes, G.Fontaine, F.Wesemael: in 'White Dwarfs', ed. G.Wegner, Springer-Verlag, 363 (1989).
5. E.Pfefferman et al.: Proc. SPIE, **733**, 519 (1986).
6. A.Wells et al.: Proc. SPIE, **1344**, 230 (1990).
7. M.A.Barstow: in 'White Dwarfs', ed. G.Wegner, Springer-Verlag, 156, (1989).
8. M.A.Barstow and J.B.Holberg: Mon.Not.R.astr.Soc., **245**, 370 (1990).

Carbon Enrichment In The Outer Layers Of Hot Helium-Rich High Gravity Stars

Klaus Unglaub, Irmela Bues
Dr. Remeis Sternwarte Bamberg, D-8600 Bamberg, Fed. Rep. of Germany

1 Introduction

The atmospheres of many White Dwarfs and Subdwarfs O (sdO's) are helium-rich with traces of heavier elements. Since these stars are in an evolutionary stage after the core helium burning, at least carbon should be present with significant abundance in the innermost regions. This is why we investigated in detail which carbon abundance is to be expected in the atmospheres of the stars in an equilibrium state of sedimentation, ordinary diffusion and selective radiative forces.

2 Methods and Results

Models of the outer, non-degenerate envelope have been computed in hydrostatic, thermal and diffusive equilibrium. They consist of the elements helium and carbon only, the gravities are log g = 8.0 (cgs) for the White Dwarfs and log g = 6.0 for the sdO's, the effective temperatures are 50000 K, 30000 K and 50000 K, 40000 K, respectively. The total stellar mass is $M_* = 10^{30}$ kg, this is about half a solar mass. A detailed computation of monochromatic absorption coefficients for each sort of particles permits the computation of the radiative forces due to line absorption and continuum absorption as well as the Rosseland mean value of the absorption coefficient. The temperature distribution is obtained with the diffusion approximation. A common property of all models is a thin outer helium convection zone. As convection is incompatible with diffusion, the integration of the hydrostatic equilibrium equations for each kind of particles (see e.g. Montmerle and Michaud, 1976, Unglaub and Bues, 1990) and the diffusion approximation starts from the inner boundary of the convection zone. The ratio carbon over helium $(C/He)_0$ by number at this depth is the independent parameter, the corresponding values for temperature and gas pressure are taken from a model atmosphere computed in advance.

Fig.1 shows results for models with log g = 8.0 and $T_{eff} = 50000$ K. The computation starts at a gas pressure log $P_G = 5.92$ and a temperature of 65770 K. This corresponds to a Rosseland mean optical depth $\tau_R = 3.12$. The number ratios at this depth point are taken $(C/He)_0 = 10^{-3}$, 10^{-4}, 10^{-5}, $0.9 * 10^{-5}$ and $0.8 * 10^{-5}$, respectively. The five curves in Fig.1 show the run of the abundance ratio C/He with gas pressure for these models. For $(C/He)_0 = 10^{-3}$ (model 1) and $(C/He)_0 = 10^{-4}$ (model 2) the number ratios C/He increase inwards to a value of about $(C/He) = 10^5$ at log $P_G = 10^{12}$.

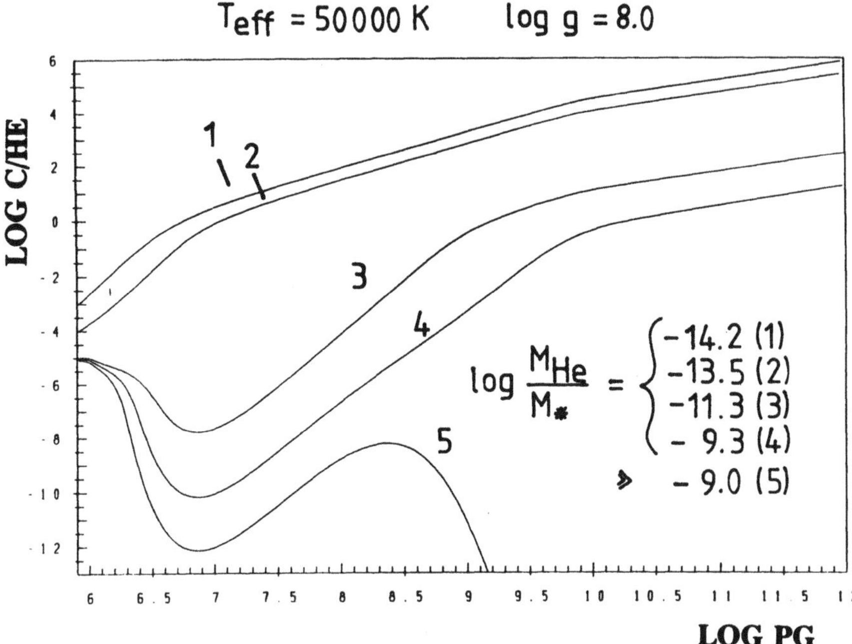

Fig. 1. Abundance C/He by number as a function of the gas pressure for models with T_{eff} = 50000 K, log g = 8.0 (cgs). The ratios C/He_0 at the inner boundary of the outer helium convection zone are 10^{-3} (1), 10^{-4} (2), 10^{-5} (3), $0.9 * 10^{-5}$ (4), $0.8 * 10^{-5}$ (5). The ratios helium layer mass to total stellar mass are indicated.

The prevailing mechanisms responsible for the change of abundances are sedimentation and ordinary diffusion. The carbon abundance is too large for the radiative forces to become effective. It is well known that the radiative forces decrease with increasing abundance due to saturation effects. In these cases the masses of the total helium layer are only $7.0 * 10^{-15}$ and $3.0 * 10^{-14}$ stellar masses, respectively. This means that for models with T_{eff} = 50000 K and log g = 8.0 an atmospheric carbon abundance of 10^{-4} or even more can only be expected for almost photospheric helium layers. The carbon abundances expected for larger helium masses can be deduced from the curves 3, 4 and 5 in Fig.1. The values $(C/He)_0 = 10^{-5}$ (model 3), $0.9 * 10^{-5}$ (model 4) and $0.8 * 10^{-5}$ (model 5) at the lower boundary of the helium convection zone vary slightly for these three models. In all cases the carbon abundance is low enough for the radiative forces on the carbon ion C^{3+} to become very effective. In the equilibrium state this leads to a negative concentration gradient C/He inwards. Both effects, sedimentation and ordinary diffusion, are necessary to balance the strong radiative forces. But at a depth with log $P_G = 6.8$ the ion C^{4+} becomes the prefered ionization stage of carbon. Yet the radiative forces acting on C^{4+} are vanishingly small. So in equilibrium sedimentation and ordinary diffusion must act in the opposite direction. This leads to an increasing carbon abundance inwards. At a depth of about log $P_G = 8.5$ the ionization stage of carbon changes from C^{4+} to C^{5+} and the effect of the radiative forces increases again. But in the models 3 and 4 the carbon abundance at this depth is already too large for the radiative forces to become effective. In the case of model 5 however, the low carbon abundance leads to very strong radiative forces and, as a consequence, to a steep negative concentration gradient. The total helium masses are $4.6 * 10^{-12}$ (model 3) and $5.3 * 10^{-10}$ (model 4) stellar masses, respectively, whereas for model 5 the helium mass is much larger than 10^{-9} stellar masses. The carbon abundance of model 5 is low enough for the radiative forces to prevent carbon from sinking. This means that an atmospheric carbon abundance of $(C/He)_0$ is a minimum value which could even then be expected if there were no carbon in the stellar interior and the star had only caught some carbon by accretion.

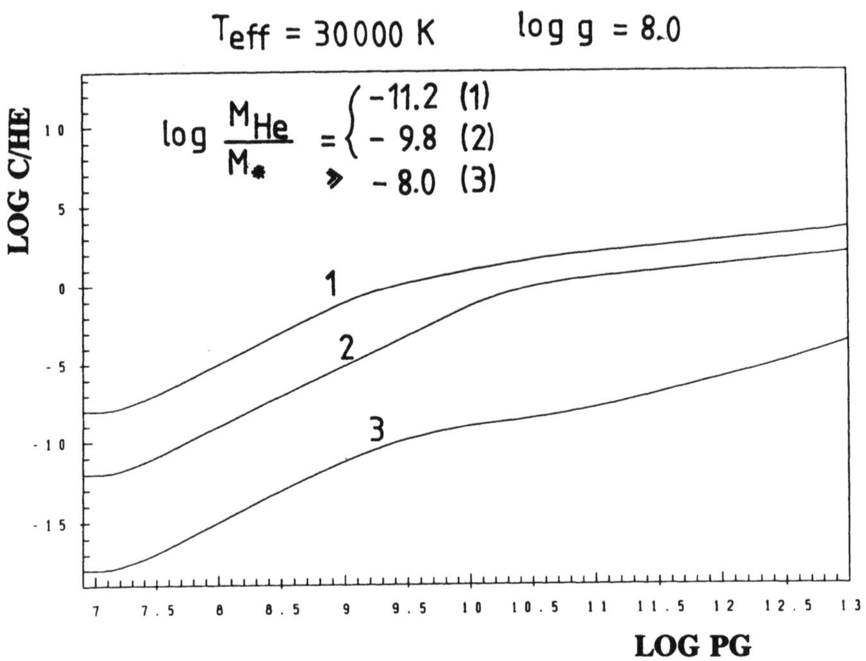

Fig. 2. The same as Fig.1 for models with T_{eff} = 30000 K, log g = 8.0, $(C/He)_0 = 10^{-8}$ (1), 10^{-12} (2), 10^{-18} (3).

Fig. 2 shows the ratio C/He as a function of gas pressure for models with T_{eff} = 30000 K and log g = 8.0. In all models C/He increases inwards. This means that radiative forces are not effective here, even for the model with $(C/He)_0 = 10^{-18}$. Therefore no detectable carbon enrichment of the atmosphere can be expected. Carbon abundances $(C/He)_0 > 10^{-8}$ are only possible for helium masses smaller than 10^{-11} M_*. For more details and results see Unglaub (1991).

References

Montmerle.T., Michaud,G.: 1976, Astrophys.J. Suppl. Ser. **31**, 489

Unglaub, K., Bues, I.: 1990, in "White Dwarfs", ed. G. Vauclair and E. Sion, NATO ASI Ser. C, p. 267, Kluwer Academic Publishers

Unglaub, K.: 1991, PhD Thesis, Erlangen 1991

Prospective EUVE Observations of Hot White Dwarfs

David. S. Finley
Center for EUV Astrophysics, University of California, Berkeley

Extreme ultraviolet (EUV, 70–900 Å) spectroscopic observations of a number of hot white dwarfs will be performed in 1992–1995 with the EUV Explorer satellite. A description of the properties of hot white dwarfs and of the capabilities of the EUVE spectrometer may be found in the white dwarf review in this volume (Koester and Finley 1991). My purpose in this presentation is to briefly sketch some examples of how EUV spectroscopy may contribute to our understanding of hot white dwarfs.

Homogeneous vs. stratified atmospheres. DA white dwarfs which show traces of helium in their spectra can do so by two very different means. The helium may be mixed with the hydrogen, due to accretion or to radiative acceleration. Or, the hydrogen layer may be sufficiently thin that the hydrogen becomes transparent in the short EUV wavelengths. EUV spectroscopy can unambiguously determine the atmospheric structure. This is demonstrated in Figure 1.

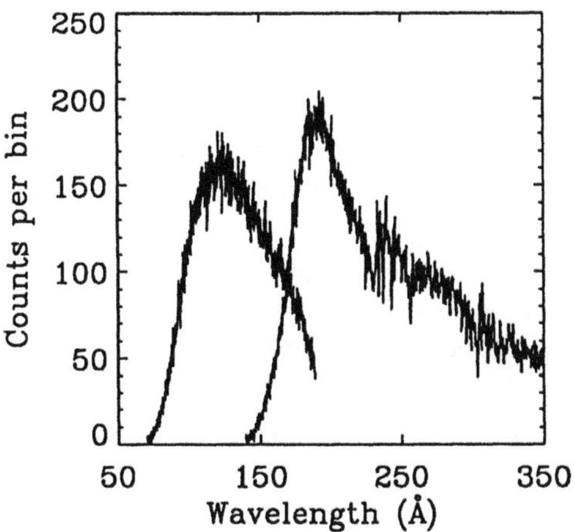

Figure 1a. Simulated spectrum of homogeneous H/He DA white dwarf, T_{eff} = 55,000 K, log g = 7.5, n(He)/n(H) = 1×10^{-5}, V = 14. Exposure time is 40,000 seconds. $N_{\mathrm{H}} = 3 \times 10^{18}$ cm^{-2}.

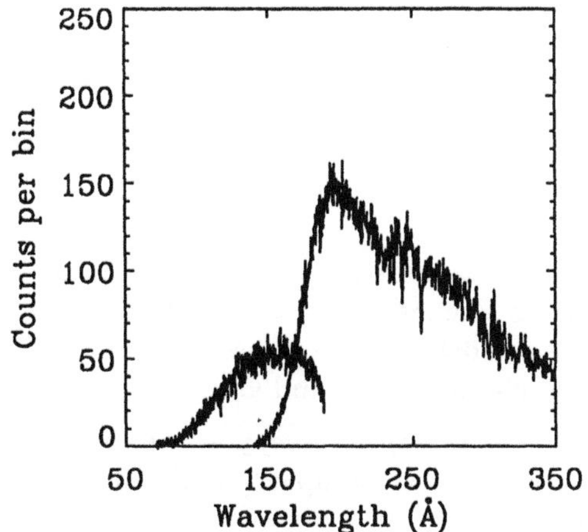

Figure 1b. Simulated spectrum of stratified 55,000 K DA white dwarf, log g = 7.5, H layer mass is 8×10^{-14} $M_{\odot}$, $N_{\mathrm{H}} = 2.8 \times 10^{18}$ cm^{-2}, V = 14.3.

The two models were chosen to have 228 Å jumps of comparable magnitude, and the V magnitude and column were adjusted slightly so that the continua longward of 228 Å matched. The continua shortward of 228 Å are absolutely irreconcilable, with the stratified model providing significantly less flux than the uniform model at all wavelengths. Furthermore, the appearance of the He II Lyman line series is quite different in the two cases, due to the increased pressure broadening in the stratified case.

Trace metals in hot DA white dwarfs. Feige 24, which has been observed spectroscopically in the EUV by *EXOSAT*, cannot be modeled using either uniform H/He or stratified models. Vennes et al. (1989) have shown that the *EXOSAT* spectrum can be fit reasonably well using a number of trace elements. However, the *EXOSAT* spectrometer resolution of only $\lambda/\Delta\lambda \sim 30$ was insufficient to resolve any of the characteristic lines or edges. The EUVE spectrometer, however, will have a resolution of ~300, allowing the determination of the trace element abundances in many cases. Figure 2a shows the input model spectrum, calculated by Vennes (1991), showing numerous lines and edges from the different species. Figure 2b shows the simulated EUVE count rate spectrum, in the medium and long wavelength spectrometers, for a 100,000 second observation.

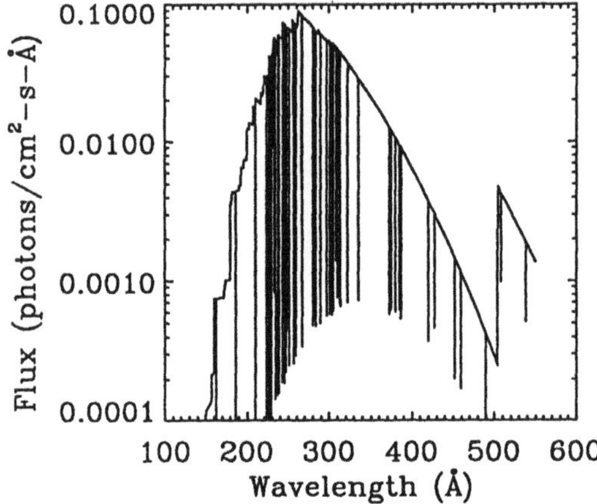

Figure 2a. Model spectrum for Feige 24 (from Vennes). T_{eff} = 55,000 K, log g = 7.2, model contains traces of He, C, N, O, Ne, Na, Si, S, Ar, and Ca.

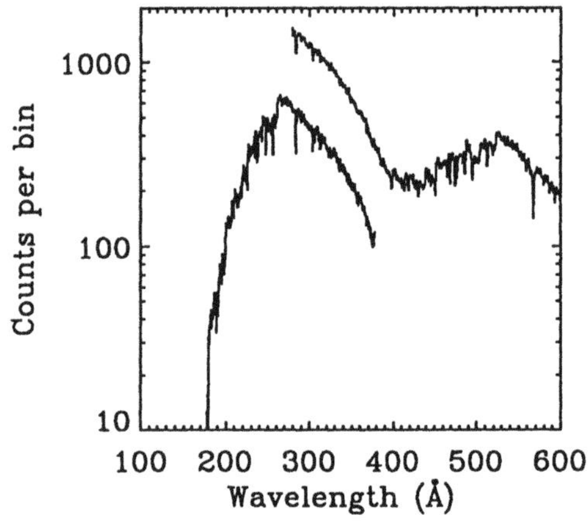

Figure 2b. Simulated EUVE count spectrum for 100,000 second observation, smoothed with 3-bin boxcar.

EUV spectroscopy at the limit. We also investigated whether faint sources would be good spectroscopy candidates. One case considered was for a 60,000 K pure H DA white dwarf at a distance of ~370 pc. The assumed HI column was 1 × 10^{20}. That star would be undetectable in the ROSAT WFC Be/C (S2) or EUVE Al/C bandpasses, and would be just above the detectability threshold in the Lexan/B bandpass for either instrument. The very narrow count spectrum (extending only from 75 to about 110 Å) would peak just below 100 Å due to the interstellar cutoff. The counting statistics would be such that even for such a distant and heavily absorbed source, a temperature and column could be derived with reasonable accuracy.

The next case was that of a hot DA white dwarf with a rather low H mass stratified atmosphere (perhaps analogous to the hot DA GD 246). As seen in Figure 3, due to the strong cutoff shortward of the He II ionization edge at 228 Å, this object would be at best marginally detectable in either photometric survey. Nonetheless, the simulated count spectrum for a 40,000 second observation is seen to be quite adequate for determining the physical parameters of this object under these observing conditions.

SPECTRA OF INTERACTING BINARY WHITE DWARF STARS

J.-E. Solheim
Institute of Mathematical and Physical Sciences
University of Tromsø, Norway

Abstract

IUE spectra of 4 known Interacting Binary White Dwarf (IBWD) stars show spectra with no hydrogen, only helium and some traces of metals. From the continuum spectra it is possible to model a disk component and a hot central object.

Introduction

One of the possible final stages of close binary star evolution is a system in which a carbon–oxygen white dwarf is accreting helium from a degenerate low mass companion. We know today 4 systems, which have only helium and some traces of metals in their spectra, and show sign of mass transfer by exposing a He-disk spectrum or flickering in their light curves. These systems, named Interacting Binary White Dwarf stars (IBWD) [1] are AM CVn (= HZ–29) [2], V803 Cen (= AE–1) [3], GR Boo (= PG 1346+082) [4] and GP Com (= G21–29) [5].

Of the 4 objects, AM CVn varies a few per cent around a mean value of $B = 13.9$. It may be compared with a cataclysmic variable such as a nova like object or a dwarf nova in constant outburst. V803 Cen and CR Boo varies like dwarf novae between a high state of $B \sim 13.5$ and a low state of $B \sim 17.2$, but unlike the dwarf novae, they spend most of their time in the high state.

The continuum spectra and models

A comparison of UV-continuum spectra for dwarf nova in quiescence and outburst, has shown that a majority of 80 percent of the dwarf novae during outburst have identical UV flux distribution within narrow limits, best described by the continuum spectra of B2–3 V–III stars of temperature 20000±2000 K [6]. A comparison of the UV-continuum flux for the IBWDs with the dwarf novae [7] shows that, except for GP Com which has a redder spectrum, the 3 other IBWDs have a much bluer spectrum than the dwarf novae.

There are basically three mechanisms for producing a steeper spectrum than we observe for the dwarf novae: a higher mass transfer rate, lower inclination, or the presence of a hotter central object. The line profiles indicate that the inclination is low, since P-Cygni profiles are not observed. Even inclination zero does not give the spectral slope observed. The absolute magnitudes of the objects are uncertain, but indicate that $\dot{M}$ is of the order $10^{-9}\ M_\odot\ yr^{-1}$. The most likely explanation of the steep spectra in

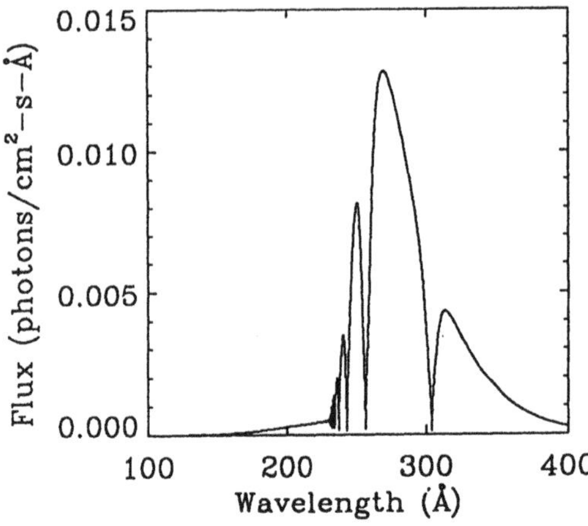

Figure 3a. Stratified DA model atmosphere, T_{eff} = 60,000 K, log g = 7.5, V = 15, H layer mass = 5 × 10^{-15} $M_{\odot}$. N_H = 5 × 10^{18} cm^{-2}.

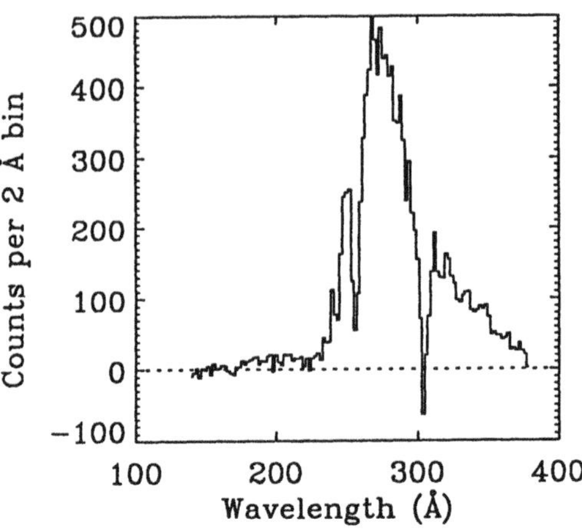

Figure 3b. Simulated count spectrum for 40,000 second observation.

DO white dwarfs. DO white dwarfs suffer from a number of observational disadvantages with respect to EUV spectroscopy. Strong absorption edges exist at 504 and 228 Å. Unless they are very hot, the stellar photon spectra in the EUV peak longward of the 504 Å edge. Of course, if the interstellar neutral hydrogen column exceeds a few 10^{18}, the long wavelength flux will be eliminated. Accordingly, we investigated the visually brightest DO white dwarf now known, PG 1034+001. Its V magnitude is 13.2, which would place it at around 75 pc, if its radius is typical. The estimated temperature is about 80,000 K. A usable spectrum will be obtainable only if it is hotter than 70,000 K and the column is less than about 2 × 10^{19} cm^{-2}.

Conclusions. Spectroscopy with EUVE will provide useful results for all types of hot white dwarfs. The sensitivity of the spectrometer is such that essentially all white dwarfs detected in the all-sky survey can be observed spectroscopically. Thus, the number of white dwarf spectroscopy candidates could be of the order of a few hundred. However, there may be few hot helium white dwarfs for which EUV spectroscopic observations will be possible. The spectra shown here (except for Feige 24) were calculated using a model atmosphere code which was kindly provided by Detlev Koester. This work was funded by grants from NASA.

References

Koester, D. and Finley, D.S., 1991, Proceedings of International CCP7 Workshop: *Atmospheres of Hot Stars*, Springer, Berlin, in press.

Vennes, S., 1991, private communication.

Vennes, S., Chayer, P., Fontaine, G., Wesemael, F., 1989, *ApJ*, **336**, L25.

the UV is that we observe flux from a hot central objct, which may be the white dwarf in the system or a boundary layer created by accretion onto the white dwarf.

To match the observed UV and UBV continuum observed for V803 Cen, we have varied $\dot{M}$, the inner (r_{in}) and outer radius (r_{out}) of the disk, and added one single source in the center with a minimum size of a He white dwarf of one solar mass to match the hottest part of the spectrum.[1] The mass of the white dwarf is kept constant, as is a zero inclination. A higher inclination will make the spectrum flatter, and the $\dot{M}$ has to be increased to compensate. A possible fit to the UV spectrum and UBV observations is shown in figure 1 with parameters given in table 1.

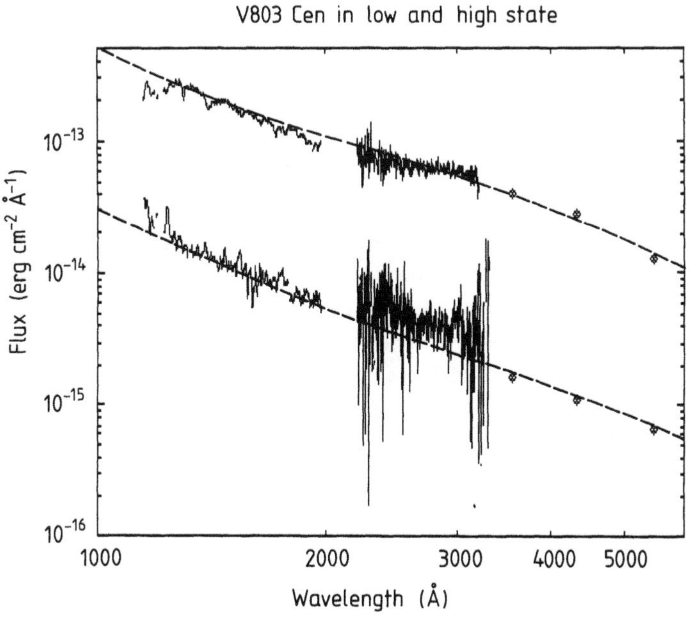

Figure 1. An example of disk model with a central hot body fit to the continuous spectrum of V803 Cen in the low and the high state. Parameters are given in table 1.

Figure 1 demonstrates that the parameters chosen make a reasonable good fit, both for the high and the low state. In the low state we observe an excess flux between 2500 and 3000 Å. Because of the long exposures needed for the IUE short wavelength spectrum, the long wavelength spectrum was taken at a different time when the system may have changed. The best fit for the low state is a small disk with a hole and a hot white dwarf with temperature of 65000 K observed in this hole. In the high state the disk is hotter and bigger. The white dwarf is still present but the hottest part of the spectrum is dominated by a larger object of temperature 150000 K, which we interprete as the boundary layer created by increased accretion when the system is bright. The cut off in the visual part of the spectrum shows that we observe the outer part of the disk and there is no trace of a secondary object, which may then be a low mass white dwarf [1].

[1]Note: The composite spectrum is constructed from black body spectra of appropriate temperature and area.

Table 1: Parameters for V803 Cen disk and central object(s).

	Low state	High state
Mass transfer rate ($\dot{M}_\odot$/year)	10^{-10}	10^{-8}
Outer disk radius r_{out} (m)	10^8	3.2×10^8
Outer disk temperature $T_{eff_{out}}$ (K)	6100	8400
Inner disk radius r_{in} (m)	2×10^7	5×10^7
Inner disk temperature $T_{eff_{in}}$ (K)	18000	31500
Radius of the central object (m)	6.9×10^6	1.3×10^7
Temperature of the central object: T_{eff} (K)	65000	150000
Type of central object	White Dwarf	Boundary Layer

Conclusions

UV-spectra give information about a hot central source and the expanding or contracting disk. Models for non-stationary helium disks need to be developed to explain CR Boo and V803 Cen variations. High resolution UV spectra, which are scheduled to be taken with IUE, will give more information about outflow processes from the inner disk/boundary layers and help us identify circumbinary matter related to earlier possible nova like outbursts. They may also help us determine abundances which are critical for understanding the past history of the systems.

Further investigation of these objects may tell us whether they are progenitors of single DB white dwarfs or doomed for a more violent future as supernovae.

Acknowledgements

This research is supported by grants from the Norwegian Research Council for Sciences and Humanities. The assistance from the IUE – Vilspa staff and the use of ULDA archives is greatly appreciated.

References

1. Nather, R.E.: in Interacting Binaries, P.P. Eggelton and J. Pringle (eds.), NATO ANSI, series C, Vol. 150, Reidel, 349 (1985) .

2. Solheim, J.-E., Robinson, E.L., Nather, R.E., and Kepler, S.O.: Astron. Astrophys. **135**, 1 (1984).

3. O'Donoghue, D. and Kilkenny, D.: Mon. Not. R. astr. Soc. **236**, 319 (1989).

4. Wood, M.A., Winget, D.E., Nather, R.E., Hessman, F.V., Liebert, J., Kurtz, D.W., Wesemael, F., and Wegner, G.: Ap. J. **313**, 757 (1987).

5. Nather, R.E., Robinson, E.L., and Stover, R.J.: Ap. J. **244**, 269 (1981).

6. la Dous, C.: Astron. Astrophys., in press (Vilspa IUE preprint no 54.) (1991).

7. Solheim, J.-E.: in Evolutionary Processes in Interacting Binary Stars, R. Polidan (ed.), Kluver, in press (1992).

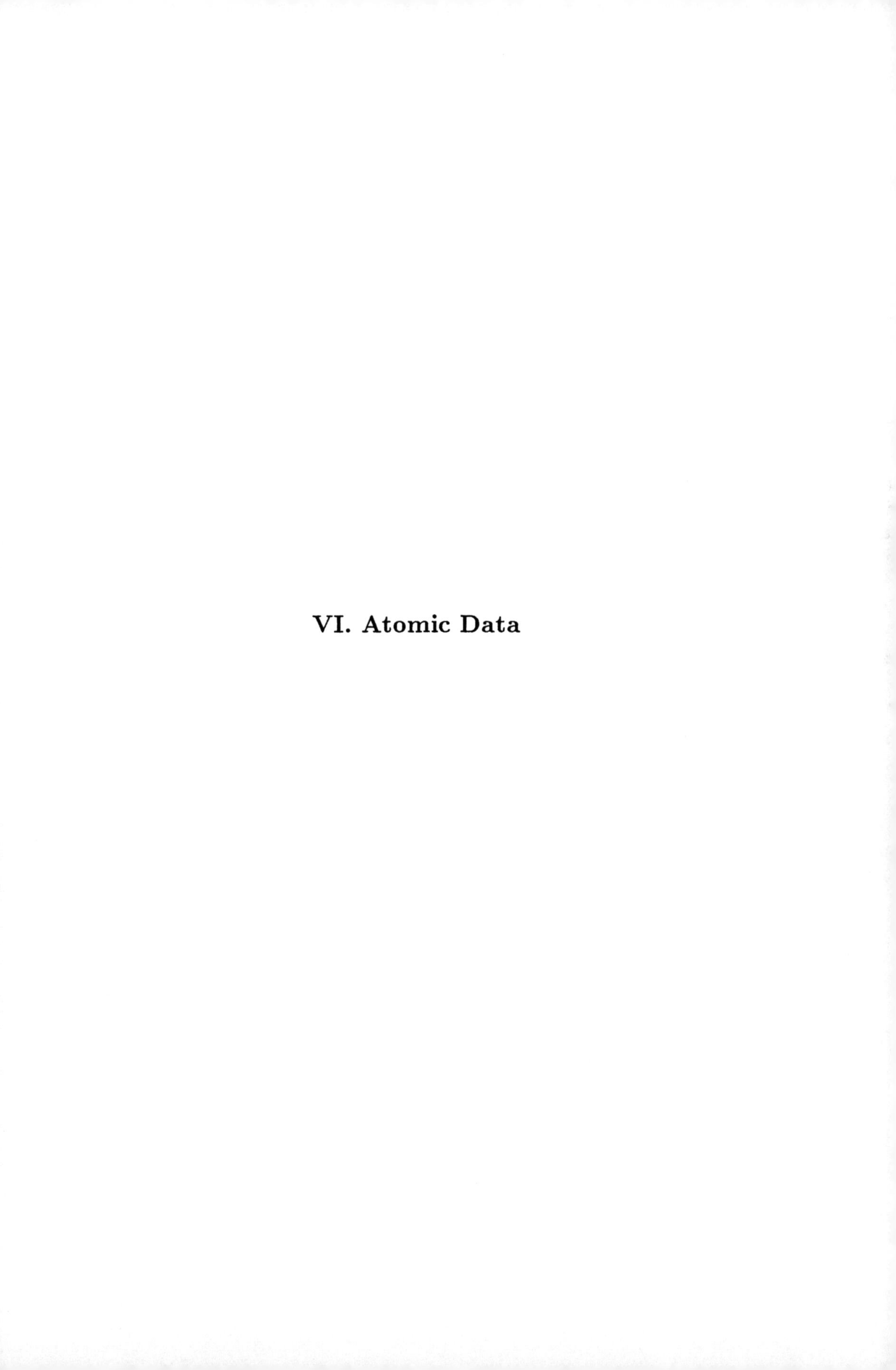

VI. Atomic Data

The Opacity Project – A Review

Keith Butler

Institut für Astronomie und Astrophysik der Universität München, Scheinerstraße 1, W–8000 München 80, Germany.

Abstract: The aims and methods of the international collaboration known as the **Opacity Project** are described. Results from all aspects of the work are presented, culminating in pulsation models for Cepheid variables which are in agreement with observation.

1 Introduction

Opacities are ubiquitous for the analysis and modelling of astrophysical spectra. For many years the Los Alamos opacities of Cox (1965) and Huebner (1985) have been the most used and useful available. Many outstanding astrophysical problems have been solved through their application. However, for many years there was a discrepancy regarding Cepheid models such that the masses derived observationally from the ratio of the fundamental period to that of the first overtone were significantly different from those produced by the models. Simon (1982) suggested a solution to this problem. He arbitrarily increased his model opacities at appropriate parameter values and showed that the conflict could then be resolved. The question as to whether such increases in the opacity (factors of order two to three were necessary) were feasible was answered in the negative by Magee *et al.* (1984). The **Opacity Project** was then conceived as an international collaboration in order to answer this question definitively. This of course meant it was necessary to calculate *accurate* occupation numbers, radiative data and line profiles for all ions of astrophysical interest and under conditions appropriate for stellar envelopes. What has been done in each of these areas is the subject of the next three sections, while the opacities obtained by putting all three aspects together and the pulsation models calculated using these numbers are presented in section 5. The final section gives some details of the future plans for the collaboration.

2 Occupation numbers

As is well known, a cut-off is needed in the calculation of partition functions which otherwise diverge. One approach is to use the "physical picture" in which the interactions are described using the many-body quantum theory (e.g. Ebeling *et al.* 1976). Hummer and Mihalas (1988), Mihalas *et al.* (1988) and Däppen *et al.* (1988) on the other hand describe the plasma in terms of its constituent molecules etc. , the "chemical picture". They use the method of free energy minimization together with an occupation probability formalism to solve the problem for temperatures $\leq 10^7$ K and densities $\leq 10^{-2}$ gcm^{-3}. The use of the occupational probability method implies that the partition function can be written

$$Z_s = \sum_i w_{is} g_{is} \exp(-\frac{E_{is}}{kT})$$

and w_{is} is the occupation probability. This is a quantity which decreases rapidly with increasing n (principal quantum number). In fact, Hummer and Mihalas adopt

$$\ln w_{is} = -\left(\frac{4\pi}{3V}\right)\left\{\sum_\nu N_\nu (r_{is} + r_{1\nu})^3 + 16\left[\frac{(Z_s+1)e^2}{\chi_{is} K_{is}^{\frac{1}{2}}}\right]^3 \sum_{\alpha \neq e} N_\alpha Z_\alpha^{\frac{3}{2}}\right\}.$$

The first term is due to perturbations by neutral particles which are treated as hard spheres while the influence of the charged particles is described by the second term which is a modification of the formula given by Unsöld (1948). Z_α is the net charge of a particle of species α (the electrons are excluded from the sum), N_ν, N_α are the total numbers of neutral and charged particles in volume V respectively, r_{is} is the radius of a particle in a state i and χ_{is} is the binding energy of such a particle. Lastly, K_{is} is a quantum-mechanical correction term such that

$$K_n = \frac{16}{3}\left(\frac{n}{n+1}\right)^2 \frac{n+\frac{7}{6}}{n^2+n+\frac{1}{2}}.$$

It should be noted that w_{is} can also be written in terms of the microfield distribution $P(F)$, e.g. the Holtsmark distribution as

$$w_{is} = \int_0^{F_c} dF P(F)$$

where the critical field strength F_c is given by the condition that the highest Stark level of quantum number n has the same energy as the lowest Stark level of quantum number n+1, viz

$$F_c(n) = Z^3 \frac{2n+1}{6n^4(n+1)^2}.$$

Hummer and Mihalas write the free energy F as the sum of four terms

$$F = F_1 + F_2 + F_3 + F_4.$$

The terms account for the following interactions:

F_1 the translational free energy of classical point particles;

F_2 the free energy of internal excitation (includes Z_s);

F_3 an ideal gas of partially degenerate electrons;

F_4 the free energy of the Coulomb interactions.

F is a function of temperature, volume and particle numbers and is minimised with respect to all N_s subject to number and charge conservation. The choices made by Hummer and Mihalas mean that all terms are analytically differentiable and hence that this non-linear system of equations can be linearised and solved in an efficient manner.

3 Radiative Data

It was clear from the beginning that in order to calculate all the necessary radiative data, methods would have to be chosen which were fast and required a minimum amount of human intervention. Electron scattering methods were used since they had already been shown to be capable of producing moderate amounts of radiative data although with much effort on the part of the authors concerned. This situation was altered drastically when Seaton (1985) showed how to find bound states fully automatically using the R-Matrix method (Burke *et al.*, 1971). Using this approach, wavefunctions for Fe^{6+} for example are found by solving the $Fe^{7+}+e$ scattering problem. It is then guaranteed that the Fe^{6+} wavefunctions are as good (or as bad) as the Fe^{7+} *target* wavefunctions. The latter are found using configuration interaction structure codes such as CIV3 (Hibbert, 1975) or SUPERSTRUCTURE (Eissner *et al.*, 1974). However, for ions such as those of iron the structure problem is not at all easy as can be seen fom fig. 1 (Saraph and Storey, 1991). Here the energy levels of Fe^{6+} are shown in a schematic way. The lines contributing most to the opacity come from levels in the 3d4s and 3d4p configurations, thus they should be well described. However, 3d4p overlaps with $3p^5 3d^3$ so at least some of these states should be included to give an accurate representation of the 3d4p states. The $3p^5 3d^3$ on the other hand are far too numerous and extend much too high in energy to be included in their entirity. Thus a compromise has to be found. This is something that requires much thought and effort and is the most time-consuming part of the calculations (at least from the human point of view).

The R-Matrix method is, in principle, straightforward. The problem is split into two parts. For radial distance $r > a$, exchange between the scattered and target electrons is negligible, the problem is simple and Coulomb functions with perturbation theory may be used. For $r < a$, all wavefunctions are expanded on a fixed set of basis functions. Because a is finite this is an eigenvalue problem which can be solved once and for all by diagonalising the Hamiltonian matrix. This is energy independent but care must be taken to ensure that the basis set is sufficiently complete to describe all energies of interest. Matching the solutions at $r = a$ then gives the physical solutions as a function of energy.

There are further advantages to be gained by the use of the R-Matrix method. Since a fixed basis set is used the radial integrals necessary for the evaluation of the radiative data can be calculated once and for all for the inner region while for the outer region, integrals over pairs of Coulomb functions can be obtained rapidly. The resulting program package has been described by Berrington *et al.* (1987).

Of course, it is important to be able to judge how accurate the data are. This is not unproblematical since the number of accurate data available for comparison is small. The calculated energies may be compared with observed values. Generally the agreement is better than 10%. Oscillator strengths may be compared. For He I the agreement is excellent (Table 1) while for Fe V, the *internal* consistency of the data is shown by the good agreement of the values derived from the length and velocity formulations (fig. 2). For exact wavefunctions, the two formulations give the same result. Since, the length value gives more weight to the outer part of the wavefunction while the velocity operator is more sensitive to the inner part the agreement between the two values demonstrates that these two regions are self-consistent (though they may

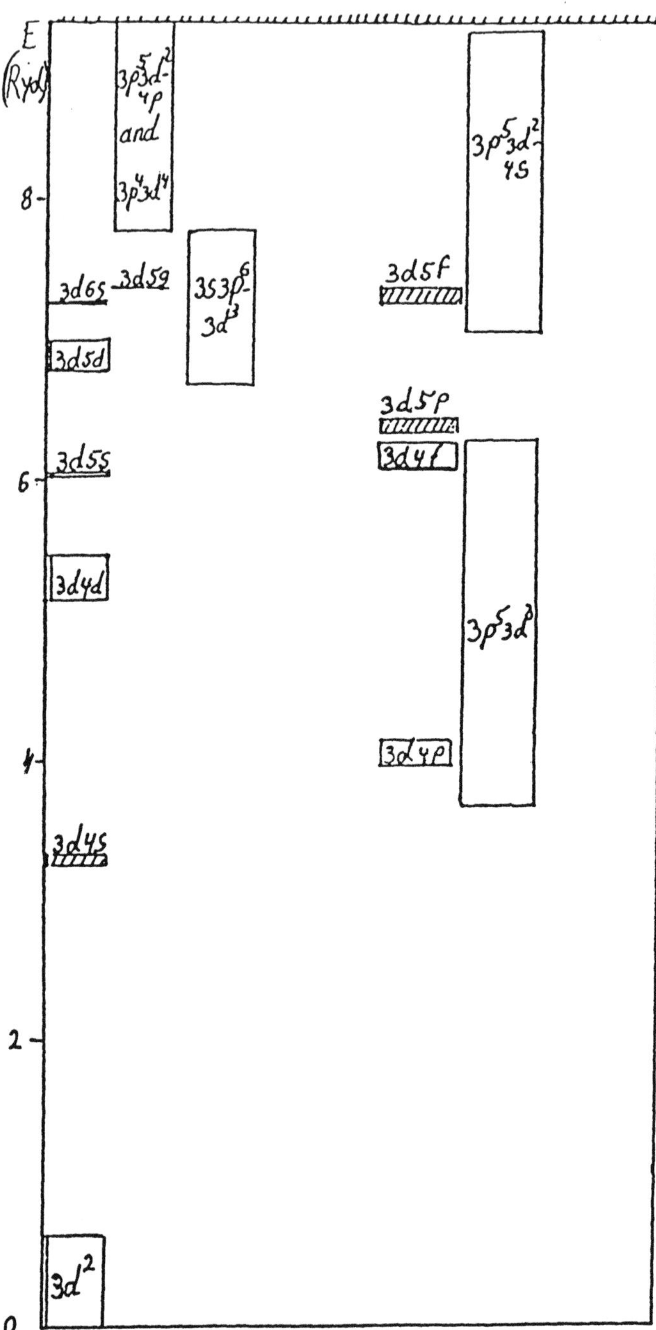

Figure 1: Schematic energy level structure of Fe^{6+} (Saraph and Storey, 1991)

Table 1: Comparison of |gf| values from Fernley *et al.* (1987) (upper entries) with those of Schiff *et al.* (1971).

	$Z = 2$ (He I)			
	Final state			
Initial state	$2\,^1P^o$	$3\,^1P^o$	$4\,^1P^o$	$5\,^1P^o$
$1\,^1S$	0.281 1	0.074 34	0.030 28	0.015 24
	0.276 2	0.073	0.030	0.015
$2\,^1S$	0.362 1	0.159 5	0.051 29	0.023 19
	0.376 4	0.151 4	0.049	0.02
$3\,^1S$	0.140 1	0.602 6	0.153 7	0.053 04
	0.145 4	0.626	0.144	0.05
$4\,^1S$	0.025 14	0.297 0	0.824 7	0.157 5
	0.025 8	0.306	0.85	0.15
$5\,^1S$	0.009 412	0.054 22	0.460 0	1.042
	0.009 6	0.055	0.47	1.1

both be in error). Finally, in a few cases the ground state photoionisation cross sections may be compared. This is done for He I and O I in figs. 3 and 4 respectively. As can be seen the agreement is excellent for He I. The 'hump' in the O I experimental cross section remains unexplained. A larger calculation undertaken by Bell *et al.* (1989) also lacks such a feature.

Since the calculations are at their worst at the neutral end of an isoelectronic sequence, such impressive agreement for such cases is good evidence for the accuracy of the data for the rest of the sequence.

Fig. 5, taken from the work of Tully *et al.* (1991), demonstrates the influence of the PEC (photoexcitation) resonances. Note the logarithmic scale though. The same authors have also shown clearly the decreasing influence of resonances with increasing ionic charge (fig. 6). Most resonances correspond to two (or more) electron transitions which have relatively small oscillator strengths. If however the photon energy is such that a core electron can be excited while the outer electron remains unaffected the corresponding oscillator strength is large being the f-value for the resonance transition of the next ion. For the Be I sequence for example, the $2sn\ell \rightarrow 2pn\ell$ resonances are orders of magnitude larger than the other resonances and for C III greatly perturb the regularity of the resonance pattern.

Results for individual sequences have been appearing in J. Phys. B at irregular intervals since 1987. It is planned to collect all the papers in a single volume to be published by the Institute of Physics in the not too distant future. Furthermore, Mendoza and Cunto (1991) have written a read only database capable of providing fast access to approximately 1 Gb of data. The program, **TOPBASE** currently runs under VM/CMS although a Unix version is planned.

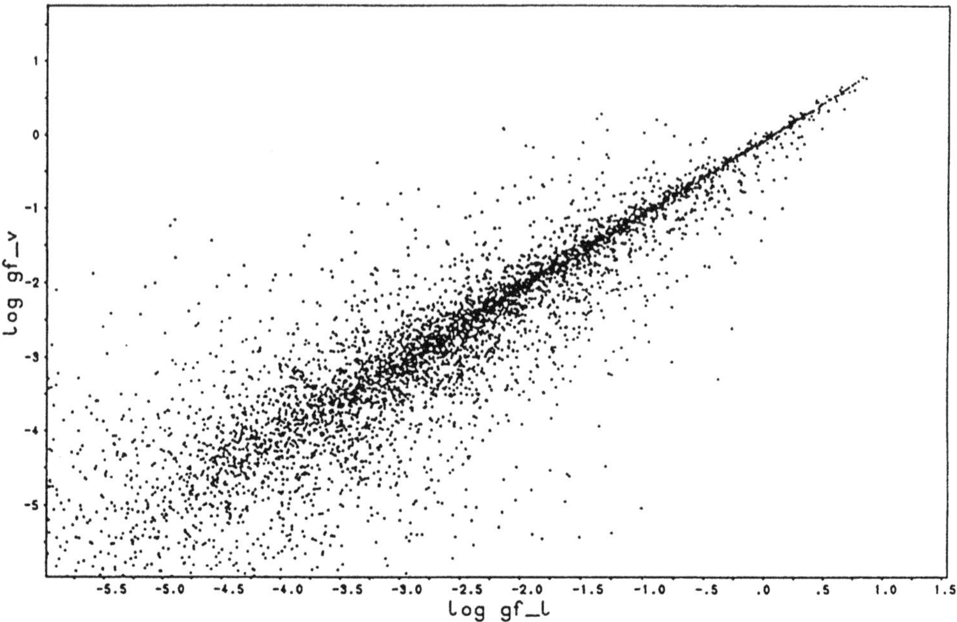

Figure 2: Comparison of length and velocity gf values for Fe^{4+} (Butler, 1992)

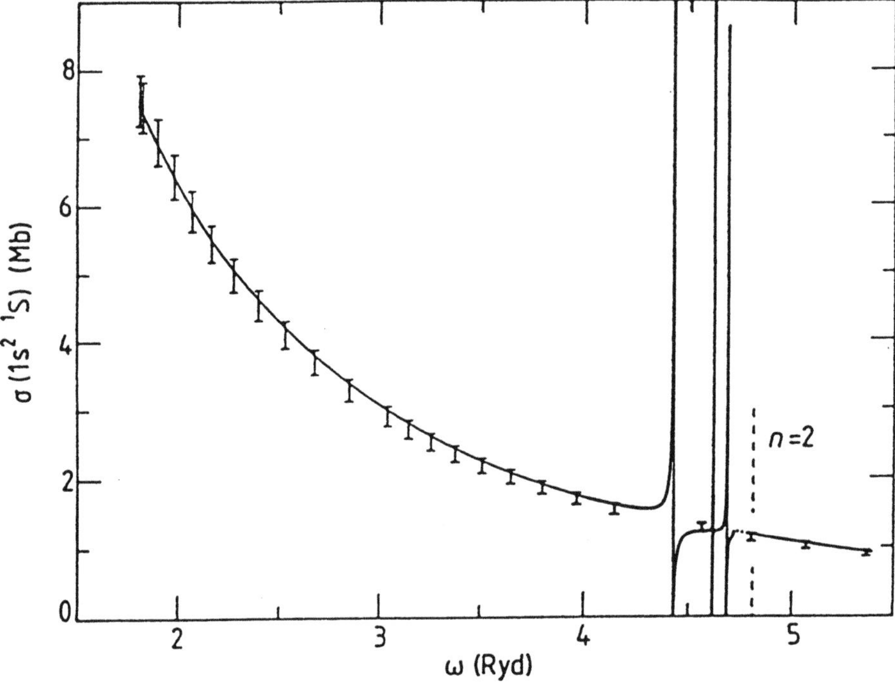

Figure 3: Comparison of the He I ground state photoionization cross section of Fernley *et al.* (1987) with the experimental data of West and Marr (1976).

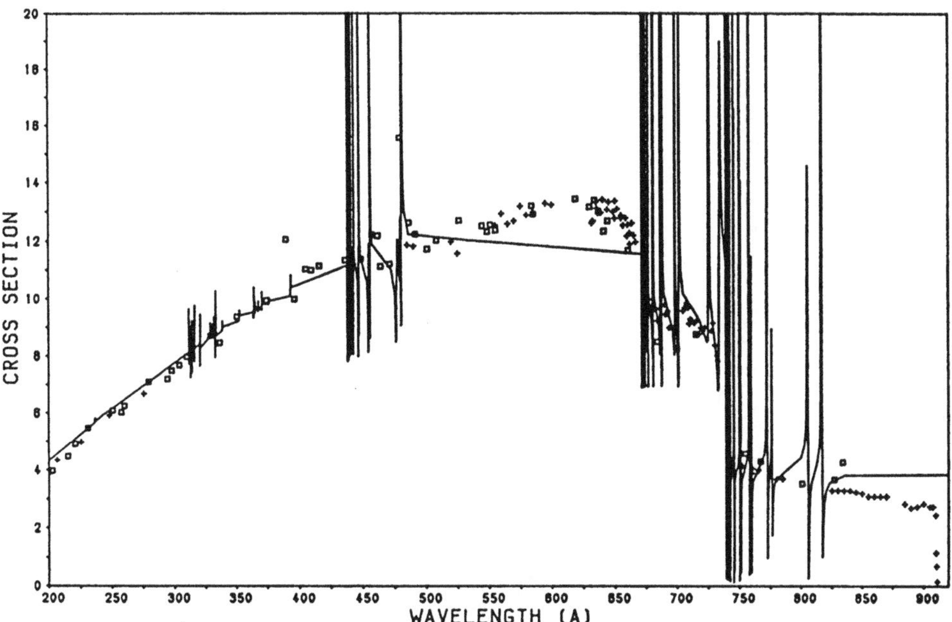

Figure 4: Comparison of the O I ground state photoionization cross section of Butler and Zeippen (1990) with the experimental data of Samson and Pareek (1985) (absolute values, □) and Angel and Samson (1988) relative values, +)

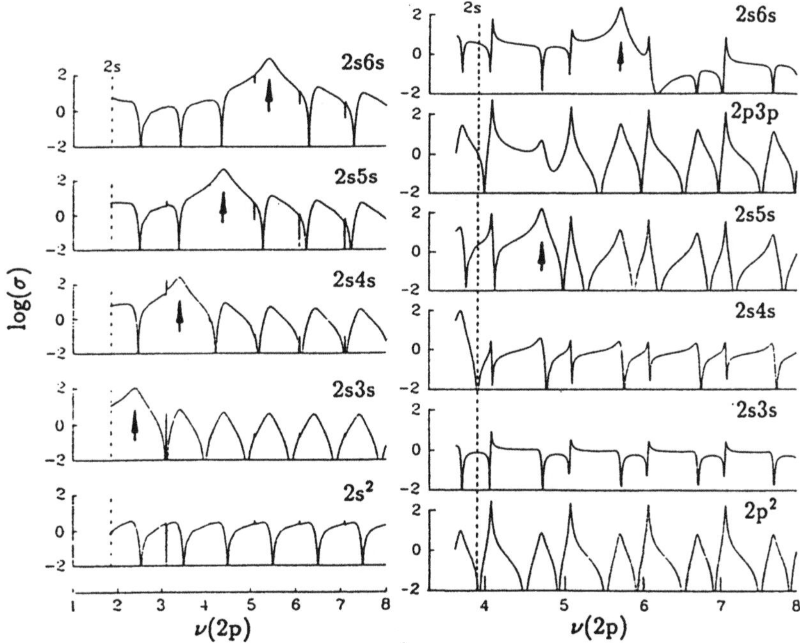

Figure 5: Details of PEC resonances in Be I and C III from Tully *et al.* (1991).

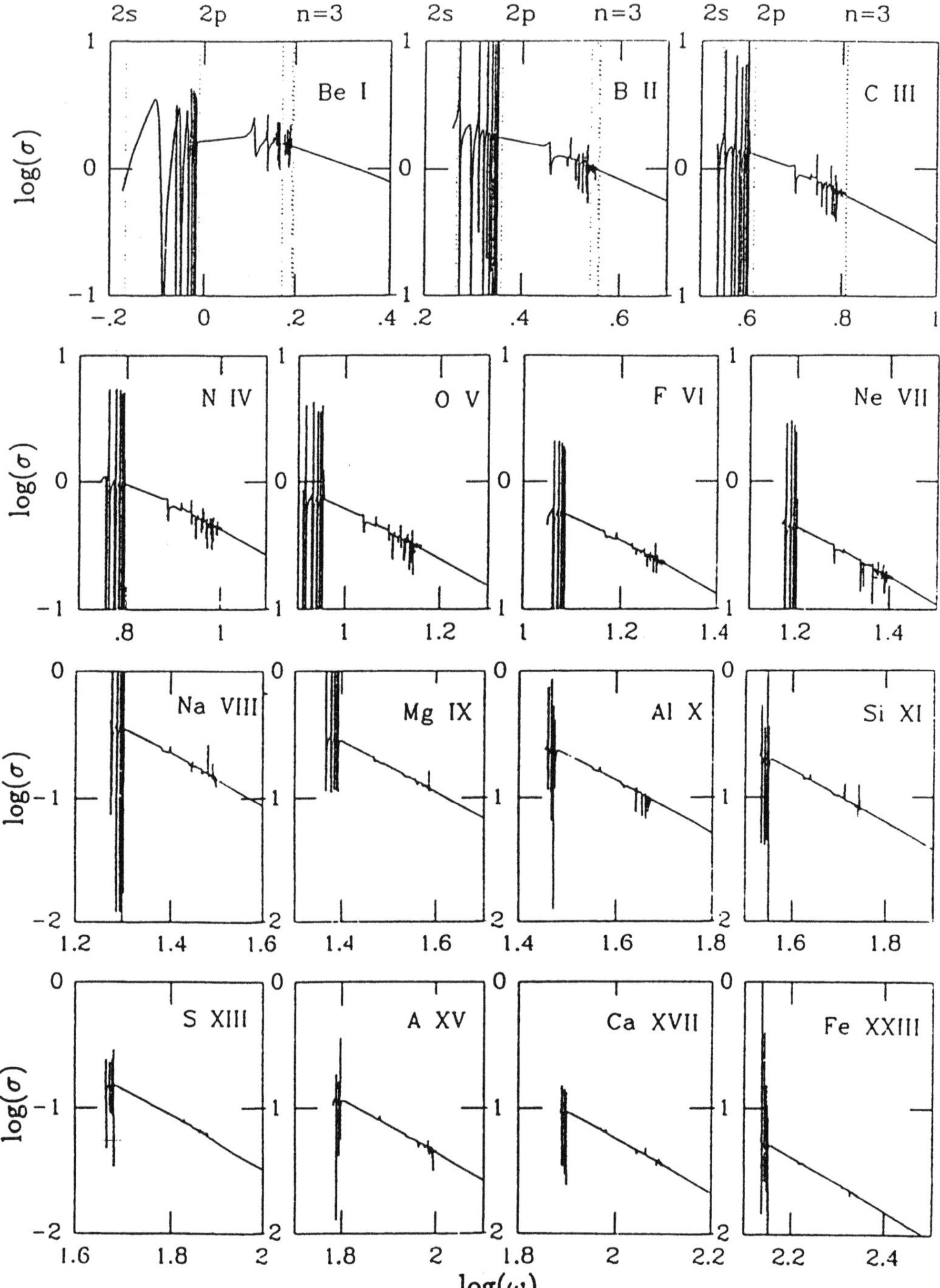

Figure 6: Cross sections for photoionization of Be-like ions from the $2s^2\ {}^1S$ ground state from Tully *et al.* (1991).

4 Line profiles

4.1 Quadratic Stark

Seaton (1987) derives expressions for the line broadening parameters due to electron broadening in the impact approximation. The final expression for the line profile is

$$\phi(\omega) = \frac{\gamma/\pi}{(\omega + x - \omega_0)^2 + \gamma^2}$$

in terms of the angular frequency ω relative to the frequency at line centre ω_0. The line width γ and displacement x are given in terms of the dimensionless *line broadening collision strength* Ω via the relations

$$\begin{aligned}
\gamma + i\,x &= a_0^3 N_e \Gamma \\
\Gamma &= 2\sqrt{\frac{\pi}{kT}}\Upsilon(T) \\
\Upsilon(T) &= \int_0^\infty \Omega(\varepsilon)\exp\left(-\frac{\varepsilon}{kT}\right)\, d\varepsilon
\end{aligned}$$

where ε is the electron energy, N_e the electron density and a_0 the Bohr radius. Ω itself is given by the complicated formula

$$\begin{aligned}
\Omega(\varepsilon) &= \sum_S \sum_{L\pi} \sum_{L'\pi'} \sum_{\ell\ell'} P(L, L', L_a, L_b, \ell, \ell') \\
&\times \{\delta(\ell,\ell') - S_a(ESL\pi;\ell,\ell')S_a^*(E'SL'\pi';\ell,\ell')\}\,.
\end{aligned}$$

Here P is a combination of Racah coefficients dependent on the angular momenta involved where the transition occurs from state a to state b while S_a, S_b are scattering matrices from the electron scattering problem. They describe the asymptotic behaviour of the scattered wavefunction. Such quantities can be calculated within the R-Matrix framework.

Seaton (1988) has carried out a number of these calculations. Since it is impractical to obtain such data for all the millions of lines required for the opacity calculations he has fitted the data to provide collisional broadening parameters good to a factor of two. The fit reads

$$Re\ \Upsilon_{ab}(T) = G(T)\{U_a(T) + U_b(T)\}$$

with

$$\begin{aligned}
g_p U_p(T) &= \sum_q S(p,q)\exp\left(-\frac{\Delta_{pq}}{kT}\right) \\
G(\text{fit}) &= 6.3 - \frac{5.9}{1+z}
\end{aligned}$$

$S(p,q)$ is the radiative line strength for the p $\rightarrow$ q transition, z being the charge on the target ion and g_p the statistical weight of the state p. The quantity Δ_{pq} is defined so that

$$\Delta_{pq} = \begin{cases} E_q - E_p - \delta & (E_q - E_p - \delta) > 0 \\ 0 & (E_q - E_p - \delta) < 0 \end{cases}$$

The Es are the excitation energies of the respective levels while δ is a further fit parameter intended to take the resonant contribution into account.

4.2 Linear Stark

The linear Stark effect is relevant when the field strengths are greater than the energy level splittings. This is the case not only for hydrogenic ions but also for any ion provided that n is sufficiently large, one problem being to define what "sufficiently large means". Seaton (1990) has approached the problem in a novel way, the main object being to obtain the line profiles quickly. He uses the quasi-static theory for the ions and the impact theory for the electrons. This means that the resultant profiles are not valid at the line centre because here the motions of the ions play a rôle but we note that this is irrelevant for the opacity work. Seaton interpolates between the one-perturber approximation in the wings which gives

$$\phi(u) = \frac{\gamma(u)/2\pi}{u^2}$$

and the impact approximation in the core

$$\phi(u) = \frac{\gamma(0)/2\pi}{u^2 + \left(\frac{1}{2}\gamma(0)\right)^2}$$

by putting

$$\phi(u) = \frac{\gamma(u)/2\pi}{u^2 + \left(\frac{1}{2}g(u)\right)^2}$$

with

$$g(u) = u\int_u^\infty \frac{\gamma(u')}{u'^2}du'$$

for the electron broadening. He calculates the matrix γ quantum mechanically using the Coulomb-Bethe approximation but avoids the matrix inversion implied in the above formulae by using only the diagonal elements. He ensures that the wings are reproduced correctly by renormalising the matrix of the diagonal elements $\bar{\gamma}$ so that $\mathbf{D}^\dagger\bar{\gamma}\mathbf{D} = \mathbf{D}^\dagger\gamma\mathbf{D}$. Finally, including ion-broadening and line dissolution leads to the expression

$$\begin{aligned}\phi\left(|\omega-\omega_0|\right) &= \sum_j^1 \Delta_j\phi_j\left(|\omega-\omega_0|\right)W(n') + \\ &\quad \sum_j^2 \Delta_j \int_0^{F_c} dFP(F)\left\{\phi\left(|\omega-\omega_0-a_jF|\right)+\phi\left(|\omega-\omega_0+a_jF|\right)\right\}.\end{aligned}$$

The summations 1 and 2 are over the unshifted and shifted components respectively and $\Delta_j = |D(j)|^2$ while a_j is the plasma parameter. The calculation of a single profile requires 0.006 s on a Cray X-MP machine. Following Däppen *et al.* (1987), Seaton computed hydrogen opacities based the occupation number theory of Hummer and Mihalas and using these line profiles. These are compared with the experimental data of Wiese *et al.* (1972) in fig. 7. The agreement is excellent. It must be borne in mind that this is a stringent test of both the line broadening parameters and the occupation numbers.

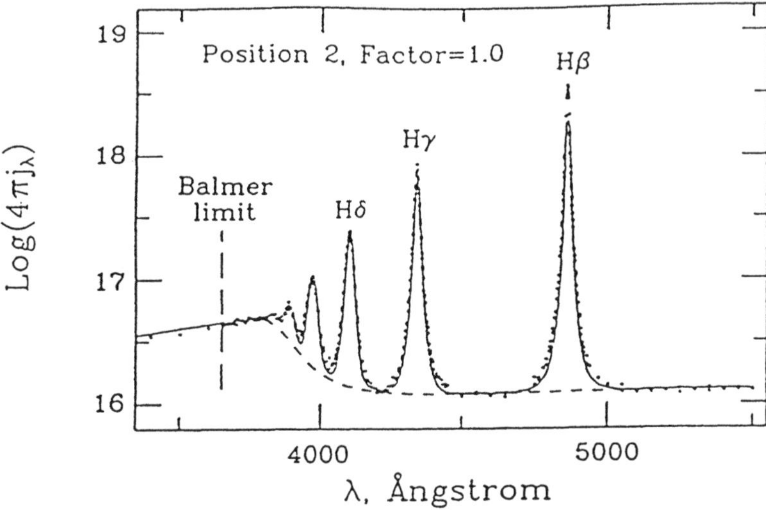

Figure 7: Emissivity of a hydrogen plasma. Full curve: theory (Seaton, 1990). Points: experiment (Wiese *et al.* 1972).

5 Opacities and Pulsation Models

Although the calculation of the opacities given the radiative data, occupation numbers and broadening parameters is in principle a straightforward exercise, it does require a great deal of effort in the organisation of the data to ensure efficient use of computing facilities. Seaton (priv. comm.) has computed a number of opacities for a mixture corresponding to the Anders and Grevesse (1989) abundances. A sample is shown in fig. 8 together with results from the Livermore group (Iglesias and Rogers, 1991). The agreement is extremely gratifying.

Following on from this, Kanbur (priv. comm.) has computed a grid of pulsation models for Cepheid variables using linear pulsation theory. His results may be seen in fig. 9 together with the observational material. Once more the agreement is highly satisfactory.

6 Outlook

The first opacities should appear on the 'open market' in the period when this volume goes to press. It should be noted that the Livermore group have already made opacities for a solar mixture available (Iglesias and Rogers, 1991). While the **Opacity Project** is coming to an end, it has already been decided to embark on a follow-up project to calculate collisional cross sections for ions of astrophysical interest and to include fine structure effects in the cases already treated. Particular attention will be paid to the iron group elements.

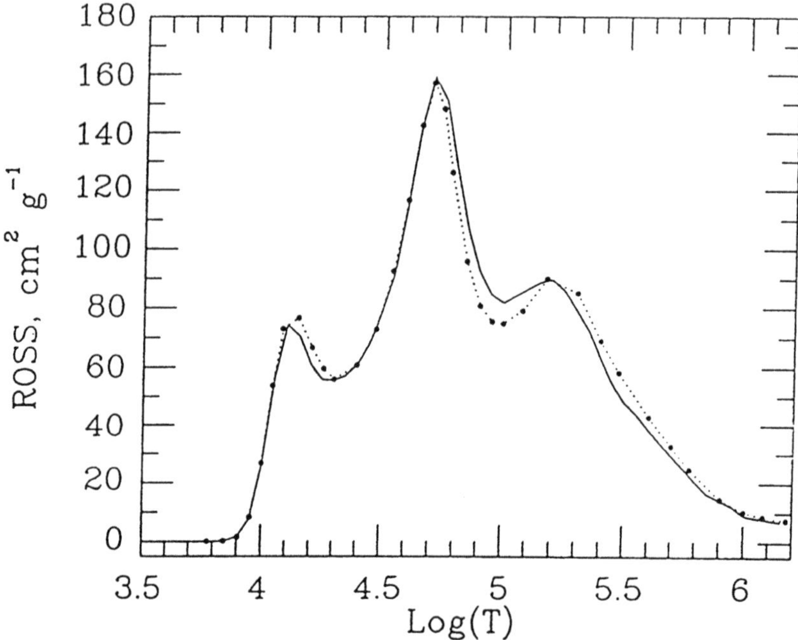

Figure 8: Opacities for a solar mixture from Seaton (priv. comm.) (full curve) and Iglesias and Rogers (1991) (dotted curve).

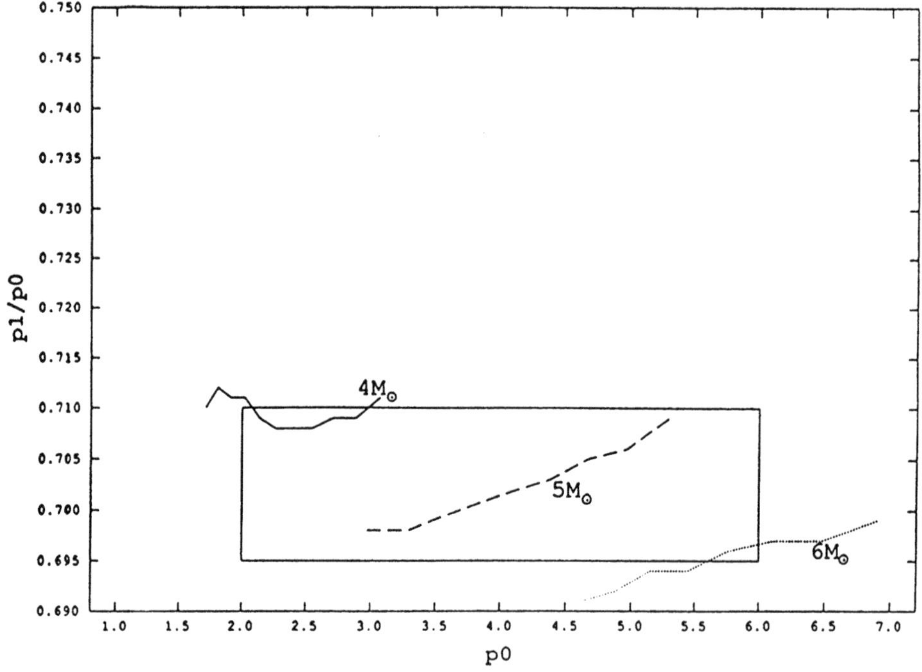

Figure 9: Comparison of pulsation theory models (Kanbur, priv. comm.) (curves) and the observations (boxed area).

Acknowledgements

It is a pleasure to thank all my colleagues for their help and advice over the years. In particular I should thank Mike Seaton and Shashi Kanbur for providing their data before publication. The calculations were carried out on the Cyber machines of the Leibniz Rechenzentrum and the Cray Y-MP of the Bayerische Akademie der Wissenschaft.

References

Angel, G.C. and Samson, J.A.R., 1988, *Phys. Rev. A*, **38**, 5578.
Anders, E. and Grevesse, N., 1989, *Geochim. Cosmochim. Acta*, **53**, 197.
Bell, K.L., Burke, P.G., Hibbert, A. and Kingston, A.E, 1989, *J. Phys. B: At. Mol. Phys.*, **22**, 3197.
Berrington, K.A., Burke, P.G., Butler, K., Seaton, M.J., Storey, P.J., Taylor, K.T. and Yu Yan, 1987, *J. Phys. B: At. Mol. Phys.*, **20**, 6379.
Burke, P.G., Hibbert, A. and Robb, W.D., 1971, *J. Phys. B: At. Mol. Phys.*, **4**, 153.
Butler, K., 1992, *Astron. Astophys.*, in preparation.
Butler, K. and Zeippen, C.J., 1990, *Astron. Astrophys.*, **234**, 569.
Cox, A.N., 1965, in *Stars and Stellar Systems* ed. Aller, L.H. and Laughlin, D.B. (Chicago:Chicago Univerity Press) p 195.
Däppen, W., Mihalas, D., Hummer, D.G. and Mihalas, B.W., 1988, *Astrophys. J.*, **332**, 261.
Däppen, W., Anderson, L. and Mihalas, D., 1987, *Astrophys. J.*, **319**, 195.
Ebeling, W., Kraeft, W.D. and Kremp, D., 1976, *Theory of Bound States and Ionization Equilibrium in Plasmas and Solids*, (Berlin:Akademie).
Eissner, W., Jones, M. and Nussbaumer, H., 1974, *Comput. Phys. Commun.*, **8**, 270.
Fernley, J.A., Taylor, K.T. and Seaton, M.J., 1987, *J. Phys. B: At. Mol. Phys.*, **20**, 6457.
Hibbert, A., 1975, *Comput. Phys. Commun.*, **9**, 141.
Huebner, W.F., 1985, in *Physics of the sun* ed. Sturrock, P., Holzer, T., Mihalas, D and Ulrich, R. (Dordrecht:Reidel) p 33.
Hummer, D.G. and Mihalas, D., 1988, *Astrophys. J.*, **331**, 794.
Iglesias, C.A. and Rogers, F.J., 1991, *Astrophys. J.*, **371**, L73.
Magee, N.H., Merts, A.L. and Huebner, W.F., 1984, *Astrophys. J.*, **283**, 264.
Mendoza, C. and Cunto, W., 1991, *IBM Venezuela: Doc. No. CSC-02-91*, **Topbase: VM/CMS User's Guide**
Mihalas, D., Däppen, W. and Hummer, D.G., 1988, *Astrophys. J.*, **331**, 815.
Samson, J.A.R. and Pareek, P.N., 1985, *Phys. Rev. A*, **31**, 1470.
Saraph, H.E. and Storey, P.J., 1991, *Journal de Phys.*, **1, C1**, ed. Zeippen, C.J. and Le Dourneuf, M., p169.
Schiff, B, Pekeris, C.L and Accad, Y., 1971, *Phys. Rev. A*, **4**, 885.
Seaton, M.J., 1985, *J. Phys. B: At. Mol. Phys.*, **18**, 2111.
Seaton, M.J., 1987, *J. Phys. B: At. Mol. Phys.*, **20**, 6363.
Seaton, M.J., 1988, *J. Phys. B: At. Mol. Phys.*, **21**, 3033.
Seaton, M.J., 1990, *J. Phys. B: At. Mol. Phys.*, **23**, 3255.

Simon, N.R., 1982, *Astrophys. J.*, **260**, L87.
Tully, J.A., Seaton, M.J. and Berrington, K.A., 1991, *Journal de Phys.*, **1**, **C1**, ed. Zeippen, C.J. and Le Dourneuf, M., p169.
Unsöld, A., 1948, *Zs. Ap.*, **24**, 355.
West, J.B. and Marr, G.V., 1976, *Proc. R. Soc. A*, **349**, 397.
Wiese, W.L, Kelleher, D.E. and Paquette, D.R., 1972, *Phys. Rev.*, **A6**, 1132.

The Lyman α line wing and application for synthetic spectra of DA white dwarfs

N.F.Allard [1] and D. Koester [2]

[1] Observatoire de Paris-Meudon
Département Atomes et Molécules en Astrophysique
92195 Meudon Principal Cedex, France
[2] Department of Physics and Astronomy
Louisiana State University
Baton Rouge, LA70803, USA

Summary

New theoretical absorption profiles are included in stellar atmosphere codes and used to predict synthetic spectra for DA white dwarfs of intermediate temperatures (20000 to 8000 K). These new calculations offer a unique opportunity to determine accurate effective temperatures and surface gravities for the variable ZZ Ceti stars.

1 Introduction

Satellite features in the far red wing of Lyman α were observed in one of the first spectra obtained with the IUE (International Ultraviolet Explorer) satellite (Greenstein 1980), and subsequently in many others (Wegner 1984, Koester *et al.* 1985, Nelan and Wegner1985, Holm *et al.* 1985). A successful identification as satellites caused by collisions with neutral (1600 Å) or ionized (1400 Å) hydrogen, however, was not achieved until 1985 (Nelan and Wegner 1985; Koester *et al.* 1985). The theoretical spectra published by these authors showed qualitative agreement with the observations, which confirmed the identification, but was not good enough for a quantitative determination of atmospheric parameters.

Recently new theoretical calculations have been performed using the unified theory of (Anderson and Talman 1956) , which takes into account the effect of multiperturber collisions. Results for H-H collisions were given in (Allard and Kielkopf 1991). These calculations have been extented for the H-H^+ collisions. The method used for these calculations is essentially the same as that used in (Allard and Kielkopf 1991) with one exception: instead of expanding the autocorrelation function in powers of density, we here use a different method based on a suggestion by (Royer 1971). The autocorrelation function is split into a "locally averaged part", which gives rise to the line core, and an "oscillating part" responsible for the wing of the profile These absorption coefficients are then used to calculate synthetical spectra for DA white dwarfs in the range of effective temperatures from 20000 to 8000 K.

2 Synthetic spectra for DA white dwarfs

Figures 1 - 2 show two typical white dwarf UV spectra from the atlas by (Wegner and Swanson 1991), compared to two synthetic spectra. The model spectra have not been fitted in detail, we have only taken the closest approximation from a sparse grid of models for log g = 8. It is clear that an almost perfect fit will be possible when T_{eff} and log g are varied.

Acknowledgements:
Part of this work was supported by grants from the National Aeronautics and Space Administration and from the National Science Foundation (D.K.). N.A. wants to express her thanks for support received from NATO and for the hospitality of Dr. J. Kielkopf at the University of Louisville. D.K. is very grateful for the kind hospitality of the Département Atomes et Molécules en Astrophysique and Dr. N. Feautrier during his stay at the Observatoire de Paris-Meudon in the summer of 1991.

References

Allard,N.F., Kielkopf,J.F.: 1991, *Astron.Astrophys.* **242**, 133
Anderson,P.W., Talman,J.D.: 1956, *Proc.Conf.Broadening of Spectral Lines*, Bell Telephone System Technical Publications, No.3117, 29, MurrayHill,N.J.
Greenstein,J.L.: 1980, *Astrophys.J.* **241** , L89
Holm,A.V., Panek,R.J., Schiffer III,F.H., Bond,H.E., Kemper,E., Grauer,A.D.: 1985, *Astrophys.J.* **289**, 774
Koester,D., Weidemann,V., Zeidler-K.T.,E.-M., Vauclair,G.: 1985, *Astron.Astrophys.* **142**, L5
Nelan,E.P., Wegner,G.: 1985, *Astrophys.J.* **289**, L31
Royer,A.: 1971, *Phys.Rev.A* **3**, 2044
Wegner,G.: 1984, *Astron.J.* **89**, L1050
Wegner,G., Swanson,R.: 1991, *Astrophys.J.Suppl.* **75**,507

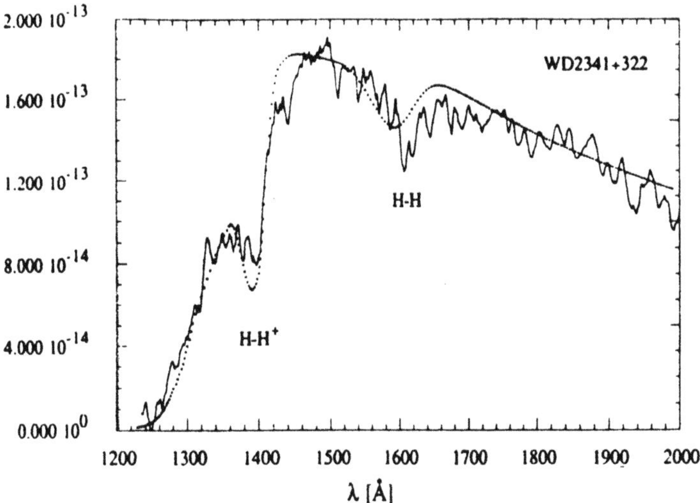

Fig. 1. IUE spectrum of WD2341+322 and a theoretical model for 13000 K, log g = 8.0.

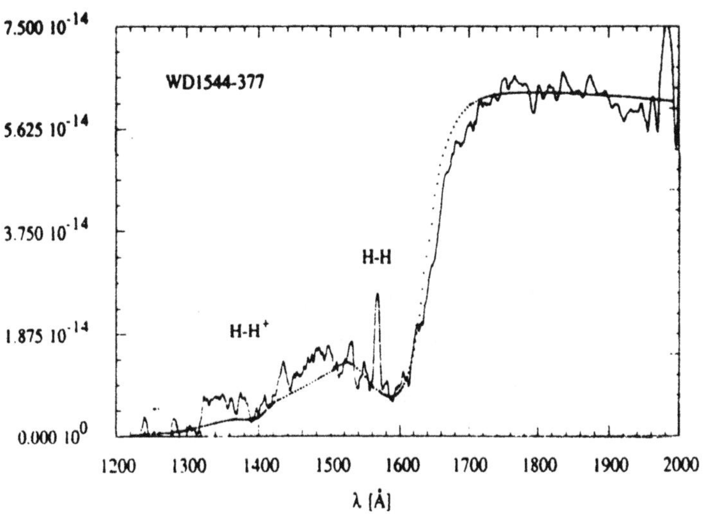

Fig. 2. IUE spectrum of WD1544-377 and a theoretical model for 11000 K, log g = 8.0.

Ion-Atom Complexes and the Absorption of Radiation in Stellar Plasma

A. A. Mihajlov [1], M. S. Dimitrijević [2]

[1]Institute of Physics, P. O. Box 57, 11001 Beograd, Yugoslavia
[2]Astronomical Observatory, Volgina 7, 11050 Beograd, Yugoslavia

Abstract: In order to provide the relevant absorption coefficients for the interpretation of the continuum absorption spectra in stellar atmospheres, the processes $H_2^+ + \hbar\omega \rightarrow H + H^+$ and $H + H + \hbar\omega \rightarrow H + H^+$ have been considered together, as well as the same processes in the helium case. We present here the corresponding coefficients for the conditions of stellar atmospheres.

1 Introduction

The significance of the combined study of the processess of the photodissociation

$$H_2^+ + \hbar\omega \rightarrow H + H^+ \tag{1a}$$

and the absorption of electromagnetic radiation by collisional ion-atom complexes:

$$H + H^+ + \hbar\omega \rightarrow \begin{cases} H + H^+ \\ H^+ + H \end{cases} \tag{1b}$$

has been demonstrated recently by Mihajlov and Dimitrijević (1986), for the conditions characteristic for stellar plasma. The simple method for the determination of corresponding absorption coefficients in the infrared and visible spectral range, proposed in the mentioned article, is applicable not only in the case of (1a) and (1b) processess but for a more numerous class of atomic systems. For the application of this method, the potentials (as a function of the internuclear distance R) for low lying energy states of molecular ions and the corresponding dypole matrix elements are needed.

2 Results and Discussion

The absorption coefficients $K_\omega^{(ab)}$ for H_2^+ and He_2^+ case as a function of λ and T for conditions in white dwarf atmospheres are presented in Figs. 1-2. From the Figs. 1-2

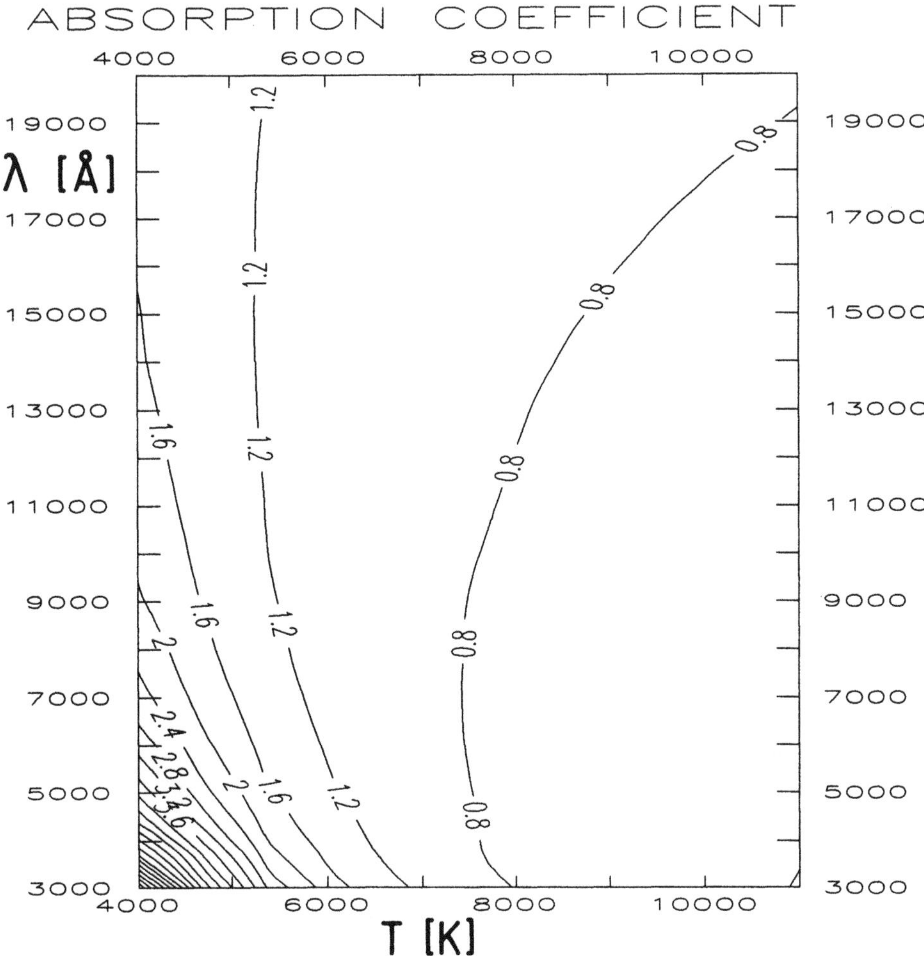

Fig. 1. The absorption coefficient $\times 10^{39}$ $[cm^{-1}]$ as a function of $T\,[K]$ and $\lambda\,[\AA]$. The case of H_2^+.

one can see that the processes 1a and 1b as well as 2a and 2b must be treated together when the process 1a or 2a are taken into account since their contributions are comparable. Moreover, the process 1b (2b) becomes more significant towards the infrared part of the spectrum. The total absorption coefficient shows a weak dependence on λ and T except in the region of low temperatures and towards the UV part of the spectrum, where a significant increase of $K_{\omega}^{(ab)}$ exists. The details of calculations as well as the additional numerical results will be published in Mihajlov and Dimitrijević (1992).

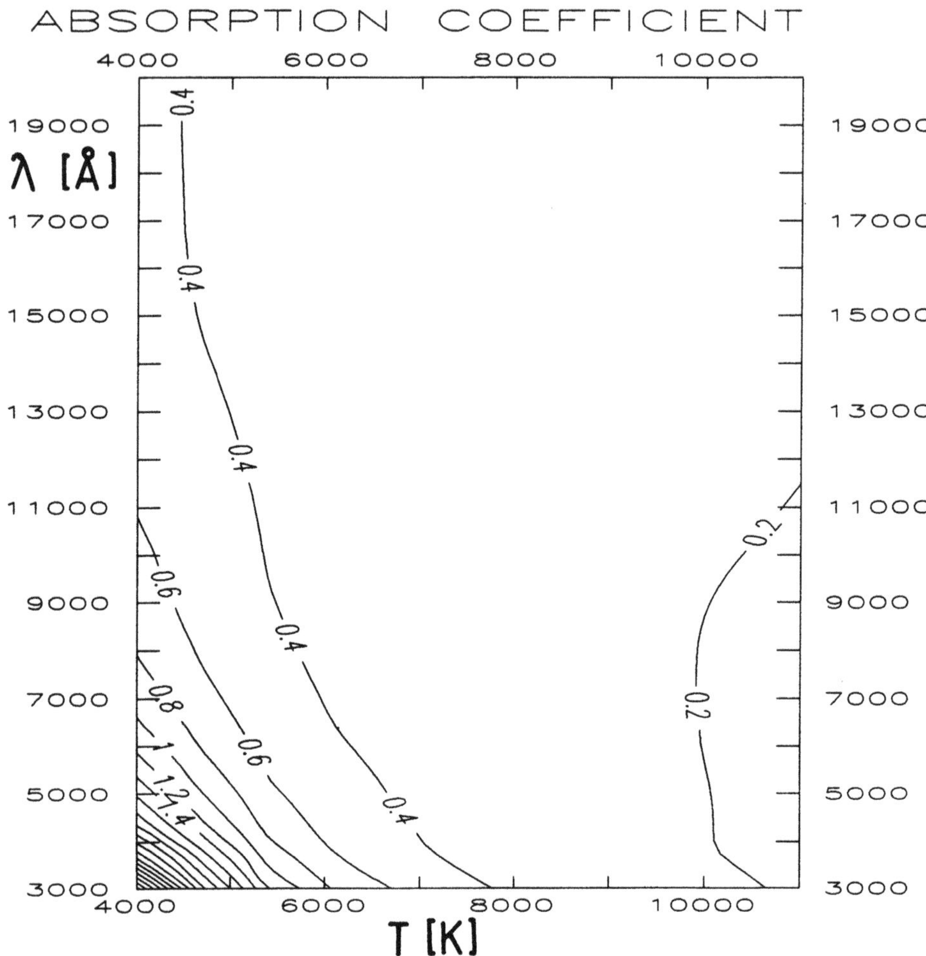

Fig. 2. The absorption coefficient $\times 10^{39}$ $[cm^{-1}]$ as a function of $T\,[K]$ and $\lambda\,[\mathring{A}]$. The case of He_2^+.

References

Ermolaev, A.M., Mihajlov, A.A.: 1991, J.Phys.B **24**, 155.

Ermolaev, A.M., Mihajlov, A.A., Popović, M.M.: 1989, in Proc. XIX Int.Conf.Phen.Ioniz.Gases, Beograd, ed. by J.M.Labat (Belgrade: University of Belgrade Press) p. 656.

Mihajlov, A.A., Dimitrijević, M.S.: 1986, Astron.Astrophys. **155**, 319.

Mihajlov, A.A., Dimitrijević, M.S.: 1992 to be published.

Ion-Atom Complexes and the Recombination in Stellar Plasma

A. A. Mihajlov [1], N. N. Ljepojević [2] and M. S. Dimitrijević [3]

[1]Institute of Physics, P. O. Box 57, 11001 Beograd, Yugoslavia
[2]Department of Applied Mathematics and Theoretical Physics, University of Cambridge, Cambridge CB3 9EW, UK
[3]Astronomical Observatory, Volgina 7, 11050 Beograd, Yugoslavia

Abstract: We show that for the study of recombination of ions and electrons in weakly ionized low temperature hydrogen plasmas, the processess $H + H^+ + e \rightarrow H + H^*(n)$ and $H_2^+ + e \rightarrow H + H^*(n)$ must both be considered since their contributions are comparable. A simple method for the calculation of the corresponding rate coefficients is presented. We also present the results of our calculations for $H^*(n)$ excited to the level of the principal quantum number $n = 4$.

1 Introduction

We will demonstrate in this contribution the significance of the combined study of the following processess

$$A_2^+ + \mathrm{e} \rightarrow A + A^*(n) \tag{1a}$$

$$A + A^+ + \mathrm{e} \rightarrow \begin{cases} A + A^*(n) \\ A^*(n) + A \end{cases} \tag{1b}$$

These processes are one of the ways for recombination of ions and electron in astrophysical plasmas. In the above equations, A denotes atom, A^+ and A_2^+ atomic and molecular ions, $A^*(n)$ atom excited to the level with the principal quantum number n, and e denotes electron. We will limit ourselves in the present contribution to the hydrogen case, and will present here a simple method for the calculation of corresponding rate coefficients for the case $n > 4$, derived within semiclassical approximation, using the resonant energy transfer model within electronic components of the atomic system considered (Mihajlov and Janev, 1981) and results obtained in Mihajlov and Dimitrijević (1986). The rate coefficients are calculated supposing that the following conditions are fulfilled: (i) Energy distribution function of free atoms and ions is Maxwellian with the temperature T_{a}; (ii) Energy distribution of bound states of molecular ions A_2^+ is Boltzmannian with the temperature T_{b}; (iii) Energy distribution of the free electrons is Maxwellian with the temperature T_{e}.

2 Theory

Starting from the results obtained in Mihajlov and Janev (1981) and, Mihajlov and Dimitrijević (1992) we have the recombination rate coefficients $K_n^{(a)}$ or $K_n^{(b)}$ in the form

$$K_n^{(a,b)}(T_a,T_e) = C_n(T_e) \int_{\varepsilon_n}^{\varepsilon_{max}} \frac{e^{-\varepsilon kT_e}}{\varepsilon} R_\varepsilon^4 e^{-U_1(\varepsilon)/kT_a} X^{(a,b)}(\varepsilon,T_a) d\varepsilon \qquad (2)$$

$$C_n(T_e) = \frac{16\pi^2}{3\sqrt{6}} n^{-3}\gamma^{-1}(kT_e)^{-3/2}\Gamma(3/2)exp(\frac{|\varepsilon_n|}{kT_e}), \qquad (3)$$

where $|\varepsilon_n| = 1/(2n^2)$ and $X^{(a,b)}$ is given in Dimitrijević and Mihajlov (1986). All quantities in the Eq. (2) are in the atomic units. The internuclear distance R_ε as a function of ε is determined from

$$\varepsilon = U_2(R_\varepsilon) - U_1(R_\varepsilon) \qquad (4)$$

assuming that the electron recombination occurs in the direct vicinity of the resonant internuclear distance R_ε. The value ε is the energy difference between the final (repulsive) (U_2) and initial (attractive) (U_1) electronic states of the ion-atom subsystem (molecular ion), $U_1(\varepsilon)$ is the energy of the initial state of $A_2^+(U_1(R) < 0)$ and ε_n the ionization energy of the atomic state $A^*(n)$. The γ in Eq. (3) is 1 for H and 1.345 for He. The maximal energy, ε_{max} in the Eq.(2) corresponds to the maximal difference in the Eq (4). In the case of hydrogen, the R_ε corresponding to ε_{max} is equal to zero (in Mihajlov and Ljepojević (1982), the calculations were performed under the assumption that $\varepsilon_{max} = \infty$, since the corresponding analytical approximations for the expression under the integral were used). The total rate coefficient for process (1) is given by the expression

$$K_n(T_a,T_e) = K_n^{(a)}(T_a,T_e) + K_n^{(b)}(T_a,T_e) \qquad (6)$$

The presented method is correct when in the output channels for (1a) and (1b) reactions, the interaction between $A^*(n) + A$ and $A^+ + A^-$ system terms may be neglected. In the hydrogen case, the electron afinity is $\varepsilon^{(-)}$ =0.75 eV (Massey, 1976). Therefore, the proposed method is correct for $n > 4$ in this case. However, the $n = 4$ case might be included since the interaction of $H^*(n = 4) + H$ and $H^+ + H^-$ system terms occurs for $R \geq 10$ a.u. In the case of $n = 4$ system, this region is not of importance. Consequently we may propose the presented method for $n \geq 4$ in the hydrogen case.

3 The Hydrogen ($n = 4$) Case

We present here as an example, calculations for $n = 4$ in the hydrogen case. In the case of $H_2^+, U_{1,2}$ correspond to the Σ_g^+ and Σ_u^+ electron states and were taken from Bates et al. (1953). Results for $K_n(T_a,T_b)$ and $K_n^{(a)}(T_a,T_e)/K_n(T_a,T_e)$ as a function of T_a and T_e are presented in Figs. 1 and 2. One can see that in plasma conditions recombination process on collisional quasi-molecular complexes (1a) and dissociative recombination process (1b) are competitive. The presented method offers possibility for the determination of their relative participation and for the calculation of corresponding rate coefficients.

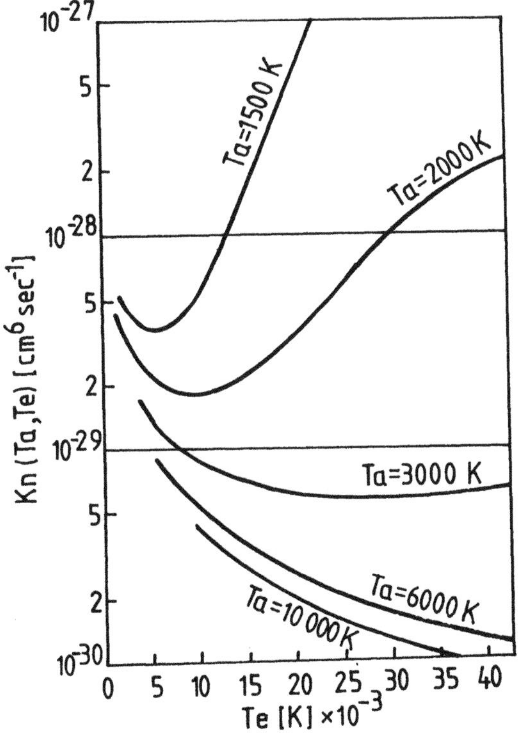

Fig. 1. Recombination rate coefficient $K_n(T_a, T_e)$ for $n = 4$ hydrogen case as a function of atomic temperature T_a and electron temperature T_e.

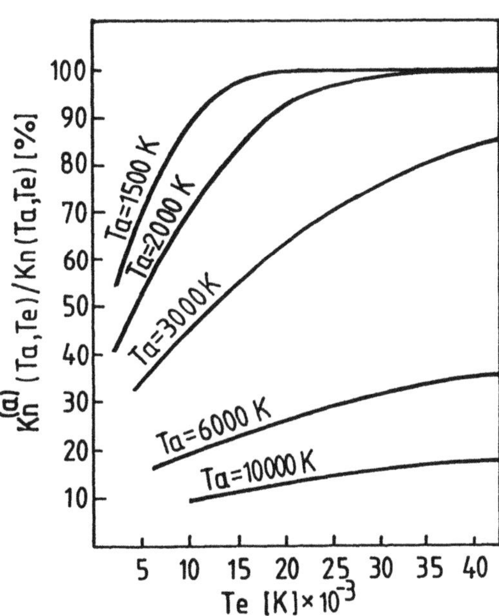

Fig. 2. Relative participation of process 1a $[K_n^{(a)}(T_a, T_e)/K_n(T_a, T_e)]$ for $n = 4$ hydrogen case as a function of atomic temperature T_a and electron temperature T_e.

References

Bates,D.R., Ledsham,K., Stewart,A.L.: 1953, Phil. Trans. Soc. London **246**, 215.

Janev,R.K., and Mihajlov,A.A.: 1980, Phys.Rev.A **21**, 819.

Massey,H., 1976, Negative Ions, Cambridge University Press, Cambridge, London, New York, Melbourne.

Mihajlov,A.A., and Dimitrijević,M.S.: 1986, Astron.Astrophys. **155**, 319.

Mihajlov,A.A., and Dimitrijević,M.S.: 1992, Astron.Astrophys. in press.

Mihajlov,A.A., and Janev,R.K.: 1981, J.Phys.B **14**, 1639.

Mihajlov,A.A., and Ljepojević,N.N., 1982, in Symp.Phys.Ioniz. Gases, Dubrovnik, 1982, Contrib.Papers, Inst.Phys.Univ. Zagreb, 385.

Stark Broadening Parameters for Spectral Lines of Multicharged Ions in Stellar Atmospheres: C IV, N V, O VI Lines and Regularities within an Isoelectronic Sequence

Milan S. Dimitrijević [1], Sylvie Sahal-Bréchot [2]

[1]Astronomical Observatory, Volgina 7, 11050 Beograd, Yugoslavia
[2]Laboratoire Astrophysique, Atomes et Molécules Département Atomes et Molécules en Astrophysique, Unité associée au C.N.R.S. No 812 Observatoire de Paris-Meudon, 92190 Meudon, France

Abstract: Using a semiclassical approach, we have calculated recently, electron-, proton-, and ionized helium-impact line widths and shifts for 39 C IV, 30 N V and 30 O VI multiplets. This comprehensive set of data has been used for the investigation of Stark broadening parameter regularities within isoelectronic sequences.

1 Introduction

For the investigations of hot star atmospheres it is of interest to know Stark broadening data for multiply charged ion lines as C IV, N V, OVI (see e.g. Werner, Heber and Hunger, 1991). In order to provide a method for quick interpolation of new data along an isoelectronic sequence it is of interest to investigate if a sufficiently regular behaviour of Stark broadening parameters along such a sequence exists. Moreover, Stark broadening of spectral lines has been taking a new interest in astrophysics (Seaton, 1987), owing to the recent development of researches on the physics of stellar interiors: in subphotospheric layers, the modellisation of energy transport needs the knowledge of radiative opacities and thus, certain atomic processes must be known with accuracy.

The present paper concerns C IV, N V and O VI lines. Beyond the interest for the stellar atmospheres investigation and the modellisation of stellar interiors, the knowledge of C IV, N V and O VI Stark broadening parameters is of great importance for a number of problems in astrophysics and plasma physics, since they have a high cosmical abundance and are present as impurity in many laboratory plasma sources. In order to provide reliable data for the mentioned lines broadened by collisions with all important charged perturbers in stellar plasmas, we have calculated electron-, proton-, and ionized helium-impact line widths and shifts for 39 C IV (Dimitrijević et al, 1991ab), 30 N V

(Dimitrijević et al, 1992a) and 30 O VI multiplets (Dimitrijević and Sahal-Bréchot, 1992b), using the semiclassical-perturbation formalism (Sahal-Bréchot, 1969ab). This is a part of an effort to provide reliable Stark broadening data for stellar plasma research (see the review on up to now performed calculations for He I, Na I, K I, F I, Be II, Mg II, Ca II, Sr II, Ba II, Si II, Ar II, Ga II, Ga III and several lines of other light elements, in Dimitrijević and Sahal-Bréchot, 1991).

2 Results and Discussion

The obtained results were used to investigate the behaviour of Stark broadening parameters within the isoelectronic sequencee in order to examine the use of such behaviour for the interpolation of new data of interest for the stellar plasma investigations.

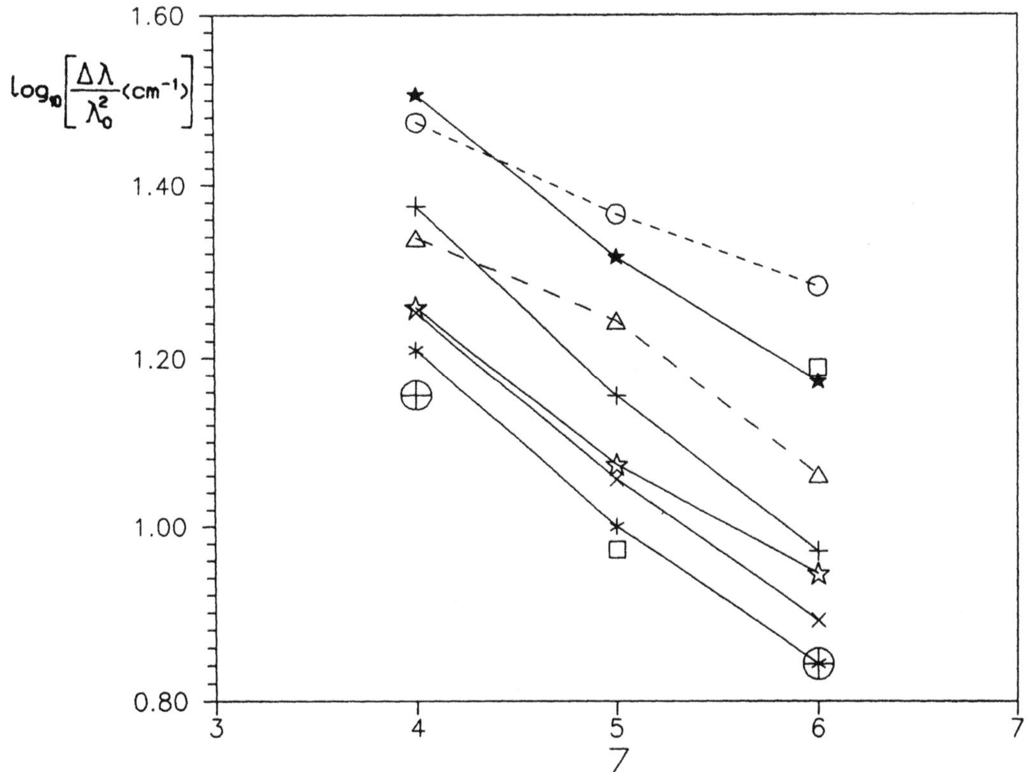

Fig. 1. Comparison between the experimental Stark full half width for $3s^2S_{1/2} - 3p^2P^0_{1/2}$ C IV, N V and O VI lines (◯) (Böttcher et al, 1988), with different Stark width (W in Å) calculations at the electron density $N = 1.8 \times 10^{18} cm^{-3}$ and T=145.000 K.
△ - present semiclassical calculations;
⊕ - quantum mechanical strong coupling calculations (Seaton, 1987);
× - modified semiempirical method (Dimitrijević and Konjević, 1980), calculated by Böttcher et al (1988);
+ - simplified semiclassical approach (Griem, 1974), calculated by Böttcher et al (1988);
∗ - quasiclassical Gaunt factor approach (Hey and Breger, 1980a,b), calculated by Böttcher et al (1988);
☆ - quasiclassical Gaunt factor approach (Hey and Breger, 1980c, 1982), calculated by Böttcher et al (1988);
⋆ - simplified approach of Baranger (1962), calculated by Böttcher et al (1988);
□ - estimate based on regularities (Purić et al, 1987, 1988), calculated by Böttcher et al (1988).

In Fig. 1 our results are compared with experimental data (Böttcher et al 1988) and with other calculations for 3s - 3p transition in the C IV, N V, O VI segment of Li isoelectronic sequence. We can see that a regular behaviour exist but additional experimental and theoretical work for the investigated case is needed as well as the extension to the other members of Li isoelectronic sequence.

References

Baranger,M.: 1962, in Atomic and Molecular Processes, ed. by D. R.Bates, Academic Press, New York.

Böttcher,F., Breger,P., Hey,J.D., Kunze,H.–J.: 1988, Phys.Rev. **38**, 2690.

Dimitrijević,M.S., and Konjević,N.: 1980, JQSRT **24**, 451.

Dimitrijević,M.S., and Sahal-Bréchot,S.: 1991, J.Phys.IV **1**, C1-111.

Dimitrijević,M.S.,Sahal–Bréchot,S.,and Bommier,V.:1991a,Astron. Astrophys.Suppl.Series **89**, 581.

Dimitrijević,M.S.,Sahal–Bréchot,S.,and Bommier,V.:1991b, Bull.Obs.Astron. Belgrade **144** in press.

Dimitrijević,M.S., and Sahal–Bréchot,S.: 1992a, Astron. Astrophys. Suppl.Series, in press.

Dimitrijević,M.S., and Sahal–Bréchot,S.: 1992b, Astron. Astrophys. Suppl.Series, in press.

Griem,H.R.: 1974, Spectral Line Broadening by Plasmas, Academic Press, New York.

Hey,J.D., and Breger,P.: 1980a, JQSRT **24**, 349.

Hey,J.D., and Breger,P.: 1980b, JQSRT **24**, 427.

Hey,J.D., and Breger,P.: 1980c, JQSRT **23**, 311.

Hey,J.D., and Breger,P.: 1982, S.Afr.J.Phys. **5**, 111.

Purić,J., Djeniže,S., Srećković,A., Platiša,M., Labat,J.: 1988, Phys.Rev.A **37**, 498.

Purić,J., Srećković,A., Djeniže,S., Platiša,M.: 1987 Phys.Rev.A **36**, 3957.

Sahal–Bréchot,S.: 1969a, Astron.Astrophys. **1**, 91.

Sahal–Bréchot,S.: 1969b, Astron.Astrophys. **2**, 322.

Seaton,M.J.: 1987, J.Phys.B **20**,6363.

Werner,K., Heber,U., and Hunger,K.: 1991, Astron.Astrophys. **244**, 437.

On Stark Line Shifts in Spectra of Very Hot Stars

Vladimir Kršljanin and Milan S. Dimitrijević

Astronomical Observatory, Volgina 7, 11050 Belgrade, Yugoslavia

Abstract: Starting from semiclassical calculations of Stark broadening parameters, Stark line shifts of several important UV lines of C IV, Si IV, N V and O VI in spectra of very hot main sequence and high-gravity stars have been estimated and possibilities for their observational confirmation have been discussed.

1. Introduction

Lines in spectra of very hot stars often reveal asymmetries and shifts, mainly caused by dynamics of stellar envelopes, mass loss, plasma waves, convective motions and gravitation. At the same time, collisions of emitting ions with charged particles in stellar atmospheres produce broadening and wavelength shift of spectral lines (Stark broadening and shift). Each layer in stellar atmosphere with its temperature and pressure conditions gives different Stark width and shift of a spectral line, and, as a result of radiative transfer, emerging flux profile is not only broadened, but also shifted and asymmetric. The influence of this effect is different for each spectral line, depending on its atomic parameters, atomic structure of emitting ion, and on perturber density and temperature distributions in line formation region. In most cases its contribution to the asymmetries and the shifts is much smaller as compared to the contribution caused by large scale motions in a stellar atmosphere. However, for a relatively "quiet" stellar atmosphere line asymmetries and shifts produced by Stark broadening could be of simmilar order of magnitude as "dynamical" ones. Consequently, it is necessary to investigate it in detail.

Influence of pressure broadening on asymmetries and shifts of solar lines (often reduced due to opposite signs of Stark and hydrogen-impact line shifts) has been discussed in *e.g.* Vince and Dimitrijević (1989) or Kršljanin *et al.*(1991). Kršljanin (1989a) demonstrated the behaviour of pressure shifts of two FeI lines in the spectra of A-G main sequence stars, finding its increase with effective temperature (Stark shifts were calculated according to approximative approach of Dimitrijević and Konjević (1986)).

In an extensive work (Kršljanin, 1989b), based on the modified semiempirical approach for electron-impact widths (Dimitrijević and Konjević, 1980) and shifts (Dimitrijević and Kršljanin, 1986) of ion lines, the contribution of Stark broadening to the observed line asymmetries and shifts in spectra of a large variety of very hot stars (includig white dwarfs) has been estimated and discussed. Pronounced Stark broadening effects have been found for the majority of metal lines in the spectra of subdwarfs and white dwarfs. Even in the spectra of main sequence stars observed line shapes and shifts can not be explained in detail without knowledge of accurate Stark broadenig parameters, particularly in the case of lines formed in photosphere.

Behaviour of pressure caused line asymmetries and shifts along the main sequence have been examined in Kršljanin and Marković-Kršljanin (1990), emphasizing that it is qualitatively simmilar (on both sides of granulation boundary) with the behaviour of the observed line patterns (blueshifted wings of the majority of cool star lines near the granulation boundary became redshifted since Stark broadening in atmospheres of hot stars predominates the hydrogen-impact broadening).

2. Results and Discussion

Stark broadening parameters for all most important C IV, Si IV, N V and O VI lines (see Dimitrijević and Sahal-Bréchot (1991) and references therein) have been calculated using the semiclassical-perturbation formalism (Sahal-Bréchot, 1969a,b).

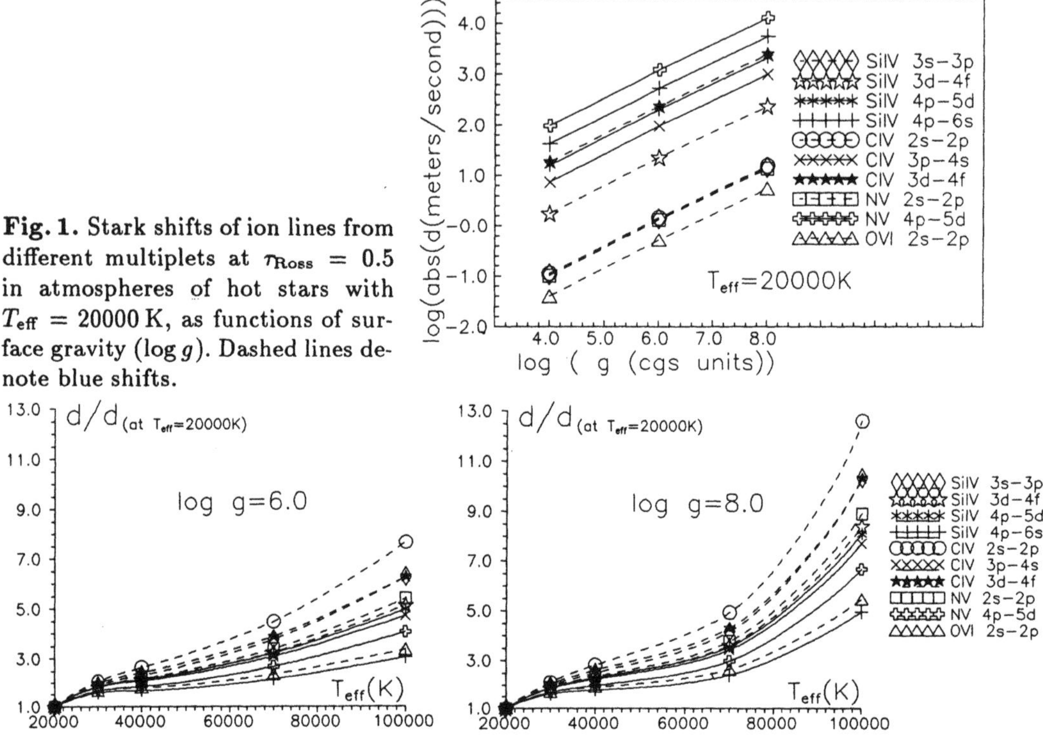

Fig. 1. Stark shifts of ion lines from different multiplets at $\tau_{\rm Ross} = 0.5$ in atmospheres of hot stars with $T_{\rm eff} = 20000$ K, as functions of surface gravity ($\log g$). Dashed lines denote blue shifts.

Fig. 2. Stark shifts (relative to the ones for stars with $T_{\rm eff} = 20000$ K) of ion lines from different multiplets at $\tau_{\rm Ross} = 0.5$ in atmospheres of hot high-gravity stars as functions of effective temperature. Dashed lines denote blue shifts.

Here are presented Stark shifts of a number of important UV lines of the mentioned ions in the line formation region of the atmospheres of O and B main sequence stars (Kurucz, 1979), subdwarfs and hot DA white dwarfs (Wesemael *et al.,* 1980). Figs. 1 and 2 show Stark shifts of these lines (in velocity units) as functions of surface gravity and effective temperature. Very prominent spectral features, UV resonant doublets of C IV, N V, O VI and Si IV, despite their high sensitivity to pressure broadening, exhibit very small (and very simmilar) Stark shifts. Other (photospheric) lines show observable (in some cases very large) Stark shifts. Influence of Stark broadening on their assymetries is ilustrated in Fig.3.

Due to known difficulties in absolute measurements of stellar spectral line shifts, a possible way of searching for an observational evidence of important Stark broadening contribution to these shifts is to estimate additional wavelength differeneces between two lines due to Stark shifts. A list of estimated relative Stark shifts for seven pairs of multiplets is given in Table 1. In the case of high-gravity stars these shifts typicaly range from several hundred meters per second for subdwarfs to several kilometers per second for white dwarfs. For the typical DA parameters (Koester *et al.*, 1979), influence of the Stark line shift to the stellar mass determined from the gravitational red shift (in solar mass units) is estimated as 0.0189 d[km/s] (Kršljanin, 1989b). In our small list, there are two pairs of multiplets (Nos. 3 and 7) with almost equal wavelengths and substantial relative shifts. In such cases, "dynamical" or gravitational shifts can be accurately distinguished and estimated.

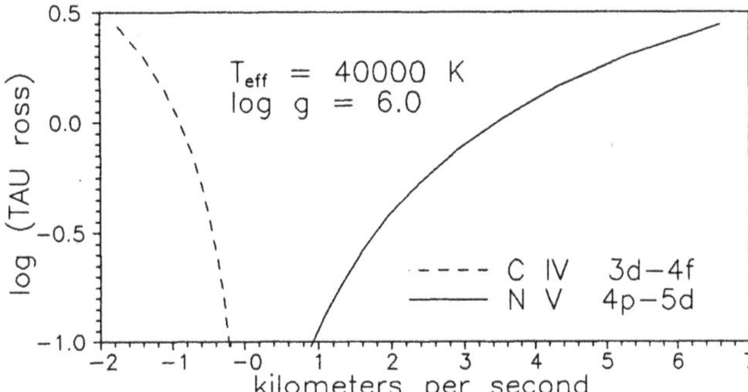

Fig. 3. Stark shifts of two multiplets in atmosphere of a hot subdwarf.

Table 1. Relative Stark shifts [km/s] of 7 pairs of multiplets at $\tau_{\rm Ross} = 0.5$

No.	WAVE-LENGTH	TRANSITION	log g	EFFECTIVE TEMPERATURE [K] 20000.	30000.	40000.	70000.	100000.
1	1169.0	CIV 3d-4f	4.0	0.025	0.034	0.025		
	1198.6	CIV 3d-4p	6.0	0.332	0.655	0.794	1.25	2.01
			8.0	3.50	7.01	8.94	14.6	34.7
2	1169.0	CIV 3d-4f	4.0	0.060	0.070	0.047		
	1230.0	SiIV 4p-6s	6.0	0.780	1.33	1.52	2.10	3.16
			8.0	8.21	14.2	17.1	24.6	54.3
3	1230.0	SiIV 4p-6s	4.0	0.034	0.035	0.022		
	1230.0	CIV 3p-4s	6.0	0.444	0.674	0.697	0.87	1.20
			8.0	4.67	7.20	7.84	10.2	20.4
4	1230.0	SiIV 4p-6s	4.0	0.042	0.044	0.029		
	1240.2	NV 2s-2p	6.0	0.541	0.850	0.901	1.17	1.66
			8.0	5.69	9.10	10.1	13.8	28.3
5	1230.0	CIV 3p-4s	4.0	0.007	0.009	0.006		
	1240.2	NV 2s-2p	6.0	0.097	0.176	0.204	0.300	0.460
			8.0	1.02	1.90	2.29	3.51	7.90
6	1367.6	SiIV 4p-5d	4.0	0.016	0.020	0.014		
	1396.7	SiIV 3s-3p	6.0	0.206	0.382	0.445	0.665	1.03
			8.0	2.17	4.09	5.01	7.75	17.6
7	1549.1	CIV 2s-2p	4.0	0.095	0.106	0.073		
	1549.1	NV 4p-5d	6.0	1.24	2.05	2.30	3.33	5.06
			8.0	13.0	21.9	25.8	38.9	86.9

References

Dimitrijević M.S., Konjević N., 1980, JQSRT 24, 451

Dimitrijević M.S., Konjević N., 1986, A&A 163, 297

Dimitrijević M.S., Kršljanin V., 1986, A&A 165, 269

Dimitrijević M.S., Sahal-Bréchot S., 1991, this volume

Koester D., Schulz,H., Weidemann,V., 1979, A&A 76, 262

Kršljanin V., 1989a, in: Solar and Stellar Granulation, eds. R.J.Rutten, G.Severino, Kluwer, Dordrecht, p.91

Kršljanin V., 1989b, Ion Lines Stark Shifts in Spectra of Hot Stars, Publ.Obs.Astron.Belgrade No.37

Kršljanin V., Marković-Kršljanin S., 1990, in: XV SPIG Contributed Papers, ed. D.Veža, Inst.Phys.Univ., Zagreb, p.370

Kršljanin V., Vince I., Erkapić S., 1991, in: The Sun and Cool Stars: activity, magnetism, dynamos, eds. I.Tuominen *et al.*, Lect.Not.Phys. 380, Springer-Verlag, Berlin, p.277

Kurucz R.L., 1979, ApJS 40, 1

Sahal-Bréchot S., 1969a, A&A 1, 91

Sahal-Bréchot S., 1969b, A&A 2, 322

Vince I., Dimitrijević M.S., 1989, in: Solar and Stellar Granulation, eds. R.J.Rutten, G.Severino, Kluwer, Dordrecht, p.93

Wesemael F., Auer L.H., Van Horn H.M., Savedoff M.P., 1980, ApJS 43, 2

VII. Model Atmospheres and Radiative Transfer

Accelerated Lambda Iteration

Ivan Hubeny

Universities Space Research Association, NASA Goddard Space Flight Center,
Code 681, Greenbelt, MD 20771, USA

Abstract: Accelerated Lambda Iteration, or ALI, methods are reviewed. An emphasis is given to the critical evaluation of various methods, analysing their physical and mathematical meaning, and recommending the most advantageous methods to interested non-specialists who consider applying these methods to solving actual line formation and model stellar atmosphere problems.

1 Introduction

The recent dramatic advances in fast numerical methods of solving the radiative transfer problem have spurred impressive progress in stellar spectroscopic diagnostics. We are now able to model stellar spectra with an unprecedented degree of realism and accuracy. Transfer solutions taking into account hundreds of atomic energy levels and several hundreds to thousands transitions are now no longer exceptional.

A central role in these developments is played by a group of methods referred to as *Accelerated Lambda Iteration* (ALI) methods. There have been several excellent review papers published recently. Rybicki (1991) reviews the recent progress, while Kalkofen (1987; his paper and the entire first part of the book) extensively reviews the earlier approaches. Also, there are many papers, not intended as reviews, which nevertheless contain excellent discussions and analyses from very diverse viewpoints, and are highly recommended to the reader (Rybicki and Hummer 1991; Puls 1991; Klein *et al.* 1989; Puls and Herrero 1988; Olson and Kunasz 1987), to name only few.

Is there then any reason for yet another review paper on ALI methods? The organizers of the workshop apparently believe so. I order to fulfill my task, I will try to make this review useful by concentrating on two issues. First, I would like to provide a historical overview of the ALI approaches, tracing the roots of their development. Second, since I do not have any personal method at stake, I will try to critically evaluate the usefulness of the individual methods from the point of view of their potential user. In business terminology, this paper is intended not as a commercial, but rather as a consumer's report.

2 The Problem

2.1 Formulation

Let us demonstrate the basis of the problem with the following simple case, namely the standard two-level atom problems in a homogeneous, static medium. The problem consists of solving simultaneously two basic equations. The first one is the radiative transfer equation, written as

$$\mu \frac{dI_{\mu\nu}}{d\tau_\nu} = I_{\mu\nu} - S_\nu \quad , \tag{1}$$

where $I_{\mu\nu}$ is the specific intensity of radiation, and S_ν the source function. The specific intensity depends on three variables, the frequency ν, the directional cosine μ, and the (monochromatic) optical depth τ_ν. Traditionally, ν and μ are written as subscripts. Let us further assume no overlapping continuum and complete frequency redistribution (CRD), in which case the source function is independent of frequency, $S_\nu \equiv S$. The formal solution of (1) may be written as

$$I_{\mu\nu} = \Lambda_{\mu\nu}[S] \quad , \tag{2}$$

where Λ operates on the quantity within [ ]. Other forms are

$$J_\nu = \Lambda_\nu[S], \quad \text{where} \quad J_\nu = \frac{1}{2}\int_{-1}^{1} I_{\mu\nu} d\mu \quad \text{and} \quad \Lambda_\nu = \frac{1}{2}\int_{-1}^{1} \Lambda_{\mu\nu} d\mu \quad , \tag{3}$$

and

$$\bar{J} = \bar{\Lambda}[S], \quad \text{where} \quad \bar{J} = \int J_\nu \phi_\nu d\nu \quad \text{and} \quad \bar{\Lambda} = \int \Lambda_\nu \phi_\nu d\nu \quad , \tag{4}$$

where ϕ_ν the (normalized) absorption profile. J_ν and $\bar{J}$ are mean intensity and frequency-averaged mean intensity, respectively, The same terminology should apply to different Λ operators, although this distinction is often missing in the literature. In the following, we will omit the bar in $\bar{\Lambda}$ and will simly write Λ.

The second basic equation is the equation of statistical equilibrium, which for a two-level atom may simply be written as an expression for the line source function,

$$S = (1-\epsilon)\bar{J} + \epsilon B \quad , \tag{5}$$

where ϵ is the collisional destruction probability, and B the Planck function.

For multilevel atoms, one may formally express the solution as

$$S_{ij} = S_{ij}(\boldsymbol{n}) \qquad \text{and} \qquad \boldsymbol{n} = \boldsymbol{n}(\bar{J}_{ij}) \quad , \tag{6}$$

where $\boldsymbol{n}$ is the vector of level populations, and subscripts ij indicate transition $i \rightarrow j$.

Equations (5) and (6) demonstrate mathematically what is clear on physical grounds: the problem arises in the *coupling* of the physical variables. The fact that (1) is a differential equation implies coupling of *depths*. Moreover, the source function couples all *frequencies* and *angles*, thus making the simple linear differential equation (1) a more complicated integro-differential equation. Finally, for multi-level problems, there is the additional coupling of source functions in different *transitions* via the statistical equilibrium condition.

Numerically, the coupling means matrices, and solving a coupled problem numerically means inverting large matrices. The realization that coupling is extremely important and finding the means with which to deal with it were the major achievements of the 1960's and 1970's, while the realization that some part of the coupling could be treated iteratively and some part globally, thus increasing enormously the computational speed of the process, was the major achievement of the late 1970's and the 1980's, and this trend will certainly continue in the 1990's.

2.2 Lambda Operator and its Matrix Representation

The Lambda operator plays a central role in the whole theory, so some explanatory comments are in order:

i) Equations (2) – (4) involve Λ written as an operator. In fact, it does not have to be constructed explicitly. Λ is normally thought of as a *process* of explicitly calculating the intensity from the source function. Any method of solving the transfer equation, as for instance by the familiar Feautrier (1964) difference-equation method, may be used.

ii) If Λ is to be understood as an explicit operator, it should be stressed that since it acts on a continuous function of position (the source function), it is generally represented by an infinite matrix. However, any practical problem involves some kind of discretization, and Λ appears as a finite matrix. Nevertheless, we shall still refer to the discretized form as an "exact" Λ operator.

iii) What is the meaning of the Λ matrix? To clarify this, let us first rewrite (2), (3), or (4) as

$$J_d = \sum_{d'=1}^{D} \Lambda_{dd'} S_{d'} \quad , \tag{7}$$

where d denotes the depth index. Equation (7) may be written for the monochromatic specific intensity $I_{\mu\nu}$, or mean intensity J_ν, or $\bar{J}$; using the appropriate Λ. Equation (7) may be thought of as the discrete representation of, for instance, the formal solution for the mean intensity (Mihalas 1978), $J_\nu(\tau_\nu) = (1/2)\int_0^\infty dt_\nu S_\nu(t_\nu) E_1(|t_\nu - \tau_\nu|)$, E_1 being the first exponential integral; this representation is not useful because exponential integrals are rather costly to compute.

A more advantageous way to express Λ is the following. Let us take, quite formally, all elements of the source function vector to be zero except the i-th element which is taken to be 1, $S_d = \delta_{di}$. Then

$$\begin{pmatrix} J_1 \\ J_2 \\ \vdots \\ J_D \end{pmatrix} = \begin{pmatrix} \Lambda_{11} & \Lambda_{12} & \dots & \Lambda_{1D} \\ \Lambda_{21} & \Lambda_{22} & \dots & \Lambda_{2D} \\ \vdots & \vdots & \ddots & \vdots \\ \Lambda_{D1} & \Lambda_{D2} & \dots & \Lambda_{DD} \end{pmatrix} \times \begin{pmatrix} 0 \\ \vdots \\ 1 \\ \vdots \end{pmatrix} = \begin{pmatrix} \Lambda_{1i} \\ \Lambda_{2i} \\ \vdots \\ \Lambda_{Di} \end{pmatrix} \quad . \tag{8}$$

In other words, the i-th column of the Λ matrix is a solution of the transfer equation with the source function given as a unit pulse function. Physically, the i-th column of Λ

therefore describes how the puls which originated at the i-th depth point spreads over all depths.

iv) Another, but related, way to construct the (monochromatic) Λ matrix explicitly is to start with the Feautrier form of (1),

$$\mu^2 \frac{d^2u}{d\tau_\nu^2} = u - S \quad , \tag{9}$$

where $u = (I_{\mu\nu} + I_{-\mu\nu})/2$. Its discretized form is written as

$$-A_d u_{d-1} + B_d u_d - C_d u_{d+1} = S_d \quad , \tag{10}$$

or, in the matrix form (omitting subscripts ν and μ),

$$\boldsymbol{T}\boldsymbol{u} = \boldsymbol{S} \quad , \tag{11}$$

where $\boldsymbol{T}$ is a tridiagonal matrix with elements A, B, C (for details refer, for instance, to Mihalas 1978). The relation between $\boldsymbol{T}$ and Λ is then

$$(\boldsymbol{T})^{-1} = (\Lambda_{\mu\nu} + \Lambda_{-\mu\nu})/2 \quad . \tag{12}$$

The fact that matrix $\boldsymbol{T}$ is tridiagonal may be used to advantage for constructing the explicit Λ operator (Rybicki and Hummer 1991; see Sect.4.3).

v) Equation (7) clearly shows the following trivial, but nontheless important fact, namely that a *diagonal* Λ operator means a *local* Λ operator; in other words $J(\tau) = \Lambda(\tau)S(\tau)$, where $\Lambda(\tau)$ has now the meaning of a single number.

vi) Equations (10) – (12) help to clarify another important point, namely that the elements of the Λ matrix are functions of the optical depth differences. This does not matter for the two-level problems since the optical depth is the basic depth variable which is not changed during the iteration process. But it does matter for multilevel problems, where the individual mean optical depths for the individual transitions, being dependent on the corresponding lower level population, *change* during the iteration process. For these problems, the "exact" Λ only means exact on the current, approximate, optical depth scale.

3 Ordinary versus Accelerated Lambda Iteration

Combining (4) and (5), we obtain in the standard simplest case (two-level, no continuum, CRD)

$$S = (1-\epsilon)\Lambda[S] + \epsilon B \quad . \tag{13}$$

Since Λ is a linear operator, (13) may in fact by solved in a single step. However, due to the above mentioned coupling, the matrix representing Λ may be enormous, and therefore a direct solution may be impractical. Another concern is that even if Λ is simple enough to warrant direct solution in a few individual cases, it may still be too time-consuming in problems requiring solution of (13) millions of times, as may easily occur in hydrodynamical simulations. For all these reasons, developing simpler, iterative methods is highly desirable.

3.1 Ordinary Lambda Iteration

The simplest possible iteration scheme is the notorious "ordinary" lambda iteration. Denoting the n-th iterate of the source function as $S^{(n)}$, we may write

$$S^{(n+1)} = (1-\epsilon)\Lambda[S^{(n)}] + \epsilon B \quad , \tag{14}$$

or, denoting

$$\bar{J}^{\mathrm{FS}} = \Lambda[S^{(n)}] \qquad \text{and} \qquad S^{\mathrm{FS}} = (1-\epsilon)\bar{J}^{\mathrm{FS}} + \epsilon B \quad , \tag{15}$$

then the basis of the lambda iteration consists in

$$S^{(n+1)} - S^{(n)} = S^{\mathrm{FS}} - S^{(n)} \quad . \tag{16}$$

The usefulness of (16) will become apparent later, when comparing to (19). The superscript FS stands for Formal Solution, which means that one solves (1) with a *known*, prescribed source function. The advantage of (14) or (16) can be seen immediately: all the coupling is treated iteratively; one performs only (cheap) formal solutions of the transfer equation, for one frequency-angle point at a time. Unfortunately, this method fails in most cases of interest (i.e., when scattering is important). The reasons for this failure are well understood (see, e.g., Mihalas 1978): the iterations correspond to following successive photon scatterings in the medium. If the number of scatterings is large, which is the usual case in the astrophysical applications (i.e. small ϵ and large optical depth), an impractical number of iterations would be required to get the converged solution. And worse yet, Lambda iteration tends to *stabilize*, not to *converge* the solution, so that a smallness of relative corrections is no guarantee of convergence (for a very illustrative example, see Auer 1991; his Fig.1). An illuminating mathematical discussion is given by Olson, Auer, Buchler (1986).

3.2 Accelerated Lambda Iteration

In a seminal paper Cannon (1973a) introduced into astrophysical radiative transfer theory the idea of "operator splitting", long known in numerical analysis. The idea consists of writing

$$\Lambda = \Lambda^* + (\Lambda - \Lambda^*) \quad , \tag{17}$$

where Λ^* is an appropriately chosen ***approximate lambda operator***. The iteration scheme (14) may then be rewritten to read

$$S^{(n+1)} = (1-\epsilon)\Lambda^*[S^{(n+1)}] \;+\; (1-\epsilon)(\Lambda-\Lambda^*)[S^{(n)}] \;+\; \epsilon B \quad , \tag{18}$$

or

$$S^{(n+1)} - S^{(n)} = [1-(1-\epsilon)\Lambda^*]^{-1}[S^{\mathrm{FS}} - S^{(n)}] \quad . \tag{19}$$

The last equation is particularly instructive. It shows that iteration is driven, similarly as the ordinary lambda iteration, by the difference between the old source function and the new source function obtained by formal solution. However, unlike the ordinary lambda iteration, this difference is amplified by the "acceleration operator" $[1-(1-\epsilon)\Lambda^*]^{-1}$.

To gain more insight, let us consider a diagonal (i.e. local) Λ^* operator. The appropriate Λ^* has to be chosen such as $\Lambda^*(\tau) \rightarrow 1$ for large τ (see Sect. 4). Since in typical cases $\epsilon \ll 1$, the acceleration operator indeed acts as a large amplification factor. This interpretation was first introduced by Hamann (1985) and Werner and Husfeld (1985), who also coined the term "Accelerated Lambda Iteration" (ALI). The acronym ALI is also sometimes understood to mean "Approximate Lambda Iteration". Other terms for ALI are *Operator Perturbation* (Kalkofen 1987), or *Approximate-Operator Iteration* (AOI; Castor *et al.* 1991). Finally, the term "accelerated Λ iteration" should not be confused with "acceleration of convergence", discussed in 4.5.

4 Linear Problems

Equation (18) provides us with the basis for understanding the requirements on constructing the Λ^* operator. First, we see that for $\Lambda^* = 0$ we recover the ordinary lambda iteration, with its disadvantage of a slow convergence; while for $\Lambda^* = \Lambda$ we recover the exact method, which is costly. So, in order that Λ^* brings an essential improvement over both methods, it has to satisfy the following requirements: i) it has to incorporate all the *essential* properties of the exact Λ operator; but at the same time, ii) it must be easy (and cheap) to invert. These requirements are generally incompatible, therefore the construction of the optimum Λ^* is a delicate matter, as the following section will demonstrate.

4.1 An Overview

Table 1 summarizes basic papers which introduced, in one way or another, a new type of Λ^* operator. The list does not contain the application–minded papers, or papers containing just very slight variants of existing approaches. The list is by no means exhaustive, and I apologize in advance to all whose work was omitted here.

Generally speaking, there were three milestone papers. The first one, as already mentioned, is the paper by Cannon (1973a; also 1973b). Although the methods themselves did not survive, they introduced the basic idea. The Λ^* operator (or, rather, procedure, since Λ was fully implicit) was represented by replacing integrals over angles and frequencies by a low-order quadrature formulae; the "exact" Λ was then represented by higher-order quadratures. Historically, his methods were introduced during the period of greatest development of complete linearization, and were thus largely overlooked by most radiative transfer specialists, who enthusiastically pursued the newly found ability to calculate, for the first time, realistic multilevel line transfer problems.

The second milestone is a brilliant paper by Scharmer (1981). Interestingly enough, its main thrust is that he combined results of two seminal papers from roughly the same epoch, which both were more or less overlooked in their time. The first is Cannons's paper, and the second is a paper by Rybicki (1972), where the extremely important concept of *core saturation* was introduced. This concept, the Scharmer method, and its subsequent modifications by the Kiel group, will be discussed in more detail in the next section.

From the historical perspective, Cannon's approach was mostly *mathematically* motivated, while Scharmer's and subsequent variants were more *physically* motivated. The basic "physics" ingredient – core saturation – brought both a deeper understanding of

Table 1. History of Λ^*

Reference	Construction of Λ^*
Cannon 1973a	Angular quadrature of low order
Cannon 1973b	Frequency quadrature of low order
Scharmer 1981	Core saturation + Eddington–Barbier relations
Hamann 1985	Core saturation for moving atmospheres in CMF
Werner, Husfeld 1985 (WH)	Scharmer; also purely local (core saturation)
Hamann 1986	Empirically modified core saturation (CMF)
Werner 1986	Empirically modified core saturation
Hempe, Schönberg 1986	Approximate Newton–Raphson (ANR) operator for moving atmospheres in CMF
Olson, Auer, Buchler 1986 (OAB)	Approximate diagonal of exact Λ
Olson, Kunasz 1987 (OK)	Exact diagonal or tri–diagonal of exact Λ (by the short characteristics method)
Puls, Herrero 1988	Analogy of OAB for spherical atmospheres
Klein *et al.* 1989	Double–splitting iteration
Hillier 1990	Multiband ANR for moving atmospheres in CMF
Steiner 1991	Adaptation of a multi–grid method to ALI
Rybicki, Hummer 1991	Exact diagonal or multiband part of exact Λ within the Feautrier formalism
Puls 1991	Analogy of OAB for moving atmospheres in CMF

CMF = comoving–frame formalism

the problem, as well as numerical advantages. However, Scharmer's approach possessed an unfortunate drawback, namely that one had to choose: either to use a parameter–free operator (which was still relatively complicated and costly to invert), or to take a much simpler one, which nevertheless involved an arbitrary, adjustable parameter, γ, or even more free parameters. Unfortunately, the speed of convergence was found to depend rather dramatically on the adopted value of γ, which generally depends on the problem at hand. Therefore, a good deal of experimentation was necessary in some cases to find adequate parameter values. An interesting paper by Puls and Herrero (1988) examines this situation in retrospect, using more modern approaches to demonstrate that in some cases it would have been nearly impossible to find an optimum value of the core saturation parameter γ by the simple method of trial and error.

What was needed to improve the situation was further progress in *mathematics*, and this is precisely what was achieved in the third milestone paper, Olson, Auer, and Buchler (1975 – OAB), who were the first to demonstrate mathematically that a nearly "optimum" Λ^* is simply the diagonal of the exact Λ operator. The art of constructing Λ^* then went from the realm of physics to the realm of mathematics and numerics, and will very likely remain there.

In the following subsections, I will discuss the individual ALI methods in more detail. Due to the space limitations, I have to leave out the double–splitting iteration method by Klein *et al.* (1989; which is related, but not identical, to ALI approaches), and the multi–grid method of Steiner (1991). The methods dealing specifically with velocity fields will be considered in Sect. 6.

4.2 Core Saturation and Related ALI Methods

The core saturation method was introduced by Rybicki (1972) and was extensively reviewed by Rybicki (1984). Very briefly, the method is based on the observation that the specific intensity saturates to the source function for large enough monochromatic optical depths; this frequency region is called the *core*:

$$I_\nu(\tau_\nu) = S_\nu(\tau_\nu), \quad \nu \in \text{core} \quad (\tau_\nu \geq \gamma) \quad , \tag{20}$$

where γ is the above mentioned adjustable free parameter, basically of the order of unity (typically 2 – 5). In the original variant of core saturation, no special assumptions are made about the specific intensity in the wing ($\tau_\nu < \gamma$), which may be determined essentially exactly.

In the context of ALI, Scharmer (1981) used (20) together with an approximation for the wing component, based on generalizing the idea of Eddington–Barbier relations, namely

$$I_\nu(\tau_\nu) = S_\nu(\tau_\nu + \gamma'), \quad \nu \in \text{wing} \quad , \tag{21}$$

where again γ' is a parameter of the order of unity. Written explicitly, Scharmer's operator (angle-averaged, monochromatic) is defined by

$$\Lambda^*_\nu\,[S(\tau_\nu)] = \begin{cases} S(\tau_\nu) & \text{for } \tau_\nu \geq \gamma; \\ (1/2)S(\tau_\nu = \gamma'), & \text{for } \tau_\nu < \gamma. \end{cases} \tag{22}$$

Scharmer also suggested an improved, parameter–free operator (written here for semi–infinite atmospheres)

$$\Lambda^*_{\mu\nu} = \begin{cases} S(\tau_{\mu\nu} + 1), & \text{for } \mu > 0; \\ (1 - e^{-\tau_{\mu\nu}})\, S\,(\tau_{\mu\nu}/(1 - e^{-\tau_{\mu\nu}}) - 1), & \text{for } \mu < 0. \end{cases} \tag{23}$$

where $\tau_{\mu\nu} = \tau_\nu/\mu$, which yields a faster convergence, but is significantly more complicated than (22). Nevertheless, it became the basis for the subsequent multilevel formulation (Scharmer 1984; Scharmer and Carlsson 1985).

Hamann (1985) and Werner and Husfeld (1985 – WH) used the simpler Scharmer operator (22) for formulating comoving–frame (CMF), and static multilevel problems, respectively. However, they also used an even simpler variant of (22), namely

$$\Lambda^*_\nu\,[S(\tau_\nu)] = \begin{cases} S(\tau_\nu), & \text{for } \tau_\nu \geq \gamma; \\ 0, & \text{for } \tau_\nu < \gamma. \end{cases} \tag{24}$$

We note that in this case one may write a particularly simple and instructive expression for the frequency averaged Λ^*, which is now a *local* (diagonal) operator

$$\Lambda^*\,[S(\tau_\nu)] = N_C(\tau_\nu)S(\tau_\nu), \tag{25}$$

where $N_C = \int_{\text{core}} \phi_\nu d\nu$ is the so–called core normalization, in the nomenclature of Rybicki (1984). Here we see an intimate connection to the escape probability theory, since the frequency averaged escape probability may be well approximated by $(1-N_C)/2$ (for more details, see Rybicki 1984).

As stressed by WH, operators (24) or (22) are very advantageous numerically, since they are either purely local (24), or upper–triangular (22), which yields a uni–directional flow of information from the bottom to the top of the atmosphere (provided that γ' is chosen so that $\gamma' \geq \gamma$). But, as it was already pointed out, the basic disadvantage of this method is its dependence on the choice of parameter γ. The limit $\gamma \to \infty$ corresponds to the ordinary lambda iteration. The smaller the value of γ, the larger the amplification of corrections – see (19), but too small a γ yields overcorrections and can cause divergence.

Another uncertainty of these approaches is their treatment of continua. Strictly speaking, the concept of core saturation applies only to lines. Although it may be easily extended to continua as well, the experience has revealed convergence problems, particularly when dealing with very optically thick UV continua. To cope with these problems, Hamann (1986) and Werner (1986) have introduced a modification of (22) or (24),

$$\Lambda^*_\nu \; [S(\tau_\nu)] = S(\tau_\nu)(1 - e^{-\tau_\nu/\beta}), \quad \text{for } \tau_\nu \geq \gamma, \tag{26}$$

where β is another free parameter of the order of unity to several tens.

4.3 OAB, or Why Do We Need to Care about Eigenvalues?

Any iterative scheme can be written in the form

$$\boldsymbol{x}^{(n+1)} = \boldsymbol{F} \cdot \boldsymbol{x}^{(n)} + \boldsymbol{x}^{(0)} \quad , \tag{27}$$

where $\boldsymbol{F}$ is called the *amplification matrix*. In the case of the linear transfer problem, (18), $\boldsymbol{F}$ is given by

$$\boldsymbol{F} = [1 - (1-\epsilon)\Lambda^*]^{-1} \, [(1-\epsilon)(\Lambda - \Lambda^*)]. \tag{28}$$

The converged solution may be written $\boldsymbol{x}^{(\infty)} = \boldsymbol{F} \cdot \boldsymbol{x}^{(\infty)} + \boldsymbol{x}^{(0)}$. If we now define the relative error of the n-th iterate, $\epsilon^{(n)} = \boldsymbol{x}^{(n)} - \boldsymbol{x}^{(\infty)}$ and expand it in terms of eigenvectors of $\boldsymbol{F}$,

$$\epsilon^{(n)} = \sum c_i \boldsymbol{u}_i \quad , \tag{29}$$

then the relative error of the $(n+1)$-th iterate may therefore be written

$$\epsilon^{(n+1)} = \boldsymbol{F} \cdot \epsilon^{(n)} = \sum \lambda_i c_i \boldsymbol{u}_i \quad , \tag{30}$$

where λ_i are the eigenvalues and $\boldsymbol{u}_i$ the eigenvectors of the amplification matrix.

Now we see why the eigenvalues of the amplification matrix are so critical: the problem converges only if $\lambda_{max} \equiv \max|\lambda_i| < 1$. Moreover, after several iterations the eigenvector with the largest eigenvalue will dominate, so $\epsilon^{(n+1)} \to \lambda_{max}\epsilon^{(n)}$. Therefore, the smaller λ_{max}, the faster the convergence will be. This is the first important achievement of the OAB paper, namely providing an objective criterion by which to judge the quality of the Λ^* operator. The second, and most basic achievement of OAB was that they showed that a nearly optimum Λ^* is simply the diagonal of the exact Λ. The mathematical background for this assertion is provided by Gerschgorin's theorem, which states that if we subtract off the diagonal, then the largest eigenvalue is bounded by the sum of the off diagonal elements of the rows of the matrix (OAB; for a thorough discussion, see Puls and Herrero 1988).

As discussed in Sect 2.2, the construction of the diagonal of Λ is equivalent to solving the transfer problem D times, D being the total number of depth points, and may therefore be time-consuming. Three different ways of constructing the diagonal of Λ were suggested.

i) First, OAB themselves suggested an approximate evaluation of the diagonal. They do not present explicit expressions but only the procedure for how the approximate diagonal may be evaluated. Working out their prescription, and adopting the Feautrier scheme for simplicity (an alternative possibility is to use the Hermite scheme of Auer, 1976, which in fact was done by OAB), we obtain for the angle–averaged OAB Λ^* operator (see also Puls and Herrero, 1988),

$$\Lambda_\nu^* = \int_0^1 d\mu \left[1 + \left(\frac{1-e^{-\Delta\tau_d}}{\Delta\tau\Delta\tau_d} + \frac{1-e^{-\Delta\tau_{d-1}}}{\Delta\tau\Delta\tau_{d-1}}\right)\right]^{-1} , \tag{31}$$

where $\Delta\tau_d = (\tau_{d+1} - \tau_d)/\mu$, and $\Delta\tau = (\Delta\tau_d + \Delta\tau_{d-1})/2$ (subscript ν was omitted). Once again, we stress that no adjustable parameter is required. Moreover, OAB, and, subsequently, Puls and Herrero (1988), demonstrated that the maximum eigenvalue of the amplification matrix corresponding to (31) is comparable to the core–saturation scheme (diagonal Hamann or Werner–Husfeld operator) with its optimum γ (which, of course, is not known *a priori*).

ii) Olson and Kunasz (1987 – OK) suggested a way by which to calculate the diagonal elements of Λ *exactly*, using the so–called *short characteristics* method. This method uses (8) to evaluate Λ, and then expresses the intensity at some mesh point through intensity in the neighboring mesh points by means of the formal solution of the transfer equation. The diagonal of Λ is then (in their linear approximation)

$$\Lambda_\nu^* = 1 - \frac{1}{2}\int_0^1 d\mu \left(\frac{1-e^{-\Delta\tau_d}}{\Delta\tau_d} + \frac{1-e^{-\Delta\tau_{d-1}}}{\Delta\tau_{d-1}}\right) . \tag{32}$$

OK introduced another important improvement in allowing Λ^* to be not only the diagonal, but also the tridiagonal or even higher–order bands of the exact Λ. They tested the tridiagonal operator numerically, and found a considerable acceleration of the convergence with respect to the diagonal form.

iii) To date, the last word in the problem of constructing diagonal, and higher order bands, of the true Λ was spoken by Rybicki and Hummer (1991). They used a formalism based on (9) - (12), employing a very efficient algorithm for inverting a tridiagonal matrix. They demonstrated that the entire set of the diagonal elements of Λ can be found with an order of D operations. This feature makes it the method of choice, since it avoids computing costly exponentials, a problem inherent to both the previous approaches (OAB, OK).

4.4 Approximate Newton–Raphson Operator

Finally, I will briefly mention methods based on the so-called Approximate Newton–Raphson (ANR) operator. The idea was first suggested by Hempe and Schönberg (1986), who used a diagonal part of the exact NR operator, and later extended by Schönberg and Hempe (1986) for multilevel atoms, and extended to tridiagonal and even pentadiagonal operator by Hillier (1990). The approach is conceptually very close to the usual ALI approach. The idea is to expand source function linearly around the current estimate (written here for simplicity for a two-level atom),

$$(S^0 + \Delta S) = (1-\epsilon)(\bar{J}^0 + \Delta\bar{J}) + \epsilon B \quad , \tag{33}$$

and express $\Delta\bar{J}$ through ΔS by means of the Jacobi matrix, as in any ordinary Newton–Raphson approach,

$$(S_d^0 + \Delta S_d) = (1-\epsilon)\left[\bar{J}_d^0 + \sum_{d'} \frac{d\bar{J}_d}{dS_{d'}} \Delta S_{d'}\right] + \epsilon B \quad . \tag{34}$$

But, in this case we know from (7) that $d\bar{J}_d/dS_{d'} = \Lambda_{dd'}$. In other words, the NR operator for this case is equal to the Λ operator and, consequently, any approximation for the Newton–Raphson operator is equivalent to the corresponding approximation for the true Λ operator. Hempe and Schönberg (1986) used the diagonal part, therefore their method (in the linear case) should be quite analogous to OAB.

4.5 Acceleration of Convergence

Acceleration of convergence has recently become an important ingredient of the ALI methods. It was reviewed extensively by Auer (1987; 1991), so I present here only a very brief summary of the basic ideas.

First, as follows from (27) – (30), in any iteration method where the $(n+1)$-th iterate is evaluated by means of the previous one, convergence may only be *linear*, and therefore generally slow. However, taking into account information from earlier iterates, one may find faster schemes. The first application of this idea was accomplished by Hamann (1981), who used the so-called Aitken extrapolation scheme to accelerate the ordinary Lambda iteration. Buchler and Auer (1985) and OAB introduced into ALI methods a powerful acceleration technique by Ng (1974), which was subsequently used by many authors. More recently, Klein *et al.* (1989) used another acceleration technique called Orthomin, developed originally by Vinsome (1976), and found it to be even more efficient than the Ng acceleration.

The general expression for the accelerated estimate of the solution of (27) is written

$$\boldsymbol{x} = \left(1 - \sum_{m=1}^{M} \alpha_m\right)\boldsymbol{x}^n + \sum_{m=1}^{M} \alpha_m \boldsymbol{x}^{n-m} \quad , \tag{35}$$

where the coefficients α are determined either by residual minimization (Ng), or minimization with respect to a set of conjugate vectors (Orthomin).

5 Multilevel Problems

5.1 An Overview

Most of basic methods discussed in the previous section (Cannon; Scharmer; OAB; OK). have been developed for the linear, two–level atom problem. However, any practical application requires considering realistic multilevel atoms, and the problem becomes inherently non–linear.

There are three basic ways to deal with this non–linearity. The first one is to linearize (called category *I* in the following), while the second one avoids linearization, for instance by means of ***preconditioning*** of the statistical equilibrium equations (category *II*). The third one also avoids linearization, by means of the equivalent–two–level atom (ETLA) approach (category *III*). The first way, linearization, actually comes in two different flavors: one may either first use ALI to formulate source function in all transitions (i.e. via equation (6) in our terminology), and then linearize the resulting set of equations (category *Ia*), or first linearize the basic equations and then use the idea of ALI to simplify the resulting equations (category *Ib*).

Table 2 summarizes basic papers which deal with the multilevel problem. It concentrates on papers which have introduced new ideas, and is by no means complete. Again, application–oriented papers papers are excluded. I include my paper (Hubeny 1975) only for the sake of historical completeness, since it seems to be the first attempt to work out ALI–related ideas (Cannon's operator) for multilevel problems. Unfortunately it was never tested numerically. The Table is self–explanatory; the applications of ALI to the treatment of expanding atmospheres is further discussed in Sect 6.1.

Table 2. History of Multilevel ALI

Reference	Categ.	Geom.	Dynam.	Equations	Comments
Hubeny 1975	Ib	PP	St	SE	Cannon + CL
Scharmer 1984	III	PP	St,O	SE	operator (23)
Scharmer, Carlsson 1985	Ia	PP	St,O	SE	operator (23)
Werner, Husfeld 1985	Ia,II	PP	St	SE	operator (22),(24)
Hamann 1986	Ia	Sph	CMF	SE	operator (26)
Werner 1986	Ia	PP	St	SE,HE,RE	model atmospheres
Schönberg, Hempe 1986	Ib	Sph	CMF	SE	diagonal ANR
Pauldrach, Herrero 1988	II	PP,Sph	St	SE	using 2 previous iterations
Werner 1989	Ia	PP	St	SE,HE,RE	tridiagonal OK operator
Klein *et al.* 1989	III	PP	St	SE	double–splitting
Hamann, Wessolowski 1990	Ia	Sph	CMF	SE,RE	OAB for continua
Hillier 1990	Ib	Sph	CMF	SE,RE	multiband ANR
Rybicki, Hummer 1991	II	PP	St,O	SE	systematic preconditioning
Hamann *et al.* 1991	Ia	Sph	CMF	SE,RE	Broyden inversion
Dreizler, Werner 1991	Ia	PP	St	SE,HE,RE	Broyden inversion
Puls 1991	II	Sph	CMF	SE	unified ALI + Sobolev

Notes: Geometry: PP=plane-parallel; Sph=spherical. Dynamics: St=static, O=observer's frame (stated explicitly), CMF=comoving frame. Equations (solved together with radiative transfer): SE=statistical, HE=hydrostatic, RE=radiative equilibrium. CL=complete linearization.

5.2 Linearization versus Preconditioning

To illustrate the basic problems of applying ALI in multilevel problems, let us first write down the expression for the radiative rates. For simplicity, let us consider only lines; the treatment of continua is analogous. The net transition rate for the line $l \rightarrow u$ is

$$R_{ul}^{\mathrm{net}} = n_u A_{ul} - (n_l B_{lu} - n_u B_{ul}) \bar{J}_{lu} \quad , \tag{36}$$

where A and B are the corresponding Einstein coefficients. The basic ALI equation, (18), gives for $\bar{J}_{lu}$

$$\bar{J}_{lu} = \Lambda^*[S^{\mathrm{new}}] + (\Lambda - \Lambda^*)[S^{\mathrm{old}}] \quad . \tag{37}$$

Here the second term, which may be written as $\Delta \bar{J}_{lu}^{\mathrm{old}}$, is known from the previous iteration. However, the first term contains S^{new} which, according to (6), is a complicated, and generally non–linear function of all populations, $S^{\mathrm{new}} = S^{\mathrm{new}}(\boldsymbol{n})$. A linearization thus seems to be necessary.

If, however, the total source function is given by the line source function S_{lu} (i.e., the case of non-overlapping lines and no background continuum), and if the Λ^* is local, then the core contributions cancel, and the net rate may be written

$$R_{ul}^{\mathrm{net}} = n_u A_{ul}(1 - \Lambda_{ul}^*) - (n_l B_{lu} - n_u B_{ul}) \Delta \bar{J}_{lu}^{\mathrm{old}} \quad , \tag{38}$$

which is equivalent to subtracting analytically a large number of scatterings in the line core (the very idea underlying the concept of core saturation). This is what is meant by preconditioning. The basic advantage of (38) is that the radiative transition rates are known, and the statistical equilibrium equations may be solved *directly*, without any linearization. Equation (38) was already worked out in the original paper (Rybicki 1972); other variants were considered in Rybicki (1984), and a systematic study of preconditioning was presented by Rybicki and Hummer (1991). They showed that with a careful use of preconditioning, linearization is avoided in two more cases: a local Λ^* with background continuum; and even a non–local Λ^* with background continuum!

Another method which avoids linearization, albeit in a somewhat different way, is that of Pauldrach and Herrero (1988), who however need to consider source functions from the two previous iterations.

Nevertheless, there are cases where linearization seems to be unavoidable. One involves the case of strong overlap of lines (although Rybicki and Hummer in the 1991 paper promised a new development with a more extensive use of preconditioning). Another, and perhaps the most important case, is the application of ALI to constructing model stellar atmospheres, where the transfer equation is solved simultaneously with the constraints of hydrostatic, radiative, as well as statistical equilibrium.

The usual way of solving set of non-linear equations is by applying the Newton–Raphson method. This may be rather time consuming because each iteration requires to set up and to invert the Jacobi matrix of the system. The matrix inversion usually consumes the dominant part of the time, although setting up the matrices may also contribute significantly (in particular, for large number of levels). Recently, the Kiel group (Hamann, Koesterke, Wessolowski 1991; Dreizler and Werner, 1991; Koesterke, Hamann, Kosmol 1991) suggested a very promising method, called the *Quasi–Newton*, or *Broyden*, or *Least Change Secant Method*, which, in principle, is a multi–dimensional variant of the Regula Falsi method. This method basically prescribes how to calculate an *inverted* matrix of the system at the i-th iteration directly from the *inverted* matrix at the previous iteration. One thus avoids both setting up a new matrix and inverting it, leading to a considerable savings of computer time.

6 Non–classical Problems

6.1 Expanding Atmospheres

Practically any method mentioned in Sect.4, developed for static problems, may include velocity fields using an observer's frame formulation; several of them mention this extension explicitly (Scharmer; Rybicki and Hummer). However, the observer's frame formalism becomes impractical for velocities larger than few times thermal velocity.

Line transfer problems in stellar winds, where i) one has to adopt the spherical rather that plane geometry; and ii) the ratio of the expansion to thermal velocity is of the order of hundreds; the comoving–frame (CMF) formalism is traditionally used (Mihalas 1978). Its basic drawback is that it may only be used for monotonic velocity fields; all the following approaches thus pössess this limitation. The application of ALI to the CMF transfer problem was pioneered by Hamann (1985), who has extended the idea of core saturation to spherically expanding atmospheres, and used already discussed Λ^* operator (24). The approach was subsequently extended to multilevel atoms (Hamann 1986), and to the simultaneous solution of the statistical and radiative equilibrium (Hamann and Wessolowski 1990); a review is presented by Hamann *et al.* (1991).

Another group of methods are the ANR methods – see 4.4. They also use the CMF formulation. Schönberg and Hempe (1986) presented a multilevel formulation, and Hillier (1990) solves the multilevel statistical equilibrium together with radiative equilibrium in CMF.

Recently, Puls (1991) developed a parameter–free Λ^* operator for expanding atmospheres in the CMF formulation, which represent an extension of OAB to moving atmospheres. Moreover, he has shown that this purely local operator is intimately related to the Sobolev approximation. This has an additional advantage, namely that the traditional Sobolev approach and the new ALI approach may be arbitrarily mixed in the system of statistical equilibrium. This approach thus shows a great promise for future non-LTE calculations of the "unified" (photosphere + wind) model stellar atmospheres.

6.2 Multi–Dimensional Geometry

Finally, I will very briefly summarize multi–dimensional ALI approaches. Cannon (1976) extended his approach and developed a multi–dimensional formalism, where Λ^* was given by the corresponding 1-D solution. Nordlund (1984) has proposed an ALI–based technique for 3-D problems by approximating the angle couplings in three dimensions by evaluating the error of the source function by means of a series of ray–by–ray 1-D transfer solutions. But word of caution about this type of approaches was recently expressed by Castor *et al.* (1991), who demonstrated that in order to achieve a desired accuracy, one may be faced with a necessity to deal with an enormous number of angles. Castor *et al.* also presented a generalization of the Klein *et al.* 's double–splitting method to two dimensions. Finally, Kunasz and Olson (1988) developed an extension of OK for 2-D geometries.

7 Conclusion

Finally, here is the promised consumer's report:

a) Applying ALI to linear (two-level or equivalent-two-level) problems:
- use Λ^* given as a band matrix of the exact Λ;
- for its numerical evaluation, use the Rybicki and Hummer (1991) procedure;
- the tridiagonal Λ^* yields faster convergence, but is sometimes not as stable as the diagonal Λ^*; a good strategy seems to be to apply the diagonal operator in the initial stages of calculations, and to switch to the tridiagonal form later, when corrections are already smaller (Rybicki and Hummer 1991);
- use acceleration of convergence, Ng's or Orthomin. Ng's appears to be simpler to code, but Orthomin tends to accelerate the solution better. Again, both acceleration methods may yield instabilities or even divergences when applied too early in the iteration process.

b) For non–linear problems (multilevel problems; stellar atmospheres):
- use Λ^* as in the previous item;
- precondition the statistical equilibrium;
- if the problem allows, use preconditioning to obtain *linear* statistical equilibrium after applying Λ^*;

c) Line formation in monotonic velocity field:
- Puls' (1991) method appears to be very promising.

ACKNOWLEDGEMENTS. I wish to thank Bruce Altner and Steve Shore for their careful reading of the manuscript.

References

Auer, L.H. (1976): JQSRT **16**, 931.
Auer, L.H. (1987): in *Numerical Radiative Transfer*, ed. by W. Kalkofen (Cambridge Univ. Press, Cambridge), p. 101.
Auer, L.H. (1991): in *Stellar Atmospheres: Beyond Classical Models*, ed. by L. Crivellari, I. Hubeny, and D.G.Hummer, NATO ASI Series C 341 (Kluwer, Dordrecht), p. 9.
Buchler, J.R., Auer, L.H. (1985): in *Proc. 2nd Intern. Conf. and Workshop on Radiat. Properties of Hot Dense Matter*, ed. by J. Davis *et al.* (World Scientific, Singapore), p.48.
Cannon, C.J. (1973a): JQSRT **13**, 627
Cannon, C.J. (1973b): ApJ **185**, 621.
Cannon, C.J. (1976): A & A **52**, 337.
Castor, J.I., Dykema, P.G., Klein R.I. (1991): in *Stellar Atmospheres: Beyond Classical Models*, ed. by L. Crivellari, I. Hubeny, and D.G.Hummer, NATO ASI Series C 341 (Kluwer, Dordrecht), p. 49.

Dreizler, S., Werner, K. (1991): in *Stellar Atmospheres: Beyond Classical Models*, ed. by L. Crivellari, I. Hubeny, and D.G.Hummer, NATO ASI Series C 341 (Kluwer, Dordrecht), p. 155.
Feautrier, P. (1964): C. R. Acad. Sci. Paris **258**, 3189.
Hamann, W.-R. (1981): A & A **93**, 353.
Hamann, W.-R. (1985): A & A **148**, 364.
Hamann, W.-R. (1986): A & A **160**, 347.
Hamann, W.-R., Wessolowski, U. (1990): A & A **227**, 171.
Hamann, W.-R., Koesterke, L., Wessolowski, U. (1991): in *Stellar Atmospheres: Beyond Classical Models*, ed. by L. Crivellari, I. Hubeny, and D.G.Hummer, NATO ASI Series C 341 (Kluwer, Dordrecht), p. 69.
Hempe, K., Schönberg, K. (1986): A & A **160**, 141.
Hillier, D.J. (1990): A & A **231**, 116.
Hubeny, I. (1975): Bull. Astron. Inst. Czechosl. **26**, 38.
Kalkofen, W. (1987): in *Numerical Radiative Transfer*, ed. by W. Kalkofen (Cambridge Univ. Press, Cambridge), p. 23.
Klein, R.I., Castor, J.I., Greenbaum, A., Taylor, D., Dykema, P. (1989): JQSRT **41**, 199.
Koesterke, L., Hamann W.-R., Kosmol, P. (1991): A & A (in press).
Kunasz, P.B., Olson, G.L. (1988): JQSRT **39**, 1.
Mihalas, D. (1978): *Stellar Atmospheres* (Freeman, San Francisco).
Ng, K.C. (1974): J. Chem. Phys. **61**, 2680.
Nordlund, Å (1984): in *Progress in Stellar Spectral Line Formation Theory*, ed. by J.E. Beckman and L. Crivellari, NATO ASI Series C 152 (Reidel, Dordrecht), p. 215.
Olson, G.L., Auer, L.H., Buchler, J.R. (1986): JQSRT **35**, 431 (OAB).
Olson, G.L., Kunasz, P.B. (1987): JQSRT **38**, 325 (OK).
Pauldrach, A., Herrero, A. (1988): A & A **199**, 262.
Puls, J. (1991): A & A **248**, 581.
Puls, J., Herrero, A. (1988): A & A **204**, 219.
Rybicki, G.B. (1972): in *Line Formation in the Presence of Magnetic Fields*, ed. by R.G.Athay, L.L.House, G.Newkirk (High Altitude Observatory, Boulder), p. 145.
Rybicki, G.B. (1984): in *Methods in Radiative Transfer*, ed. by W. Kalkofen (Cambridge Univ. Press, Cambridge), p. 21.
Rybicki, G.B. (1991): in *Stellar Atmospheres: Beyond Classical Models*, ed. by L. Crivellari, I. Hubeny, and D.G.Hummer, NATO ASI Series C 341 (Kluwer, Dordrecht), p. 1.
Rybicki, G.B., Hummer, D.G. (1991): A & A **245**, 171.
Scharmer, G.B. (1981): ApJ **249**, 720.
Scharmer, G.B. (1984): in *Methods in Radiative Transfer*, ed. by W. Kalkofen (Cambridge Univ. Press, Cambridge), p. 173.
Scharmer, G.B., Carlsson, M. (1985): J. Comput. Phys. **59**, 56.
Schönberg, K., Hempe, K. (1986): A & A **163**, 151.
Steiner, O. (1991): in *Stellar Atmospheres: Beyond Classical Models*, ed. by L. Crivellari, I. Hubeny, and D.G.Hummer, NATO ASI Series C 341 (Kluwer, Dordrecht), p. 19.
Vinsome, P.K.W. (1976): in *Proc. of the 4th Symp. on Reservoir Simulation* (Soc. of Petroleum Engineers, Boulder), p.149.
Werner, K. (1986): A & A **161**, 177.
Werner, K. (1989): A & A **226**, 265.
Werner, K., Husfeld, D. (1985): A & A , **148**, 417 (WH).

Instabilities in Hot-Star Winds: Basic Physics and Recent Developments

Stanley P. Owocki [1]

[1]Bartol Research Institute, University of Delaware
Newark, DE 19716 USA

Abstract: The winds of the hot, luminous, OB stars are driven by the line-scattering of the star's continuum radiation flux. Several kinds of observational evidence indicate that such winds are highly structured and variable, and it seems likely that a root cause of this variability is the strong instability of the line-driving mechanism. This paper reviews the basic physics of the linear instability and summarizes results from numerical simulations of its nonlinear evolution. Particular emphasis is placed on the dynamical importance of the diffuse, scattered radiation field, and on recent methods for incorporating such scattering effects into the numerical simulations. I also summarize recent preliminary results on synthetic UV line, Hα, IR continuum spectra in dynamical wind models with extensive structure.

1 Introduction

Hot-star winds are believed to be driven by line-scattering of the star's continuum radiation field (Lucy and Solomon 1970; Castor, Abbott, and Klein 1975; hereafter CAK). Indeed, modern theoretical models based on this line-driving mechanism (Friend and Abbott 1986; Pauldrach, Puls, and Kudritzki 1986, hereafter PPK) predict wind speeds and mass-loss rates that are now in quite good agreement with observationally inferred values. However, such smooth, steady wind models seem inherently incapable of explaining several other observational characteristics of these stars, including: nonthermal radio emission (Abbott, Bieging, and Churchwell 1981, 1984); enhanced thermal infrared emission (Abbott, Telesco, and Wolff 1984); UV lines with variable, discrete absorption components (DAC) (Lamers, Gathier, and Snow 1982) or black absorption troughs (Lucy 1982a); and soft X-ray emission (Harnden et al. 1979; Seward et al. 1979). Together these suggest a wind with a high degree of structure and variability. One quite promising explanation for such structure is that it is a direct consequence of a strong instability in the line-driving mechanism itself.

The present paper will review the basic physics of this instability, describe recent developments that enable inclusion of scattering effects in simulations of its nonlinear evolution, and summarize preliminary results on synthesis of various observational signatures in unstable wind models with extensive structure. The emphasis is chosen to supplement and complement several other recent reviews on both the linear instability (Rybicki 1987) and on attempts to model its nonlinear evolution (Owocki 1990, 1991a; Castor 1991). The Castor review is particularly recommended to the reader for its succinct yet quite comprehensive discussion of many subtle aspects not covered here.

2 Physics of Driving by Line-Scattering

The basic physics of line-driving and its instability were already recognized in the early work of Milne (1926), and in the original paper by Lucy and Solomon (1970) that suggested line-driving as a mechanism for hot-star winds. The reason for instability is quite simple: a small-scale increase in radial flow speed Doppler-shifts the frequency for local line-absorption out of the shadow of underlying material, leading to a increased radiative force which tends to further increase the flow speed. However, formal linear stability analyses (MacGregor et al. 1979; Carlberg 1980; Lucy 1984; Owocki and Rybicki 1984, 1985) have since illuminated several additional, more subtle effects that under various circumstances can reduce or even eliminate this instability. The following represents an attempt to explain physically the origin of both the instability and the effects that tend to reduce it. The discussion centers on graphical illustrations of the formation of various components of the radiative force (Figures 1 and 2).

2.1 The Direct and Diffuse Force from a Single Line

The radiative acceleration $g_{\rm rad}$ (i.e., force-per-unit mass) arising from line-scattering depends on a frequency integral over the line profile of the specific intensity I,

$$g_{\rm rad}(r) = \left(\frac{2\pi\kappa v_{\rm th}}{c^2}\right)\int_{-1}^{1} d\mu\,\mu \int_{-\infty}^{\infty} dx\,\phi(x-\mu u(r))I(x,\mu,r). \tag{1}$$

Here κ is the line opacity, ϕ is the line-profile function, x is the observer's frame line-frequency (measured from line-center in units of Doppler-width $\Delta\nu_D \equiv \nu v_{\rm th}/c$), and $u(r) = v(r)/v_{\rm th}$ is the velocity in units of the thermal speed $v_{\rm th}$; $I(x,\mu,r)$ is the specific intensity at frequency x along a direction that makes an angle $\cos^{-1}\mu$ with the radial direction at the local radius r.

For the following physical discussion, it is sufficient to consider a "two-stream" approximation in which $\mu = \pm 1$, so that radiation moves only along a radial ray in either the forward (+) or backward (−) direction. The intensity then takes the form,

$$\begin{aligned} I(x,r) &= I_{\rm dir} + I_{\rm diff}{}^{\pm} \\ &= I_* e^{-t_+(x,r)} + \int S(r')e^{-|t-t'|}dt' \\ &\approx I_* e^{-t_+(x,r)} + S_o(r)\left(1 - e^{-t_\pm(\pm x,r)}\right), \end{aligned} \tag{2}$$

where we have split the intensity into a *direct* component that originates from the underlying star I_*, plus forward and backward scattered, *diffuse* components $I_{\rm diff}{}^{\pm}$. The last equality represents a "Smooth Source Function" (SSF) approximation (Owocki

1991b). It is based on the idea that the source function $S(r')$ is given by a frequency- and angle-integrated mean intensity, and so is less sensitive to local variations in, e.g. the velocity, than the optical depth terms $t(x,\mu,r')$; thus by replacing $S(r') = S_o(r)$, where S_o is to be estimated (e.g. from Sobolev theory), the formal solution integral is carried out analytically. The optical depths are given by

$$t_{\pm}(x,r) \equiv \int \kappa\rho(r')\phi(\pm(x-u))dr', \tag{3}$$

where ρ is the density and the integral is understood always to be outward, i.e. either from R_* to r for t_+ and from r to ∞ for t_-.

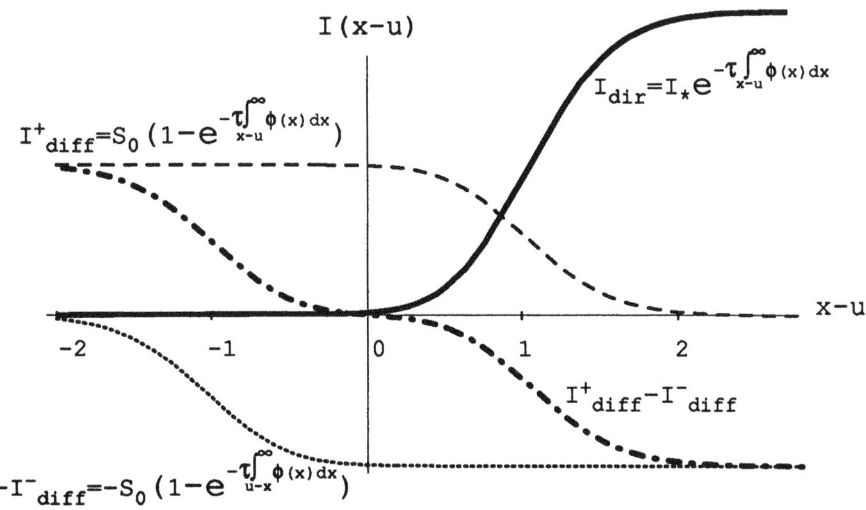

Fig. 1. **Various components of the intensity vs. comoving frame frequency $x-u$, for the Sobolev case of smooth, supersonic flow.**

In a smooth, supersonic wind, the variation in the integrand of eqn. (3) is dominated by the Doppler-shift associated with changes in the velocity $u(r')$; we may thus use the ***Sobolev approximation*** to transform the optical depth spatial integral into a locally evaluated integral over the comoving frame frequency $x' \equiv x - u(r')$,

$$t_{\pm}(x,r) \approx \tau \int_{\pm(x-u)}^{\infty} \phi(x')dx', \tag{4}$$

where the Sobolev optical depth $\tau \equiv \kappa\rho L$ represents a collection of spatial variables that are assumed to be approximately constant over a Sobolev length $L \equiv (du/dr)^{-1}$. Applying (4) to (2) we see that the intensity components have the form illustrated in Figure 1,

$$I_{\rm dir}(x,r) = I_* e^{-\tau \int_{x-u(r)}^{\infty} \phi(x')dx'}. \tag{5}$$

$$I_{\rm diff}{}^{\pm} = S_o \left(1 - e^{-\tau \int_{\pm(x-u)}^{\infty} dx'\, \phi(x')}\right). \tag{6}$$

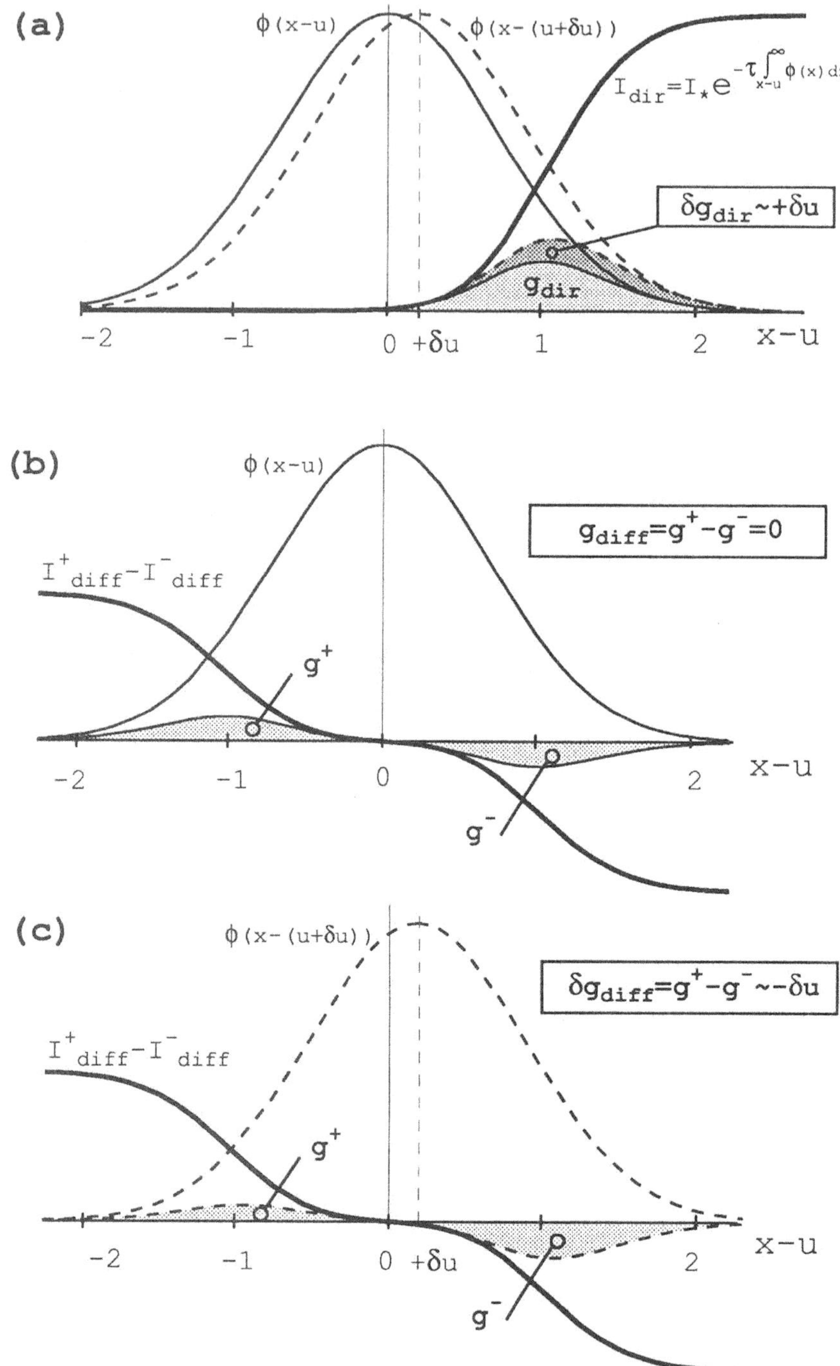

Fig. 2. Graphical illustration of the formation of various force components, proportional here to the shaded areas.

The line-force is given by frequency integration of this intensity over the line profile (eqn. (1)). Figures 2 illustrate this graphically by combining plots of the profile and intensities to form shaded areas proportional to various components of the force. In Figure 2a, for example, the light shaded area is proportional to the direct force. Analytically, in this Sobolev case this direct force component has the simple form,

$$g_{\text{rad}}(r) = g_{\text{thin}} \frac{1 - e^{-\tau}}{\tau}, \tag{7}$$

where $g_{\text{thin}} = 2\pi\kappa v_{\text{th}} I_*(R_*/cr)^2$ is the line-force in the optically thin limit $\tau \ll 1$. For a strong line $\tau \gg 1$ we obtain the optically thick form $g_{\text{thick}}(r) = g_{\text{thin}}/\tau \propto (du/dr)$. Note that the latter is independent of the opacity κ, depending instead on the velocity gradient du/dr.

The net diffuse force component is given by frequency integration of the net flux, which is proportional to the difference $I_+ - I_-$. Figure 2b illustrates that the positive and negative contributions, represented by the shaded areas above and below the x-axis, exactly cancel. This illustrates the well-know result (Lucy 1971; Castor 1974) that the diffuse force *vanishes* in the Sobolev approximation.

2.2 Response to Velocity Perturbations

Let us now consider the effect of a small-amplitude velocity perturbation of the form $\delta u \propto e^{ikr}$. First, if the perturbation wavelength $\lambda = 2\pi/k$ is large compared to the Sobolev length L, then the Sobolev analysis applies to the perturbation too, and so the perturbed force $\delta g_{\text{dir}} \propto \delta u' \propto ik\delta u$ (Abbott 1980). In this case the perturbed force and velocity are 90° out of phase, and so no net work is done on the perturbation, implying that it must be stable. On the other hand, in the opposite limit $\lambda \ll L$, one can show that the perturbation is optically thin, in the sense that one may ignore the perturbations in the optical depth t_+. The perturbed force in this case is thus just computed from the Doppler shift of the line-profile (MacGregor, Hartmann, and Raymond 1979; Carlberg 1980). As illustrated by the dashed curves in Figure 2a, a positive velocity perturbation increases the overlap between the profile and direct intensity; since then $\delta g_{\text{dir}} \propto +\delta u$, we now find instability. From a full computation for arbitrary k, Owocki and Rybicki (1984) derived a simple law that "bridges" these two limits,

$$\frac{\delta g_{\text{dir}}}{\delta v} \approx \Omega \, \frac{ikL}{2 + ikL}. \tag{8}$$

The instability growth rate is $\Omega \approx g_{\text{dir}}/v_{\text{th}} \approx v/L$, and this implies a cumulative growth of order $e^{\int \Omega dt} \approx e^{u}$, which can exceed e^{100} in these highly supersonic winds!

However, as first discussed by Lucy (1984), this instability can be substantially reduced, or even eliminated, by the *line-drag* effect of the perturbed *diffuse* force. Figure 2c illustrates the physical origin of this effect: For the diffuse radiation a positive velocity perturbation now results in both an increase in the negative force from the blue side of the line, and a decrease of the positive force from the red side of the line. Together these yield a net perturbed diffuse force of the form, $\delta g_{\text{diff}} \propto -\delta u$, which thus tends to offset the instability of the direct term. The degree of reduction depends on the magnitude of

S_o/I_*, with complete cancellation for $S_o = I_*/2$, which turns out to be approximately true at the wind base. (Recent work even shows how nonzero thermalization can lead to $S_o > I_*/2$ and thus to a net damping; Feldmeier and Owocki 1991). Away from the wind base, the combined effects of spherical expansion and the shrinking of the solid angle subtended by the stellar core quickly reduce the relative importance of the drag term. At $r = 1.5R_*$, for example, the net growth rate is about half that obtained from a pure-absorption analysis (Owocki and Rybicki 1985), but this still implies a strong instability, with cumulative growth now about 50 (vs. 100) e-folds. In practice this means, of course, that any small, but finite amplitude perturbation would quickly become nonlinear in the wind. Efforts to determine the possible forms of this nonlinear structure are discussed below (see §3.2).

Rybicki et al. (1990) showed that this line-drag effect also tends to strongly dampen *horizontal* velocity perturbations (for which the countervailing direct term is much weaker). This implies that wind velocity variations should be predominantly in the radial direction, and so in this sense might be reasonably simulated within a simple 1-D, spherically symmetric model. Of course, in general there will still be horizontal phase variations that will require a 2-D or 3-D treatment.

2.3 The Force from a Line Ensemble

In hot-star winds a large number of lines contribute to the driving. A major advance of the CAK model was to develop a formalism for efficiently including the cumulative effect of a large ensemble of lines, which they effectively assumed to have a flux-weighted number distribution that is a power law in opacity, i.e., $N(\kappa){\sim}\kappa^{\alpha-2}$, where $0< \alpha <1$. Assuming the lines do not significantly overlap, the cumulative force can then be computed by integrating the above force expressions over this number distribution. If one also assumes validity of the Sobolev approximation, integration of eqn. (7) over $N(\kappa)$ yields the CAK line-force

$$g_{\rm rad} \approx g_{\rm dir} \approx g_{\rm CAK} = g_{{\rm thin},\kappa_o} \frac{\Gamma(\alpha)}{(1-\alpha)\tau_o^\alpha} = g_{{\rm thin},\kappa_o} \frac{\Gamma(\alpha)}{1-\alpha} \left(\frac{1}{\kappa_o \rho} \frac{du}{dr} \right)^\alpha , \tag{9}$$

which here is normalized to the optically thin force for a line with opacity κ_o, where $\kappa_o N(\kappa_o) \equiv 1$. (This normalization is related to the CAK constant k by $(\kappa_o/\kappa_e)^{1-\alpha} = (1-\alpha)kc/v_{\rm th}\Gamma(\alpha)$, where κ_e is the electron scattering opacity.)

If instead of the full Sobolev approximation, we assume the SSF approximation, integration over the line ensemble yields

$$g_{\rm dir}(r) = \frac{g_{{\rm thin},\kappa_o}\Gamma(\alpha)}{\kappa_o^\alpha} \int_0^1 dy \int_{-\infty}^{\infty} dx\, \phi(x - \mu_y u(r)) \eta_+^{-\alpha}(x, \mu_y, r) \tag{10}$$

and

$$g_{\rm diff}(r) = \frac{g_{{\rm thin},\kappa_o}\Gamma(\alpha)}{\kappa_o^\alpha} \frac{r^2 S_o}{R_*^2 I_*} \int_0^1 dy \int_{-\infty}^{\infty} dx\, \phi(x - \mu_y u(r)) \left[\eta_-^{-\alpha}(x, \mu_y, r) - \eta_+^{-\alpha}(x, \mu_y, r) \right] . \tag{11}$$

These expressions also represent a generalization of the above two-stream treatment to now include an integration over a set of rays that intercept the stellar disk at impact parameters $\sqrt{y}R_*$, and that have direction cosines $\mu = \mu_y = \sqrt{1 - y(R_*/r)^2}$ relative to the local radial direction at a radius r. In practice, it turns out that using just a single-ray quadrature with $y = 0.5$ yields a direct force that is typically within a few percent of the full, finite-disk value. (It differs significantly, however, from the point-star result, computed with a single ray at $y = 0$.) For computational economy, this single-ray approximation is nominally used for the diffuse term as well.

Note that the integration over κ has resulted in the optical depth $t_\pm$ of individual lines (eqn. (3)) being replaced by profile-weighted mass-column-depths $\eta_\pm \equiv t_\pm/\kappa$. When these are computed from the full depth integration rather than from the Sobolev approximation, then expressions (10) and (11) provide a practical way to incorporate the dynamics of line-driving and its instability into numerical simulations (see §3.2). A great advantage of the SSF approach lies in the very similar forms of these two expressions, requiring very little extra computation to obtain the diffuse as well as direct force. The computational cost is in fact dominated by calculation of the $\eta_\pm$ over a grid of depths and frequencies; but this is still far less (by a factor ~100) than what would be required for a full, direct transfer solution in a sufficient sample of driving lines.

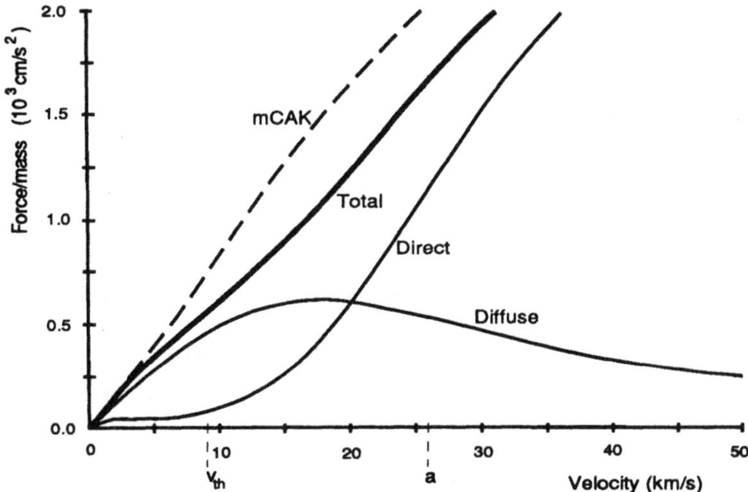

Fig. 3. The variations of various force components vs. flow velocity in a smooth wind model.

Figure 3 shows how these direct and diffuse forces vary with flow velocity near the base of a smooth, steady-state wind. In the supersonic portion, each of these force components approaches the appropriate Sobolev limit. For the diffuse force, this again means $g_{\rm diff} \approx 0$, a result which follows here from the near symmetry of the η_+ and η_- terms. For the direct term, we find $g_{\rm dir}(r) \approx g_{\rm mCAK}$, where "mCAK" stands for CAK modified to take account of finite-disk effects (PPK).

In the region near the sonic point ($v \approx a$), the flow is being rather abruptly transformed from a near hydrostatic balance to strongly driven, with the velocity gradient accordingly going from shallow to steep. The shallowness of the subsonic side implies a reduction in $\eta_+^{-\alpha}$, and thus in the direct force. But the steepness on the supersonic side implies an enhancement in the $\eta_-^{-\alpha}$ term, and this net asymmetry now leads to a net, positive diffuse force. Figure 3 shows however that the total force is still such that $g_{\rm tot} \equiv g_{\rm dir} + g_{\rm diff} \approx g_{\rm CAK}$! As discussed by both Castor (1991) and Owocki (1991b), the reason is that the next order corrections to the Sobolev approximation have the character of an *viscosity* for the diffuse term, but an *anti*-viscosity for the direct term. Cancellation of these means that the *Sobolev force is accurate to one higher order for scattering lines than for absorption lines.* This appears to be one important reason why Sobolev wind models like CAK are in such good agreement with models based on a full comoving frame transfer solution (PPK).

3 Models of the Nonlinear Evolution

3.1 Heuristic Models of a Wind with Embedded Shocks

Given the supersonic nature of the wind outflow and the large instability growth rate, it seems inevitable that, whatever its detailed form, the nonlinear wind structure arising from this instability will include shocks. Several of the attempts to study the consequences of this instability have thus centered on determining the properties of these assumed shocks. (See review by Castor 1987).

The first models along these lines (e.g., Lucy and White 1980; Lucy 1982b) assumed the wind contains a quasi-periodic train of *forward* shocks. Relatively fast flow exposed to a nearly unshadowed stellar flux is strongly driven, and so is pushed against relatively slow flow, which is shadowed by the fast material and only weakly driven. A forward shock thus forms that sweeps up and accelerates this slower material, thereby adding it to the faster flow. A crucial point is that the fast material represents *post-shock* flow, with its associated *high density* and (at least initially) *high temperature.* To maintain the structure, this material must quickly cool and reform the driving ions that line-absorb the radiative momentum.

The question of the ionization and energy balance behind the shock was analyzed by Krolik and Raymond (1985) for a single, forward shock that propagates outward through the wind. For massive winds, they inferred that the cooling would be sufficiently rapid to allow the driving ions to reform and so maintain the structure. For low-density winds, however, they argue that the shock amplitude would be limited by the requirement that the cooling time be less than the flow time.

MacFarlane and Cassinelli (1989) studied the evolution of wind shocks with a phenomenological numerical-hydrodynamics model in which an initially smooth, slow wind is disrupted by a sudden increase in the driving force, thereby accelerating the wind to a much higher speed. For parameters chosen to give initial and final state terminal speeds of 500 km/s and 2500 km/s, collision between the fast and slow flow forms a forward/reverse pair of shocks of velocity amplitude ~1000 km/s each. This turns out to be just what's needed to reproduce the observed X-ray properties of their model star, τ Sco.

Another heuristic wind-structure model worth mentioning is Mullan's (1984, 1986) application of the well-known solar wind phenomena of "corotating interaction regions" to the case of hot-star winds. The winds here again contain dense shells bounded by forward/reverse shock pairs, but now this is not thought of as arising from the line-driven instability, but rather from an assumed azimuthal variation in the wind flow speed and the interaction between fast and slow wind streams that results from the stellar rotation. Mullan argues that the resulting shocks and dense shells can give rise to the observed X-ray emission and discrete absorption components in UV lines. Although there is no independent evidence that O star winds have such an azimuthal stream structure, the effect of rotation in laterally extending any structure forming from the wind instability may turn out to be quite similar.

3.2 Radiation-Hydrodynamics Simulations

The last few years have seen the first attempts to carry out direct numerical simulations of the nonlinear evolution of this instability. Although there have been recent initiations of independent efforts by the Munich group and by Waldron et al. (1991), most of the detailed results so far have been from a code developed by J. Castor and myself, and so it is these that I will focus on here. The efforts so far have had to make some severe simplifications, including:

1. Isothermality.
2. 1-D, radial flow tube with spherical area expansion.
3. No line overlap; fixed, constant opacities with number distribution $N(\kappa) \propto \kappa^{\alpha-2}$.
4. Pure-absorption or Smooth-Source-Function line-force.

The latter two of these have already been discussed above, but let us first briefly consider what's involved in the former two. Isothermality means that one assumes that any heat by, e.g. shocks, is quickly cooled by radiative losses over a region that is narrow compared to any competing scales. The validity of this assumption depends on the strength of any shocks, the wind mass loss rate, and the competing scale. For a shock of velocity jump Δv, the column depth for cooling is roughly $N_{cool} \approx 7 \times 10^{17}(\Delta v/100\,\mathrm{km\,s^{-1}})^4 cm^{-2}$ (Castor 1987, 1991); the column between shocks scales as $N_{shock} \approx 5 \times 10^{22}(\dot{M}_6/r_{12}^2)\Delta t_4$, where $\dot{M}_6$ is the mass loss rate in $10^6 M_\odot/yr$, r_{12} is the radius in $10^{12} cm$, and Δt is the time interval between shocks in $10^4 s$. Comparison indicates that isothermality is not too bad an assumption for the massive winds of luminous OB supergiants (e.g. ζPup), but that it may break down for strong shocks in the less dense winds on the main sequence (e.g. τ Sco). In either case, however, it will eventually be necessary to include a full energy balance to derive X-ray luminosities from the shocks. Work along these lines is underway; see Cooper and Owocki (1991) for some initial results.

The restriction to 1-D means that important effects, such as rotation, cannot yet be included, despite observational evidence that it may play an important role (e.g. Prinja 1988). 1-D does not necessarily imply spherical symmetry, since the models can in principle be thought of as applying to individual radial streams; but since interaction between these streams is ignored, there is no way to determine the lateral coherence scale. Unfortunately, the computational requirements of the radiative force means that extensions of instability simulation to 2-D are probably still a few years away.

3.2.1 Results for Pure-Absorption Models

The initial approach of Owocki, Castor, and Rybicki (1988; hereafter OCR) was simply to ignore the diffuse, scattered radiation, and compute the force from numerical integration of the frequency-dependent line *absorption* of the direct radiation from the star. Perhaps their most important result (see OCR figures 4 and 5) is that periodic perturbations, initially of a small amplitude ($\delta v/a \sim 0.01$) when introduced at the photospheric wind base, are amplified in the wind into a series of high-speed, low-density rarefactions that terminate in *reverse* shocks. This is quite different from the Lucy (1982b) model, for which *forward* shocks were assumed to abruptly accelerate ambient wind material as it is rammed by a high-speed, dense, strongly driven flow. Instead the reverse shocks here act to *decelerate* high-speed, rarefied flow as it impacts slower material that has been compressed into dense clumps.

The highest speed waves actually contain only a very little material (See, e.g. OCR figure 10.) Thus, although previous discussions of this instability have tended to focus on the velocity variations, a perhaps more important consequence is the tendency for the wind to become highly *clumped.* OCR found clumping factors (defined as the ratio of the mean squared density over the square of the mean density) of up to 6. For computional economy, these initial simulations assumed an artificially enhanced wind temperature, $T_{\rm wind} = 10T_*$, and, because of the enhanced gas pressure, this turns out to limit somewhat the clumping. Subsequent models with $T_{\rm wind} = T_*$ have even narrower, denser clumps, with clumping factors of 10 or more. As discussed below (§3.2.3), this implies a substantial enhancement in emission diagnostics that depend on density squared, e.g., IR emission, Hα.

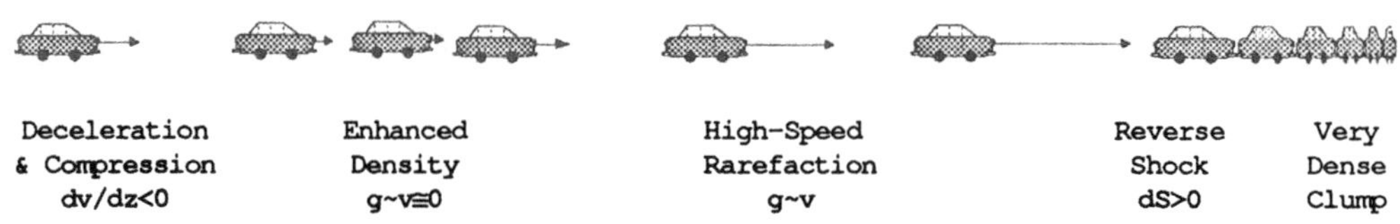

Fig. 4. Traffic analogy for the formation of various types of structure in unstable winds. The lengths of the arrows denote velocity.

This preponderance of high-speed rarefactions, reverse shocks, and dense clumps appears to be a quite robust result. The basic explanation is really quite straightforward: To satisfy mass continuity, a flow generally tends to become more rarefied in its highly accelerated parts, and denser in its slowly moving parts. The same kinematic effect occurs, for example, in traffic flow, and Figure 4 illustrates that the structures in both traffic and unstable winds are really quite analogous. In both cases an additional important factor is the tendency for the acceleration to vary inversely with density. To obtain the high-speed, high-density structure envisioned in earlier heuristic wind models (§IIIa) would require that the acceleration increase with density, which is quite the opposite of the actual scaling of the line-force (cf. eqns. (9)-(11)).

3.2.2 Results for SSF Scattering Models

With development of the SSF formalism described in §2.3 it became practical to include some of the most important dynamical effects of scattering with only a modest (~10%) timing increase over what is required pure-absorption simulations.

One particularly important feature of the SSF form of the diffuse force is that it naturally incorporates the Lucy line-drag effect discussed in §2.3. SSF wind models thus tend to be notably less unstable than corresponding pure-absorption models. The latter often exhibit an *intrinsic variability* which persists even in the absence of explicit perturbation (Owocki, Poe, and Castor 1990; Owocki 1990, 1991a). This is apparently closely related to a *degeneracy* in the steady-state solutions (Poe, Owocki, and Castor 1990); since there is no unique steady-state solution to which the time-dependent flow can relax, it never settles down. Owocki (1991b) showed that addition of the SSF diffuse force has the dramatic effect of effectively *eliminating* this intrinsic variability. This may just reflect the reduction in instability by the line-drag effect; but an additional factor may be the role of the diffuse radiative viscosity in "breaking" the solution degeneracy (Owocki and Zank 1991) In any event, unperturbed SSF models approach a steady-state which is in excellent agreement with the (finite-disk) CAK model.

Although scattering effects apparently eliminate this intrinsic or *absolute* instability, such winds still remain *advectively* unstable, albeit at a somewhat reduced rate. This means that perturbations introduced at the flow base can still become strongly amplified in the wind. As in the pure-absorption case, when structure does develop it is still dominated by dense clumps that are separated from high-speed rarefactions by reverse shocks. The height at which structure forms depends on details of assumptions about the amplitude of the perturbation and the form of the smooth source function S_o. The original formulation of the SSF method assumed a form appropriate to optically thin scattering, $S_o/I_* = (1-\mu_*)/2$, where $\mu_* \equiv \sqrt{1-(R_*/r)^2}$. Near the stellar surface $S_o/I_* \rightarrow \frac{1}{2}$, which is what's required for the diffuse drag term to completely cancel the instability (see §2.2 and Figure 2); but away from the star, the fall off in S_o is too rapid, and this causes the instability growth rate to rise from zero faster than it should according to the full analysis of Owocki and Rybicki (1985). Work is underway to develop alternative forms that improve upon this and also take into account the potentially stabilizing effects of nonzero photon thermalization (Feldmeier and Owocki 1991). Preliminary results indicate that it may actually be quite difficult to generate much structure below about $r \approx 1.5R_*$, but that above about $r \approx 2R_*$ substantial structure seems almost unavoidable.

3.2.3 Generation of Synthetic Diagnostics

One area of much current effort is toward the synthesis of observational diagnostics from the simulation models, with the ultimate aim of comparing these with the various types of datasets (X-ray, UV, optical, IR, radio) mentioned in the Introduction as evidence for highly structured winds. The complexity of the wind models leads to many difficulties not encountered in the usual smooth, steady case, and for each type of diagnostic developing methods to deal with these constitutes its own challenging research project. I have already mentioned work underway regarding X-rays. In the following I will briefly describe preliminary results of some recent work done in collaboration between myself and members of the Munich group (J. Puls and P. Najarro) on synthesis of UV resonance lines, Hα, and the IR continuum.

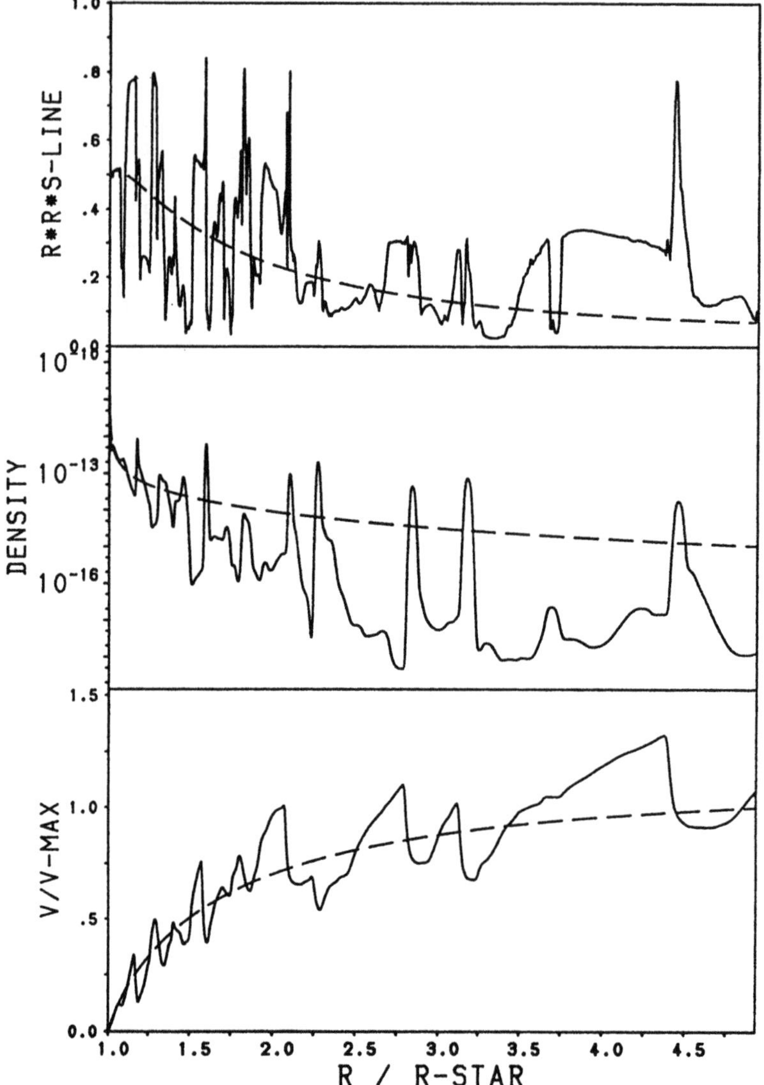

Fig. 5. Snapshot of the wind structure versus height in an SSF pure-scattering wind model that is perturbed at its base by a sound wave of amplitude $\delta\rho/\rho = 0.25$ and period 10,000 s. From bottom up, the panels show the radial variation of velocity, density, and source function for line synthesis. The dashed curves are the same quantities in an analogous smooth, CAK/Sobolev model. The stellar parameters, the same as assumed by OCR, are for a generic O supergiant.

The simulation models for these initial calculations were purposely chosen to contain substantial structure, in order both to challenge and test the methods and to obtain clear effects. They thus define an extreme rather than a "best-fit" structure model. They are based on the SSF method with the optically thin scattering source function, and assume a strong driving at the wind base with a periodic sound wave of amplitude $\Delta\rho/\rho = \Delta v/a = 0.25$ and period 10,000 s. The lower two panels of Figure 5 show a shapshot of the density and velocity variation vs. radius at a typical instant in time (~60,000 s) long enough for the base perturbations to have propagated into the wind and been amplified into the characteristic dense clumps, reverse shocks, etc. The dashed curves denote the same quantities in a corresponding CAK model.

The top panel shows the source function (actually $r^2 S$) used in the synthesis of a strong ($\tau \approx 100$) UV line. This is obtained using the method of Rybicki and Hummer (1978) for Sobolev line-transport in a wind with a nonmonotonic velocity law. Although this properly takes into account the coupled interaction of multiple resonance layers, it still assumes that the local photon escape is dominated by frequency shift associated with the velocity gradient. Given the existence here of strong *density* gradients as well, this may not be justified, and work in progress aims ultimately to solve the full transfer problem using the Accelerated Lambda Iteration methods reviewed by Hubeny in this volume. Note nonetheless that even the source function in Figure 4 is far from smooth, as was assumed in the dynamics, and so eventually the dynamical role of source function gradients should also be investigated.

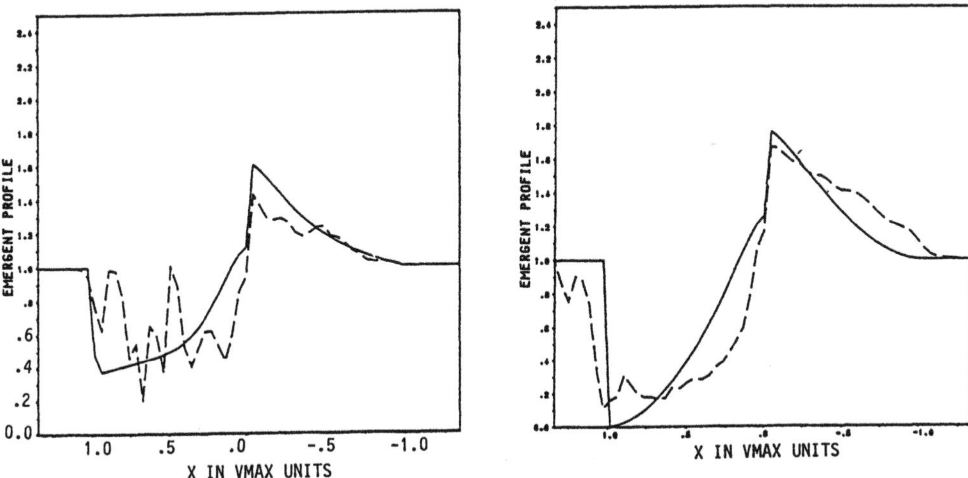

Fig. 6. Synthetic UV line-profiles in a smooth (solid curves) and a structured (dashed curves) wind model. The left and right panels are respectively for marginally optically thick and very optically thick lines, i.e. $\tau \approx 1$ and $\tau \approx 100$ in the smooth wind.

Figure 6 shows samples of the resulting profiles for a marginally optically thick line ($\tau \approx 1$ in the smooth model; left) and a very optically thick line ($\tau \approx 100$; right) (Puls and Owocki, work in progress). For comparison, the profiles in the smooth model are plotted as well. The extensive wind structure is clearly reflected in the profiles, particularly in the absorption part, and there are many features reminiscent of observed spectra, e.g. discrete absorption components and extended blue edges. In these 1-D, spherically symmetric models, however, the dense "clumps" are "shells", and this tends to exaggerate such features. There is even noticeable structure in the emission component, which is generally not observed. Indeed, the results suggest that the lack of observed variability in emission might ultimately be used to place an upper limit on the lateral scale of structure. This would nicely complement information from the DAC's, which require the structure to cover a substantial fraction of the stellar disk and so already place a lower limit on this lateral scale.

Line-synthesis calculations have also been carried out for Hα. For the smooth CAK model, the profile remains in absorption, but in the clumped model, it exhibits a variable emission, which is often just what's observed (Ebbetts 1982). This is so even though the time-averaged mass loss rates of the two models are the same. The reason is, of course, that formation of Hα depends on the density *squared*, and, as discussed above, this is enhanced in these structured models by a clumping factor of 10 or more.

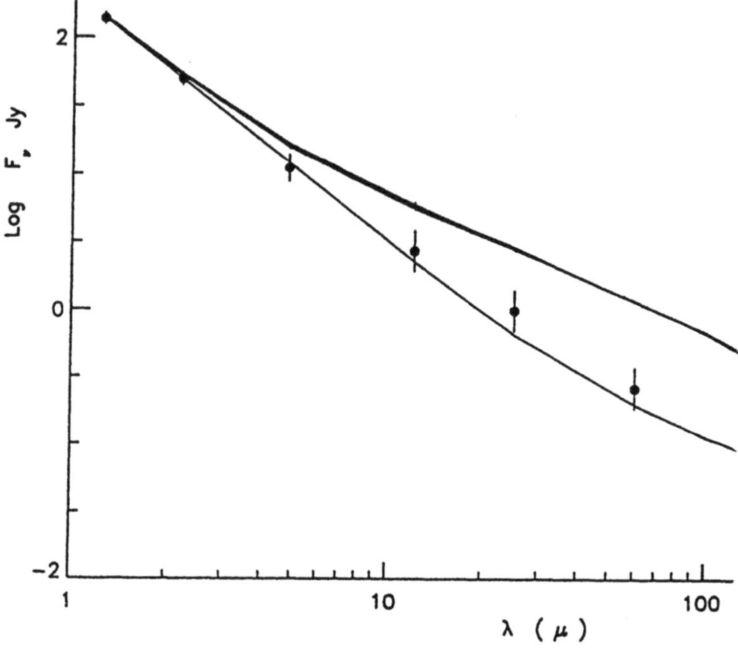

Fig. 7. IR continuum flux vs. wavelength for smooth and structured wind models of ζ *Pup*. The points are the observed data.

A similar enhancement is found for the flux in the IR continuum (Najarro and Owocki, work in progress). Figure 7 compares the wavelength dependence of the IR flux in a smooth model (lower curve) with that in a structured model at several instants in time (upper curves). The stellar parameters in this case were those for ζ *Pup*, and the wind parameters represent a best fit to the observed UV lines. The points plotted denote the observed fluxes (Lamers et al. 1986); note that these are slightly higher than predicted by the smooth model, but well below the strongly enhanced emission of the clumped model. As mentioned above, this structured model is probably too extreme, and work is underway to develop a more reasonable "best-fit" clumped model.

Finally, let us briefly return to the X-ray signatures of such unstable models. Although detailed calculations are still in progress, it now seems clear that such dynamical models of wind structure also have many favorable characteristics for explaining the available X-ray data. The range in shock strengths $\Delta v \approx 100 - 1000\,\mathrm{km/s}$ is roughly what's needed to obtain the soft and hard X-ray component observed in X-ray spectra, and the small fraction of strongly shocked material implies that a similarly small fraction $\lesssim 10^{-3}$ of the total flow kinetic energy $\dot{M}v_{\infty}^2/2 \approx 10^{-3}\,L_*$ goes into shock heating, which agrees roughly with the observed X-ray luminosity scaling $L_{\mathrm{x}} \approx 10^{-7}\,L_*$. Both aspects seem to favor such models over the previous forward shock picture.

4 Summary

Let us conclude by summarizing some of the most important points:

1. In the line-driven instability, only a very little material is actually accelerated to high speed; for most of the mass, the dominant effect is clumping. Perhaps we should call this the *line-driven clumping instability*?
2. The diffuse radiation plays an important role in reducing the instability, especially near the wind base. However even small base perturbations are inevitably amplified to give substantial structure beyond about $r \gtrsim 2R_*$.
3. Efforts to synthesize X-ray, UV, optical, and IR observations from structured models have begun, and preliminary results show some encouraging possibility of explaining various features observed in these datasets.
4. Much work remains to relax various simplifications, including the source function smoothness, isothermality, and 1-D radial flow.

This work was supported in part by NSF grant AST 88-14580 and NASA grant NAGW-1487. Many of the computations were carried out using an allocation of supercomputer time from the San Diego Supercomputer Center. I thank A. Fullerton and G. Cooper for helpful comments, and J. Puls and P. Najarro for permission to use Figures 5-7 prior to publication.

References

Abbott, D. A. 1980, *Ap. J.*, **242**, 1183.
Abbott, D. C., Bieging, J. H., and Churchwell, E. 1981, *Ap. J.*, **250**, 645.
Abbott, D. C., Bieging, J. H., and Churchwell, E. 1984, *Ap. J.*, **280**, 671.
Abbott, D. C., Telesco, C. M., and Wolff, S.C. 1984, *Ap. J.*, **279**, 225.
Carlberg, R. G. 1980, *Ap. J.*, **241**, 1131.
Castor, J. I. 1974, *Mon. Not. R. Astr. Soc.*, **169**, 279.
Castor, J. I., Abbott, D. C., and Klein, R. I. 1975, *Ap. J.*, **195**, 157. (CAK)
Castor, J. I. 1987, in *Instabilities in Luminous Early Type Stars*, ed. by H.J.G.L.M. Lamers and C.W.H. de Loore (Reidel, Dordrecht), p.159.
Castor, J. I. 1991, in *Stellar Atmospheres: Beyond Classical Models*, L. Crivellari, I. Hubeny and D. Hummer, eds., (Kluwer, Dordrecht), 221.

Cooper, R. G., and Owocki, S. P. 1991, in *Nonisotropic and Variable Outflows from Stars*, L. Drissen, C. Leitherer, and A. Nota, eds., (Astron. Soc. Pacific, San Francisco), in press.
Ebbets, D. 1982, *Ap. J. Supp.*, **48**, 399.
Feldmeier, A., and Owocki, S. P. 1991, *Ap. J.*, , in preparation.
Friend, D. B., and Abbott, D. C. 1986, *Ap. J.*, **311**, 701.
Harnden, F. R. *et al.* 1979, *Ap. J. (Letters)*, **234**, L51.
Krolik, J. H., and Raymond, J. C. 1985, *Ap. J.*, **298**, 660.
Lamers, H. J. G. L. M., Gathier, R., and Snow, T. P. 1982, *Ap. J.*, **258**, 186.
Lamers, H. J. G. L. M., Waters, L. B., Wesselius , P. R. 1984, *Astron. Astrophys.*, **134**, L17.
Lucy, L. B. 1971, *Ap. J.*, **163**, 95.
Lucy, L. B. 1982a, *Ap. J.*, **255**, 278.
Lucy, L. B. 1982b, *Ap. J.*, **255**, 286.
Lucy, L. B. 1984, *Ap. J.*, **284**, 351.
Lucy, L. B., and Solomon, P. M. 1970, *Ap. J.*, **159**, 879.
Lucy, L. B., and White, R.L. 1984, *Ap. J.*, **241**, 300.
MacGregor, K. B., Hartmann, L., and Raymond, J. C. 1979, *Ap. J.*, **231**, 514.
Milne, E. A. 1926, *Mon. Not. R. Astr. Soc.* **86**, 459.
MacFarlane, J. J. and Cassinelli, J. P. 1989, *Ap. J.*, **347**, 1090.
Mullan, D. J. 1984, *Ap. J.*, **283**, 303.
Mullan, D. J. 1986, *Astron. Astrophys.*, **165**, 157.
Owocki, S. P. 1990, in *Reviews of Modern Astronomy*, **3**, (Springer, Berlin), 98.
Owocki, S. P. 1991a, in *Wolf-Rayet Stars and their Interrelations to Other Massive Stars in Galaxies*, IAU Symposium 143, K. Van der Hucht and B. Hidayat, eds., Kluwer, Dordrecht, 155.
Owocki, S. P. 1991b, in *Stellar Atmospheres: Beyond Classical Models*, L. Crivellari, I. Hubeny and D. G. Hummer, eds., (Kluwer, Dordrecht), 235.
Owocki, S. P., and Rybicki, G. B. 1984, *Ap. J.*, **284**, 337.
Owocki, S. P., and Rybicki, G. B. 1985, *Ap. J.*, **299**, 265.
Owocki, S. P., and Zank, G. P. 1991, *Ap. J.*, **368**, 491.
Owocki, S. P., Castor, J. I., and Rybicki, G. B. 1988, *Ap. J.*, **335**, 914. (OCR)
Owocki, S. P., Poe, C. H., and Castor, J.I., 1990, in *Properties of Hot Luminous Stars*, ed. C. Garmany, (Astron. Soc. Pacific, San Francisco), p. 283.
Pauldrach, A., Puls, J., and Kudritzki, R. P. 1986, *Astr. Ap.*, **164**, 86. (PPK)
Poe, C. H., Owocki, S. P., and Castor, J. I. 1990, *Ap. J.*, **355**, 199.
Prinja, R. K. 1988, *M.N.R.A.S.*, **231**, 21P.
Rybicki, G. B. 1987, in *Instabilities in Luminous Early Type Stars*, ed. by H.J.G.L.M. Lamers and C.W.H. de Loore (Reidel, Dordrecht) p.175.
Rybicki, G. B., Owocki, S. P., and Castor, J. I. 1990, *Ap. J.*, **349**, 274.
Rybicki, G. B. and Hummer, D. G. 1978, *Ap. J.*, **219**, 654.
Seward, F. D. *et al.* 1979, *Ap. J. (Letters)*, **234**, L55.
Sobolev, V. V. 1960, *Moving Envelopes of Stars*, (Harvard University Press, Cambridge).
Waldron, W. L., Klein, L., and Altner, B. 1991, in *Nonisotropic and Variable Outflows from Stars*, L. Drissen, C. Leitherer, and A. Nota, eds., (Astron. Soc. Pacific, San Francisco), in press.

Radiation-Driven Wind Theory: Not (Yet?) Working

Werner Schmutz [1], Daniel Schaerer [2]

[1]Institut für Astronomie, ETH Zürich, CH-8092 Zürich, Switzerland
[2]Observatoire de Genève, Ch. des Maillettes 51, CH-1290 Sauverny, Switzerland

Abstract: Line-blanketed non-LTE models for spherically expanding atmospheres have been constructed, including multiple scattering and line overlap and without an imposed core-halo structure. They have been used to evaluate the radiation force in the winds of two Of stars, ζ Pup and R84. We find for both stars that radiation forces are not sufficient to drive the observed mass-loss rates. For ζ Pup the disagreement is less than a factor of two, and could in principle be explained by uncertainties in the interpretation of the observations. However, for R84 the difference is unequivocally due to a failure of the radiation-driven wind theory in its present form. Having found a case where the theory does not work we are inclined to interpret the problems for ζ Pup as real, and therefore we suggest that the present theory of radiation-driven winds is not sufficient for a complete understanding of the observed winds of hot stars.

Introduction

The theory of radiation-driven stellar winds is quite successful in explaining the observed mass-loss properties of hot stars (Howarth and Prinja 1989). Since radiation pressure was first suggested as the driving force by Lucy and Solomon (1970), and following the formulation as a quantitative theory by Castor, Abbott, and Klein (1970, CAK), there have been many papers refining the evaluation of the radiation force. The most important improvements have been introduced by Abbott (1982), Panagia and Macchetto (1982), Abbott and Lucy (1985), Friend and Abbott (1986), Pauldrach (1987), and Puls (1987). Today, quantitative predictions of mass-loss rates and terminal velocity for individual stars are in most cases within the (usually large) uncertainties of the observations. However, it is suspected that there are still systematic discrepancies between theory and reality (Groenewegen et al. 1989; Blomme 1990; Blomme 1991; Lamers, Schuiling, and Leitherer 1991). The claimed disagreement is that the theory predicts lower mass-loss rates and higher terminal velocities than observed. It is not clear what the suspected differences imply; is it just that the theory is not yet good enough, or do we omit a significant contributor to the driving force in addition to radiation pressure? In this contribution we present evidence for the second possibility.

Method

To construct a model we proceed as follows. First, we adopt a "known" velocity structure and mass-loss rate. The hydrogen and helium populations, together with a first guess of the depth-dependent radiation field, are then calculated with the code of Hamann and Schmutz (1987). This atmosphere code treats the line radiation transfer in the comoving frame and solves the coupling between transfer and statistical equilibrium equations with the approximate lambda iteration technique (Hamann 1987). Then we calculate separately the ionization and excitation states of metals (which depend on the radiation field obtained with the atmosphere code) by using the approximate formulae of Schmutz (1991). With all level populations specified a formal solution of the radiation transfer equations is performed with Monte Carlo simulation. During the Monte Carlo calculation intensity-weighted mean opacities are evaluated for broad wavelength bands (typically 100 Å). These mean opacities are then fed back into the atmosphere code for the recalculation of populations together with a new radiation field, and the process is iterated to convergence. The final opacities and radiation field automatically give the radiation force for the assumed dynamical structure.

The unique feature of our code is the method that allows to treat line-blanketing in a non-LTE code (Schmutz 1991). By the nature of our method this determination of the radiation force includes automatically the effects of multiple scattering, line overlap, continuum contribution, and changing energy distribution of the flux as a function of depth. Thus, our evaluation of the radiation force accounts for more effects than any previous treatment. New are the inclusion of the contribution to the force from continuum absorption and that we account for changing energy distribution of the flux as a function of depth due to absorption and emission processes. Both effects are important mainly in the optically thick part of the atmosphere but it is essential to include them since our inner model-boundary is at large optical depths. In the optically thin part multiple scattering and line overlap are of importance. Multiple scattering has been studied first by Panagia and Macchetto (1982). Friend and Castor (1983) added line overlap but assumed the lines to be randomly distributed in frequency. Abbott and Lucy (1985) and Puls (1987) based their calculations on realistic line lists and for the optically thin part of the wind our method is in practice equivalent to their approach.

In order to evaluate the radiation force in the form of a CAK force multiplier ($M(r) = k(r)t^{\alpha(r)}$, where $t = \rho/\frac{dv}{dr}$), we then calculate a second atmosphere with the mass-loss rates reduced by a factor of two but with otherwise identical stellar parameters and with the same velocity structure. The pair of radiation forces from two corresponding models allows us to determine the CAK parameter $k(r)$ and $\alpha(r)$ as a function of radius. We then follow CAK to solve the hydrodynamic equation. Note, that in contrast to the approach of Pauldrach, Puls, and Kudritzki (1986), who introduced a modified force multiplier in order to account for "non-radially streaming" photons, the parameters $k(r)$ and $\alpha(r)$ determined in this way have all effects built in. This results in a strong depth dependence of $k(r)$ and $\alpha(r)$, i.e. the two parameters are not constants, or almost constant, as they are in the usual approach.

The resulting radiation forces depend on the ionization equilibrium and level excitation. For the metals our approach is clearly inferior to the extensive calculations of Pauldrach (1987). However, we will show below that the calculated emergent spectrum agrees well with the observed energy distribution. This fact signifies that the used ionization and excitation conditions are close to the true situation in the continuum formation region. For the outer part of the wind there is no direct check possible. Since the approximations are more likely to deteriorate in regions which are remote from the star the calculated radiation forces may be less accurate with increasing distance from the star. Bearing in mind this caveat we give less weight to disagreements in the regions where the wind reaches its terminal velocity. Instead, we concentrate our investigation on the inner parts of the wind, i.e. on the acceleration zone and on predicted mass-loss rates.

The stellar models

The philosophy of our approach is that we first calculate semi-empirical model atmospheres, i.e. we adopt "observed" velocity laws, before we solve the hydrodynamic equation for the density structure. The reason is that we want to test if the radiation force is sufficient to support the "observed" wind structure. We have constructed model atmospheres for two Of stars: HD 66811 (ζ Pup) and HDE 269227 (R84). The adopted stellar parameters of the two models are given in Schaerer and Schmutz (this workshop). The stellar radius and temperature of ζ Pup are taken from Bohannan et al. (1990), the mass-loss rate from Leitherer and Robert (1991), and the velocity law from Groenewegen et al. (1989). The parameters of R84 are those given in Schmutz et al. (1991) except for the effective temperature, which was revised by Schmutz (1991). The calculations presented in this contribution differ from the work of Schmutz (1991) in that the treatment of absorption during the line-scattering event has been improved. This results in a stronger line-blanketing effect around 1600 Å, caused by numerous Fe IV lines. The calculated spectrum and observed energy-distribution are compared in Figure 1.

Discussion

The calculated radiation forces are compared with the forces required to drive the observed stellar winds in Schaerer and Schmutz (this workshop, Figs. 1 and 3). For both models the calculated radiation forces are smaller than the forces necessary to drive the adopted wind. It is not obvious what these differences imply since there are plenty of possible explanations.

On the observational side the most uncertain quantity is the velocity law. In order to investigate the possibility that we get too small a radiation force because we adopted a wrong (e.g. too steep) velocity law, we solve the hydrodynamic equations using the force multipliers obtained from the model calculations. Adopting a mass of 37 $M_{\odot}$ for ζ Pup (from the observed $\log g = 3.5$) and of 18 $M_{\odot}$ for R84 (from the mass–luminosity relation of helium stars) we obtain solutions with mass-loss rates of $2.8\,10^{-6}$ $M_{\odot}$/yr and $\approx 1\,10^{-6}$, respectively. (The terminal wind velocities are 2400 km s^{-1} and some 1000 km s^{-1}.) For ζ Pup the difference to the observed mass-loss rate, which is $5\,10^{-6}$ $M_{\odot}$/yr, is

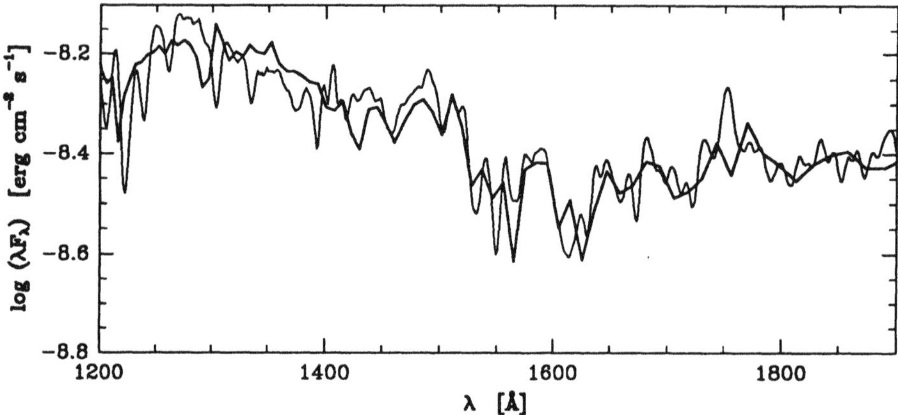

Fig. 1. Comparison of the calculated flux (thick line) with the dereddend IUE spectrum of R84 (thin line). The good agreement of the theoretical energy-distribution with the observed one indicates that the calculated ionization and excitation conditions are roughly correct.

less than a factor of two and could be (almost) accounted for by a different interpretation of the observed radio flux.[1] However, for R84 there is so much force missing that we have to conclude that it cannot be due to imperfectness of the atmosphere model: The theory of radiation driven winds in its present form is not able to explain theoretically the observed wind of this star. Because we have found an example where we are convinced that there must be an additional mechanism acting, we are inclined to interpret the problems in the case of ζ Pup as indicative that also in Of stars there is an additional contribution to the driving force.

Conclusion

There is no question that for "normal" O stars the dominant mechanism that drives their winds is radiation pressure. However, when investigating stars with winds of increasing strength, there is the tendency that a "force gap" is continuously opening. The difference is getting larger from Of stars, to Of/Wolf-Rayet transition type stars (like R84), WN L, WN E, and to WC stars. It could well be that the mechanism driving the Wolf-Rayet

[1] During the workshop we pointed out that we obtain considerably smaller force multipliers for ζ Pup than Gabler et al. (1989) and suspected that this is due to the combined effects of multiple scattering, line overlap, and that we do not use the core halo approximation. After the workshop a preprint by Kudritzki et al. (1991) came to our attention where they present new results for ζ Pup. They predict a mass-loss rate of $3\,10^{-6}$ $M_{\odot}$/yr. If we solve the hydrodynamic equation with their stellar parameters but our force multipliers we obtain $2.6\,10^{-6}$ $M_{\odot}$/yr. Thus, our predicted mass-loss rate is basically in agreement with their result. However, in contrast to our interpretation, they claim agreement with the observations. This is possible because the interpretation of the observed radio flux depends on what is assumed for the helium ionization equilibrium in the region where the radio flux is formed. Kudritzki et al. (1991) assume He^{++} to be dominant while we follow the observational evidence of Leitherer and Robert (1991) that helium is only once ionized.

winds is also present in O and that its contribution is non-negligible at least for Of type stars. The "yet?" in the title of this contribution refers to the fact that improvements of the radiation driven wind theory are still possible, e.g. inclusion of the effects of a finite intrinsic line profile (see Schaerer and Schmutz, this workshop). We cannot exclude that the winds of all these stars are radiation driven.

Acknowledgements: We are grateful to Ian Howarth for editorial assistance. D. S. acknowledges a travel grant by the Astronomy Commission of the Swiss National Academy of Sciences.

References

Abbott, D.C., 1982, ApJ 259, 282
Abbott, D.C., Lucy, L.B., 1985, ApJ, 288, 679
Blomme, R., 1990 A&A 229, 513
Blomme, R., 1991 A&A 246, 199
Bohannan, B., Voels, S.A., Hummer, D.G., Abbott, D.C., 1990, ApJ 365, 729
Castor, J.I., Abbott, D.C., Klein, R.I., 1975, ApJ 195, 157 (CAK)
Friend, D.B., Abbott, D.C., 1986, ApJ 311, 701
Friend, D.B., Castor, J.I., 1983, ApJ 272, 259
Gabler, R., Gabler, A., Kudritzki, R.P., Puls, J., Pauldrach, A., 1989, A&A 226, 162
Groenewegen, M.A.T., Lamers, H.J.G.L.M., Pauldrach, A.W.A., 1989, A&A 221, 78
Hamann, W.-R., 1987. In: W. Kalkofen (ed.) Numerical Radiative Transfer. Cambridge University Press, p. 35
Hamann, W.-R., Schmutz, W., 1987, A&A 174, 173
Howarth. I.D., Prinja, R.K. 1989, ApJS 69, 527
Kudritzki, R.-P., Hummer, D.G., Pauldrach, A.W.A., Puls, J., Najarro, F., Imhof, J., 1991, MPA Preprint 585
Lamers, H.J.G.L.M., Schuiling, E., Leitherer, C., 1991 (in preparation)
Leitherer, C., Robert, C., 1991, ApJ 377, 629
Lucy, L.B., Solomon, P., 1970, A&A 159, 879
Panagia, N., Macchetto, 1982, A&A 106, 266
Pauldrach, A., 1987, A&A 183, 295
Pauldrach, A.W.A., Puls, J., Kudritzki, R.-P., 1986, A&A 164, 86
Puls, J., 1987, A&A 184, 227
Schmutz, W., 1991. In: Crivellari, L., Hubeny, I., Hummer, D.G. (eds.) NATO ASI Series C, Vol. 341, Stellar Atmospheres: Beyond Classical Models. Kluwer Academic Publishers, p. 191
Schmutz, W., Leitherer, C., Hubeny, I., Vogel, M., Hamann, W.-R., Wessolowski, U., 1991, ApJ 372, 664

Radiation-Driven Wind Theory: The Influence of Turbulence

Daniel Schaerer [1], Werner Schmutz [2]

[1]Observatoire de Genève, Ch. des Maillettes 51, CH-1290 Sauverny, Switzerland
[2]Institut für Astronomie, ETH Zürich, CH-8092 Zürich, Switzerland

Abstract: Line-blanketed non-LTE models for spherically expanding atmospheres without core-halo approximation, including multiple scattering and line overlap, are used to evaluate the radiation pressure in stellar winds of Of and Ofpe/WN9 stars. The enhancement of the radiative forces due to microturbulence (line-broadening) is discussed.

Introduction and Method

Schmutz and Schaerer (this workshop, Paper I) presented evidence that radiation pressure alone might not provide enough force to drive the observed stellar winds of hot stars. The aim of this contribution is to investigate the role of turbulence. Turbulence could enhance the acceleration in two ways: First, turbulence itself provides an outward directed force (turbulent pressure) and second, turbulence widens the line profiles which results in an increase of the radiation pressure. In this pilot study we only explore the second contribution.

The procedure we follow is that we adopt stellar parameters and "observed" velocity laws and mass-loss rates (see paper I). Given the physical structure of the atmosphere we calculate line-blanketed non-LTE models for spherically expanding atmospheres using Monte Carlo (MC) simulations. We then compare the resulting radiative forces with those required to drive the adopted stellar winds of an Of star (ζ Pup) and an Ofpe/WN9 (R84).

In paper I we have evaluated the forces in the narrow-line limit. In order to study turbulence, as observed in O–stars (e. g. by Groenewegen et al. 1989), and its influence on the line-forces we have to account for the finite line widths. This is achieved in the MC code by trapping the photons in the broadened line at a frequency displacement of one dopplerwidth. Then possible multiple resonance scattering is replaced by single scattering and the angular redistribution is chosen. The photons are then released at the frequency which corresponds to the optical depth of $\tau = 1$ for strong lines.

Results

ζ Puppis:
The stellar parameters taken from Bohannan et al. (1990) and the wind parameters according to Groenewegen et al. (1989) are the following: $logL = 5.9L_\odot$, $T = 42000K$, $log\dot{M} = -5.3M_\odot yr^{-1}$, $v_\infty = 2200kms^{-1}$, $\beta = 0.7$. Figure 1 shows the forces in the narrow-line limit and compares them with the force required for $M = 33\ M_\odot$. It can be seen that radiation pressure, without any line broadening, is not sufficient to drive the stellar wind of ζ Pup. Only at distances $r > 10\ R_\star$, far behind the sonic point, we obtain sufficiently large forces.

In Figure 2 we examine the forces obtained with a depth-dependent turbulence $v_{turb} = min(0.7v, v_\infty/2)$. With such large a turbulence, the force increases significantly due to line-broadening but does not match the value required.

The Ofpe/WN9 star HDE 269227 (R84):
The parameters given by Schmutz et al. (1991) are: $logL = 5.7L_\odot$, $T = 28500K$, $log\dot{M} = -4.61M_\odot yr^{-1}$, $v_\infty = 400kms^{-1}$, $\beta = 1$. The force obtained in the narrow-line limit is shown in Fig. 3. We also show the CAK force (as calculated by Schmutz et al. 1991) and the force required to drive the adopted wind. One sees that the CAK force largely overestimates the MC force. Note that, as for ζ Pup, the total energy transfer to the wind is large enough to reproduce the adopted mass loss, but its distribution is not able to drive the observed wind.

Figure 4 gives the "largest" force obtained assuming a (unrealistically large) depth-dependent turbulence. Considering line-broadening alone the conclusions are the same as for ζ Pup. However it is important to recognize the importance of turbulent pressure close to $R_\star$, since this has been neglected so far.

Conclusion

Line-blanketed models, including multiple scattering and line overlap, show radiative forces calculated in the narrow-line limit to be insufficient to drive the adopted (observed) wind structure of Of and Ofpe/WN9 stars.

Line-broadening due to turbulence increases the forces significantly but even unrealistically large values do not provide enough force. The contribution of turbulent pressure (cf. Blomme et al. 1991), neglected in our study, will be the object of future work.

Acknowledgements: D. S. acknowledges a travel grant by the Astronomy Commision of the Swiss National Academy of Sciences.

References

Blomme, R., Vanbeveren, D., Van Resbergen, W., 1991, A&A 241, 479
Bohannan, B., Voels, S.A., Hummer, D.G., Abbott, D.C., 1990, ApJ 365, 729
Groenewegen, M.A.T., Lamers, H.J.G.L.M., Pauldrach, A.W.A., 1989, A&A 221, 78
Schmutz, W., Leitherer, C., Hubeny, I., Vogel, M., Hamann, W.-R., Wessolowski, U., 1991, ApJ 372, 664
Schmutz, W., Schaerer, D., these proceedings, (Paper I)

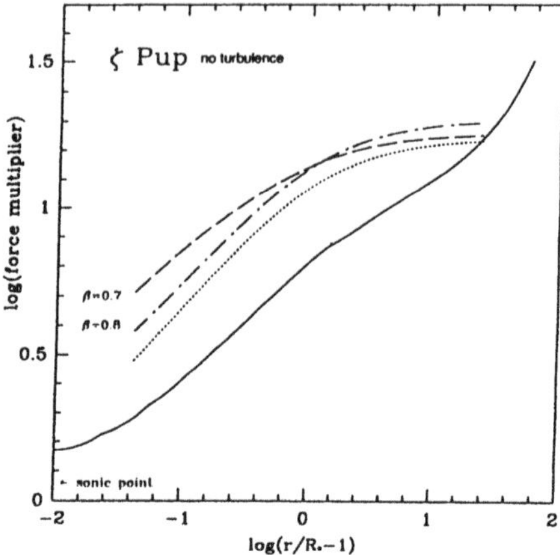

Fig. 1. Comparison of the obtained force (narrow-line limit) with the required force for $M = 33\ M_\odot$. The bold line shows the MC force multiplier (fm). The fm required to maintain the assumed $\beta = 0.7$ (dashed) resp. 0.8 (dash-dotted) velocity field is given. Using a higher luminosity ($logL = 6.$), the required force gives the dotted line. One clearly sees the "missing force" of a factor of approximately 2 up to $r = 4\ R_\star$.

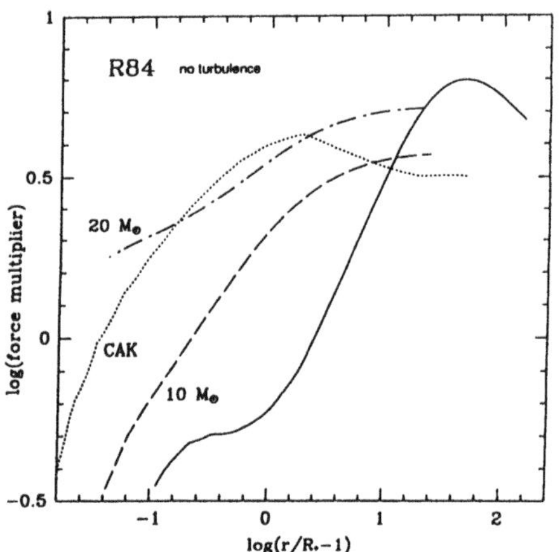

Fig. 3. Same as Fig. 1. The bold line shows the MC force multiplier (fm). The required fm is given for $M = 10\ M_\odot$ and $20\ M_\odot$. The dotted line gives the CAK force multiplier obtained by Schmutz et al.

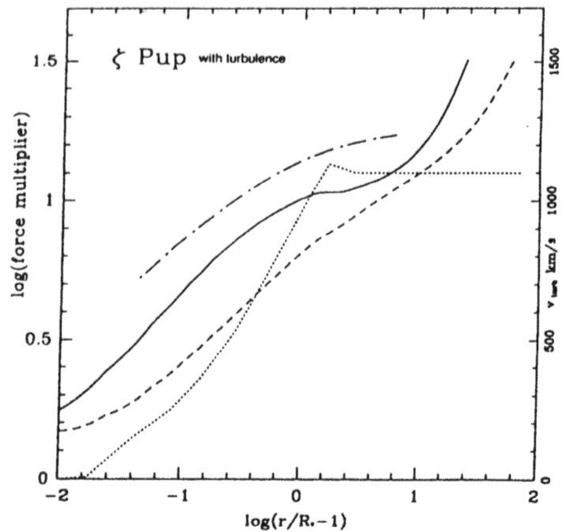

Fig. 2. Comparison of the obtained force including turbulence (line-broadening) with the required force. The bold line shows the MC force multiplier (fm) obtained with the depth-dependent turbulence given by the dotted line (cf. Groenewegen et al., 290 $km\ s^{-1}$). The required fm is given by the dash-dotted line. The dashed line gives the force in the narrow-line limit, shown in Fig. 1

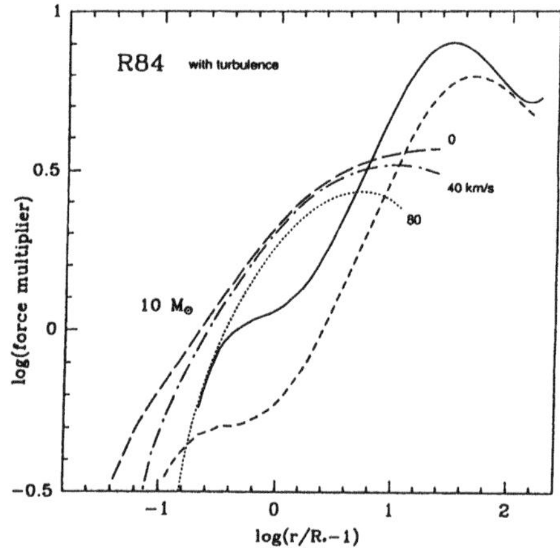

Fig. 4. Same as Fig. 2. The bold line shows the MC fm adopting outward increasing turbulence. The required fm is given for $v_{turb} = 0$ (long dashed), 40 (dash-dotted) and 80 $km\ s^{-1}$ (dotted) taking turbulent pressure into account. The short dashed line gives the force in the narrow-line limit shown in Fig. 3.

APPLICATION OF THE ETLA APPROACH IN THE COMOVING FRAME TO THE STUDY OF WINDS IN HOT STARS

M. Perinotto[1] and C. Catala[2]

1) Dipartimento di Astronomia e Scienza dello Spazio
Università di Firenze, Largo E. Fermi 5, 50125 Firenze, Italy

2) Observatoire de Paris, Meudon Unité de Recherche Associée
au CNRS 264, F-92195 Meudon Principal Cedex, France

Abstract

Among the various methods developed for studying the formation of line profiles in an expanding atmosphere, the one based on the direct solution of the coupled comoving-frame transfer and statistical equilibrium equations with multilevel, multi-ion model atoms in the equivalent-two-level-atom (ETLA) scheme (Mihalas and Kunasz, 1978), offers particular advantages over other methods. It handles correctly the transonic wind region and converges relatively easily. We have adapted a version of this code extensively used by one of us for Herbig Ae/Be stars (e.g. Catala and Kunasz, 1987), for application to hot stars. Appending a wind model to a photosphere calculated with the non-LTE TLUSTY code of Hubeny (1989), we have computed the CIV line profiles in a test case calculated by Pauldrach et al. (1990) with the entirely different technique, called "unified atmosphere-wind theory". The effect of the turbulence across the winds appears to be important to reproduce observed profiles in hot stars.

1. The ETLA Method Used in this Work

In semi-empirical methods, line profiles are calculated assuming a given structure of the wind (mass loss rate, velocity law, turbulence). They are particularly adequate to investigate 'observationally' the relevant physical quantities, without limitations imposed by the chosen mechanism of production of the wind, and with the minimum of parameters.

The main properties of the ETLA model used are: a) the structure of the photosphere is calculated with the TLUSTY non-LTE code of Hubeny (1989); b) a wind model with a given mass loss rate is appended on top of the photosphere with a smooth transition between the two structures at the optical depth of 2/3 at a given wavelength; c) the velocity law in the wind is specified; d) the density follows from the continuity equation; e) the statistical equilibrium and radiative transfer equations are solved for each bound-bound transition in the comoving frame of the flow with multilevel, multi-ion model atoms in the equivalent two-level-atom (ETLA) scheme (Mihalas and Kunasz, 1978); f) a Doppler profile is assumed for the intrinsic shape of the lines, with a width defined by the thermal broadening and a constant turbulent velocity across the wind; g) multiplets are treated as single transitions.

The advantages of the method are: i) a consistent treatment of lines and continua over the full range of the flow velocities from the transonic region up to the high terminal velocities; ii) a smooth overlapping of the results of the photospheric calculations with the wind models; the transfer equation is solved consistently through the wind and photosphere, down to a sufficiently deep layer to ensure thermalization of radiation at all wavelengths; iii) a relatively fast convergence.

2. Preliminary Application to the CIV Lines and Results

For the present application to the CIV lines, the model atom includes the ions CIII, CIV, and CV. A total of 6 levels are taken into account with 7 transitions treated "explicitely", i.e. the corresponding radiative rates are recalculated at each ETLA iteration. Background sources and sinks of radiation, including H, He, HeI and HeII; N, O, Ne, Si, and their ions, are assumed to be in LTE, with normal abundances.

Our test calculations were performed with a wind model approaching the C3 case of Pauldrach et al. (1990, thereafter PKPB), which refers to a model star, intended to resemble a main sequence star of 60 solar masses, $T_{eff} = 35{,}480$ K, log $g = 3.4$ (cgs), He/H = 0.1, log (L/Lo) = 5.91. The terminal velocity of this model star is calculated by PKPB to be of 2650 km/s and the mass loss rate to be of $M = 1.63\ 10^{-6} M_0\ yr^{-1}$.

Our prime goal was to investigate the influence of each parameter of the model on the resulting profile. As expected, the dependence of the profile on the velocity law is significant. A detailed comparison of the computed profiles with the observed profiles of several lines (CIV, NV, NIV, SiIV, OVI, Hα) should therefore provide a powerful diagnostic of the shape of the velocity law in this type of wind.

One interesting aspect is the role of the turbulence in the wind, which is expected to be important in the winds of hot stars (e.g. Owocki, 1991) and in fact has been found to be important in many OB stars (Groenwegen and Lamers, 1990).

In Fig. 1 we show the effect of increasing the turbulent velocity V_{turb}, from 45 km/s to 200 km/s, with a velocity law close to the standard one with $\beta = 2$. The changes are quite significant, with the new strongly 'turbulent' profile probably better resembling the real OB stars, because of its lower emission component relative to the less turbulent case. (cf. PKPB where no 'turbulence' was included in the theoretical calculations and the observed profiles have in general less emission than the calculated ones). We therefore conclude that the inclusion of some type of turbulent velocity in the model is necessary for a realistic comparison with observed profiles.

References

Catala, C. and Kunasz, P.B. 1987, Astron, Astrophys. 174, 158.

Groenewegen., M.A.T. and Lamers, H.J.G.L.M. 1990, Astron. Astrophys. Suppl. 79, 359.

Hubeny, I. 1989, Comput. Phys. Commun. 52, 103.

Mihalas, D. and Kunasz, P.B. 1978, Ap.J. 219, 635.

Owocki, S.P. 1991, in "Stellar Atmospheres: Beyond Classical Models", Eds. L. Crivellari, I. Hubeny and D.G. Hummer, (Kluwer: The Netherland), p. 235.

Pauldrach, A.W.A., Kudritzki, R.P., Puls, J. and Butler, K. 1990, Astron. Astrophys. 228, 125 (PKPB).

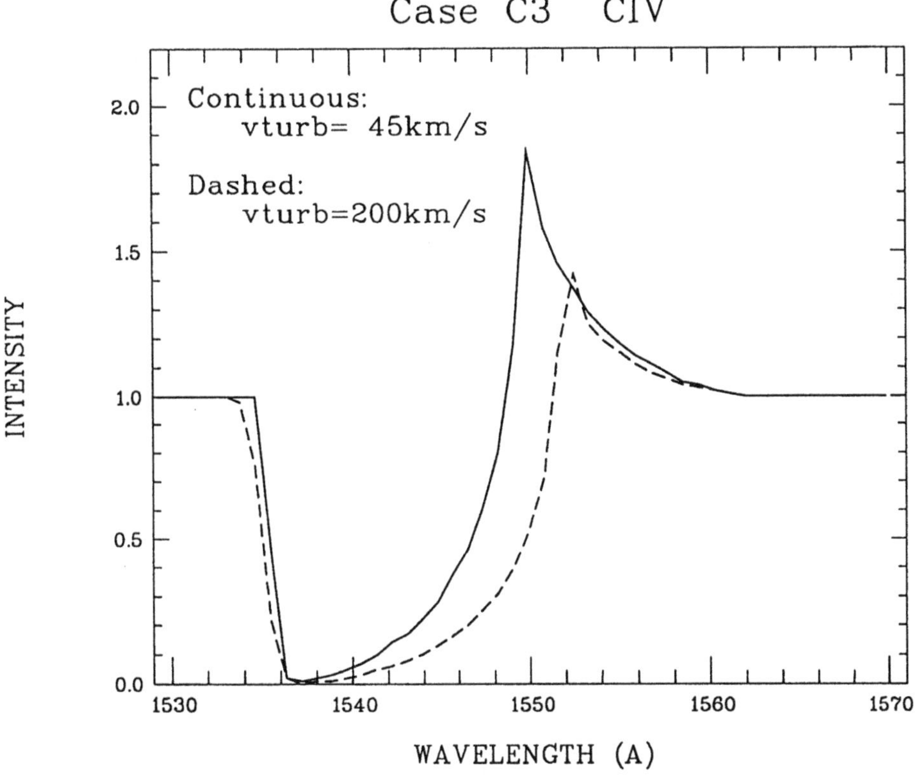

Fig 1. Role of the turbulent velocity (V_{turb} = 45 and 200 km/s) in the calculation of the CIV profile.

Interactive LTE Spectrum Synthesis

A.E. Lynas-Gray

Department of Physics, University of Oxford, Keble Road, Oxford, OX1 3RH, England.

Abstract: Local Thermodynamic Equilibrium (LTE) is known to be a useful simplifying assumption when computing line-blanketed model stellar atmospheres and synthetic spectra. Element abundance adjustment can optimise agreement between a synthetic and an observed spectrum, providing a useful technique for abundance determination when spectrum lines are blended. Interactive abundance adjustment techniques are proposed as a means of minimising the effort involved in using LTE spectrum synthesis to obtain abundances for several elements in a large sample of stars.

1 Introduction

The limitations of synthetic spectra computed with the assumption of LTE are discussed by Mihalas and Athay (1973). More recently, Kurucz (1991) demonstrates the importance of line-blanketing by showing that the emergent solar flux can only be reproduced (using LTE) when blanketing from 58 000 000 lines is included. Progress on non-LTE line-blanketing is reviewed by Anderson and Grigsby (1991) who concede that, for the present at least, synthetic LTE spectra are in better agreement with observation even in the case of main sequence early B stars. LTE is therefore a useful simplifying assumption, and LTE spectrum synthesis a suitable technique for stellar abundance determination in most cases.

The study of abundance gradients in the Galaxy, for example, requires analyses for a large number of stars. Cayrel et al. (1991) present a method for determining fundamental stellar parameters and metalicities using a grid of synthetic spectra, its application to a large number of stars being straightforward. Interactive LTE spectrum synthesis, as presented in this paper, is proposed as an alternative strategy; an advantage is that individual metal abundances and isotope ratios are obtainable.

2 Method

Inspection of an observed spectrum gives an approximate effective temperature (T_{eff}), a surface gravity (log g) and abundances; it is easier and more precise if a spectrum of another star (for which these quantities are similar and known) is also available for comparison. Good initial estimates for T_{eff} and log g are desirable. Moon and Dworetsky (1985) present a method for T_{eff} and log g determination from uvbyβ photometry. If the stellar energy distribution is known, together with an infrared monochromatic flux, an accurate T_{eff} is obtainable (Blackwell and Shallis 1977). Except in the case of late-type stars where few if any lines of ionised species are observed, ionisation ratios may be used to derive log g once T_{eff} is known.

With initial estimates for T_{eff}, log g and abundances, a line-blanketed LTE model stellar atmosphere may be computed using (for example) ATLAS6 (Kurucz 1979) or MARCS (Gustafsson et al. 1975). Standard techniques summarised by Cayrel et al. (1991) would then give a synthetic spectrum for comparison with observation and subsequent refinement. An element whose abundance is to be interactively adjusted is selected; it must be a minor constituent of the stellar atmosphere (not hydrogen or helium) and not appreciably associated into molecules. For each wavelength and model stellar atmosphere depth point, separate line opacity summations are computed for the adjustable abundance element along with its ionised forms (L_λ') and all remaining species (L_λ''). Adding each L_λ' and L_λ'' to the corresponding continuous opacity enables the optical depth scale to be derived, and radiative transfer equation solved, at each wavelength point to yield a synthetic spectrum.

For interactive and repeated element abundance adjustment, it is essential to minimise computation. Brault and White's (1971) deconvolution techniques are used to remove instrumental and rotational broadening from an observed spectrum; these do not then have to be convolved with the synthetic spectrum each time it is computed. After dividing by a previously calculated continuum flux, a normalised synthetic spectrum can be graphically compared with a deconvolved observed spectrum.

A weak line attributable to the adjustable abundance element, or one of its ionisation stages, is chosen with a graphics cursor; it must lie on the linear part of the curve of growth so that the equivalent width is directly proportional to abundance, or the species number density at each model stellar atmosphere depth point. The synthetic spectrum profile desired for a chosen line is defined with a graphics cursor using the observed profile as a guide. Not only can a new species abundance be obtained by scaling the previous abundance with the equivalent width ratio, but scaling previous values of L_λ' with the equivalent width ratio appropriately updates them; this is because all ionisation fractions are fixed in LTE, through the Saha equation, by temperature and electron pressure at each model atmosphere depth. Revised L_λ' values are thereby obtained without the expense of recomputing Voigt functions; it only works if the species is a minor stellar atmosphere constituent (otherwise Voigt damping parameters are altered significantly) and not appreciably associated into molecules. A new synthetic spectrum is now directly obtainable as before.

Several iterations will be required before an optimum synthetic spectrum is obtained, with each element (other than hydrogen or helium) selected for interactive abundance adjustment in turn. It is important to reduce uncertainties, by obtaining optimum synthetic spectra with different model stellar atmospheres and projected rotation speeds (v sin i). Once final values for T_{eff}, log g, v sin i and abundances are known, they should be used to initiate a synthetic spectrum calculation as a consistency check.

3 Program

A new program (STARSPEC) was developed to implement interactive LTE spectrum synthesis. Improved algorithms for Voigt function evaluation (Lether and Wenston 1990, 1991) enable, in principle, synthetic spectra to be calculated to a user-determined numerical precision dependent upon the observed spectrum to be analysed. Accurate oscillator strengths for most transitions of astrophysical interest are becoming available through Opacity Project work (Seaton 1987) and attention now needs to be devoted to numerical precision in synthetic spectrum calculations. It is also essential to be able to include different isotopic compositions and allow for stratification in the stellar atmosphere.

Related scalars and vectors, such as those required to specify a LTE stellar atmosphere, are logically arranged into single vectors for passing as arguments to subroutines. Of crucial importance to minimising memory utilisation is the provision of a workspace vector in the main program whose length is the maximum required by any subroutine; this vector is used in many different contexts. Only quantities needed for subsequent parts of the calculation are stored outside the workspace vector.

Optimum utilisation of the central processing unit (cpu) is achieved by minimising the number of subroutine calls, which requires loops to be placed inside subroutines; this is especially important in the case of Voigt function evaluation. In order to ensure efficient operation on vector processors, inner loops are written so that they vectorise. The best performance on CRAY computers will be achieved by unrolling critical outer loops following Dongarra and Eisenstat (1984).

The program is controlled through user-supplied commands and choices made with a graphics cursor. Help information for each command, and its qualifying parameters, is built into the code. Script files containing several commands can be used for interactive or batch mode operation.

4 Illustration

Van der Waals damping constants are estimated (where necessary) in STARSPEC using the impact approximation, as summarised by Warner (1967). Effective principal quantum numbers are obtained from energy differences deduced from spectroscopic data; this is adequate for simple atoms if the optical electron has a principal quantum number which exceeds that of all the core electrons. The Van der Waals damping constant is also dependent upon the mean relative speeds of the radiating atom and its perturbers; this is thermal in origin and therefore dependent upon masses.

The isotopic composition of a Van der Waals perturber therefore influences damping constants. As an illustration of the capabilities of STARSPEC, Figure 1 shows C I λ4269Å ($^1P^o - {}^1D$) profiles computed using an ATLAS6 model atmosphere; the stronger profile is a result of a exaggerated increase in the deuterium fraction. Deuterium atoms will have lower mean relative speeds with respect to radiating atoms, and are therefore less effective as Van der Waals perturbers at a given temperature. Replacing hydrogen with deuterium results in reduced damping; the same effect could be achieved by reducing pressure through a lowering of surface gravity.

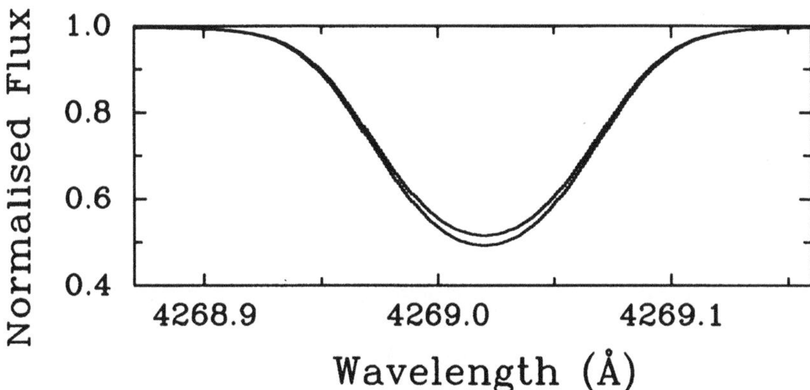

Fig. 1. The influence of hydrogen isotopic composition on Van der Waals broadening as seen in the C I λ4269.2Å line computed for $T_{eff} = 6000K$, $\log g = 4.0$ and abundances by number of N(H) = 0.899, N(He) = 0.100, N(C) = 0.001. The weaker profile is for D/H = 0 and the stronger for D/H = 2.

5 Summary

The ratio of line to continuous opacity is critical to the calculation of synthetic spectra; line opacities are dependent on oscillator strengths and damping constants which are often uncertain. Interactive spectrum synthesis enables a judgement to be made as to which lines deserve a higher weight when choosing an optimum fit. Approximations to the true physical conditions in a stellar atmosphere, atomic data quality and finite computational resources will limit what can be achieved in practice.

A brief description of the method employed for interactive spectrum synthesis has been presented. Planned developments to the code are the inclusion of Stark broadening, H I and He I line profiles. Extensive tests will include comparisons of spectra, synthesised with Holweger & Müller's (1974) empirical solar model atmosphere and Grevesse's (1984) solar abundances, with the Solar Flux Atlas of Kurucz et al. (1984). It is anticipated that the STARSPEC Fortran source, along with a detailed description, will be submitted to Computer Physics Communications.

References

Anderson, L., Grigsby, J.A. (1991): "Line Blanketing Without LTE: The Effect on Diagnostics for B-Type Stars", in Stellar Atmospheres: Beyond Classical Models, ed. by L. Crivellari *et al.*, NATO ASI Series C: Mathematical and Physical Sciences **341**, 365.

Blackwell, D.E., Shallis M.J. (1977): "Stellar Angular Diameters from Infrared Photometry. Application to Arcturus and Other Stars; with Effective Temperatures" in Mon. Not. R. astr. Soc. **180**, 177.

Brault, J.W., White, C.R. (1971): "The Analysis and Restoration of Astronomical Data via the Fast Fourier Transform" in Astron. Astrophys. **13**, 169.

Cayrel, R., Perrin, M.N., Barbuy, B., Buser, R. (1991): "A Grid of Synthetic Spectra for the Determination of Effective Temperature, Gravity and Metallicity of F, G, and K Stars I. Description of the Method", in Astron. Astrophys. **247**, 108.

Dongarra, J.J., Eisenstat, S.C. (1984): "Squeezing the Most out of an Algorithm in CRAY Fortran", in ACM Trans. Math. Software **10**, 219.

Grevesse, N. (1984):"Accurate Atomic Data and Solar Photospheric Spectroscopy", in Physica Scripta **T8**, 49.

Gustafsson, B., Bell, R.A., Eriksson, K., Nordlund, Å. (1975): "A Grid of Model Atmospheres for Metal-Deficient Giant Stars I", in Astron. Astrophys. **42**, 407.

Holweger, H., Müller, E.A. (1974): "The Photospheric Barium Spectrum: Solar Abundance and Collision Broadening of Ba II Lines by Hydrogen", in Solar Phys. **39**, 19.

Kurucz, R.L. (1979): "Model Atmospheres for G, F, A, B and O Stars", in Astrophys. J. Suppl. **40**. 1.

Kurucz, R.L. (1991): "New Opacity Calculations", Stellar Atmospheres: Beyond Classical Models, ed. by L. Crivellari *et al.*, NATO ASI Series C: Mathematical and Physical Sciences **341**, 441.

Kurucz, R.L., Furenlid, I., Brault, J., Testerman, L. (1984): "Solar Flux Atlas from 296 to 1300nm", Sunspot, New Mexico: National Solar Observatory.

Lether, F.G., Wenston, P.R. (1990): "An Algorithm for the Numerical Computation of the Voigt Function", in Applied Mathematics and Computation **35**, 277.

Lether, F.G., Wenston, P.R. (1991): "The Numerical Computation of the Voigt Function by a Corrected Midpoint Quadrature Rule for $(-\infty, \infty)$", in Journal of Computation and Applied Mathematics **34**, 75.

Mihalas, D., Athay, R.G. (1973): "The Effects of Departures from LTE in Stellar Spectra", in Ann. Rev. Astron. Astrophys. **11**, 187.

Moon, T.T., Dworetsky, M.M. (1985): "Grids for the Determination of Effective Temperature and Surface Gravity of B, A and F Stars using uvbyβ Photometry", in Mon. Not. R. astr. Soc. **217**, 305.

Seaton, M.J. (1987): "Atomic Data for Opacity Calculations: I. General Description", in J. Phys. B: At. Mol. Phys. **20**, 6363.

Warner, B. (1967): "Some Effects of Pressure Broadening in Solar and Stellar Curves of Growth" in Mon. Not. R. astr. Soc. **136**, 381.

Calculations of Line Positions in the Presence of Magnetic and Electric Fields in White Dwarf Spectra

Susanne Friedrich [1], Roland Östreicher [2], Hanns Ruder [1]

[1] Lehr- und Forschungsbereich Theoretische Astrophysik, Universität Tübingen, W-7400 Tübingen, Fed. Rep. of Germany
[2] Landessternwarte, W-6900 Heidelberg, Fed. Rep. of Germany

Abstract: We present first results of our line position calculations in the presence of strong magnetic and electric fields, as one can find in the dense atmospheres of magnetic White Dwarfs. The problem was solved by second order perturbation theory with hydrogenic wavefunctions calculated in a pure magnetic field and arbitrary orientations of the electric and magnetic fields.

1 The Problem

For DA White Dwarfs with temperatures above 10000 K Stark broadening is the dominant line broadening mechanism for hydrogen. In the atmosphere there are electric fields of about $1{\cdot}10^7$ V/m caused by free electrons and ions.

In the presence of magnetic fields the degeneracy among levels of the same principal quantum number is removed. Thus whenever the magnetic field is strong enough dominating the Stark interaction via inter-l splitting, the electric field leads only to an additional shift of the line component.

2 The Solution

We calculate these shifts by second order perturbation theory for arbitrary orientations between the electric and magnetic fields (Tworz 1989). The unperturbed diamagnetic hydrogen wavefunctions were derived by matrix diagonalization in a pure magnetic field (Zeller 1990). As shown by quantum mechanical calculations for parallel magnetic and electric fields our calculations are valid up to electric field strength of $1{\cdot}10^6$ V/m. Depending on the temperature of the model atmosphere this is a typical value in the outer layers of a White Dwarf atmosphere.

In cylindrical coordinates and atomic units the Hamiltonian of the Hydrogen atom in homogenous outer fields is given by

$$H = \frac{1}{2}p^2 - \frac{1}{r} + \beta L_z + \frac{1}{2}\beta^2\rho^2 + F_\perp x + F_\parallel z \tag{1}$$
$$= H_0 + H_1 + H_2 \ , \tag{2}$$

where H_1 describes the magnetic and H_2 the electric field. The eigenvalue E_j of the Hydrogen atom in a magnetic field

$$(H_0 + H_1)|\text{"diamagnetic"}, j\rangle = E_j|\text{"diamagnetic"}, j\rangle \tag{3}$$

is calculated by huge matrix diagonalization, and the energy correction $\triangle E_j$, due to electric fields,

$$\triangle E_j = \sum_{j \neq i} \frac{|\langle\text{"diamagnetic"}, j|H_2|\text{"diamagnetic"}, i\rangle|^2}{E_j - E_i} \ . \tag{4}$$

by second order perturbation theory. The energy of the eigenstate j is than given by

$$E_j{}^{(2)} = E_j + \triangle E_j \ . \tag{5}$$

3 Results

In Fig.1 and Fig.2 we show the wavelength shifts due to the electric field of two transitions. The solid line describes the behavior in a pure magnetic field, the dashed line in a magnetic and electric field. The transitions are characterized by their zero field quantum numbers.

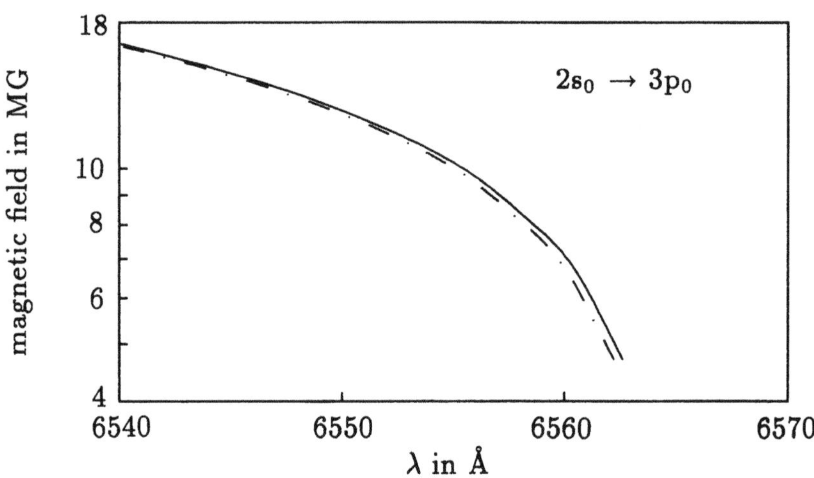

Fig. 1. Wavelength shifts due to the electric field for the transition $2s_0 \rightarrow 3p_0$. The solid line describes the behavior in a pure magnetic field, the dashed line in a magnetic and electric field

The shifts are very small - about 1 Å - or even negligible as Fig.2 shows. Thus they may only be detectable for stationary lines ($\frac{d\lambda}{dB} \approx 0$), e.g. observed in the spectrum of Grw+70°8247.

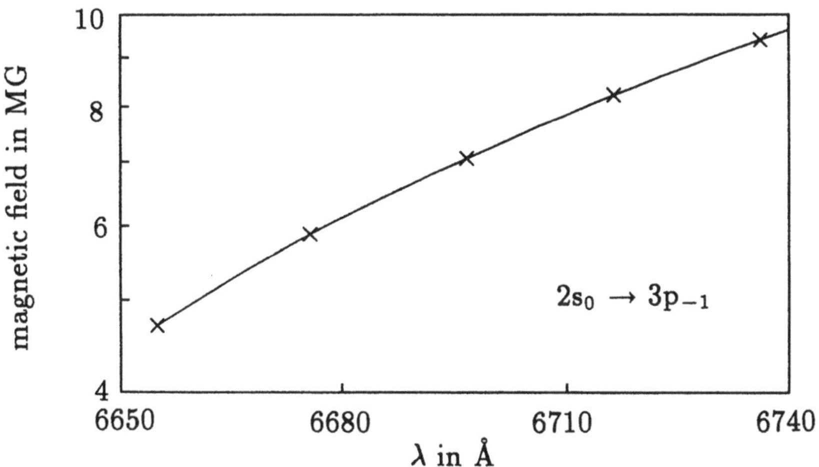

Fig. 2. Same as Fig.1 but for the transition $2s_0 \rightarrow 3p_{-1}$. There is no wavelength shift due to the electric field

References

Tworz, J. (1989): private communication
Zeller, G. (1990): PhD thesis, Fakultät für Physik der Universität Tübingen

Treatment Of Strong Magnetic Fields In Very Hot Stellar Atmospheres

Dieter Engelhardt, Irmela Bues
Dr. Remeis Sternwarte Bamberg, D-8600 Bamberg, Fed. Rep. of Germany

Abstract: Polarized radiative transfer in a hydrogen-rich model atmosphere of an effective temperature of 50000 K and a magnetic field strength of B=1000 Tesla must be investigated in the NLTE scheme. The four transfer equations for each Zeeman transition $\triangle m = 0, +1, -1$ coupled by the interaction of the polarized beam and the electron, are connected to the rate equations in terms of energy. Within the framework of the Dirac equation as a transfer equation the four equations decouple including magnetooptical parameters. Application of the Greens's function method leads to a formal solution, which is inverted numerically.

1 Transfer equation of polarized light

Polarized light can be completely described by the Lorentz vector potential A, summed over photon number. The time coordinate is related to the energy of the light beam, the space coordinates are connected to three projections of the photon spin. The variation of A as a wave along a relativistic line element can be written in terms of the Dirac equation, which ensures relativistic isochronic invariance and Hermiticity of the interaction matrix. The photon-electron coupling is described by a current density $\hat{H}_{int} = -ie\Psi^{+}\gamma_0\gamma_\mu A^\mu \Psi$ (Sakurai 1967), which is added to the usual Hamiltonian density of the Dirac equation. Ψ denotes a four component spinor wave function of the electron, γ represents the Dirac matrices and e is the coupling constant. The resulting Dirac equation is solved by covariant perturbation theory in the interaction picture. The transition probability of an initial state $|i>$ to a final state $<f|$ reads in the S-matrix formalism for a process of first order (Sakurai):

$$|S_{if}|^2 = |<f|e \int_{-\infty}^{\infty} dx^4 \Psi^{+}\gamma_0\gamma_\mu A^\mu \Psi |i>|^2. \tag{1}$$

The main advantage of the interaction picture is the fact, that a plane wave solves the the perturbed Dirac equation. Thus the photon wave function

$$\Psi_{Photon} = \frac{1}{\sqrt{2}} \begin{pmatrix} \chi \\ \underline{\sigma}\chi \end{pmatrix} exp(\frac{i}{\hbar}(+\underline{p}\underline{x} - Et)), \tag{2}$$

where $\underline{\sigma}$ denote the Pauli matrices and χ a two component spinor $(1,0)^{+}$, is a solution of the Dirac equation in question. In other words, this is the Stokes vector for which the interaction matrix $\gamma_0\gamma_\mu A^\mu$ is diagonal. For non relativistic and non polarized electrons without spin-spin interaction and a dipole photon wave the scattering process is still described by the structure of this interaction matrix. Since the interaction matrix is idempotent the argument is independent of the order of the perturbation. We are dealing with a massless particle. Therefore the Dirac spinors of the photons span a two dimensional vector space. The Zeeman-effect is described by three times one equations.

The two chiralities of the photon are split due to the magnetic field into left and right circularly polarized light and a projection, where no spin is observed. We can evaluate the fully decoupled transfer equation after application of density matrix theory (v.Neumann 1927) and get two linear Lorentz-rotations $I^{\triangle m}_{Stokes} = R(-\phi) * R^{+}(\theta) * I^{\triangle m}$ of the uncoupled transfer equation

$$\frac{dI^{\triangle m}}{N^{\triangle m} ds} = \frac{1}{2}(diag(2 \quad 0 \quad 1 \quad 1) + diag(0 \quad 0 \quad +1 \quad -1))(\kappa^{\triangle m} I^{\triangle m} - \eta^{\triangle m} I_0^{\triangle m}) \quad (3)$$

to the usual (see Sidlichovsky 1974) Stokes parameters. The transformations are bijective for $\triangle m = \pm 1$ and for $\triangle m = 0$ for $\theta \epsilon]0, 2\pi[$. The transformation concerning θ reads:

$$R(\theta) = \begin{pmatrix} 1 & 0 & 0 & 0 \\ 0 & 1 & 0 & 0 \\ 0 & 0 & \frac{1}{\sqrt{2}} & -\frac{i}{\sqrt{2}} \\ 0 & 0 & \frac{i}{\sqrt{2}} & -\frac{1}{\sqrt{2}} \end{pmatrix} * \begin{pmatrix} w_0\sqrt{2} & w_1\sqrt{2} & 0 & w_3\sqrt{2} \\ w_0\sqrt{2} & -w_1\sqrt{2} & 0 & -w_3\sqrt{2} \\ 0 & 0 & 2w_0 & 0 \\ 0 & -2w_3 & 0 & 2w_1 \end{pmatrix}, \quad (4)$$

where the angular dependence is written in Euler coordinates:

$$N^{\pm} w^{\pm} = \left(\tfrac{1+c_\theta}{4}, \ \tfrac{-s_\theta^2}{4}, \ 0, \ \tfrac{\pm c_\theta}{2} \right), \quad N^0 w^0 = s_\theta^2 \left(\tfrac{1}{2} \ , \tfrac{1}{2} \ , 0 \ , 0 \right)$$

$$N^{\triangle m} = \begin{cases} \frac{1+c_\theta^2}{2} & \text{for } \triangle m = \pm 1 \\ s_\theta^2 & \text{for } \triangle m = 0 \end{cases} . \quad (5)$$

The transformation for ϕ is well known (Sidlichovsky), mixing the linear polarizations. The first diagonal absorption matrix of (3) represents the usual (House, Steinitz 1975) real transfer equation. The second is added, because the eigenvalues of the complex transfer matrix are real. Furthermore we neglect the restriction of a Planck like source function and use the constant additive vector$I_0^{\triangle m}$ for spontaneous transitions. We assume the boundary conditions of both zero transport equations to be zero and transform the solution back to the transfer equation of the usual Stokes parameters (I,Q,U,V), which reads

$$\frac{d}{ds}(I, Q, U, V)^T = \sum_{\triangle m} N^{\triangle m}(w_0, w_1 - iw_3, w_2, w_3 + iw_1)^{\triangle m, T} \ (\kappa I - \eta)^{\triangle m}. \quad (6)$$

The solution for real absorption consists of one energy equation for each m dipole photon. In this picture the polarization is simply generated by an absorption coefficient, which is different for each $\triangle m$ transition. The part of $\kappa^{\triangle m}$, which is equal for each $\triangle m$ results directly in natural light. The Stokes parameters are generated by the angular dependence of the Stokes projectors.

2 Numerical Treatment

The atmosphere is divided into various cells to form a numerical grid. Each cell consists of hot pure hydrogen plasma. The magnetic field lines pass through each cell homogeneously, forming a centred dipole on the whole star. Light is propagating from the edges, passing the cell or being scattered at least once. The propagation of the beam is written in terms of a two point boundary problem, a formal solution reads as

$$u(\tau) = u(\tau') * exp(-|\tau - \tau'|) + \int d^3\tau' \frac{exp(-|\tau - \tau'|)}{|\tau - \tau'|} S(\tau'). \quad (7)$$

where u is the mean intensity of the incoming and outcoming radiation. For nearest neighbour coupling and one dimension this can be implemented as a tridiagonal matrix inversion. The rate equations are incorporated, in the sense of the operator perturbation theory (Cannon 1973), by the source function, which is divided into a local interaction and a long range interaction proportional to the intensity. The departure from the local source function is determined numerically by the rate equations. The programming has been done for radiative transfer up to five hundred depth points, the H model atom of the rate equations has seven n levels. Scalar Thomson scattering is added. The equation of state is considered to be constant. The dipole model was done on a 500x500 grid, solving the transfer equation for 25 depth points, as plotted in figure (1) for the variation of circularly polarized light for various magnetic angles.

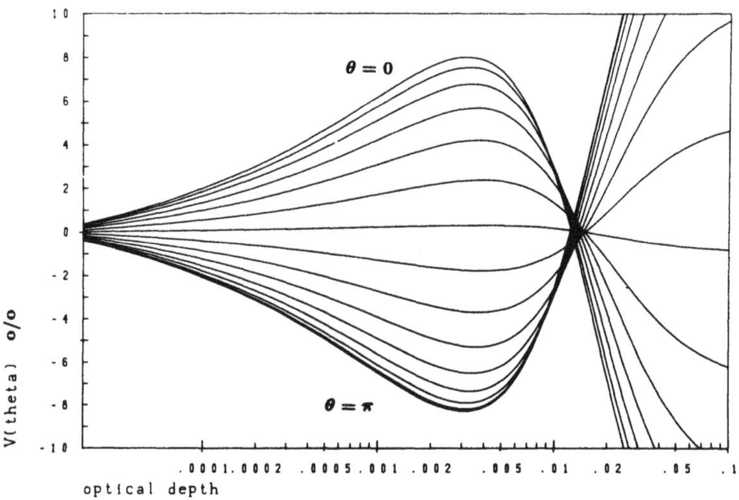

Fig. 1. Circular polarization is beamed by magnetic poles for $\triangle$ m =1 and $0 < \theta < 2\pi$

3 Future projects

Bound free opacities, magnetic Thomson scattering or Landau level scattering should be incorporated. The framework of the Dirac equation should be useful for the formulation of a consistent set of transport equations for various photon-electron interactions, dependent on the spin of the photon. Furthermore we plan to extend the program to treat conditions of hot stars with strong magnetic fields.

References

Cannon, C.J.: 1973, Astrophys.J., **185** ,621
House, L.L., Steinitz, R.: 1975, Astrophys. J.,**195**, 235
Landau, L.D., Lifschitz E.M.: 1986, "Quantenelektrodynamik"
v.Neumann, J.: 1927, Göttinger Nachrichten, 1,No 10,11 reprinted in Neumann,J., 1961, "Logic, Theory of Sets and Quantum Mechanics", ed. A.H.Taub, Pergamon Press, New York
Sakurai, J.J.: 1967,"Advanced Quantum Mechanics", Addison-Wesley, Redwood City
Sidlichovsky, M.: 1974, Bull.Astr.Institute Czechoslovakia, **25**, 198
Unno W.: 1956, Publ. Astron. Soc. Japan, **8**,108

Blends, frequency grids and the Scharmer scheme

M.J. Stift

Institut für Astronomie, Türkenschanzstraße 17, A-1180 Wien, Austria

"This much is already known: for every sensible line of straightforward statement, there are leagues of senseless cacophonies, verbal jumbles and incoherences."

Jorge Luis Borges: *The Library of Babel*

1 Generalising the Scharmer scheme to blends

The use of operator perturbation methods such as the Scharmer scheme (Scharmer & Carlsson, 1985) makes it possible to solve problems involving complex atomic and atmospheric models. Extensions to the original implementation by Carlsson (1986) are numerous and the present contribution is intended to discuss the particular problems associated with the treatment of blends in the context of Carlsson's code. Adhering closely to the notation employed by Scharmer & Carlsson, I assume complete redistribution and use a purely diagonal lambda operator. With the symbols $n_{ij} = (n_i/n_j)^*, \epsilon_{ij} = e^{-h\nu_{ij}/kT}$, and $Z_{ij} = n_i - n_j\, n_{ij}\, \epsilon_{ij}$ the set of linear equations for the δn's takes the form

$$
\begin{aligned}
-E_i = &+ \delta n_i \sum_{j>i} \left\{ +C_{ij} + 4\pi \iint \frac{\alpha_{ij}}{h\nu_{ij}} I \, d\nu d\mu \right\} \\
&+ \delta n_i \sum_{j<i} \left\{ +C_{ij} + 4\pi \iint \frac{\alpha_{ji}}{h\nu_{ji}} \left[n_{ji}\, \epsilon_{ji} \left(I + \frac{2h\nu_{ji}^3}{c^2} \right) \right] d\nu d\mu \right\} \\
&+ \sum_{j>i} \delta n_j \left\{ -C_{ji} - 4\pi \iint \frac{\alpha_{ij}}{h\nu_{ij}} \left[n_{ij}\, \epsilon_{ij} \left(I + \frac{2h\nu_{ij}^3}{c^2} \right) \right] d\nu d\mu \right\} \\
&+ \sum_{j<i} \delta n_j \left\{ -C_{ji} - 4\pi \iint \frac{\alpha_{ji}}{h\nu_{ji}} I \, d\nu d\mu \right\} \\
&+ 4\pi\lambda \sum_r \delta n_r \left\{ \sum_{j>i} \iint \frac{\alpha_{ij}}{h\nu_{ij}} \frac{Z_{ij} Y_r}{\kappa} d\nu d\mu - \sum_{j<i} \iint \frac{\alpha_{ji}}{h\nu_{ji}} \frac{Z_{ji} Y_r}{\kappa} d\nu d\mu \right\}
\end{aligned}
$$

with

$$
\kappa = \kappa_c + \sum_l \sum_{m>l} \alpha_{lm} Z_{lm}
$$

and

$$Y_r = \sum_{l<r} \alpha_{lr}\, n_{lr}\, \epsilon_{lr} \left(I + \frac{2h\nu_{lr}^3}{c^2} \right) - \sum_{l>r} \alpha_{rl}\, I.$$

The error term is calculated from

$$\begin{aligned} E_i = &+ n_i \sum_{j>i} \left\{ +C_{ij} + 4\pi \iint \frac{\alpha_{ij}}{h\nu_{ij}}\, I\, d\nu d\mu \right\} \\ &+ n_i \sum_{j<i} \left\{ +C_{ij} + 4\pi \iint \frac{\alpha_{ji}}{h\nu_{ji}} \left[n_{ji}\, \epsilon_{ji} \left(I + \frac{2h\nu_{ji}^3}{c^2} \right) \right] d\nu d\mu \right\} \\ &+ \sum_{j>i} n_j \left\{ -C_{ji} - 4\pi \iint \frac{\alpha_{ij}}{h\nu_{ij}} \left[n_{ij}\, \epsilon_{ij} \left(I + \frac{2h\nu_{ij}^3}{c^2} \right) \right] d\nu d\mu \right\} \\ &+ \sum_{j<i} n_j \left\{ -C_{ji} - 4\pi \iint \frac{\alpha_{ji}}{h\nu_{ji}}\, I\, d\nu d\mu \right\}. \end{aligned}$$

2 The frequency quadrature

The accuracy of the final solution is solely determined by the precision with which the error term is calculated and great care has to be taken to ensure adequate frequency and angle quadrature. It is often assumed that any grid point distribution that leads to correct integration of the Voigt function is appropriate for this purpose. From Fig. 1 however it transpires that even in the rather simple case of Carlsson's (1986) Ca 2 atomic model in conjunction with the VAL3C atmospheric model the functions under the double integrals behave in a strongly angle-dependent and somewhat counterintuitive way and that their frequency-dependency is not directly related to that of the Voigt function. Taking the optimum diagonal lambda operator proposed by Olson et al. (1986) it is easy to show that at large optical depths the integrand can be proportional to the *inverse* of the profile function. I find that the judicious distribution of quadrature points necessary for a satisfactory numerical frequency integration is a non-trivial problem.

An explicit grid equation similar to the one employed by Stift (1985) may be used to ensure a good quadrature point distribution. One of the countless possible approaches – albeit probably not the optimum one – consists in taking the depth- and angle-dependent integrands $r_{d,\mu}(\nu)$ (normalised to an amplitude range between -1 and $+1$ at each standard optical depth τ_d separately in order to give equal weight to all atmospheric layers) to define

$$S(\nu_i) = \max_{d,\mu} |r_{d,\mu}(\nu_i) - r_{d,\mu}(\nu_{i-1})|$$

with $S(\nu_1) = 0$. Adding up the individual values of $S(\nu_i)$ results in the discrete function

$$\Gamma(\nu_i) = \sum_{k=1}^{i} S(\nu_k).$$

Finally the range in Γ is divided into the desired number of intervals and the frequency quadrature points are determined by interpolation – using rational splines on account of their favourable stability properties (Späth 1983) – in the inverse relation $\nu(\Gamma)$ at points equidistant in Γ.

3 Results and conclusions

Figure 2 illustrates how the proposed grid equation works in the case of a moving atmosphere. The frequency resolution of $r_{d,\mu}(\nu)$ is displayed at three different iteration stages, for all atmospheric layers and three values of μ. The original grid obviously is not optimum but as the iterations proceed the points concentrate near strong gradients of the (changing) integrand leading to a satisfactory resolution of all features, reflected in fuzzier images. It is important that gradients in the integrands be resolved everywhere in the atmosphere; grid equations based on a minimax criterion are the natural choice. It should be noted that the Feautrier technique – which cannot be applied to blends – is not the best choice for moving atmospheres in conjunction with adaptive grids because quadrature points are wasted in the symmetric distribution about the unshifted central frequency of a line.

The increase in computing time is of the order of 4% if the grid is updated after 8 iterations. In exchange the method ensures a good distribution of the frequency points without a costly trial and error procedure. Although the improvement in the final solution due to the adaptive grid is negligible for most static atmospheres, the situation is quite different in the more interesting cases, e.g. moving cepheid atmospheres. Here various reasonably looking fixed grids yield discrepant results: occupation numbers differ by up to 7% depending on the precise integration range and vary in excess of 3% with the core extension (using 101 frequency points per transition). While the use of a grid equation is recommended for all moving atmospheres I anticipate that adaptive frequency grids will play a vital role in the treatment of blends as outlined in the first section.

Acknowledgements: This work was supported by the Austrian Fonds zur Förderung der wissenschaftlichen Forschung, project number P7252-PHY. Thanks go to Dr. M. Carlsson, Dr. E.A. Dorfi and Dipl.Ing. M. Feuchtinger.

References

Carlsson, M. 1986, Uppsala Astron. Obs. Rep. No. 33
Olson, G.L, Auer, L.H., Buchler, J.R. 1986, JQSRT 35, 431
Scharmer, G.B., Carlsson, M. 1985, J. Comput. Phys. 59, 56
Späth, H. 1983, Spline-Algorithmen zur Konstruktion glatter Kurven und Flächen, R. Oldenbourg, München
Stift, M.J. 1985, MNRAS 217, 55

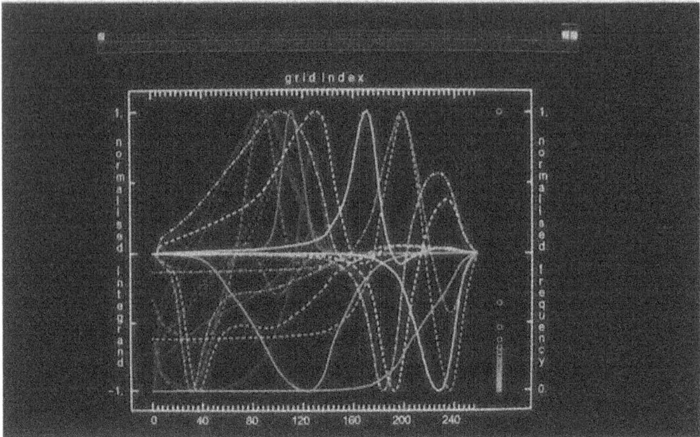

Fig. 1. Run with frequency of the depth- and angle-dependent integrand $r_{d,\mu}(\nu)$ in the radiative error terms for the Ca K line in Carlsson's example. The frequency grid index is taken as abscissa, the integrands are normalised to unity amplitude separately at each optical depth and angle point. Colours correspond to depth and range from red (deepest layers) to violet (uppermost layers); full, dashed and dashed-dotted lines correspond to $\mu = 0.047, 0.500, 0.953$ respectively. The green circles to the right show the frequency distribution of the grid points.

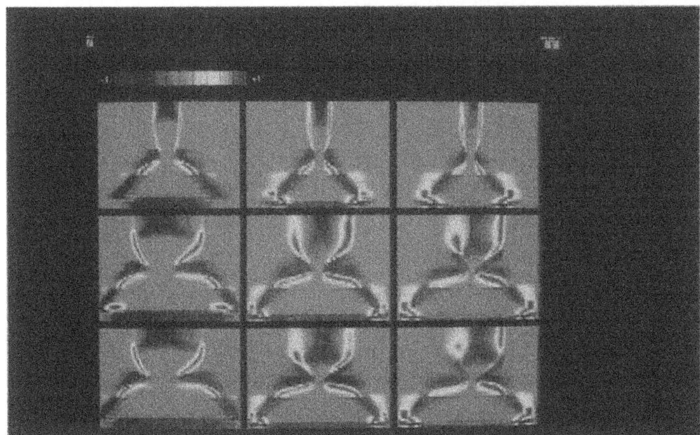

Fig. 2. Run with frequency and angle of the depth-dependent normalised integrand $r_{d,\mu}(\nu)$ for the same problem as in Fig. 1 but with an artificial velocity field superimposed (increasing monotonically from zero in the deepest to 10 km/s in the uppermost layers). In the individual squares the frequency grid index is the horizontal coordinate, the optical depth grid index the vertical coordinate; violet corresponds to -1, red to $+1$ (see the top of the picture). Iterations proceed from top to bottom, the values of μ (see Fig. 1) increase from left to right.

MULTILEVEL NON-LTE RADIATIVE TRANSFER USING EXACT COMPLETE AND DIAGONAL OPERATORS

Joseph J. MacFarlane
Department of Astronomy and Fusion Technology Institute
University of Wisconsin, Madison, Wi 53706 USA

We have investigated the convergence properties of Λ-acceleration methods for non-LTE radiative transfer problems in planar and spherical geometry. Matrix elements of the "exact" Λ-operator were used to accelerate convergence to a solution in which both the radiative transfer and atomic rate equations are simultaneously satisfied. Convergence properties of 2-level and multilevel atomic systems have been investigated for methods using: (i) the complete Λ-operator, and (ii) the diagonal of the Λ-operator. We found that the convergence properties for the method utilizing the complete Λ-operator are significantly better than those of the diagonal Λ-operator method, often reducing the number of iterations needed for convergence by a factor of between 2 and 7. However, the overall computational time required for large scale calculations — that is, those with many atomic levels and spatial zones — is typically a factor of a few larger for the complete Λ-operator method, suggesting that the approach should be restricted to problems in which convergence is especially difficult. In addition, we have shown that for problems with spherical symmetry the convergence properties found when using a Λ-operator based on contributions from only one hemisphere (thereby neglecting contributions from symmetry points) are only marginally worse than those when using the exact diagonal. A detailed description of this study in presented in Ref. 1.

---

[1]MacFarlane, J. J., 1991, submitted to Astronomy and Astrophysics

IRON LINE BLANKETING IN NLTE MODEL ATMOSPHERES FOR O STARS: FIRST RESULTS

S. DREIZLER and K. WERNER
Institut für Theoretische Physik und Sternwarte der Universität Kiel
Olshausenstr. 40, 2300 Kiel, Germany

Absorption lines of highly ionized iron (Fe III–Fe VII) dominate the UV spectra of many hot stars. They have been identified in IUE spectra of subdwarf O stars (e.g. HD 49798, BD +28°4211, BD +75°325; Bruhweiler et al. 1981; Dean & Bruhweiler 1985), as well as in central stars of PNe (e.g. LSS 1362, NGC 7293; Schönberner & Drilling 1985). Most recently, Feibelman & Bruhweiler (1990) announced the detection of Fe VII lines in PG 1159 type degenerate stars (PG 1159-035 and K 1-16) and in the hottest He-rich (DO-) white dwarfs (PG 1034+001 and KPD 0005+5106).

These findings seriously underline the necessity to account for iron line blanketing when calculating model atmospheres for O stars. Due to the high effective temperatures this has to be done under NLTE conditions and, for this reason, adequate models were lacking up to now. We present the first NLTE model atmospheres blanketed by some ten thousands of Fe lines. We study the blanketing effects onto the atmospheric structure, onto the emergent line and continuum fluxes, and we examine the consequences for metal abundance determinations.

The numerical approach to the NLTE metal line blanketing problem is an adoption of Anderson's (1989) method along with an opacity sampling technique to treat the huge number of lines. We have implemented this method into the framework of our Accelerated Lambda Iteration (ALI) code (Werner 1986).

The iron group elements have (due to a partially filled 3d subshell) so many levels and line transitions that the employment of "classical" model atoms is impossible. Instead of correlating real levels with model atom levels, we follow Anderson's (1989) statistical approach. All real levels l within an interval around a peak in the distribution of the statistical weight over the excitation energy are grouped together into a model band L, assuming Boltzmann statistics for the population within the band. Each of these model bands L is treated as a single NLTE level with suitably averaged energies E_L

$$E_L = \sum_{l \epsilon L} E_l g_l e^{-E_l/kT} \Big/ \sum_{l \epsilon L} g_l e^{-E_l/kT}$$

and statistical weights g_l^* and G_L

$$g_l^* = g_l e^{(E_L - E_l)/kT} \qquad \text{and} \qquad G_L = \sum_{l \epsilon L} g_l^* \, .$$

The photon cross section of a transition σ_{LU} between model bands L and U consists of many (hundreds or thousands) individual transitions lu:

$$\sigma_{LU} = \frac{\pi e^2}{mc} \frac{1}{G_L} \sum_{l \epsilon L, u \epsilon U} g_l^* f_{lu} \phi(\nu_{lu} - \nu) \, .$$

The opacity of transitions within a band L are calculated in the same way and are taken into account as a background opacity. For the model atmosphere calculations it is not necessary to treat the cross sections in full frequency resolution since individual features are sufficiently dense to be sampled (Fig. 1). The sampled cross sections are computed in advance of the model atmosphere calculations, with a fixed temperature for the Doppler broadening and with two electron densities (0 and $1.0 \cdot 10^{16}\,\mathrm{cm}^{-3}$) for the collisional broadening. The model atmosphere code interpolates between the two cross sections. The quality of the sampling procedure is checked by integration of $\sigma_{LU}(\nu)$ over frequency and the cross section is then renormalized by:

$$\frac{\pi e^2}{mc}\frac{1}{G_L}\sum_{lu} g_l^* f_{lu} \Big/ \int \sigma_{LU}(\nu) d\nu \ .$$

The sampling technique is adequate for line blanketing effects since the contribution of all lines to the opacity is accounted for, at least statistically. However, for detailed NLTE line formation calculations a "classical" (i.e. detailed) model atom and for spectrum synthesis the full frequency dependence of the cross sections are required.

The model atom and the cross sections are constructed using atomic data from Kurucz (1991). As a first step we designed a model atom containing Fe III–Fe VII (Tab. 1). Data for *observed* transitions have been included up to now. However, it is in principle no difficulty to consider also the *predicted* transitions, which will increase the number of line transitions by a factor of 100. This will be done in the next step, when the full Kurucz data set is available to us.

Table 1: Number of NLTE levels (NLTE), number of line transitions between model bands L and U (R_{LU}) and number of transitions between real levels l and u (R_{lu}) contributing to the R_{LU} transitions.

Ion	NLTE	R_{LU}	R_{lu}	Ion	NLTE	R_{LU}	R_{lu}
Fe III	7	25	15871	Fe IV	5	3	7897
Fe V	5	3	3670	Fe VI	6	2	814
Fe VII	7	2	117				

We constructed two NLTE model atmospheres for post-AGB stars to explore the effect of iron line blanketing in hot star atmospheres. Effective temperatures and gravities of the models are 60,000 K, $10^{4.5}$ cm/s^2 and 40,000 K, $10^{4.0}$ cm/s^2. The models contain H, He and Fe with solar abundances. Subsequent line formation calculations for nitrogen were performed in the 40,000 K model. Reference models excluding iron were also constructed. The results presented here are preliminary, since only the observed Fe lines are included in the calculations (see above). Therefore, the effects might be more drastic when all the predicted lines are also taken into account.

- Due to cooling in the outer regions and backwarming in the inner regions the temperature in the Fe line blanketed model is closer to the LTE stratification than in the unblanketed NLTE model.

- H and He line profiles become deeper and broader (Fig. 2). This might resolve the discrepancies encountered in the fit of He II and H lines in hot subdwarfs (Dreizler 1992).

- At present only a very small effect onto the nitrogen lines is found. Iron line blocking in the region of the N III 2p ^{2}P^o continuum is too weak to affect the ionisation balance. This will probably change, when the complete iron line list and other iron group elements will be implemented.

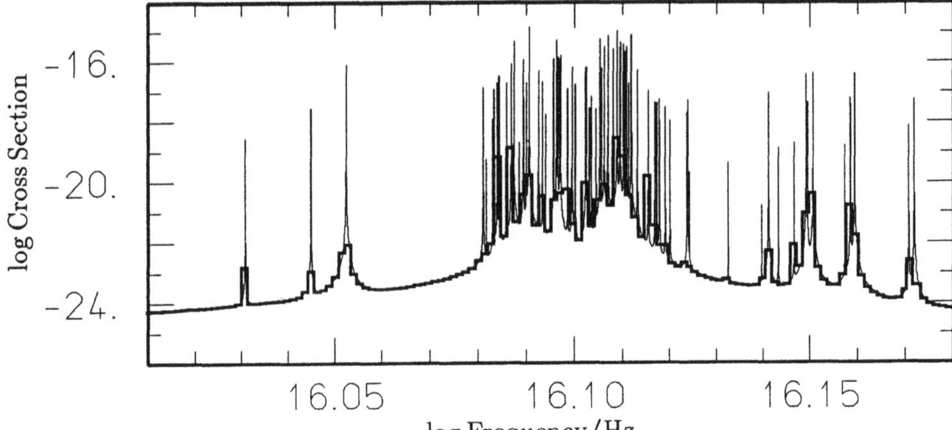

Figure 1: Complex (thin) and sampled (thick) cross section $\sigma_{1\to3}$ of Fe VII.

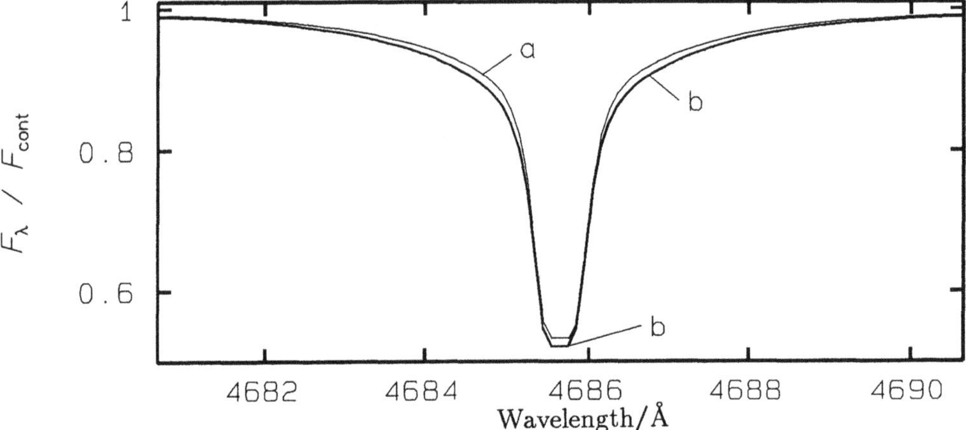

Figure 2: Comparison of He II 4686 Å line profiles at $T_{\text{eff}} = 40,000$ K, $\log g = 4.0$: **a)** H/He model and **b)** iron line blanketed model.

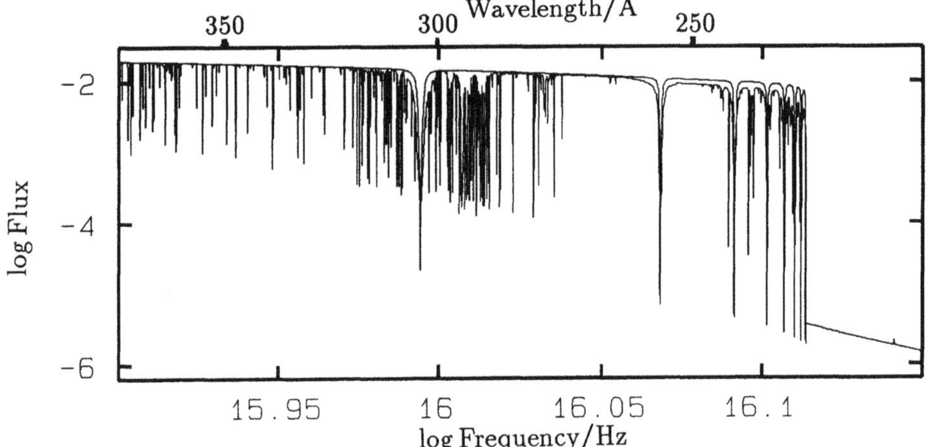

Figure 3: Flux distribution of the 60,000 K models *with* and *without* iron line blanketing, in the vicinity of the He II 228 Å edge.

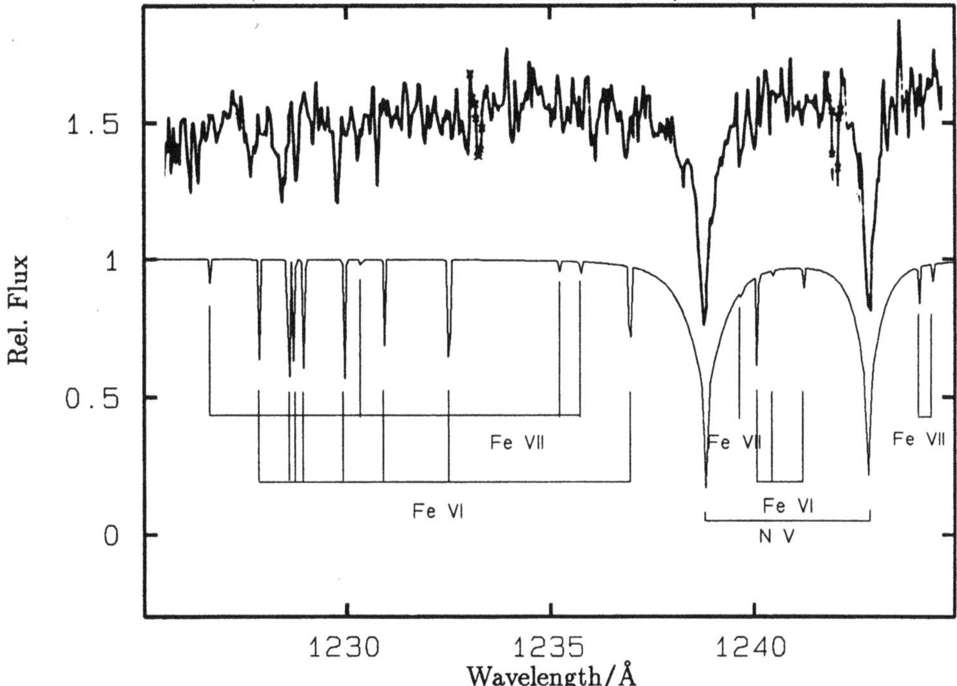

Figure 4: Relative flux of the 60,000 K iron line blanketed NLTE model around the N V 1240 Å resonance doublet compared with an IUE high resolution spectrum of the sdO star **BD +28°4211**.

The importance of iron line blanketing for the EUV flux distribution is demonstrated in Fig. 3. Significant flux blocking due to Fe VI and Fe VII lines occurs redwards of the He II ground state edge (228 Å) and should, therefore, be taken into account in analyses of EUV data.

The potential of our model calculations for the interpretation of IUE data is shown in Fig. 4. We have plotted a detail of a high resolution spectrum of the sdO star BD +28°4211, together with the synthetic spectrum of the model with $T_{\rm eff} = 60,000$ K, $\log g = 4.5$. Note that this is not meant to be a profile fit, because BD +28°4211 is probably somewhat hotter ($T_{\rm eff} \approx 80,000$ K) and has a larger surface gravity ($\log g \approx 6$). Nevertheless, the similarity between observed and calculated iron lines is striking. A detailed analysis of the sdO star BD +28°4211 is in progress.

In summary, the results of our first iron line blanketed NLTE model calculations are very encouraging and we will continue our work by incorporating more iron lines and considering other iron group elements.

Acknowledgements. This work is supported by the DFG (grant Hu 39/29-2) and by the BMFT (grant 50 OR 9007 3). We thank Lawrence Anderson for helpful discussions on his method.

References

Anderson, L.S. 1989, ApJ, 339, 558
Bruhweiler, F.C., Kondo, Y., McCluskey, G.E. 1981, ApJS, 46, 255
Dean, C.A., Bruhweiler, F.C. 1985, ApJS, 57, 133
Dreizler, S. 1992, these proceedings
Feibelman, W.A., Bruhweiler, F.C. 1990, ApJ, 357, 548
Kurucz, R.L. 1991, in Stellar Atmospheres: Beyond Classical Models, NATO ASI series C, Vol. 341, eds. L. Crivellari, I. Hubeny and D.G. Hummer, p. 441
Schönberner, D., Drilling, J.S. 1985, ApJ, 290, L49
Werner, K. 1986, A&A, 161, 177

Calculations of non-LTE radiative transfer in extended outflowing atmospheres using the Sobolev approximation for line transfer

A. de Koter
SRON Space Research Lab.
Utrecht, The Netherlands

W. Schmutz
ETH
Zürich, Switzerland

H. J. G. L. M. Lamers
SRON Space Research Lab.
Utrecht, The Netherlands

1 Introduction

We are currently developing an atmosphere code in order to investigate the structure of the instable atmospheres of Luminous Blue Variables (LBVs). In this paper we report test calculations in which we compare non-LTE level populations of a 5-level Hydrogen model atom with those of a reference code. Calculations including Helium and more complex model atoms are in preparation.

LBVs are massive stars ($M_{ZAMS} \geq 50 M_{\odot}$) in a relatively short (10^4 to 10^5 yr) evolutionary phase in between the O-Of stage and the Wolf-Rayet stage. They show photospheric and spectroscopic variability within a range of different time-scales and amplitudes. The most characteristic type of variation in LBVs is a horizontal (i.e. occurring at constant Luminosity) excursion from the left to the right (and vice versa) in the HRD which involves crossing the Humphreys-Davidson limit. Throughout such a cycle the mass-loss rates vary from the values predicted by the radiation-driven wind theory at visual minimum (i.e. left in the HRD), to much higher rates of up to about $5 \times 10^{-5} M_{\odot}$/yr at visual maximum. In addition to high mass-loss rates LBVs have low terminal velocities, typically 200 km/sec. Therefore their winds are extremely dense. Given the physical conditions of LBV atmospheres a model code must be able to treat extended atmospheres that expand with both subsonic and supersonic velocities. Electron opacities dominate and therefore, the level populations have to be evaluated in non-LTE.

The ultimate aim of our project is to obtain an understanding of the physical processes that are responsible for the variations in the LBVs.

2 Properties of the model code

The model solves the radiative transfer in the continuum and in the lines, subject to the constraint of statistical equilibrium. The model is semi-empirical in the sense that density and temperature structure can be specified in such a way that the calculated energy distribution and line profiles fit the observations.

2.1 density structure

With the mass-loss rate $\dot{M}$ as a free parameter, the density structure can be derived from a specified velocity law $v(r)$, using the equation of mass continuity. The mass continuity equation with a constant mass-loss rate, $\dot{M}(r) = \dot{M}$, is in principle valid for variations that occur on time-scales longer than typically $10 R_{\star}/v_{\infty} \simeq$ months.

We split $v(r)$ in two parts, one for the low and one for the high velocity domain. In the low velocity part, where $v \lesssim v_{sound}$, $v(r)$ corresponds to a hydrostatic density stratification in spherical geometry; in the high velocity part $v(r)$ can be specified as an arbitrary monotonically increasing function, e.g. as a β -law. We require a smooth transition from the low to the high velocity part (following Leitherer et al., (1989)).

2.2 temperature structure

The temperature structure can be chosen arbitrarily. For the first tests we have chosen the T-structure of a grey spherically extended atmosphere in LTE. If the temperature drops below a certain minimum value, T_{min}, then $T(r)$ is set to T_{min}.

2.3 chemical composition

Although for the calculation presented here the chemical composition of the model consists of Hydrogen only, the code also contains atomic data for Helium. For the atomic data we followed Wessolowski et al. (1988). The singlet and triplet states of He I are treated simultaneously, each state containing six main quantum numbers n. For $n \leq 4$ the different angular momentum states are treated as different levels, yielding eleven levels. For $n = 4$ to 6 all states with the same quantum number are treated as one level in both the singlet and the triplet system, which extends the total number of levels for He I to 17. The hydrogenic ions are each represented by ten levels combining all states of the same main quantum number. For the hydrogenic ions additional implicit levels can be included, e.g. by assuming them to be in LTE relative to their continuum.

The line radiation transfer is treated explicitly for all transitions, i.e. for allowed and forbidden (intercombination) transitions.

3 Computational Method

The main difficulty in solving non-LTE model atmospheres is that the state of the gas and the *non local* radiation field are coupled; thus the problem is fundamentally a global one. In principle a change in the population of one level of one ion somewhere in the atmosphere changes the populations of all other levels of all ions of all elements as well as the radiation field at all frequencies.

In order to solve the problem one iteratively solves radiation transfer and statistical equilibrium. As is well known, this method of lambda iteration converges extremely slowly in optically thick cases. We solve this problem using an Approximate Lambda Iteration (ALI) method. For a review of the ALI technique see Hubeny in these Proceedings.

In ALI, essentially, one decouples the local and the non-local contributions to the radiation field. When one is solving for the state of the gas, the local radiation field is expressed in terms of the populations that have not yet been solved. This last step makes the set of statistical equilibrium (SE) equations highly non-linear. Since we do not want non-linear equations we developed a technique to precondition the SE equations in such a way that they stay linear.

3.1 approximate lambda operator

The decoupling of the local and non-local contribution to the radiation field is formally described by:

$$J_\nu^i = \Lambda_\nu^\star S_\nu^i + (\Lambda_\nu - \Lambda_\nu^\star) S_\nu^{i-1} \tag{1}$$

Here $\Lambda_\nu^\star$ is the approximate lambda operator; i is the iteration number. S_ν^i is thus based on populations that have not yet been solved. If $\Lambda_\nu^\star$ is a diagonal matrix, $\Lambda_\nu^\star S_\nu^i$ is the strictly local contribution to J_ν^i and $(\Lambda_\nu^\star - \Lambda_\nu) S_\nu$ the non-local one.

We use a $\Lambda_\nu^\star$ given by:

$$\Lambda_\nu^\star = \mathrm{diag}[T_\nu^{-1}] \tag{2}$$

where T_ν is the tridiagonal transfer equation matrix. By definition, this is the best choice one can make of a strictly local $\Lambda_\nu^\star$.

To calculate $\Lambda_\nu^\star$ we use a method developed by Rybicki and Hummer (1991). They showed that, by combining forward elimination & back substitution ánd backward elimination & back substitution, this particular problem can be solved in the order of N ($\equiv$ dimension T) operations.

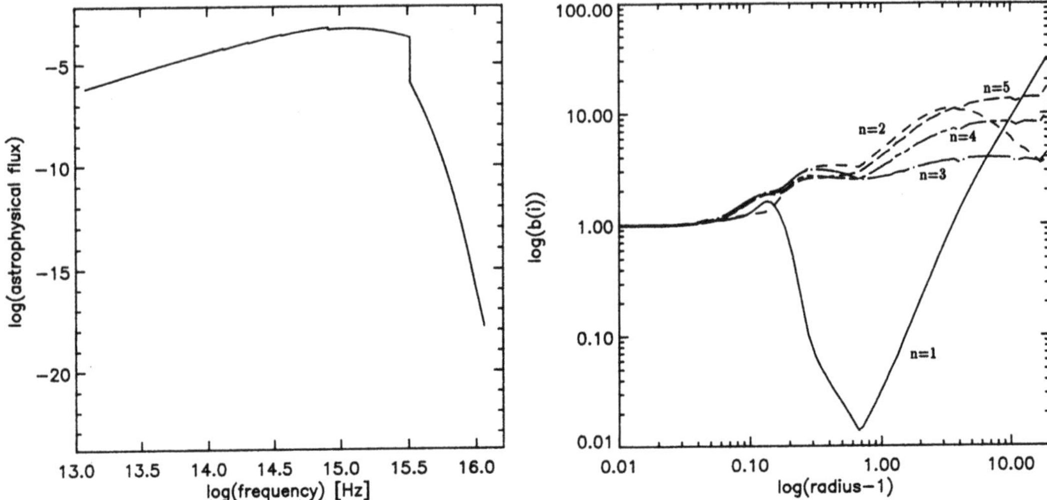

Figure 1: Energy distribution and departure coefficients of the first five Hydrogen levels of an LBV with stellar parameters characteristic for R71 or R88 at their visual minimum.

3.2 preconditioning statistical equilibrium

As stated above, the drawback of using ALI is that the set of SE equations becomes non-linear. In order to linearize the set of SE equations we follow the method developed by Pauldrach and Herrero (1988). Although they developed the method to linearize line transfer rates, we used it, at least partially, to linearize the continuum problem. With regard to the SE equation that describes the population of some level j, we split from the continuum source function, S, the part that describes the bound-free contribution from level j, thus:

$$S = \varepsilon_j^{bf} S_j^{bf} + S^{rest} \tag{3}$$

where $\varepsilon_j^{bf} \equiv \kappa_j^{bf}/(\kappa_{tot} - s_e\Lambda^\star)$. $\Lambda^\star$ enters this equation because we used equation 1 to describe the scattering term in S. For ε_j^{bf} and S^{rest} we use old population values; for S_j^{bf} we will use the new ones. The $\varepsilon_j^{bf} S_j^{bf}$ part can be incorporated in the SE equations to give a net rate with respect to $\varepsilon_j^{bf}\Lambda^\star$. The modified rate terms become:

$$R'_{j\kappa} = 4\pi \int_{\nu_o}^{\infty} \frac{\alpha_{j\kappa}}{h\nu}(\Delta J + \Lambda^\star S^{rest}) d\nu \tag{4}$$

$$\begin{aligned} R'_{\kappa j} = {} & 4\pi\left(\frac{n_j}{n_\kappa}\right)^* \int_{\nu_o}^{\infty} \frac{\alpha_{j\kappa}}{h\nu}(\Delta J + \Lambda^\star S^{rest}) e^{-\frac{h\nu}{kT}} d\nu + \\ & 4\pi\left(\frac{n_j}{n_\kappa}\right)^* \int_{\nu_o}^{\infty} \frac{\alpha_{j\kappa}}{h\nu}\frac{2h\nu}{c^2}(1 - \varepsilon_j^{bf}\Lambda^\star) e^{-\frac{h\nu}{kT}} d\nu \end{aligned} \tag{5}$$

where $\Delta J = (\Lambda - \Lambda^\star)S^{i-1}$ and $*$ denotes LTE.

Note that this method works extremely well, for instance in an atmosphere with an extremely optically thick Lyman continuum.

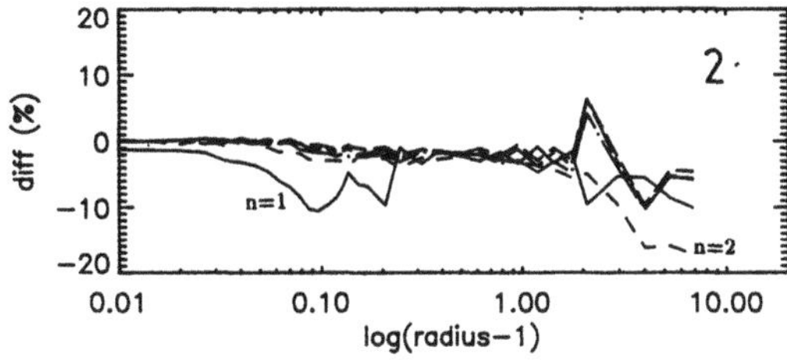

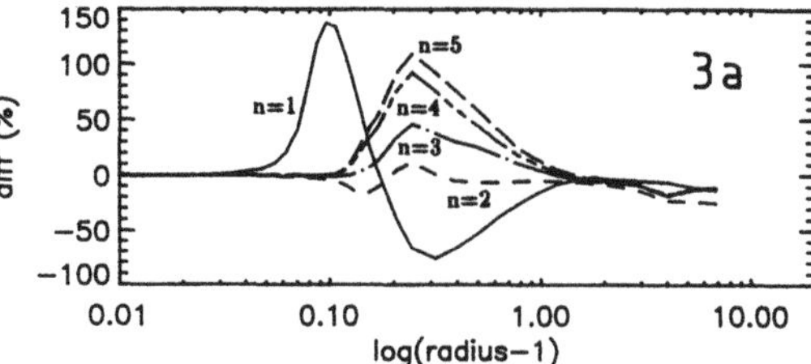

Figures 2 and 3: The percentage differences in the level population numbers obtained with the Sobolev and the CMF models are plotted against radius. In figure 2 (above) the difference is relative to a CMF calculation that uses a Doppler velocity of 0.0001 km/sec. Figure 3 shows the result relative to a CMF calculation using v_D = 15 km/sec. Continuum absorption in the resonance region is not taken into account in fig. 3a, but it is in fig 3b.

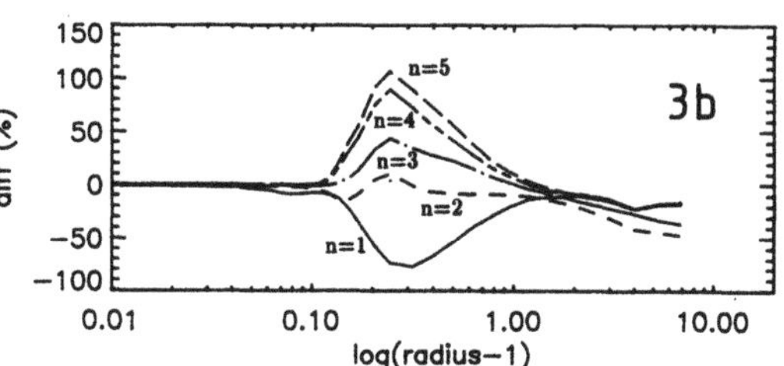

4 Validity of the Sobolev approximation for LBVs

Compared to Wolf-Rayet and O-type stars, LBVs have low velocity winds ($v_\infty \approx 200$ km/sec). We therefore investigated the validity of the Sobolev approximation for LBVs, by comparing our calculations with the results of an identical model calculation done with the atmosphere code developed by Hamann and Schumtz (Hamann et al.,(1987)), who treat the line transfer in the comoving-frame (CMF).

We took a model with parameters $\log(L/L_\odot) = 5.6$; $T_{eff} = 16000$ K; $\log(g_{eff}) = 1.3$; $v_\infty = 170$ km/sec and $\dot{M} = 10^{-6.5} M_\odot$/yr. The chemical composition is purely Hydrogen, the first five levels of which are treated in non-LTE. The model is not representative for a particular star. However, these values are close to what has been derived for the LBVs R71 and R88 at their visual minimum (Leitherer et al, 1989).

In figure 1 the emergent flux distribution and the departure coefficients of the first five Hydrogen levels are plotted.

In figure 2 we show the percentage differences, in the level population numbers obtained with the Sobolev approximation and the CMF calculations, assuming for the latter a Doppler velocity of v_D = 0.0001 km/sec. The difference is defined as:

$$\text{difference}(\%) = \frac{n_j(Sobolev) - n_j(CMF)}{n_j(CMF)} \times 100\%. \tag{6}$$

These differences give an idea of how accurately two different codes reproduce the same atmosphere. All differences below $2R_{core}$ are less than 4%, except for the ground state which deviates by a maximum of 10%. Since the intrinsic line width used in the comoving-frame calculation is less than the expansion velocities anywhere in the atmosphere, the Sobolev approximation should be valid and it is impossible to judge from this slight discrepancy which of the two codes yields the more accurate result. In spite of the fact that innumerable technical details are treated differently in the two codes the agreement between the two codes, which have been written completely independently of each other, is satisfactory.

Figure 3 shows the differences in the departure coefficients resulting from the Sobolev approximation calculation and a CMF calculation using a Doppler velocity of v_D is 15 km/sec. Obviously, due to the Sobolev approximation the differences amount to 140% for the ground level and between 15 (n=2) to 100 (n=5) % for the higher levels.

The Sobolev approximation can be improved by including the effect of absorption by the continuum within the resonance region (Hummer and Rybicki (1985) using the formalism of Puls and Hummer

(1988)). One may expect changes for populations in regions where a) the chances of escape from the resonance region or into the local continuum are about equal and additionally b) continuum processes no longer solely determine the state of the gas. These criteria are fulfilled for the ground level in the range 1.05 R_{core} (where the Balmer continuum becomes optically thin) to 1.15 R_{core} (where the Lyman continuum becomes optically thin). We found that the inclusion of continuum absorption in the Sobolev formalism improved the results considerably. For the ground level, in the region mentioned above, the differences decreased from 140% to only a few percent!

The key question one needs to ask when using the Sobolev approximation for LBVs is how it affects the diagnostic tools. Firstly, the energy distribution is ***not affected*** because the deviations relative to the CMF result occur only outside the continuum-forming region. Secondly, the strong and medium strong lines will be ***hardly affected*** because they depend on the source function or n_u/n_l, which deviates by not more than some tens of percents in a limited geometrical region. Lines arising from the ground level may be affected by the Sobolev approximation. However, not all of the observable quantities involve the ground level. Moreover, the line profiles are determined by integration over all geometrical regions. Therefore both codes will yield essentially the same line profiles for the observational transitions.

5 Conclusions

The code that we are developing for the empirical modelling of LBVs using the Sobolev approximation for the line transfer, yields accurate results that compare well with those obtained using the co-moving frame method. By including absorption by the continuum in the resonance region in addition to escape from the resonance region we get a much more accurate result for the hydrogen ground level population.

The following properties make the code extremely fast:

- the Sobolev approximation for the line transfer
- the choice of the local approximate lambda operator Λ^*, which yields the exact local contribution to the continuum radiation field
- the method of preconditioning the set of statistical equilibrium equations, which makes computationally expensive linearization unnecessary.

This means that the code is ideally suited for empirical modelling, in the sense that one can iteratively fit an atmospheric model to the observations.

References

[1] Hamann, W-R and Schmutz, W., 1987, *Astron. Astrophys.* **174** *173*

[2] Hummer, D. G. and Rybicki, G. B., 1985, *Astrophys. J.* **293**, *258*

[3] Leitherer, C.; Schmutz, W, and Abbott, D. C.: 1989, *Astrophys. J.* **346**, *919*

[4] Pauldrach, A. and Herrero, A.: 1988, *Astron. Astrophys.* **199**, *262*

[5] Puls, J. and Hummer, D. G.: 1988, *Astron. Astrophys.* **191**, *87*

[6] Rybicki, G. B., and Hummer, D. G., 1991, *Astron. Astrophys.* **245**, *171*

[7] Wessolowski, U.; Schmutz, W. and Hamann, W.-R.: 1988, *Astron. Astrophys.* **194**, *160*

3D RADIATIVE LINE-TRANSFER FOR DISK-SHAPED BE STAR ENVELOPES

W. Hummel

Astronomisches Institut, Ruhr-Universität Bochum,

Postfach 10 21 48, D-4630 Bochum 1

Be stars, defined as early-type emission-line stars of luminosity classes III-V, exhibit stellar absorption and circumstellar emission features in their spectra. The emission lines are produced in rotationally supported gaseous circumstellar disks (Struve, 1931). By high-resolution, high signal-to-noise ratio spectroscopy, important fine structure, such as shell-type absorption components or symmetric flank inflections ("winebottle-type" profiles) can be detected in Hα and Hβ emission lines (Hanuschik et al., 1988). In the present work, a new interpretation of winebottle-type emission lines is presented. The spatially implicit first-order volume technique (Adam et al., 1990) is used to solve the radiative line-transfer equation in three dimensions for a two-level-atom with complete re-distribution :

$$(\mathbf{n}\cdot\nabla) I_\lambda = \varkappa^l_\lambda S^l_\lambda + \varkappa^c_\lambda S^c_\lambda - (\varkappa^l_\lambda + \varkappa^c_\lambda) I_\lambda \quad (1)$$

Operator-splitting methods have been applied to accelerate the iteration convergence (Olson, et al., 1986). The local (diagonal) approximative Λ-operator $\Lambda^*[S^n] = f_c * S^n$ was constructed for three dimensions.

The present model consists of a central star ($T_{eff} = 20\,000$ K) radiating spherically symmetric into a cylindrically symmetric disk-like envelope, extending from $r = R_*$ ($= R_{in}$) to 20 R_* ($= R_{out}$) in the equatorial plane and from $z = -R_*$ to R_* in vertical direction. The envelope is taken to be isothermal ($T_{env} = 20\,000$ K, $\lambda_0 = 6563$ Å, $\Delta\lambda_{therm} = 22$ km s^{-1}), homogeneous ($\varkappa^l_0 =$ const.) and *Keplerian rotating* ($V_\varphi(r) = V_{\varphi crit}(r/R_*)^{-1/2}$). The scattering parameter ($\varepsilon = 0.01$), the continuous absorption coefficient ($\varkappa^c = 1/(100\,R_*)$; $\tau^c = 0.01$) and the passive continuum source function ($S^c = B_\lambda(T_{env})$) are constant in the disk. The resulting line source function S^l and J are plotted in Fig. 1 ($\tau^l = 10$); the corresponding emission line profiles F_λ ($i = \vartheta$) are plotted in Fig. 2 ($\tau^l = 10$) for different inclination angles ($\vartheta = 0^o, 15^o, 60^o, 75^o$).

Be stars often show indications for strong stellar winds giving rise to radial motions inside the envelope. As a first approach to this phenomenon, an additional *expansion velocity* $V_r = V_{r0}(r/R_*)^{-Jr}$ ($V_{r0} = 40$ km s^{-1}) is introduced. Results for F_λ ($\vartheta = 45^o, 85^o$) are given in Fig. 3.

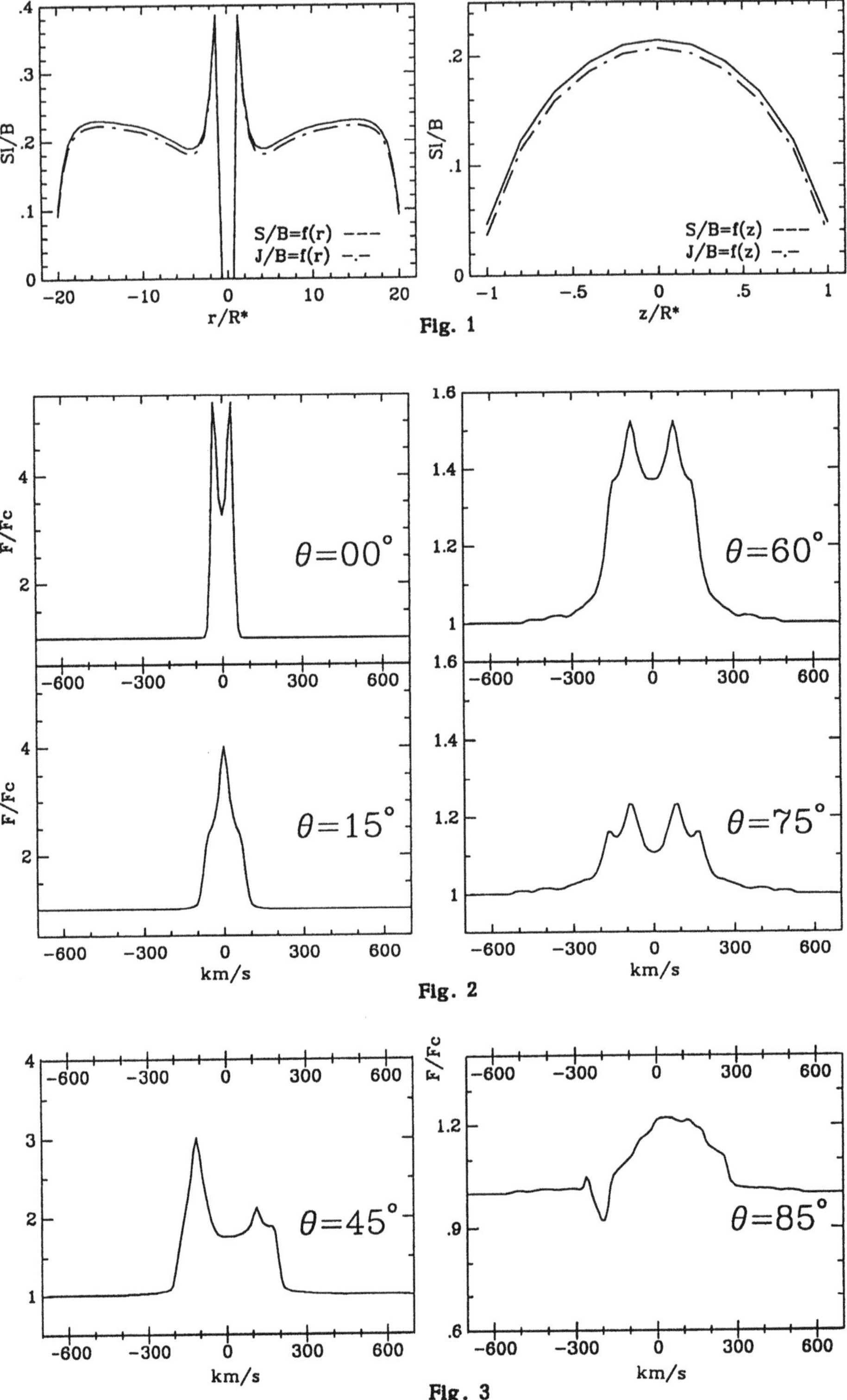

Fig. 1

Fig. 2

Fig. 3

For optically thick line radiation from *Keplerian disks*, the line source function S is increasing with r, reflecting the decreasing photon escape probability, caused by the decreasing velocity gradient: $\partial v_{\varphi}(r)/\partial r = v_{\varphi}(r)/2r$. As expected, S^l increases with τ^l, because of the increasing influence of the central star. Emission line profiles from optically thick disks are determined by the following effects :

1) The anisotropic opacity $\tau^l(\varphi)$ in directions perpendicular to the rotation axis (shear broadening) generates a deep central depression ("shell-type") in the emission line profile (Horne and Marsh, 1986). This effect vanishes for inclination angle $i = 0^0$ and is dominant for $i > 60^0$ (Fig. 2, $\vartheta = 75^0$).

2) If the line radiation F_λ ($\vartheta \approx 0^0$) is optically thick, then winebottle-type profiles (Fig. 2, $\vartheta = 15^0$) and flank inflections (Fig. 2, $\vartheta = 60^0, 75^0$) appear in the emission line profiles as a natural consequence of optically thick line radiation in a slab (Jefferies, 1955), reflecting the symmetric characteristic of the line source function S^l. The formation of this emission line feature can be explained by a Keplerian rotation broadening of the optically thick F_λ ($\vartheta = 0^0$) double-peak emission line profile. This explanation is at variance with the two-disk model (Kogure, 1969), where flank inflections of observed emission line profiles are interpreted as a consequence of two concentric disks and have been called "two-component-structure".

Rotating disks with *radially expanding motions* generate asymmetric double-peak profiles and P Cygni type profiles (Fig. 3).

ACKNOWLEDGEMENTS

Many thanks go to J. Dachs, R. Wehrse and H. Störzer for helpful discussions.

REFERENCES

Adam, J., Innes, D. E., Shaviv, G., Störzer, H., Wehrse, R. : 1990, "Theory of Accretion Disks", (eds. F. Meyer at al.), p.403, NATO ASI C, Vol **290**, Kluwer, Dortrecht

Hanuschik, R. W., Kozok, J., Kaiser, D. : 1988, Astron. Astrophys. **189**, 147

Horne, K., Marsh, T. R. : 1986, MNRAS **218**, 761

Jefferies, J. T. : 1955, Astrophys. J. **115**, 617

Kogure, T. : 1969, Astron. Astrophys. **1**, 253

Olson, G. L., Auer, L. H., Buchler, J. R. : 1986, JQSRT **35-6**, 431

Struve, O. : 1931, Astrophys. J. **73**, 94

Author index

First authors are denoted by bold figures

Lecture Notes in Physics

For information about Vols. 1–365
please contact your bookseller or Springer-Verlag

New Series m: Monographs

GPSR Compliance
The European Union's (EU) General Product Safety Regulation (GPSR) is a set of rules that requires consumer products to be safe and our obligations to ensure this.

If you have any concerns about our products, you can contact us on

ProductSafety@springernature.com

In case Publisher is established outside the EU, the EU authorized representative is:

Springer Nature Customer Service Center GmbH
Europaplatz 3
69115 Heidelberg, Germany

www.ingramcontent.com/pod-product-compliance
Ingram Content Group UK Ltd.
Pitfield, Milton Keynes, MK11 3LW, UK
UKHW061655190726
13853UKWH00008B/2226

* 9 7 8 3 6 6 2 1 3 8 8 5 4 *